A COURSE IN ORDINARY DIFFERENTIAL EQUATIONS

A COURSE IN ORDINARY DIFFERENTIAL EQUATIONS

Randall J. Swift

Stephen A. Wirkus

Chapman & Hall/CRC
Taylor & Francis Group

Boca Raton   London   New York

Chapman & Hall/CRC is an imprint of the
Taylor & Francis Group, an informa business

Chapman & Hall/CRC
Taylor & Francis Group
6000 Broken Sound Parkway NW, Suite 300
Boca Raton, FL 33487-2742

© 2007 by Taylor & Francis Group, LLC
Chapman & Hall/CRC is an imprint of Taylor & Francis Group, an Informa business

No claim to original U.S. Government works
Printed in the United States of America on acid-free paper
10 9 8 7 6 5 4 3 2

International Standard Book Number-10: 1-58488-476-2 (Hardcover)
International Standard Book Number-13: 978-1-58488-476-7 (Hardcover)

Visit the Taylor & Francis Web site at
http://www.taylorandfrancis.com

and the CRC Press Web site at
http://www.crcpress.com

To our families

Kelly, Kaelin, Robyn, Erin, and Ryley

and

Erika Tatiana and Alan Sebastian

for showing us the true concept and meaning of

∞ infinity ∞

with their tireless

patience, love, and understanding.

Preface

This book is based upon lectures given by the first author at Western Kentucky University and by both authors at California State Polytechnic University–Pomona (Cal Poly Pomona). The text is intended for a one-semester sophomore-level course in ordinary differential equations. However, there is ample material for a two-quarter sequence, as well as sufficient linear algebra in the text so that it can be used for a one-quarter course that combines ordinary differential equations and linear algebra. The focus of the text is upon applications and methods of solution, both analytical and numerical, with emphasis on methods used in the typical engineering, physics, or mathematics student's field of study.

It is the intent that students who study this book and work *most* of the problems contained in these pages will be very prepared to continue their studies in engineering and mathematics. The student whom we "typically encounter" has had one year of Calculus and is usually a major in a field other than pure mathematics. For a course that fully incorporates the linear algebra, it would be best if the student has taken a one-quarter or one-semester course in multivariable calculus.

The vast majority of our students also have *no previous experience with Matlab, Maple, or Mathematica* and we start from the basics and teach the students informed use of the relevant mathematical software. The mathematical topics and use of these software packages to shed light on the concepts are geared for these students. We have tried to provide sufficient problems of a mathematical nature at the end of each section so that even the pure math major will be sufficiently challenged. With the applied math major, engineer, or other science major in mind, we have made this text very application-based wherever possible.

Most differential equations we have encountered in practice have needed analytical approximations or numerical approximations to gain insight into their behavior. We don't feel that students use technology wisely if they simply ask the computer to solve a given problem. We thus focus on what we consider to be the basics necessary for adequately preparing a student for study in her or his respective fields, including mathematics. A unique feature of our presentation is that the relevant computer algebra code is imbedded throughout text. We present the syntax from Matlab, Maple, and Mathematica alongside the relevant theory. We feel that this provides the readers a better understanding of the theory and allows them to gain more insight into real-world problems they are likely to encounter.

Our book is traditional in its approach and coverage of basic topics in ordinary differential equations. However, we cover a number of "modern" topics that are commonly not found in a traditional sophomore-level text. For example, Chapter 2 covers direction fields, phase line techniques, and the Runge-Kutta method; Chapter 5 covers linear algebraic topics such as transformations and eigenvalues; Chapter 6 considers linear and nonlinear systems of equations from a dynamical systems viewpoint and makes use of the linear algebra insights from Chapter 5; it also includes modern applications such as epidemiological models. There are also projects included at the end of each chapter that give useful insight to past and future topics covered in the book. The topics covered in these projects include a mix of traditional, modeling, numerical, and linear algebra aspects of ordinary differential equations.

As it would seem how most texts based upon lecture notes develop, many of the examples, exercises and projects have been collected over many years for various courses taught by both authors. Many were taken from others' textbooks and papers. We have tried to give proper credit throughout this text; however, it was not always possible to properly acknowledge the original sources. It is our hope that we repay this explicit debt to earlier writers by contributing our (and their) ideas to further student understanding of differential equations.

We owe a very special thanks to Erika Camacho (Loyola Marymount University) for her help in writing the Matlab and Maple code for this book and for detailed suggestions on numerous sections. John Fay and Gary Etgen reviewed earlier drafts of this text and provided helpful feedback. We owe a big thanks to our former students David Monarres, for help in preparing portions of this book, and Walter Sosa and Moore Chung, for their help in preparing solutions. We would also like to acknowledge our Cal Poly Pomona colleagues Michael Green, Jack Hofer, Tracy McDonald, Jim McKinney, Dick Robertson, Paul Salomaa, Jenny Switkes, Karen Vaughn, and Mason Porter (Caltech) for their willingness to use draft versions of this text in their courses and their important suggestions, which improved the overall readability of the text. Our colleague Siew-Ching Pye deserves an extra special thanks for her tremendously detailed (and helpful!) critiques of the text. Mary Jane Hill assisted us with certain aspects of the text and helped in typesetting some of the chapters of the initial drafts of the book; her effort is greatly appreciated. The production and support staff at Chapman & Hall/CRC Press have been very helpful. We particularly wish to thank our project coordinator Theresa Del Forn and project editor Prudence Board. Our editor Bob Stern deserves a special thanks for believing in this project and for his guidance, advice and patience. We sincerely thank all these individuals, without their assistance this text would not have succeeded.

Randall J. Swift
Stephen A. Wirkus
Pomona, CA

A few remarks for students and professors:

This book will succeed if any fears and reservations about learning one of the three computer algebra systems used in this book are put aside. Computers are not here to supplant us, but rather they are here to help illustrate and illuminate concepts and insights that we have. Nothing is foolproof and we stress the importance of *informed use of the relevant mathematical software.* Numerical answers, although quite accurate most of the time, should always be examined carefully because computers are as smart as the programmer allows them to be. There should never be a blind trust in an answer.

It is essential that the technology that you choose—Matlab, Maple, or Mathematica—be introduced early in the class, just as it is introduced early in the book. While certain mathematical software packages may be better suited for studying differential equations, none have the versatility that the above three programs have to give insight into other areas of mathematics. The two keys to learning the programs are (1) learning the syntax and (2) learning to use the help menus to figure out some of the commands. Setting aside one class, for example, to give a brief tutorial on one of these software packages in the computer lab is a very worthwhile investment. It is by no means necessary and the typical student will be able to learn the material on his/her own by carefully following Appendix A. For reinforcement, it is crucial to include at least one technology problem with each homework assignment. The conscientious student will be well prepared to use the same software package in any upper division course in *any* branch of the mathematical sciences and its applications.

It is not necessary to bring computer demonstrations into the classroom. Both authors have taught their courses successfully without classroom demonstrations; handouts sometimes are useful, especially from the appendices. The students, for better or worse, are generally far less afraid of technology than one might expect. If students are sent to the computer lab with an assignment to do and aided with Appendix A, the vast majority will come back with satisfactory answers. Yes, you may bang your head against your desk in frustration at times, but just ask the person next to you for help and also seek the help menus and you will be able to learn Matlab, Maple, and Mathematica quite well.

Contents

List of Computer Codes

Chapter 1

Traditional First-Order Differential Equations

The study of Differential Equations began very soon after the invention of Differential and Integral Calculus, to which it forms a natural sequel. In 1676 Newton solved a differential equation by the use of an infinite series, only 11 years after his discovery of the *fluxional* form of differential calculus in 1665. These results were not published until 1693, the same year in which a differential equation occurred for the first time in the work of Leibniz (whose account of the differential calculus was published in 1684).

In the next few years progress was rapid. In 1694–1697 John Bernoulli explained the method of "Separating the Variables," and he showed how to reduce a homogeneous differential equation of the first order to one in which the variables were separable. He applied these methods to problems on orthogonal trajectories. He and his brother Jacob (after whom "Bernoulli Equation" is named; see Section 1.5.2) succeeded in reducing a large number of differential equations to forms they could solve. Integrating Factors were probably discovered by Euler (1734) and (independently of him) by Fontaine and Clairaut, though some attribute them to Leibniz. Singular Solutions, noticed by Leibniz (1694) and Brook Taylor (1715), are generally associated with the name of Clairaut (1734). The geometrical interpretation was given by Lagrange in 1774, but the theory in its present form was not given until much later by Cayley (1872) and M.J.M. Hill (1888).

Today, differential equations are used in many different fields. They can often accurately capture the behavior of continuous models or a large number of discrete objects where the current state of the system determines the future behavior of the system. Such models are called *deterministic* (as opposed to *stochastic* or *random*). The study of *nonlinear* differential equations is still a very active area of research. Although this text will consider some nonlinear differential equations, here the focus will be on the linear case. We will begin with some basic terminology.

1

1.1   Some Basic Terminology

1.1.1   Order, Linear, Nonlinear

We begin our study of differential equations by explaining what a differential equation is. From our experience in calculus, we are familiar with some differential equations. For example, suppose that the acceleration of a falling object is $a(t) = -32$, measured in ft/sec^2. Using the fact that the derivative of the velocity function $v(t)$ (measured in ft/sec) is the acceleration function $a(t)$, we can solve the equation

$$v'(t) = a(t) \quad \text{or} \quad \frac{dv}{dt} = a(t) = -32.$$

Many different types of differential equations can arise in the study of familiar phenomena in subjects ranging from physics to biology to economics to chemistry. We give examples from various fields throughout the text and engage the reader with many such applications.

It is clearly necessary (and expedient) to study, independently, more restricted classes of these equations. The most obvious classification is based on the nature of the derivative(s) in the equation. A differential equation involving derivatives of a function of one variable (ordinary derivatives) is called an *ordinary differential equation*, whereas one containing partial derivatives of a function of more than one independent variable is called a *partial differential equation*. In this text, we will focus on ordinary differential equations.

The *order* of a differential equation is defined as the order of the highest derivative appearing in the equation.

Example 1: The following are examples of differential equations with indicated orders:
   a) $dy/dx = ay$ (first order)
   b) $x''(t) - 3x'(t) + x(t) = \cos t$ (second order)
   c) $(y^{(4)})^{3/5} - 2y'' = \cos x$ (fourth order)

Our focus will be on *linear* differential equations, which are those equations that have an unknown function, say y, and each of its higher derivatives appearing in linear functions. That is, we do *not* see them as $y^2, yy', \sin y$, or $(y^{(4)})^{3/5}$.[1] More precisely, a linear differential equation is one in which the dependent variable and its derivatives appear in additive combinations of their first powers. Equations where one or more of y and its derivatives appear in

---

[1]Most of the equations we consider will involve an unknown function y that depends on x. Two other common variables used are (i) the unknown function y that depends on t and (ii) the unknown function x that depends on t, the latter being used in Example 1b.

nonlinear functions are called *nonlinear* differential equations. In the above example, only **c** is a nonlinear differential equation.

Example 2: Classify the equations as linear or nonlinear.
  a) $y'' + 3y' - x^2 y = \cos x$
  b) $y'' - 3y' + y^2 = 0$
  c) $y^{(3)} + yy' + \sin y = x^2$

The first of these equations is linear as it consists of an additive combination of y, y', and y'', each of which is raised to the first power. In contrast to this, the second equation is nonlinear because of the y^2 term. The last equation is nonlinear both because of the yy' term and the $\sin y$ term—either of these terms by itself would have made the equation nonlinear. Our study of nonlinear differential equations will focus on techniques for specific equations or on understanding the qualitative behavior of a nonlinear differential equation, since general techniques of solution are rarely applicable.

Much of this book is concerned with the solutions of linear differential equations. Thus we need to explain what we mean by a solution. First we note that any nth-order differential equation can be written in the form

$$F(x, y, y', ..., y^{(n)}) = 0, \tag{1.1}$$

where n is a positive integer. For example, $y' = x^2 + y^2$ can be written as

$$y' - x^2 - y^2 = 0.$$

Here $F(x, y, y') = y' - x^2 - y^2$. The second-order equation $y'' - 3x^2 y' + 5y = \sin x$ can be written as

$$y'' - 3x^2 y' + 5y - \sin x = 0$$

and we see that $F(x, y, y', y'') = y'' - 3x^2 y' + 5y - \sin x$.

A *solution* to an nth-order differential equation is a function that is n times differentiable and that satisfies the differential equation. Symbolically, this means that a solution of differential equation (1.1) is a function $y(x)$ whose derivatives $y'(x), y''(x), ..., y^{(n)}(x)$ exist and that satisfies the equation

$$F(x, y(x), y'(x), ..., y^{(n)}(x)) = 0$$

for all values of the independent variable x in some interval (a, b) where

$$F(x, y(x), y'(x), ..., y^{(n)}(x))$$

is defined. (Note that the solution to a differential equation does not contain any derivatives, although the derivatives of this solution exist.) The interval (a, b) may be infinite; that is, $a = -\infty$, or $b = \infty$, or both.

Example 3: The function $y(x) = 2e^{3x}$ is a solution of the differential equation

$$\frac{dy}{dx} = 3y,$$

for $x \in (-\infty, \infty)$ because it satisfies the differential equation by giving an identity:

$$\frac{dy}{dx} = 2\frac{de^{3x}}{dx} = 6e^{3x} = 3y.$$

1.1.2   Initial-Value vs. Boundary-Value Problems

We will soon see that solving a general differential equation gives rise to a solution that has constants. These constants can be eliminated by specifying the initial state of the system or conditions that the solution must satisfy on its domain of definition or "boundary." An example of the first situation is specifying the position and velocity of a mass on a spring. An example of the second is a rope hanging from two supports, given the location of these two supports.

Consider a first-order differential equation

$$\frac{dy}{dx} = f(x, y)$$

and suppose that the solution $y(x)$ was subject to the condition that $y(x_0) = y_0$. This is an example of an *initial-value problem*. The condition $y(x_0) = y_0$ is called an *initial condition* and x_0 is called the *initial point*. More generally, we have the following:

> **DEFINITION 1.1**   *An initial-value problem consists of an nth-order differential equation together with n initial conditions of the form*
>
> $$y(x_0) = a_0, \quad y'(x_0) = a_1, ..., \quad y^{(n-1)}(x_0) = a_{n-1}$$
>
> *that must be satisfied by the solution of the differential equation and its derivatives at the initial point x_0.*

Example 4: The following are examples of initial-value problems:

  a) $dy/dx = 2y - 3x$, $y(0) = 2$ (here $x = 0$ is the initial point)

  b) $x''(t) + 5x'(t) + \sin(tx(t)) = 0$, $x(1) = 0, x'(1) = 7$ (here $t = 1$ is the initial point)

(Note that the differential equation in **a** is linear, whereas the equation in **b** is nonlinear.) We define a *solution* to an nth-order initial-value problem as a function that is n times differentiable on an interval (a, b); this satisfies the given differential equation on that interval, and satisfies the n, given initial conditions with the requirement that $x_0 \in (a, b)$. As before, the interval (a, b) might be infinite.

In contrast to an initial-value problem, a *boundary-value problem* consists of a differential equation and a set of conditions *at different x-values* that the solution $y(x)$ must satisfy. Although any number of conditions (≥ 2) may be specified, usually only two are given. Rather than specifying the initial state of the system, we can think of a boundary-value problem as specifying the state of the system at two different physical locations, say $x_0 = a, x_1 = b, a \neq b$.

Example 5: The following are examples of boundary-value problems:
a) $d^2y/dx^2 + 5xy = \cos x, y(0) = 0, y'(\pi) = 2$
b) $dy/dx + 5xy = 0, y(0) = y(1) = 2$

Although a boundary-value problem may not seem too different from an initial-value problem, methods of solution are quite varied. We will focus on initial-value problems. We ask whether an initial-value problem has a unique solution. Essentially this is two questions:
1. Is there a solution to the problem?
2. If there is a solution, is it the only one?

As we see in the next two examples, the answer may be "no" to each question.

Example 6: *An initial-value problem with no solution.*
The initial-value problem

$$\left(\frac{dy}{dx}\right)^2 + y^2 + 1 = 0$$

with $y(0) = 1$ has no real-valued solutions, since the left-hand side is always positive for real-valued functions.

Example 7: *An initial-value problem with more than one solution.*
The initial-value problem

$$\frac{dy}{dx} = xy^{1/3}$$

with $y(0) = 0$ has at least two solutions in the interval $-\infty < x < \infty$. Note that the functions

$$y = 0 \text{ and } y = \frac{x^3}{3\sqrt{3}}$$

both satisfy the initial condition and the differential equation.

In the next several sections we will develop methods for finding solutions to first-order differential equations. We will then discuss existence and uniqueness of solutions in Chapter 2.

1.1.3   Plotting in Matlab, Maple, and Mathematica

This section is the first of the "technology" sections where the reader is introduced to the Mathematical Software packages of Matlab, Maple, and Mathematica. The reader should only read this section after reading Appendix A, where the introductory details and some basic commands and examples are given. Although every attempt has been made to make the syntax given in this book applicable to a range of versions/releases, certain syntax may be different for an earlier or later version of a given package. We assume the reader has *no* familiarity with any package but has access to at least one of them. From this point, we will assume the reader is familiar with the relevant section of Appendix A. In particular, the first part of Appendix A gives basic examples in Matlab, Maple, and Mathematica of how to plot a simple function such as $y = x^3 - x$. Here we will begin by considering the situation of plotting multiple functions on the same graph.

Basic Plotting: Plot the function $y = Ce^{kx}$, for C values of -1 and 1, for k values of $-2/3$ and $2/3$, and for x ranging between -2 and 2.

Computer Code 1.1:   Superimposing basic plots

Matlab, Maple, Mathematica

```
                              Matlab
>>  x=-2:.01:2;
>>  y1=-1*exp(-2*x/3); % C=-1, k=-2/3
>>  plot(x,y1,'r') %curve is solid and red
>>  hold on    % subsequent plots will be superimposed
>>  y2=1*exp(-2*x/3); % C=1, k=-2/3
>>  plot(x,y2,'g:')  %curve is dotted and green
>>  y3=-1*exp(2*x/3); % C=-1, k=2/3
>>  plot(x,y3,c--') %curve is cyan and dashed (two hyphens)
>>  y4=1*exp(2*x/3); % C=1, k=2/3
>>  plot(x,y4,'m-.')  %curve is magenta and dashdot
>>  legend('y1','y2','y3','y4')
>>  axis([-2 2 -3 3]);
>>  title('Graph of y=C*exp(k*x)');
>>  xlabel('x'); ylabel('y')
>>  hold off
```

$$\boxed{\textbf{Maple}}$$

```
> eq1:=-1*exp(-2*x/3);
> plot(eq1,x=-2..2,color=red); #curve is solid and red
> eq2:=C*exp(k*x);
> eq3:=subs(C=1,k=-2/3,eq2);
> eq4:=subs(C=-1,k=2/3,eq2);
> eq5:=subs(C=1,k=2/3,eq2);
> plot([eq1,eq3,eq4,eq5],x=-2..2,-3..3,labels=[x,y],legend=
   ["y1","y2","y3","y4"], linestyle=[1,2,3,4], color=[red,
   green,cyan,magenta],title="Graph of y=C*exp(k*x)" );
```

This last line of Maple input gives us the superimposed plots. If a legend does not show up, you may need to right click on the picture and click *Legend* $\longrightarrow$ *Show Legend*. Also, later versions of Maple allow the user to type `linestyle=DOT` instead of the `linestyle=2` (the latter syntax still is acceptable).

For Mathematica, many of the commands are entered via the basic input palette. For example, the integral sign, the exponential function, and fractions are all entered with the palette to make it aesthetically pleasing; alternatively, there are commands that could be entered directly.

$$\boxed{\textbf{Mathematica}}$$

```
Needs["Graphics`Colors`"]; (*Loads Color graphics package*)
eq1 = -e^{-2x/3}; (*both e and fraction entered from palette!*)
      (*neither is typed in from the keyboard; however, this*)
      (*could be typed in as E^(-2x/3) or E^(-2*x/3)  *)
Plot[eq1,{x,-2,2}, PlotStyle → Red];
eq2=e^{-2x/3};
eq3=-e^{2x/3};
eq4=e^{2x/3};
Plot[{eq1,eq2,eq3,eq4}, {x,-2,2}, PlotStyle→{Red, Green,
   Cyan, Magenta}, AxesLabel→{"x","y=Ce^{kx}"}]
```

In each case, we plotted the curves on the same graph. Many plotting options are available for the programs. The reader is encouraged to become familiar with Appendix A and the syntax therein.

Problems

In problems 1–12, verify that the given function is a solution to the differential equation by substituting it into the differential equation and showing that the equation holds true on the given interval. Do NOT attempt to solve the differential equation.

1. Verify that $y(x) = 2x^3$ is a solution on $(-\infty, \infty)$ to $x\dfrac{dy}{dx} = 3y$.

2. Verify that $y = 2$ is a solution on $(-\infty, \infty)$ to $\dfrac{dy}{dx} = x^3(y-2)^2$.

3. Verify that $y(x) = \dfrac{-1}{5x+4}$ is a solution on $(-4/5, \infty)$ to $\dfrac{dy}{dx} = 5y^2$.

4. Verify that $y(x) = e^x - x$ is a solution on $(-\infty, \infty)$ to $\dfrac{dy}{dx} + y^2 = e^{2x} + (1-2x)e^x + x^2 - 1$.

5. Verify that $y(x) = x^3$ is a solution on $(-\infty, \infty)$ to $\dfrac{dy}{dx} = 3y^{2/3}$.

6. Verify that $y(x) = \dfrac{-1}{x-3}$ is a solution on $(-\infty, 3)$ to $\dfrac{dy}{dx} = y^2$.

7. Verify that $y(x) = x^2 - x^{-1}$ is a solution to $x^2\dfrac{d^2y}{dx^2} = 2y$ for all $x \neq 0$.

8. Verify that $y(x) = \sin x + 2\cos x$ is a solution on $(-\infty, \infty)$ to $\dfrac{d^2y}{dx^2} + y = 0$.

9. Verify that $y(x) = x$ is a solution on $(-\infty, \infty)$ to $y'' + y = x$.

10. Verify that $y(x) = x + C\sin x$ is a solution on $(-\infty, \infty)$ to $y'' + y = x$ for any constant C.

11. Verify that $y_1(x) = e^x, y_2(x) = e^{2x}$ are both solutions on $(-\infty, \infty)$ to $y'' - 3y' + 2y = 0$.

12. Verify that $y_1(x) = e^x, y_2(x) = xe^x$ are both solutions on $(-\infty, \infty)$ to $y'' - 2y' + y = 0$.

13. Which of the following functions are solutions to the differential equation $y'' + 9y = 0$?
    a. $\sin 3x$,   b. $\sin x$,   c. $\cos 3x$,   d. e^{3x},   e. x^3

14. Which of the following functions are solutions to the differential equation $y'' + 6y' + 9y = 0$?
    a. e^x,   b. e^{-3x},   c. xe^{-3x},   d. $4e^{3x}$,   e. $2e^{-3x} + xe^{-3x}$

15. Which of the following functions are solutions to the differential equation $y'' - 7y' + 12y = 0$?
    a. e^{2x},   b. e^{3x},   c. e^{4x},   d. e^{5x},   e. $e^{3x} + 2e^{4x}$

16. Which of the following functions are solutions to the differential equation $y'' + 4y' + 5y = 0$?
    a. e^{-2x},   b. $e^{-2x}\sin 2x$,   c. $e^{-2x}\cos 2x$,   d. $\cos 2x$

17. Find values of r for which $y(x) = e^{rx}$ is a solution to $y'' + 3y' + 2y = 0$ on $(-\infty, \infty)$.

18. Find values of r for which $y(x) = xe^{rx}$ is a solution to $y'' + 4y' + 4y = 0$ on $(-\infty, \infty)$.

19. Classify the differential equations by specifying (i) the order, (ii) whether it is linear or nonlinear, and (iii) whether it is an initial-value or boundary-value problem (where appropriate).
    a. $3y'' + y = \sin x$
    b. $y'' + \sin y = 0$
    c. $y^{(3)} + (\sin x)y^{(2)} + y = x$, $y(0) = 1, y'(0) = 0, y''(0) = 2$
    d. $y' + e^x y = y^4$, $y(0) = 0$
    e. $y'' + y' - y = 0$
    f. $y'' + e^x y' + y^2 = 0$, $y(0) = 1, y(\pi) = 0$

20. Classify the differential equations by specifying (i) the order, (ii) whether it is linear or nonlinear, and (iii) whether it is an initial-value or boundary-value problem (where appropriate).
    a. $y'' - 3yy' = x$
    b. $y'' = \sin x$
    c. $y'' + 3y' = 0$, $y(0) = 1, y'(1) = 0$
    d. $y'' = 0$, $y(1) = 1, y'(1) = 2$
    e. $y'' - 4y' + 4y = 0$, $y(0) = 1, y'(0) = 1$
    f. $x^2 y'' + y' + (\ln x)y = 0$

For the following problems, graph the curves using Matlab, Maple, or Mathematica over the specified interval. **Be sure to label the curves (and plot with different linestyles), label the axes, and title the graph.**

21. Graph $\sin x, \sin(2x + 3), \sin(x + 3)$, $-2\pi \le x \le 2\pi$ on the same graph.

22. Graph $\cos x, \sqrt{2x + 3}, e^{x+3}$, $-1.5 \le x \le 4$ on the same graph.

23. Graph $\ln x, \tan(2x + 3), \arctan(x + 3)$, $0 \le x \le 5$ on the same graph. Limit the vertical viewing window to $[-5, 5]$.

24. Graph $x^3, x^3 + 3x - 4, x^2 + 4x - 5$, $-3 \le x \le 3$ on the same graph. Limit the vertical viewing window to $[-5, 5]$.

25. Graph $(x - 1)^3, (x - 1)^2(x + 1), \sin x \cos 2x$, $-3 \le x \le 3$ on the same graph. Limit the vertical viewing window to $[-5, 5]$.

26. Graph e^x, e^{-x}, e^{x+3}, $-1 \le x \le 1$ on the same graph.

27. Graph $x^2, 3x^2 + \sqrt{x}, -\sqrt{x}$, $0 \le x \le 1$ on the same graph.

28. Graph $e^{2x/3}, e^{-2x/3}, e^{x^2}, e^{-x^2}$, $-1 \le x \le 1$ on the same graph.

29. Graphically show the identity

$$\sin(2x) = 2\sin(x)\cos(x)$$

by superimposing the plots of each side of the equation. Use different symbols for each side.

30. Just as we used the computer to verify a trigonometric identity, we can also use it to observe the region where an approximation is valid. Consider the Taylor expansion of $\sin x$ about $x = 0$:

$$\sin x = x - \frac{x^3}{3!} + \frac{x^5}{5!} + \cdots .$$

For small x, the first three terms of this expansion give a good approximation of $\sin x$. Graphically find the approximate x-values where the three-term expansion gives an answer that is no more than 0.2 away from the known value.

31. Consider the Taylor expansion of e^x about $x = 0$:

$$e^x = 1 + x + \frac{x^2}{2!} + \frac{x^3}{3!} + \cdots .$$

Graphically find the approximate x-values where the first four terms of the expansion give an answer that is no more than 0.2 away from the known value.

32. Consider the function

$$f(x) = \frac{x}{1 - x - x^2}, \quad x > 1,$$

and the approximations to it

$$f_1(x) = \frac{-1}{x} + \frac{1}{x^2}, \quad f_2(x) = \frac{-1}{x} + \frac{1}{x^2} - \frac{2}{x^3}, \quad f_3(x) = \frac{-1}{x} + \frac{1}{x^2} - \frac{2}{x^3} + \frac{3}{x^4}.$$

Which of the functions is the better approximation for large x-values? Based on the pattern of the three approximation functions, can you write a function that has six terms and is a better approximation than each of the three functions given here? The approximate functions $f_i(x)$ are called *asymptotic expansions* of the original function.

1.2   Separable Differential Equations

We will now introduce the simplest first-order differential equation. Although these are the simplest class of differential equations we will encounter, they appear in numerous applications and aspects of subsequent theory. We make the following definition:

DEFINITION 1.2   *A first-order differential equation that can be written in the form*

$$g(y)\, y' = f(x) \quad \text{or} \quad g(y)\, dy = f(x)\, dx,$$

where $y = y(x)$, is called a separable differential equation.

Separable differential equations are solved by collecting all the terms involving the dependent variable y on one side of the equation and all the terms involving the independent variable x on the other side. Once this is completed (it may require some algebra), both sides of the resulting equations are integrated. That is, the equation

$$g(y)\, y' = f(x)$$

can be written in "differential form"

$$g(y)\,\frac{dy}{dx} = f(x)$$

so that treating dy/dx as a fraction, we have

$$g(y)\, dy = f(x)\, dx.$$

Here the variables are separated, so that integrating both sides gives

$$\int g(y)\, dy = \int f(x)\, dx. \tag{1.2}$$

The Method of Separation of Variables, which we just applied to (1.2), is the name given to the method we use to solve Separable Equations—it is one of the simplest and most useful methods for solving differential equations. (Incidentally, it is an important technique for solving certain classes of partial differential equations, too.)

Sometimes we will be able to solve (1.2) for y. When we can do so, we will say we can express the solution *explicitly* and will write $y = h(x)$. Other times, we will not be able to solve (1.2) or it will not be worth our time and efforts to do so. In these situations, the solution is said to be given *implicitly* by (1.2). When our solution can be written explicitly, it will be easy to plot solutions in the x-y plane, by hand or with the computer; however, when the solution is implicit, plotting solutions by hand is challenging at best. The various computer programs will allow us to view plots in the x-y plane without much additional work. We now consider a number of examples.

Example 1: Solve $y' = ky$ where k is a constant.

Writing y' as $\frac{dy}{dx}$ gives

$$\frac{dy}{dx} = ky.$$

Treating $\frac{dy}{dx}$ as a "fraction" and rearranging terms gives

$$\frac{dy}{y} = k\, dx.$$

This step will only be valid if $y \neq 0$. We note that $y = 0$ is also a solution to the original differential equation. Integrating gives

$$\int \frac{dy}{y} = \int k \, dx,$$

which is

$$\ln |y| = kx + C_1,$$

so that

$$|y| = e^{kx+C_1}.$$

This gives

$$y = \pm e^{kx} e^{C_1}.$$

Now e^{C_1} is a positive constant, so that we may let $C = \pm e^{C_1}$. In the above process, we encountered the constant solution $y = 0$, which also gives us the possibility that $C = 0$. Thus, we have

$$y = Ce^{kx}$$

as our solution, where $x \in (-\infty, \infty)$. It is also important to remember the "trick" for getting rid of the absolute values—it will come up quite often in practice! We will consider a few more examples with similar standard "tricks." Note that this solution was plotted in Section 1.1.3 in Chapter 1.

Example 2: Solve

$$\frac{dx}{dt} = e^{t-x}, \quad x(0) = \ln 2,$$

for $x(t)$.

Separating the variables gives

$$\frac{dx}{dt} = e^t e^{-x}$$

and thus

$$e^x \, dx = e^t \, dt.$$

Integrating both sides of this equation gives

$$e^x = e^t + C.$$

Solving for x, we have

$$x = \ln |e^t + C|.$$

Applying the initial condition $x(0) = \ln 2$ yields

$$\ln 2 = \ln |1 + C|, \quad \text{so that} \quad C = 1.$$

Thus

$$x = \ln(e^t + 1),$$

which is defined for all t. Note that $e^t + 1$ is always positive so that we can drop the absolute value signs. *We should also note that after integrating, we could have applied the initial condition to determine C and then proceeded to solve for x instead of first solving for x and then applying the initial condition to determine C. Both methods will result in the same final answer.* See Figure 1.1 for a plot of the solution.

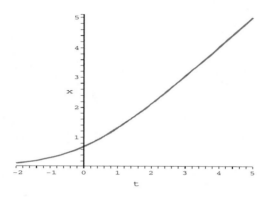

FIGURE 1.1: Plot of solution for Example 2.

If we wanted to plot the solutions in Matlab, Maple or Mathematica we could do so in a straightforward manner.

Computer Code 1.2: **Basic plots using natural log, exponential**

Matlab, Maple, Mathematica

```
                         Matlab
>>  t=-5:.1:5;
>>  y=log(exp(t)+1);
>>  plot(t,x,'LineWidth',2)
>>  xlabel('x','Fontsize',12);ylabel('x','Fontsize',12)
```

```
                              Maple
> eq1:=log(exp(t)+1);
> plot(eq1,t=-5..5,thickness=2,labels=[t,x],
    labelfont=[12,12]);
```

```
                           Mathematica
Plot[Log[e^t + 1],{t,-5,5},AxesLabel→{"x","y"}]
    (*this e is entered from the palette; use E if typing*)
```

Example 3: Solve

$$(x-4)\,y^4 - x^3\,(y^2-3)\,\frac{dy}{dx} = 0.$$

To separate variables, divide by x^3y^4, so that

$$\frac{x-4}{x^3}\,dx = \frac{y^2-3}{y^4}\,dy.$$

This simplifies to $(x^{-2} - 4x^{-3})\,dx = (y^{-2} - 3y^{-4})\,dy$. Integrating gives

$$\frac{-1}{x} + \frac{2}{x^2} = \frac{-1}{y} + \frac{1}{y^3} + C$$

as the general solution. This is definitely a case where giving the solution in an implicit representation is acceptable! We also refer the reader to the end of this section for the computer code used to plot these types of solutions with one of the software packages. There is, however, a more important idea that is illustrated by this example. Note that when we divided by x^3y^4, we implicitly assumed that $x \neq 0$ and $y \neq 0$. If we rewrite the original differential equation as

$$\frac{dy}{dx} = \frac{(x-4)y^4}{x^3(y^2-3)},$$

then one can clearly see that $y = 0$ is a solution. (That is, when $y = 0$ is substituted into both sides of the equation we get an identity for all x.) This problem shows that the separation process can lose solutions.

How can we verify that

$$\frac{-1}{x} + \frac{2}{x^2} = \frac{-1}{y} + \frac{1}{y^3} + C$$

is a solution? We need to substitute it into the differential equation as before. This will require us to find y' and we will do so with implicit differentiation. Taking the derivative of both sides of the equation gives

$$\frac{1}{x^2} - \frac{4}{x^3} = \frac{1}{y^2}y' - \frac{-3}{y^4}y'.$$

We solve for y' and then simplify the complex fraction to obtain

$$y' = \frac{y^4(x-4)}{x^3(y^2-3)},$$

which is an equivalent form of our original differential equation.

Although the separation process will work on any differential equation in the form of Definition 1.2, evaluating the integrals in (1.2) can sometimes be a daunting, if not impossible, task. As discussed in calculus, certain indefinite integrals such as

$$\int e^{x^2}\, dx$$

cannot be expressed in finite terms using elementary functions. When such an integral is encountered while solving a differential equation, it is often helpful to use definite integration by assuming an initial condition $y(x_0) = y_0$.

Example 4: Solve the initial value problem

$$\frac{dy}{dx} = e^{x^2}y^2, \quad y(2) = 1$$

and use the solution to give an approximate answer for $y(3)$.

Dividing both sides by y^2 and integrating from $x = 2$ to $x = x_1$ gives

$$\int_2^{x_1} [y(x)]^{-2}\frac{dy}{dx}\, dx = -[y(x)]^{-1}\big|_2^{x_1}$$

$$= \frac{-1}{y(x_1)} + \frac{1}{y(2)}$$

$$= \int_2^{x_1} e^{x^2}\, dx.$$

If we let t be the variable of integration and replace x_1 by x and $y(2)$ by 1, then we can express the solution to the initial value problem by

$$y(x) = \left(1 - \int_2^x e^{t^2}\, dt\right)^{-1}.$$

With an explicit solution, we often want to be able to find the corresponding y-value given any x. The right-hand side still cannot be solved exactly but can be approximated if x is given. For example, if we set $x = 3$, then the calculations using our computer packages would be as follows:

Computer Code 1.3:   **Numerically approximating a definite integral**

Matlab, Maple, Mathematica

```
                         Matlab
>>  Q=quad('exp(t.^2)',2,3);
>>  format long;
>>  y=1./(1-Q)
>>  int('exp(t.^2)',t,2,3); %works better for nicer functions
```

```
                         Maple
>  eq1:=int(exp(t^2),t=2..x);
>  Digits:=15;
>  eq2:=evalf(1/(1-subs(x=3,eq1)));
>  eq3:=evalf(int(exp(t^2),t=2..3));
```

```
                      Mathematica
N[1 - ∫₂³ e^{t²} dt]    (*integral sign entered from palette*)
```

It is sometimes the case that a substitution or other "trick" will convert the given differential equation into a form that we can solve. A differential equation of the form

$$\frac{dy}{dx} = f(ax + by + k),$$

where a, b, and k are constants, is separable if $b = 0$; however, if $b \neq 0$ the substitution

$$u(x) = ax + by + k$$

makes it a separable equation.

Example 5: Solve

$$\frac{dy}{dx} = (x + y - 4)^2$$

by first making an appropriate substitution.

We let $u = x + y - 4$ and thus $\frac{dy}{dx} = u^2$. We need to calculate $\frac{du}{dx}$. For this example, taking the derivative with respect to x gives

$$\frac{du}{dx} = 1 + \frac{dy}{dx}.$$

Substitution into the original differential equation gives

$$\frac{du}{dx} - 1 = u^2.$$

This equation is separable. We obtain

$$\frac{du}{1 + u^2} = dx$$

and integrating gives

$$\arctan(u) = x + c.$$

Thus $u = \tan(x + c)$. Since $u = x + y - 4$, we then have

$$y = -x + 4 + \tan(x + c).$$

1.2.1 Implicit Plotting in Matlab, Maple, and Mathematica

We have considered basic plotting in Matlab, Maple, and Mathematica. Now, we present an example where implicit plotting is needed. The methods presented here (as with all the code in this book) are not meant to be the most efficient, most correct, or anything else other than a relatively easy to understand method for accomplishing the given goals. The reader is *highly encouraged* to explore the programs and uncover the wealth of applications available.

For Matlab, our method of solution given below involves first creating a `meshgrid` of x-y pairs and evaluating a two-dimensional function $f(x, y)$ at each pair. This gives a two-dimensional surface and each constant value C gives the surface height at this value. The result is plotted as a contour plot.

For Maple and Mathematica, there is a command `implicitplot` which will allow us to do this plot without creating a two-dimensional surface.

Plotting a solution of Example 3 for one initial condition: Plot the solution of Example 3, rewritten here as

$$\frac{-1}{x} + \frac{2}{x^2} + \frac{1}{y} - \frac{1}{y^3} = C,$$

with initial condition $y(1) = \frac{1}{2}$.

Substitution of this IC into the solution and solving for C gives $C = -5$. Thus, we want to use our software packages to plot

$$\frac{-1}{x} + \frac{2}{x^2} + \frac{1}{y} - \frac{1}{y^3} = -5.$$

Computer Code 1.4:   **Implicit plotting of $f(x,y) = C$ for a single C-value**

Matlab, Maple, Mathematica

```
                           Matlab
>>  [X,Y]=meshgrid(-2:.011:2,-2:.011:2);
>>  Z=-1./X+2./X.^2+1./Y-1./Y.^3;
>>  contour(X,Y,Z,[-5 -5]);
>>  xlabel('x'); ylabel('y');
>>  title('Plot of Example 3 with y(1)=1/2')
```

```
                           Maple
>  eq1:=-1/x+2/x^2+1/y-1/y^3=-5;
>  with(plots): #loads package needed for implicit plotting
>  implicitplot(eq1,x=-2..2,y=-2..2,numpoints=1000,
   title="Plot of Example 3 with y(1)=1/2");
```

```
                       Mathematica
<< Graphics`ImplicitPlot`
    (*Loads package for implicit plotting*)
```
ImplicitPlot$[\frac{-1}{x} + \frac{2}{x^2} + \frac{1}{y} - \frac{1}{y^2} == -5, \{x, -2, 2\}]$
```
    (*note:  palette used to enter fractions *)
```

Now for a few comments. For Matlab, making stepsize smaller in the mesh-grid gives smoother looking cures because more points are plotted. Here 0.011 worked well. In Maple, making *numpoints* larger accomplishes the same goal. As mentioned earlier, many plotting options are available for both programs.

The second line in the above Maple code `with(plots):` tells Maple to load the `plots` package, where `implicitplot` is defined. Once the package is loaded, it will remain loaded until the session is ended or `restart;` is typed in the window. Similarly, `<< Graphics`ImplicitPlot`` in the Mathematica code loads the implicit plot package.

Plotting solutions of Example 3 for multiple initial conditions: Plot the solution of Example 3, rewritten here as $\frac{-1}{x} + \frac{2}{x^2} + \frac{1}{y} - \frac{1}{y^3} = C$, for C values of $-10, -5, -1, 0, 1, 2, 6, 50$.

Computer Code 1.5: Superimposing implicit plots, $f(x,y) = C$

Matlab, Maple, Mathematica

Matlab

```
>>  % Note:  left-hand side not defined if x=0 or y=0.
>>  % A stepsize of .011 (for example) avoids this situation.
>>  [X,Y]=meshgrid(-2:.011:2,-2:.011:2);
>>  Z=-1./X+2./X.^2+1./Y-1./Y.^3;
>>  contour(X,Y,Z,[-10 -5 -1 0 1 2 6 50]);
>>  xlabel('x'); ylabel('y');
>>  title('Plot of Example 3 with C=-10, -5, -1, 0,1,2,6,50')
```

Maple

```
> eq1:=-1/x+2/x^2+1/y-1/y^3=C;
> with(plots): #loads package needed for implicit plotting
> eq2:=subs(C=-10,eq1);
> eq3:=subs(C=-5,eq1):
> eq4:=subs(C=-1,eq1):
> eq5:=subs(C=0,eq1):
> eq6:=subs(C=1,eq1):
> eq7:=subs(C=2,eq1):
> eq8:=subs(C=6,eq1): eq9:=subs(C=50,eq1):
> implicitplot({eq2,eq3,eq4,eq5,eq6,eq7,eq8,eq9},x=-2..2,
  y=-2..2,numpoints=1000,title="Plot of Example 3
  with C=-10, -5, -1, 0,1,2,6,50");
```

Mathematica

```
<< Graphics`ImplicitPlot`
```
(*Loads package for implicit plotting*)

$eq1 = \frac{-1}{x} + \frac{2}{x^2} + \frac{1}{y} - \frac{1}{y^2} == -10$

(*As before, palette used to create fractions*)

$eq2 = \frac{-1}{x} + \frac{2}{x^2} + \frac{1}{y} - \frac{1}{y^2} == -5$

$eq3 = \frac{-1}{x} + \frac{2}{x^2} + \frac{1}{y} - \frac{1}{y^2} == -1$

$eq4 = \frac{-1}{x} + \frac{2}{x^2} + \frac{1}{y} - \frac{1}{y^2} == 0$

$eq5 = \frac{-1}{x} + \frac{2}{x^2} + \frac{1}{y} - \frac{1}{y^2} == 1$

$eq6 = \frac{-1}{x} + \frac{2}{x^2} + \frac{1}{y} - \frac{1}{y^2} == 2$

$eq7 = \frac{-1}{x} + \frac{2}{x^2} + \frac{1}{y} - \frac{1}{y^2} == 6$

$eq8 = \frac{-1}{x} + \frac{2}{x^2} + \frac{1}{y} - \frac{1}{y^2} == 50$

```
ImplicitPlot[{eq1,eq2,eq3,eq4,eq5,eq6,eq7,eq8}, {x, -2, 2}]
```

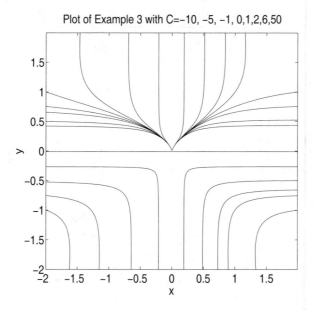

FIGURE 1.2:   Implicit plot for Example 3. The curves plotted here satisfy the implicit solution.

We note here that the C-values chosen were good for this problem, but it often takes ingenuity, experience, trial and error, or some combination of these to get a "nice" picture; see Figure 1.2. We also could have used many of the other plotting options to change the color or style of the curve. For Matlab, we specifically note the syntax for one curve is slightly different than for multiple curves. We encourage the reader to try to remember some of these details. Or maybe you should just fold this page so you will always be able to flip to it for a reference!

1.2.2   Homogeneous Equations

We have now been introduced to separable differential equations and their relative ease of solution. We will now consider a class of differential equations that can be reduced to separable equations by a change of variables.

Example 6: Consider the differential equation

$$\frac{dy}{dx} = \frac{x-y}{x+y}.$$

After a minute or so of reflection, we see that this is not a separable equation. We can, however, rewrite the equation as

$$\frac{dy}{dx} = \frac{1 - \frac{y}{x}}{1 + \frac{y}{x}} \tag{1.3}$$

so that we can isolate the fraction y/x. This suggests we consider the change of variable

$$v = \frac{y}{x}$$

or equivalently

$$y = vx.$$

Our original problem has dy/dx and thus we take the derivative of both sides of the above equation with respect to x to get

$$\frac{dy}{dx} = v + x\frac{dv}{dx}.$$

Substitution of this and $y = vx$ into (1.3) gives

$$v + x\frac{dv}{dx} = \frac{1 - v}{1 + v}.$$

Simplifying results in the separable equation

$$x\frac{dv}{dx} = \frac{1 - 2v - v^2}{1 + v},$$

and we separate its variables as

$$\frac{1 + v}{1 - 2v - v^2}dv = \frac{dx}{x},$$

and integrate to give

$$\ln|1 - 2v - v^2| = -2\ln x + C_1.$$

Exponentiation of both sides yields

$$|1 - 2v - v^2| = e^{C_1}x^{-2} = C_2 x^{-2}.$$

But, $v = y/x$ so that substitution gives

$$1 - \frac{2y}{x} - \left(\frac{y}{x}\right)^2 = \pm C_2 x^{-2} = C x^{-2}.$$

Multiplying by x^2 to clear the fraction gives

$$x^2 - 2xy - y^2 = C$$

as the implicit solution to the differential equation.

This is an example of a general method of reducing a class of differential equations to that of a separable equation. We need some terminology.

DEFINITION 1.3   *The first-order differential equation*

$$M(x,y) + N(x,y)\frac{dy}{dx} = 0$$

is said to be of homogeneous type (or homogeneous) if, when written in the derivative form

$$\frac{dy}{dx} = f(x,y),$$

there exists a function g such that $f(x,y)$ can be expressed in the form $g(y/x)$.

By classifying the equation as homogeneous, we will be able to apply the above technique in order to reduce the differential equation to one that is separable. It is sometimes not obvious that a given equation can be rewritten as a homogeneous equation. We present two examples now to help clarify this concept.

Example 7: The differential equation

$$(x^2 - 3y^2) + 2xy\frac{dy}{dx} = 0$$

is homogeneous, since the equation can be written in derivative form as

$$\frac{dy}{dx} = \frac{3y^2 - x^2}{2xy},$$

and we can rearrange this as

$$\frac{3y^2 - x^2}{2xy} = \frac{3}{2}\left(\frac{y}{x}\right) - \frac{1}{2}\left(\frac{1}{y/x}\right)$$

so that

$$\frac{dy}{dx} = \frac{3}{2}\left(\frac{y}{x}\right) - \frac{1}{2}\left(\frac{1}{y/x}\right).$$

The right-hand side is of the form $g(y/x)$ for the function

$$g(z) = \frac{3z}{2} - \frac{1}{2z},$$

and so the differential equation is homogeneous.

Example 8: The differential equation

$$\left(y + \sqrt{x^2 + y^2}\right)dx - x\,dy = 0$$

can be written as

$$\frac{dy}{dx} = \frac{y + \sqrt{x^2 + y^2}}{x}.$$

Now

$$\frac{y + \sqrt{x^2 + y^2}}{x} = \frac{y}{x} \pm \sqrt{1 + \left(\frac{y}{x}\right)^2},$$

so

$$\frac{dy}{dx} = \frac{y}{x} \pm \sqrt{1 + \left(\frac{y}{x}\right)^2}$$

which is of the form $g(y/x)$ for a function of the form

$$g(z) = z + \sqrt{1 + z^2}$$

or

$$g(z) = z - \sqrt{1 + z^2}$$

depending upon the domain of definition of the independent variable x; thus the differential equation is homogeneous.

As we mentioned, we have introduced homogeneous differential equations because they are related to separable equations; in fact, we have the following theorem which formalizes the method used in Example 1.

THEOREM 1.2.1 *If*

$$M(x,y) + N(x,y)\frac{dy}{dx} = 0 \qquad (1.4)$$

is a homogeneous equation, then the change of variables

$$y = vx$$

transforms (1.4) into a separable equation in the variables v and x.

Note that this change of variables implies that

$$y' = v + xv'$$

by the product rule.

Example 9: Solve

$$y + (x - 2y)\frac{dy}{dx} = 0.$$

We first observe that this can be rewritten as

$$\frac{dy}{dx} = \frac{y}{2y - x}.$$

Dividing numerator and denominator by x gives

$$\frac{dy}{dx} = \frac{y/x}{2y/x - 1}.$$

The right-hand side is then of the form $g(y/x)$ and making the change of variables $y = vx$ gives

$$v + x\frac{dv}{dx} = \frac{v}{2v - 1},$$

which becomes

$$x\frac{dv}{dx} = \frac{2(v - v^2)}{2v - 1}.$$

This equation is separable! Rearranging gives

$$\frac{2v - 1}{2(v - v^2)}\,dv = \frac{1}{x}\,dx,$$

and integrating both sides yields

$$-\frac{1}{2}\ln|v - v^2| = \ln|x| + C_1.$$

We then use $v = y/x$ to reintroduce the y-variable. Thus

$$-\frac{1}{2}\ln\left|\frac{y}{x} - \left(\frac{y}{x}\right)^2\right| = \ln|x| + C_1,$$

but we can let $C_1 = \ln C$ for an arbitrary constant C, so that

$$-\frac{1}{2}\ln|v - v^2| = \ln|Cx|$$

is the implicit solution. Again, we could plot these solutions for various C-values with our favorite software package by following the syntax outlined in Section 1.2.1.

Problems

Solve each of the following differential equations. Explicitly solve for $y(x)$ or $x(t)$ when possible. As an optional part, plot the solutions (whether implicit or explicit) in Matlab, Maple, or Mathematica. If no initial condition is given, choose three different C-values; if an initial condition is given, only plot the solution passing through that specific initial condition. Make sure to label the axes in your plots and choose axes that best show the behavior of the solution.

1. $4xy\,dx + (x^2 + 1)\,dy = 0$
2. $\tan x\,dy + 2y\,dx = 0$
3. $(e^x + 1)\cos y\,dy + e^x(\sin y + 1)\,dx = 0, \quad y(0) = 3$

4. $2x(y^2 + 1)\,dx + (x^4 + 1)\,dy = 0, \quad y(1) = 1$

5. $xy\,dx + (x + 1)\,dy = 0$

6. $\sqrt{y^2 + 1}\,dx = xy\,dy$

7. $(x^2 - 1)y' + 2xy^2 = 0, \quad y(\sqrt{2}) = 1$

8. $y' \cot x + y = 2 \quad y(0) = -1$

9. $y' = 10^{x+y}$

10. $x\frac{dx}{dt} + t = 1$

11. $y' = \cos(y - x)$

12. $y' - y = 2x - 3$

13. $(x + 2y)y' = 1 \quad y(0) = -2$

14. $y' = \sqrt{4x + 2y - 1}$

15. $(y + 2)\,dx + y(x + 4)\,dy = 0, \quad y(-3) = -1$

16. $8\cos^2 y\,dx + \csc^2 x\,dy = 0, \quad y(\pi/12) = \pi/4$

17. $\frac{dy}{dx} = \frac{y^3 + 2y}{x^2 + 3x}, \quad y(1) = 1$

18. $y' = e^{x^2}, \quad y(0) = 0$

19. $y' = xye^{x^2}, \quad y(0) = 1$. Explain why this differential equation guarantees that its solution is symmetric about $x = 0$.

20. Find the solution of the equations that satisfies the given conditions for $x \to +\infty$:
    a. $x^2 y' - \cos 2y = 1, \quad y(+\infty) = \frac{9\pi}{4}$
    b. $3y^2 y' + 16x = 2xy^3, \quad y(x)$ is bounded for $x \to +\infty$

21. Suppose that the population $N(t)$ of a given species (bacteria, elves, Toolie birds, college students, etc.) is not always zero and varies at a rate proportional to its current value. That is,

$$\frac{dN}{dt} = rN,$$

where $r \in \mathbb{R}$ is some measured constant proportionality factor. If the initial population is assumed to be $N(0) = N_0 > 0$, solve this exponential differential equation and discuss the behavior of the solution as $t \to \infty$ for different values of r.

22. An equivalent way of thinking of the exponential growth problem 21 is to assume the per capita growth rate, $\frac{1}{N}\frac{dN}{dt}$, is constant. That is, we assume $\frac{1}{N}\frac{dN}{dt} = r$. It is more realistic to assume that the per capita growth rate decreases as the population grows. If we assume this decrease is linear and agrees with the exponential growth model for small populations, we can write the equation

$$\frac{1}{N}\frac{dN}{dt} = r\left(1 - \frac{N}{K}\right)$$

where the left-hand side is the per capita growth rate and the right-hand side is a linearly decreasing function in N that has y-intercept r and x-intercept K. Multiplying both sides by N gives

$$\frac{dN}{dt} = r\left(1 - \frac{N}{K}\right)N,$$

which is the well-known *logistic* differential equation. If the initial population is given as $N(0) = N_0 > 0$, solve this differential equation and discuss the behavior of the solution as $t \to \infty$. From this behavior, why is K called a *carrying capacity*?

For problems 23–40, solve the homogeneous differential equation analytically. Then use the computer to plot the implicit or explicit solution for three different initial conditions.

23. $(x + y)\,dx - x\,dy = 0$

24. $\frac{dy}{dx} = \frac{y^2 + 2xy}{x^2}$

25. $2x^2\frac{dy}{dx} = x^2 + y^2$

26. $xy' - y = \sqrt{x^2 + y^2}$

27. $(x + 2y)dx - xdy = 0$

28. $(y^2 - 2xy)dx + x^2dy = 0$

29. $2x^3y' = y(2x^2 - y^2)$

30. $(x^2 + y^2)y' = 2xy$

31. $xy' - y = x\tan(\frac{y}{x})$

32. $(2x + y)dx - (4x + 2y)dy = 0$

33. $y^2 + x^2y' = xyy'$

34. $x - y + (y - x)y' = 0$

35. $(x + 4y)y' = 2x + 3y$

36. $(x - y)dx + (x + y)dy = 0$

37. $ydx = (2x + y)dy$

38. $y' = 2(\frac{y}{x+y})^2$

39. $2xdy + (x^2y^4 + 1)ydx = 0$

40. $ydx + x(2xy + 1)dy = 0$

41. A function F is called *homogeneous of degree n* if

$$F(tx, ty) = t^n F(x, y) \text{ for all } x \text{ and } y.$$

That is, if tx and ty are substituted for x and y in $F(x, y)$ and if t^n is then factored out, we are left with $F(x, y)$. For instance, if $F(x, y) = x^2 + y^2$, we note that

$$F(tx, ty) = (tx)^2 + (ty)^2 = t^2 F(x, y)$$

so that F is homogeneous of degree 2. Homogeneous differential equations and functions that are homogeneous of degree n are related in the following manner. Suppose the functions M and N in the differential equation

$$M(x,y)\,dx + N(x,y)\,dy = 0$$

are both homogeneous of the same degree n.

a. Show, using the change of variables $t = 1/x$, that

$$M\left(1, \frac{y}{x}\right) = \left(\frac{1}{x}\right)^n M(x,y),$$

which implies that

$$M(x,y) = \left(\frac{1}{x}\right)^{-n} M\left(1, \frac{y}{x}\right).$$

b. Show, using a similar calculation, that

$$N(x,y) = \left(\frac{1}{x}\right)^{-n} N\left(1, \frac{y}{x}\right),$$

so that the differential equation

$$M(x,y)\,dx + N(x,y)\,dy = 0$$

becomes

$$\frac{dy}{dx} = \frac{-M(x,y)}{N(x,y)} = -\frac{\left(\frac{1}{x}\right)^{-n} M(1, \frac{y}{x})}{\left(\frac{1}{x}\right)^{-n} N(1, \frac{y}{x})}.$$

Simplifying gives

$$\frac{dy}{dx} = -\frac{M(1, \frac{y}{x})}{N(1, \frac{y}{x})}.$$

c. Show that both numerator and denominator of the right-hand side of

$$\frac{dy}{dx} = -\frac{M(1, \frac{y}{x})}{N(1, \frac{y}{x})}$$

are in the form $g(y/x)$ and conclude that if M and N are both homogeneous functions of the <u>same</u> degree n, then the differential equation

$$M(x,y)\,dx + N(x,y)\,dy = 0$$

is a homogeneous differential equation.

42. Using the idea presented in problem 41, show that each of the equations in problems 23–40 are homogeneous.

43. Suppose that the equation $M(x, y)\,dx + N(x, y)\,dy = 0$ is homogeneous. Show that the transformation $x = r\cos\theta$, $y = r\sin\theta$ reduces this equation to a separable equation in the variables r and θ.

44. Use the method of problem 43 to solve the following equations:
    a. $(x + y)\,dx - x\,dy = 0$
    b. $(2xy + 3y^2)\,dx - (2xy + x^2)\,dy = 0$

45. a. Solve
$$\frac{dy}{dx} = \frac{y - x}{y + x}.$$

   b. Now consider
$$\frac{dy}{dx} = \frac{y - x + 1}{y + x + 5}. \tag{1.5}$$

   (i) Show that this equation is NOT homogeneous.
   How can we solve this? Consider the equations $y - x = 0$ and $y + x = 0$. They represent two straight lines through the origin. The intersection of $y - x + 1 = 0$ and $y + x + 5 = 0$ is $(-2, -3)$. Check it! Let $x = X - 2$ and $y = Y - 3$. This amounts to taking new axes parallel to the old with an origin at $(-2, -3)$.
   (ii) Use this transformation to obtain the differential equation

$$\frac{dY}{dX} = \frac{Y - X}{Y + X}.$$

   (iii) Using the solution from part **a**, obtain the solution to (1.5).

   Use the technique of problem 45 to solve problems 46–50.

46. $(2x + y + 1)dx - (4x + 2y - 3)dy = 0$

47. $x - y - 1 + (y - x + 2)y' = 0$

48. $(x + 4y)y' = 2x + 3y - 5$

49. $(y + 2)dx = (2x + y - 4)dy$

50. $y' = 2(\frac{y+2}{x+y-1})^2$

1.3   Some Physical Problems Arising as Separable Equations

Now that we have studied separable equations in detail, we consider some applications. The wide variety of application problems that we will consider all lead to equations in which variables can be separated.

1.3.1   Free Fall, Neglecting Air Resistance

We will begin this application section with an easy problem from elementary physics. This application should be very familiar.

If $x(t)$ represents the position of a particle at time t, then the velocity of the particle is given by

$$v(t) = \frac{dx}{dt}.$$

Similarly, the acceleration of the particle is

$$a(t) = \frac{dv}{dt} = \frac{d^2x}{dt^2}.$$

Thus, if we consider a particle that is in free fall, where the acceleration of the particle is due to gravity alone, we have

$$a(t) = -g.$$

Here g is assumed to be a constant and we use $-g$ as gravity acts downward. For the moment, we ignore the effects of air resistance. Thus,

$$\frac{dv}{dt} = -g,$$

which is a simple separable equation, so that

$$v(t) = -gt + c.$$

If we assume that the particle has an initial velocity v_0, so that $v(0) = v_0$, then $v(t) = -gt + v_0$. Now this gives the separable equation

$$\frac{dx}{dt} = -gt + v_0$$

which has solution

$$x(t) = \frac{-g}{2}t^2 + v_0 t + C_1.$$

If the particle has initial position x_0, then

$$x(0) = x_0$$

which gives

$$x(t) = \frac{-g}{2}t^2 + v_0 t + x_0 \tag{1.6}$$

as the position $x(t)$ of the particle in free fall, at time t.

Example 1: A man standing on a cliff 60 m high hurls a stone upward at a rate of 20 m/sec. How long does the stone remain in the air and with what speed does it hit the ground below the cliff?

Here $x_0 = 60$ and $v_0 = 20$. We take $g = 9.8$ m/sec^2. Thus,

$$x(t) = -\frac{9.8}{2}t^2 + 20t + 60$$

and

$$v(t) = -9.8t + 20.$$

The stone is in the air while $x(t) > 0$, so to find the time t that the stone is in the air, we set $x(t) = 0$ and solve for t. Using the quadratic equation,

$$t = \frac{-20 \pm \sqrt{(20)^2 - 4(-4.9)(60)}}{2(-4.9)} = -2.01, \; 6.09.$$

The stone is thus in the air for about 6.1 sec. We use this time to find the velocity upon impact:

$$v(6.1) = -9.8(6.1) + 20 = -39.78 \; \text{m/sec.}$$

1.3.2   Air Resistance

We will now consider the effects of air resistance. The amount of air resistance (sometimes called the *drag force*) depends upon the size and velocity of the object, but there is no general law expressing this dependence. Experimental evidence shows that at very low velocities for small objects it is best to approximate the resistance R as proportional to the velocity, while for larger objects and higher velocities it is better to consider it as proportional to the square of the velocity [28].

By Newton's second law $F = ma$, so that if $v(t)$ is the velocity of the object, we have

$$m\frac{dv}{dt} = F_1 + F_2$$

where F_1 is the weight of the object,

$$F_1 = mg,$$

and F_2 is the force of the air resistance on the object as it falls, so

$$F_2 = k_1 v \quad \text{or} \quad F_2 = k_2 v^2$$

where k_1, k_2 are proportionality constants. Note that $k_i < 0$ because air resistance is always opposite the velocity; see examples 2 and 3 below. We also point out that the units of k_1 and k_2 are different. In SI units, force has units of Newtons $= N = \text{kg} \cdot \text{m/sec}^2$. Thus k_1 must have the units of kg/sec. On the other hand, k_2 can be written as

$$k_2 = -\frac{1}{2}C\rho A$$

where ρ is the air density (SI units of kg/m^3), A is the cross-sectional area of the object (SI units of m^2), and C is the drag coefficient (unitless) [28].

Example 2: An object weighing 8 pounds falls from rest toward earth from a great height. Assume that air resistance acts on it with a force equal to $2v$. Calculate the velocity $v(t)$ and position $x(t)$ at any time. Find and interpret $\lim_{t\to\infty} v(t)$.

Remembering that pounds is a *force* (not a mass), we see that we need to calculate the mass of the object in order to apply Newton's second law. Using $g = 32$ ft/sec^2 gives $m = w/g = 8/32 = 1/4$. Thus by Newton's second law

$$m\frac{dv}{dt} = F_1 + F_2,$$

that is

$$\frac{1}{4}\frac{dv}{dt} = 8 - 2v.$$

This is a separable equation and can be written as

$$\frac{dv}{8 - 2v} = 4dt$$

so that upon integrating both sides we have

$$-\frac{1}{2}\ln|8 - 2v| = 4t + c.$$

Using the condition that the object fell from rest, so that $v(0) = 0$, we can determine the constant c and solve for $v(t)$. We have

$$v(t) = 4 - 4e^{-8t}$$

as the velocity of the object at any time. A graph of this velocity is shown in Figure 1.3. Figure 1.3 shows the behavior of the velocity over time. Analytically, we see that $v(t)$ approaches 4 as $t \to \infty$. This value is known as the *limiting* or *terminal velocity* of the object.

Now since $\frac{dx}{dt} = v(t)$, we have

$$\frac{dx}{dt} = 4 - 4e^{-8t}.$$

This is easily integrated to obtain $x(t) = 4t + \frac{1}{2}e^{-8t} + c$. If we take the initial position of the object as zero, so that $x(0) = 0$, then

$$x(t) = 4t + \frac{1}{2}e^{-8t} - \frac{1}{2}.$$

Computer Code 1.6:   **Basic plot with axes labeled**

Matlab, Maple, Mathematica

```
                          Matlab
>>  t=0:.01:0.5;
>>  y=4-4*exp(-8*t);
>>  plot(t,y)
>>  axis([0 0.5 0 4]);
>>  xlabel('time');
>>  ylabel('velocity');
```

```
                           Maple
>  eq1:=4-4*exp(-8*t);
>  plot(eq1,t=0..0.5,0..4,labels=[time,velocity]);
```

```
                        Mathematica
eq1[t_]=4 − 4e^{-8t}  (*e is from palette*)
Plot[eq[t], {t, 0,0.5}, AxesLabel → {"time", "velocity" }]
```

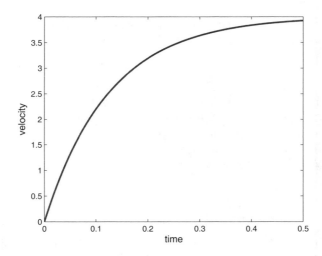

FIGURE 1.3:   Approach to terminal velocity of free-falling object of Example 2.

1.3.3   A Cool Problem

In addition to free-fall problems, separable equations arise in some simple thermodynamics applications. One such application is the following example.

Example 3: Suppose that a pie is removed from a 350°F oven and placed in a room with a temperature of 75°F. In 15 min the pie has a temperature of 150°F. We want to determine the time required to cool the pie to a temperature of 80°F, when we can finally enjoy eating it.

This example is an application of **Newton's law of cooling**, which states

> *the rate at which the temperature $T(t)$ changes in a cooling body is proportional to the difference between the temperature of the body and the constant temperature T_s of the surrounding medium.*

Symbolically we know the rate of change is the derivative and the statement is expressed as

$$\frac{dT}{dt} = k(T - T_s), \tag{1.7}$$

with the initial temperature of the body $T(0) = T_0$ and k a constant of proportionality. We observe that if the initial temperature T_0 is larger than the temperature of the surrounding T_s, then $T(t)$ will be a decreasing function of t (as the body is cooling), so $dT/dt < 0$, but $T_0 - T_s > 0$ so that the proportionality constant k must be negative. A similar analysis with $T_0 < T_s$ also gives $k < 0$. This condition on k also follows by noting that the temperature of the body will approach that of the surrounding medium as time gets large.

To solve (1.7), we seek a function $T(t)$ that describes the temperature at time t. For this equation, separating the variables we have

$$\frac{dT}{T - T_s} = k\, dt.$$

Integrating both sides of this equation gives

$$\int \frac{dT}{T - T_s} = \int k\, dt.$$

Evaluating both integrals, we obtain

$$\ln |T - T_s| = kt + C,$$

where C is the constant of integration. Exponentiating both sides and simplifying gives

$$|T - T_s| = e^{kt} e^C \implies T - T_s = \pm e^C e^{kt}.$$

Solving for the temperature, we see that

$$T(t) = C_1 e^{kt} + T_s$$

where $C_1 = \pm e^C$. We can then apply the initial condition $T(0) = T_0$, which implies $T_0 = C_1 + T_s$, so that $C_1 = T_0 - T_s$ and the solution is then

$$T(t) = (T_0 - T_s)e^{kt} + T_s. \tag{1.8}$$

We know that the temperature of the body approaches that of its surroundings and this can be seen mathematically as

$$\lim_{t \to \infty} T(t) = T_s,$$

which is true because $k < 0$.

Returning to our pie-cooling example, we see that $T_0 = 350$ and $T_s = 75$. Substituting these values in (1.8) gives

$$T(t) = 275e^{kt} + 75.$$

We still need to find k or equivalently e^k, which quantifies how fast the cooling of the pie occurs. We were given the temperature after 15 min, i.e., $T(15) = 150$. Thus

$$275e^{15k} + 75 = 150,$$

and solving for e^k gives

$$e^k = \left(\frac{3}{11}\right)^{1/15},$$

or $k = -0.08662$. Thus

$$T(t) = 275\left(\frac{3}{11}\right)^{t/15} + 75,$$

and this can be used to find the temperature of the pie at any given time. We can also calculate the time it takes to cool to any given temperature. We want to know when $T(t) = 80°\mathrm{F}$. Thus we solve

$$275\left(\frac{3}{11}\right)^{t/15} + 75 = 80$$

for t to obtain

$$t = \frac{-15\ln 55}{\ln 3 - \ln 11} \approx 46.264.$$

Thus, the pie will reach a temperature of $80°\mathrm{F}$ after approximately 46 min.

It is interesting to note that the first term in our equation for the pie temperature satisfies

$$275\left(\frac{3}{11}\right)^{t/15} > 0$$

for all $t > 0$. Thus

$$T(t) = 275\left(\frac{3}{11}\right)^{t/15} + 75 > 75.$$

The pie never actually reaches room temperature! This is an artifact of our model; we do note, however, that

$$\lim_{t \to \infty} 275 \left(\frac{3}{11} \right)^{t/15} + 75 = 75,$$

which can also be seen in Figure 1.4.

To reinforce the syntax of your favorite software package, we will present one example in each section using the packages.

Computer Code 1.7:   **Basic plot with axes adjusted**

Matlab, Maple, Mathematica

Matlab
```
>>  t=0:.1:80;
>>  y=275*(3/11).^(t/15) +75;
>>  plot(t,y)
>>  hold on
>>  y1=75*ones(size(t));
>>  plot(t,y1);
>>  axis([0 80 0 350]);
>>  hold off
```

Maple
```
> eq1:=275*(3/11)^(t/15)+75;
> plot(eq1,t=0..80,0..350);
```

Mathematica
```
Plot[275(3/11)^(t/15) + 75, {t, 0, 80} PlotRange -> {0, 350}]
```

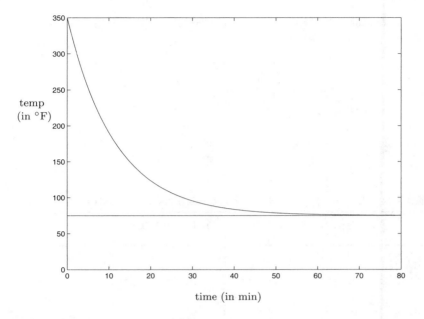

FIGURE 1.4:   Graph of pie temperature vs. time.

We present another example from forensic science of Newton's law of cooling.

Example 4: In the investigation of a homicide, the time of death is important. The normal body temperature of most healthy people is 98.6°F. Suppose that when a body is discovered at noon, its temperature is 82°F. Two hours later it is 72°F. If the temperature of the surroundings is 65°F, what was the approximate time of death?

This problem is solved as the last example. Here $T(0)$ represents the temperature when the body was discovered and $T(2)$ is the temperature of the body 2 hours later.

Thus, $T_0 = 82$ and $T_s = 65$ so that (1.8) becomes

$$T(t) = 17e^{kt} + 65.$$

Using $T(2) = 72$, we solve

$$17e^{2k} + 65 = 72$$

for e^k to find

$$e^k = \left(\frac{7}{17}\right)^{1/2}$$

so that

$$T(t) = 17 \left(\frac{7}{17}\right)^{t/2} + 65.$$

This equation gives us the temperature of the body at any given time. To find the time of death, we use the fact that the body temperature was at 98.6°F at this time. Thus we solve

$$17 \left(\frac{7}{17}\right)^{t/2} + 65 = 98.6$$

for t and find that

$$t = \frac{2\ln(1.97647)}{\ln 7 - \ln 17} \approx -1.53569.$$

This means that the time of death occurred approximately 1.53 hours before being discovered. Therefore, the time of death was approximately 10:30 a.m. because the body was found at noon.

1.3.4 Mixture Problems

Problems involving mixing typically give rise to separable differential equations. A typical mixture problem is given in the following example.

Example 5: A bucket contains 10 L of water and to it is being added a salt solution that contains 0.3 kg of salt per liter. This salt solution is being poured in at the rate of 2 L/min. The solution is being thoroughly mixed and drained off. The mixture is drained off at the same rate so that the bucket contains 10 L at all times. How much salt is in the bucket after 5 min?

Let $y(t)$ be the number of kilograms of salt in the bucket at the end of t minutes. We need to derive a differential equation for this problem and we do so by considering change in this system over a small time interval. We first find the amount of salt added to the bucket between time t and time $t + \Delta t$. Each minute, 2 L of solution is added so that in Δt minutes, $2\Delta t$ liters is added.

In these $2\Delta t$ liters the amount of salt is

$$0.3 \text{ kg/L} \times (2\Delta t) \text{ L} = (0.6\Delta t) \text{ kg.}$$

On the other hand, $2\Delta t$ liters of solution is withdrawn from the bucket in an interval Δt. Now at time t the 10 L in the flask contains $y(t)$ kilograms of salt. Then $2\Delta t$ of these liters contains approximately $(0.2\Delta t)(y(t))$ kilograms of salt if we suppose that the change in the amount of salt $y(t)$ is small in the short period of time Δt.

We have computed the amount of salt added in the interval $(t, t + \Delta t)$, as well as the amount subtracted in the same interval. But the difference between the amounts of salt present at times $t + \Delta t$ and t is $y(t + \Delta t) - y(t)$, so that we have obtained the equation

$$y(t + \Delta t) - y(t) = 0.6\Delta t - (0.2\Delta t)(y(t)).$$

We now divide by Δt and let $\Delta t \to 0$. The left side approaches the derivative $y'(t)$, and the right side is $0.6 - 0.2y(t)$. The differential equation is thus

$$y'(t) = 0.6 - 0.2y(t), \tag{1.9}$$

which can be thought of as the rate of change in the number of kilograms of salt in the bucket $y'(t)$ is equal to the rate of salt (in kg) flowing into the bucket 0.6 ($= 0.3$ kg/L $\times$ 2 L) minus the rate of salt flowing out of the bucket $0.2y(t)$.

Equation (1.9) is a separable equation and can be written as

$$\frac{dy}{0.6 - 0.2y} = dt.$$

Integrating both sides gives

$$\ln|0.6 - 0.2y| = -0.2t + c$$

so that solving for $y(t)$ we obtain

$$y(t) = 3 - Ce^{-0.2t}. \tag{1.10}$$

When t is zero, the amount of salt in the bucket is zero, that is, $y(0) = 0$. Equation (1.10) shows that when $t = 0$, we have

$$y(0) = 3 - C;$$

or $C = 3$. The value of C is now known, so that equation (1.10) becomes

$$y(t) = 3 - 3e^{-0.2t}.$$

To find y at the end of 5 min, we simply substitute $t = 5$ so that the amount of salt in the bucket is $y(5) \approx 1.9$ kg.

Problems

These problems deal with the subjects of air resistance, cooling mixing covered in this section. They also illustrate other concepts that yield separable equations, including radioactive decay problems, geometrical problems, and a few others.

In problems 1–7 it will be convenient to take the velocity to be the unknown function.

1. Assume that air resistance is proportional to the square of velocity. The terminal velocity of a 75-kg human in air of standard density is 60 m/sec. [28]. Neglecting the variation of air density with altitude, find when a man's parachute should be opened, assuming that he falls from an altitude of 1.8 km, and his parachute must open when he reaches an altitude of 0.5 km.

2. The mass of a football is 0.4 kg. Air resists passage of the ball, the resistive force being proportional to the square of the velocity, and being equal to 0.48 N when the velocity is 1 m/sec. Find the height to which the ball will rise, and the time to reach that height if it is thrown upward with a velocity of 20 m/sec. How is the answer altered if air resistance is neglected?

3. The football of the preceding exercise is released (from rest) at an altitude of 16.3 m. Find its final velocity and time of fall.

4. A ball thrown straight up climbs for 3.0 sec before falling. Neglecting air resistance, with what velocity was the ball thrown?

5. A ball dropped from a building falls for 4.00 sec before it hits the ground. If air resistance is neglected, answer the following questions:
   a. What was its final velocity just as it hit the ground?
   b. What was the average velocity during the fall?
   c. How high was the building?

6. You drop a rock from a cliff, and 5.00 sec later you see it hit the ground. Neglecting air resistance, how high is the cliff?

7. Superman is flying at treetop level near Paris when he sees the Eiffel Tower elevator start to fall (the cable snapped). His x-ray vision tells him Lois Lane is inside. If Superman is 2 km away from the tower, and the elevator falls from a height of 350 m, how long does Superman have to save Lois, and what must be his average velocity? Solve this problem with and without assuming air resistance. (Of course, Superman instantly does the calculations required in both of these problems, as he is an expert in differential equations!)

Problems 8–12 concern Newton's law of cooling.

8. At the request of their children, Randy and Stephen make homemade popsicles. At 2:00 p.m., Kaelin asks if the popsicles are frozen (0°C), at which time they test the temperature of a popsicle and find it to be 5°C. If they put the popsicles with a temperature of 15°C in the freezer at 12:00 noon and the temperature of the freezer is −2°C, when will Alan, Erin, Kaelin, Robyn and Ryley be able to enjoy the popsicles?

9. Determine the time of death if a corpse is 79°F when discovered at 3:00 p.m. and 68°F 3 hours later. Assume that the temperature of the surroundings is 60°F and that normal body temperature is 98.6°F.

10. An object cools in 10 min from 100°C to 60°C. The surroundings are at a temperature of 20°C. When will the body cool to 25°C?

11. A container holds 1 kg of water at 20°C. A 0.5 kg mass of aluminum is added at 75°C. The heat capacity of aluminum is 0.2. (To say that the specific heat capacity of aluminum is 0.2 means that 1 kg of aluminum contains as much heat as 0.2 kg of water.) In 1 minute's time the water is warmed 2°C. When did the aluminum cool by 1°C, and when did the water warm by 1°C? Assume that the water loses no heat to its surroundings.

12. A slug of metal at a temperature of 800°F is put in an oven, the temperature of which is gradually increased during an hour from $a°$ to $b°$. Find the temperature of the metal at the end of an hour, assuming that the metal warms kT degrees per minute when it finds itself in an oven that is T degrees warmer.

In problems 13–17 it is supposed that the amount of gas (or liquid) contained in any fixed volume is constant. Also, thorough mixing is assumed.

13. A 20 L vessel contains air (assumed to be 80% nitrogen and 20% oxygen). Suppose 0.1 L of nitrogen is added to the container per second. If continual mixing takes place and material is withdrawn at the rate at which it is added, how long will it be before the container holds 99% nitrogen?

14. A 100-L beaker contains 10 kg of salt. Water is added at the constant rate of 5 L/min with complete mixing, and drawn off at the same rate. How much salt is in the beaker after 1 hour?

15. A tank contains 25 lb of salt dissolved in 50 gal of water. Brine containing 4 lb/gal is allowed to enter at a rate of 2 gal/min. If the solution is drained at the same rate find the amount of salt as a function $S(t)$ of time t. Find the concentration of salt at time. Suppose the rate of draining is modified to be 3 gal/min. Find the amount of salt and the concentration at time t.

16. Consider a pond that has an initial volume of 10,000 m³. Suppose that at time $t = 0$, the water in the pond is clean and that the pond has two streams flowing into it, stream A and stream B, and one stream flowing out, stream C. Suppose 500 m³/day of water flows into the pond from stream A, 750 m³/day flows into the pond from stream B, and 1250 m³ flows out of the pond via stream C. At $t = 0$, the water flowing into the pond from stream A becomes contaminated with road salt at a concentration of 5 kg/1000 m³. Suppose the water in the pond is well mixed so the concentration of salt at any given time is constant. To make matters worse, suppose also that at time $t = 0$ someone begins dumping trash into the pond at a rate of 50 m³/day. The trash settles to the bottom of the pond, reducing the volume by 50 m³/day. To

adjust for the incoming trash, the rate that water flows out via stream C increases to 1300 m^3/day and the banks of the pond do not overflow. Determine how the amount of salt in the pond changes over time. Does the amount of salt in the pond reach 0 after some time has passed?

17. A large chamber contains 200 m^3 of gas, 0.15% of which is carbon dioxide (CO_2). A ventilator exchanges 20 m^3/min of this gas with new gas containing only 0.04% CO_2. How long will it be before the concentration of CO_2 is reduced to half its original value?

Problems 18–21 concern radioactive decay. The decay law states that the amount of radioactive substance that decays is proportional at each instant to the amount of substance present.

18. The strength of a radioactive substance decreases 50% in a 30-day period. How long will it take for the radioactivity to decrease to 1% of its initial value?

19. It is experimentally determined that every gram of radium loses 0.44 mg in 1 year. What length of time elapses before the radioactivity decreases to half its original value?

20. A tin organ pipe decays with age as a result of a chemical reaction that is catalyzed by the decayed tin. As a result, the rate at which the tin decays is proportional to the product of the amount of tin left and the amount that has already decayed. Let M be the total amount of tin before any has decayed. Find the amount of decayed tin $p(t)$.

21. A certain piece of mineral contains 100 mg of uranium and 14 mg of uranium lead. It is known that uranium loses half its radioactivity in 4.5×10^9 years and the original amount of 238 g of uranium decays to 206 g of uranium lead. Calculate the mineral's age. (Assume that when the mineral was born it contained no lead. Also, neglect the intermediate products of composition since the products to which uranium decomposes change themselves much more rapidly than uranium does.)

Problems 22–26 deal with geometric situations where the derivative arises and yields a separable equation.

22. Find a curve for which the area of the triangle determined by the tangent, the ordinate to the point of tangency, and the x-axis has a constant value equal to a^2.

23. Find a curve for which the sum of the sides of a triangle constructed as in the previous problem has a constant value equal to b.

24. Find a curve with the following property: the segment of the x-axis included between the tangent and normal at any point on the curve is equal to $2a$.

25. Find a curve with the following property: if through an arbitrary point of the curve parallels are drawn to the coordinate axes and meet these axes forming a rectangle, the area of this rectangle is divided by the curve in the ratio 1:2.

26. Find a curve such that the tangent at an arbitrary point makes equal angles with the radius vector and the polar axis (principal direction).

In problems 27–31, assume that water emerging from an aperture in a vessel has velocity $0.6 \times 2gh$ m/sec, where $g = 9.8$ m/sec^2 is the force of gravity, and h is the height of the surface of the water above the aperture.

27. A vertical cylindrical vessel has diameter $2R = 1.8$ m and height $H = 2.45$ m. How long will it take to empty the vessel through a hole in the bottom of diameter $2r = 6$ cm?

28. Answer the same question if the axis of the cylinder is horizontal and the hole is at a lowest point.

29. A cylindrical beaker with vertical axis can be half drained through a hole in the bottom in 5 min. How long would it take to empty the beaker completely?

30. A conical funnel has radius $R = 6$ cm and height $H = 10$ cm. If the opening of the funnel is a circle of diameter 0.5 cm, how long will it take to empty the entire funnel if it is initially full of water?

31. A rectangular vessel has base 60 cm × 75 cm, and height 80 cm, with an opening in the bottom 2.5 cm^2 in area. Water is being added to it at the rate of 1.8 L/sec (1 L = 1000 cm^3). How long does it take to fill the vessel? Compare the answer with the result that would be obtained on neglecting the hole in the bottom.

The remaining problems are a collection of physical examples that also give rise to separable equations.

32. *Rate of Learning.* The rate of learning is proportional to the amount learned. Suppose you are memorizing 100 numbers. Is it easier to learn the first 10 or the middle 10? (Here, the more you memorize, the longer it takes to learn each successive number.)

33. On an early Monday morning in February in rural Kentucky (not far from Western Kentucky University) it started to snow. There had been no snow on the ground before. It was snowing at a steady, constant rate so that the thickness of the snow on the ground was increasing at a constant rate. A snowplow began clearing the snow from the streets at noon. The speed of the snowplow in clearing the snow is inversely proportional to the thickness of the snow. The snowplow traveled two

miles during the first hour after noon and traveled one mile during the second hour after noon. At what time did it begin snowing?

34. A raft is being slowed down by resistance of the water, the resistance being proportional to the speed of the raft. If the initial speed was 1.5 m/sec and at the end of 4 sec was 1 m/sec, when will the speed decrease to 1 cm/sec? What total distance will the raft travel?

35. The amount of light absorbed by a thin layer of water is proportional to the amount of incident light and to the thickness of the layer. If a layer of water 35 cm thick absorbs half the light incident on its surface, what proportion of the incident light will be absorbed by a layer of water 200 cm thick?

36. A uniform extensible cord 1 m long is stretched $k \cdot f$ meters by a force of f kg. A cord of the same material 2 m long has mass P kg. If it is suspended by one end, how much is it extended under the weight of its own mass?

37. Assume the density of air is 0.0012 g/cm^3. Neglecting any variation in temperature, pressure is proportional to the density and is 1 kg/cm^2 at the earth's surface. Find the pressure as a function of the height h.

38. A closed vessel with volume V m^3 contains liquid water and air. The speed of evaporation of the water is proportional to the difference between the saturation concentration q_1 of water vapor (amount per m^3) at the given temperature, and the amount q of water vapor per m^3 actually present in the air (assume that the temperature of the air and water, and the amount of area on which evaporation occurs do not change). Initially there are m_0 grams of water in the vessel, and q_0 grams of vapor in each m^3 of air. How much water remains in the vessel at the end of t units of time?

39. The mass of a rocket, including a full chamber of fuel, is M; its net mass (without fuel) is m. The products of combustion are ejected with velocity c. If the rocket starts from rest, derive *Ciolkovskii's formula*, which gives the speed imparted to the rocket by the burning of the fuel, neglecting the resistance of the atmosphere.

1.4   Exact Equations

We will now introduce another type of differential equation. Exact equations are not separable equations and their solution will provide us important insight into several other types of differential equations.

Consider the first-order differential equation

$$\frac{dy}{dx} = f(x, y);$$

we observe that it can always be expressed in the differential form

$$M(x,y)\,dx + N(x,y)\,dy = 0$$

or equivalently as

$$M(x,y) + N(x,y)\,\frac{dy}{dx} = 0$$

and vice versa. We will now consider a type of differential equation that is not separable, but, nevertheless, has a solution. We need a definition from multivariable calculus to proceed:

DEFINITION 1.4   *Let $F(x,y)$ be a function of two real variables such that F has continuous first partial derivatives in a domain D. The total differential dF of F is defined by*

$$dF(x,y) = \frac{\partial F(x,y)}{\partial x}\,dx + \frac{\partial F(x,y)}{\partial y}\,dy$$

for all $(x,y) \in D$.

Example 1: Suppose $F(x,y) = xy^2 + 2x^3y$; then

$$\frac{\partial F}{\partial x} = y^2 + 6x^2y \quad \text{and} \quad \frac{\partial F}{\partial y} = 2xy + 2x^3$$

so that the total differential dF is given by

$$dF(x,y) = \frac{\partial F(x,y)}{\partial x}\,dx + \frac{\partial F(x,y)}{\partial y}\,dy$$

$$= (y^2 + 6x^2y)\,dx + (2xy + 2x^3)\,dy.$$

DEFINITION 1.5   *The expression*

$$M(x,y)\,dx + N(x,y)\,dy \tag{1.11}$$

is called an exact differential in a domain D if there exists a function F of two real variables such that this expression equals the total differential $dF(x,y)$ for all $(x,y) \in D$. That is, (1.11) is an exact differential in D if there exists a function F such that

$$\frac{\partial F}{\partial x} = M(x,y) \quad \text{and} \quad \frac{\partial F}{\partial y} = N(x,y)$$

for all $(x,y) \in D$.

If $M(x, y)\, dx + N(x, y)\, dy$ is an exact differential, then the differential equation

$$M(x, y)\, dx + N(x, y)\, dy = 0 \qquad (1.12)$$

is called an *exact differential equation*. Equivalently, we can use the form

$$M(x, y) + N(x, y)\, \frac{dy}{dx} = 0 \qquad (1.13)$$

as the standard form for an exact equation.

Example 2: The differential equation

$$y^2 + 2xy\, \frac{dy}{dx} = 0$$

is exact, since if $F(x, y) = xy^2$ then

$$\frac{\partial F}{\partial x} = y^2 \text{ and } \frac{\partial F}{\partial y} = 2xy.$$

Not all differential equations, however, are exact. Consider

$$y + 2x\, \frac{dy}{dx} = 0.$$

We cannot find an $F(x, y)$ so that

$$\frac{\partial F}{\partial x} = y \text{ and } \frac{\partial F}{\partial y} = 2x.$$

Numerous trials and errors may be enough to convince us that this is the case. What we really need is a method for testing a differential equation for exactness and for constructing the corresponding function $F(x, y)$. Both are contained in the following theorem and its proof.

THEOREM 1.4.1 *Consider the differential equation*

$$M(x, y) + N(x, y)\, \frac{dy}{dx} = 0 \qquad (1.14)$$

where M and N have continuous first partial derivatives at all points (x, y) in a rectangular domain D. Then the differential equation (1.14) is exact in D, **if and only if**

$$\frac{\partial M(x, y)}{\partial y} = \frac{\partial N(x, y)}{\partial x} \qquad (1.15)$$

for all (x, y) in D.

Remark: The proof of this theorem is rather important, as it not only provides a test for exactness, but also a method of solution for exact differential equations.

Proof: To prove one direction of the theorem, we first suppose the differential equation (1.14) is exact in D and show that (1.15) must hold as a result. If (1.14) is exact, then there is a function F such that

$$\frac{\partial F}{\partial x} = M(x,y) \quad \text{and} \quad \frac{\partial F}{\partial y} = N(x,y).$$

So

$$\frac{\partial^2 F}{\partial y \partial x} = \frac{\partial M}{\partial y} \quad \text{and} \quad \frac{\partial^2 F}{\partial x \partial y} = \frac{\partial N}{\partial x}$$

by differentiation. Now we have assumed the continuity of the first partials of M and N in D, so that

$$\frac{\partial^2 F}{\partial y \partial x} = \frac{\partial^2 F}{\partial x \partial y}.$$

This means that

$$\frac{\partial M}{\partial y} = \frac{\partial N}{\partial x},$$

which is the same as (1.15).

To prove the other direction, we assume (1.15) and show that (1.14) must be exact. (Proving this direction will also show us how to construct the solution for a given exact equation.) Thus, we assume

$$\frac{\partial M}{\partial y} = \frac{\partial N}{\partial x}$$

and find an F so that

$$\frac{\partial F}{\partial x} = M(x,y) \quad \text{and} \quad \frac{\partial F}{\partial y} = N(x,y). \tag{1.16}$$

It is clear that we can find an F that satisfies either of these equations, but can we find an F that satisfies both? Let's proceed and see what happens. Suppose that F satisfies

$$\frac{\partial F}{\partial x} = M(x,y).$$

We can integrate both sides of this equation to get

$$F(x,y) = \int M(x,y)\, dx + \phi(y) \tag{1.17}$$

where $\int M(x,y)\, dx$ is the partial integration with respect to x holding y constant. Note that our "constant" of integration, $\phi(y)$, is a function but is a function of y only (it might also include an additive constant, but definitely no x). This is because the expression $\partial F/\partial x$ would result in the loss of any "only y functions." Now we need to find an $F(x, y)$ that satisfies both equations in (1.16). We thus need to make sure the $F(x, y)$ in (1.17) also satisfies $\frac{\partial F}{\partial y} = N(x, y)$. We calculate $\partial F / \partial y$ by differentiating (1.17) with respect to y:

$$\frac{\partial F}{\partial y} = \frac{\partial}{\partial y} \int M(x, y) \, dx + \frac{d\phi(y)}{dy}.$$

Equating with $N(x, y)$ gives

$$N(x, y) = \left(\frac{\partial}{\partial y} \int M(x, y) \, dx \right) + \phi'(y),$$

where $\phi'(y) = d\phi(y)/dy$. Solving for $\phi'(y)$ gives

$$\phi'(y) = N(x, y) - \frac{\partial}{\partial y} \int M(x, y) \, dx.$$

Since $\phi(y)$ is a function of only y, it must also be the case that $\phi'(y)$ is a function of only y. We can see this by showing

$$\frac{\partial}{\partial x} \left(N(x, y) - \frac{\partial}{\partial y} \int M(x, y) \, dx \right) = 0.$$

Evaluating the left-hand side and simplifying give

$$\frac{\partial}{\partial x} \left(N(x, y) - \frac{\partial}{\partial y} \int M(x, y) \, dx \right) = \frac{\partial N}{\partial x} - \frac{\partial^2}{\partial x \partial y} \int M(x, y) \, dx$$

$$= \frac{\partial N}{\partial x} - \frac{\partial^2 F}{\partial x \partial y} \quad \text{(by noting what } F \text{ is)}$$

$$= \frac{\partial N}{\partial x} - \frac{\partial^2 F}{\partial y \partial x} \quad \text{(by continuity)}$$

$$= \frac{\partial N}{\partial x} - \frac{\partial^2}{\partial y \partial x} \int M(x, y) \, dx$$

$$= \frac{\partial N}{\partial x} - \frac{\partial M}{\partial y}$$

$$= 0,$$

where the last equality holds since we have assumed that

$$\frac{\partial N}{\partial x} = \frac{\partial M}{\partial y}.$$

What this means is that

$$N(x,y) - \frac{\partial}{\partial y} \int M(x,y)\,dx$$

cannot depend on x since its derivative with respect to x is zero. Hence,

$$\phi(y) = \int \left(N(x,y) - \frac{\partial}{\partial y} \int M(x,y)\,dx \right) dy$$

and thus

$$F(x,y) = \int M(x,y)\,dx + \phi(y)$$

is a function that satisfies both

$$\frac{\partial F}{\partial x} = M(x,y) \quad \text{and} \quad \frac{\partial F}{\partial y} = N(x,y).$$

Thus,

$$M(x,y) + N(x,y)\frac{dy}{dx} = 0$$

is exact in D. $\square$
   In short, the criterion for exactness is (1.15):

$$\frac{\partial N}{\partial x} = \frac{\partial M}{\partial y}.$$

If this equation holds, then the differential equation is exact. If this is not true, the differential equation is not exact.

Example 3: We considered the differential equation

$$y^2 + 2xy\frac{dy}{dx} = 0 \tag{1.18}$$

earlier. We see that

$$M(x,y) = y^2 \quad \text{and} \quad N(x,y) = 2xy.$$

Thus,

$$\frac{\partial M}{\partial y} = 2y = \frac{\partial N}{\partial x},$$

so that the differential equation is exact. On the other hand,

$$y + 2x\frac{dy}{dx} = 0 \tag{1.19}$$

gives $M(x,y) = y$ and $N(x,y) = 2x$ so that

$$\frac{\partial M}{\partial y} = 1 \neq 2 = \frac{\partial N}{\partial x}.$$

Hence $y + 2x\frac{dy}{dx} = 0$ is not exact.

Example 4: Consider the differential equation

$$(2x\sin y + y^3 e^x) + (x^2\cos y + 3y^2 e^x)\frac{dy}{dx} = 0.$$

So

$$M(x,y) = 2x\sin y + y^3 e^x \quad\text{and}\quad N(x,y) = x^2\cos y + 3y^2 e^x;$$

hence

$$\frac{\partial M}{\partial y} = 2x\cos y + 3y^2 e^x = \frac{\partial N}{\partial x}.$$

Thus the differential equation is exact.

Remark: The test for exactness applies to equations in the form

$$M(x,y) + N(x,y)\frac{dy}{dx} = 0. \tag{1.20}$$

If the left-hand side is an exact differential, then we can solve the exact differential equation (1.20) by finding a function $F(x,y)$ so that

$$\frac{\partial F(x,y)}{\partial x}\,dx + \frac{\partial F(x,y)}{\partial y}\,dy = 0.$$

More simply, using the total differential, we obtain $dF(x,y) = 0$. Thus,

$$F(x,y) = C$$

is a solution to (1.20).

THEOREM 1.4.2 *Suppose the differential equation*

$$M(x,y) + N(x,y)\frac{dy}{dx} = 0$$

is exact. Then the general solution of this differential equation is given implicitly by

$$F(x,y) = C,$$

where $F(x,y)$ is a function such that

$$\frac{\partial F}{\partial x} = M(x,y) \quad\text{and}\quad \frac{\partial F}{\partial y} = N(x,y).$$

Remark 1: As with separable and homogeneous equations, the constant in Theorem 1.4.2 is determined by an initial condition.

Remark 2: We have an explicit form for $F(x,y)$, namely,

$$F(x,y) = \int M(x,y)\,dx + \phi(y),$$

where $\phi(y) = \int \left(N(x, y) - \dfrac{\partial}{\partial y} \int M(x, y) \, dx \right) dy$. This form, however, is not always useful. We will see by example how to solve exact differential equations.

Remark 3: We integrated $\partial F / \partial x = M$ and substituted this into $\partial F / \partial y = N$. We instead could have solved $\partial F / \partial y = N$ first (by integrating with respect to y and obtaining a "constant" $\psi(x)$) and then substituted into $\partial F / \partial x = M$. The resulting F is the same but would be written

$$F(x, y) = \int N(x, y) \, dy + \psi(x), \qquad (1.21)$$

where $\psi(x) = \int \left(M(x, y) - \dfrac{\partial}{\partial x} \int N(x, y) \, dy \right) dx$. See problem 14 at the end of this section.

Example 5: Show that

$$(3x^2 + 4xy) + (2x^2 + 2y) \frac{dy}{dx} = 0$$

is exact and then solve it by the methods discussed in this section.

We have

$$M(x, y) = 3x^2 + 4xy \quad \text{and} \quad N(x, y) = 2x^2 + 2y$$

so that the equation is exact, since

$$\frac{\partial M}{\partial y} = 4x = \frac{\partial N}{\partial x}.$$

Our goal is to find an $F(x, y)$ that simultaneously satisfies the equations

$$\frac{\partial F}{\partial x} = M(x, y) \quad \text{and} \quad \frac{\partial F}{\partial y} = N(x, y).$$

That is, F must satisfy

$$\frac{\partial F}{\partial x} = 3x^2 + 4xy \quad \text{and} \quad \frac{\partial F}{\partial y} = 2x^2 + 2y.$$

Integrating $\partial F / \partial x$ with respect to x gives

$$F(x, y) = \int (3x^2 + 4xy) \, dx$$

$$= x^3 + 2x^2 y + \phi(y).$$

This same F must also satisfy $\partial F / \partial y = N$ and we then have

$$2x^2 + \phi'(y) = \frac{\partial F}{\partial y} = 2x^2 + 2y.$$

Thus, $\phi'(y) = 2y$. Integrating with respect to y gives

$$\phi(y) = y^2 + C_0$$

so that

$$F(x, y) = x^3 + 2x^2 y + y^2 + C_0.$$

Thus, a one-parameter family of solutions is given by

$$x^3 + 2x^2 y + y^2 = C.$$

We now solve an exact equation by first integrating with respect to y; see Remark 3 above.

Example 6: Show that

$$(2x \cos y + 3x^2 y) + (x^3 - x^2 \sin y - y) \frac{dy}{dx} = 0$$

is exact and solve it subject to the initial condition $y(0) = 2$. Plot the solution.

We have

$$M(x, y) = 2x \cos y + 3x^2 y \quad \text{and} \quad N(x, y) = x^3 - x^2 \sin y - y.$$

The equation is exact because

$$\frac{\partial M}{\partial y} = 3x^2 - 2x \sin y = \frac{\partial N}{\partial x}.$$

Now we find an $F(x, y)$ so that

$$\frac{\partial F}{\partial x} = M(x, y) \quad \text{and} \quad \frac{\partial F}{\partial y} = N(x, y).$$

This time we will integrate $\partial F/\partial y = N$ with respect to y. Thus

$$F(x, y) = \int N(x, y) \, dy$$

$$= \int (x^3 - x^2 \sin y - y) \, dy$$

$$= x^3 y + x^2 \cos y - \frac{y^2}{2} + \psi(x).$$

This must also satisfy $\partial F/\partial x = M$. Calculating $\partial F/\partial x$ gives

$$\frac{\partial F}{\partial x} = 3x^2 y + 2x \cos y + \psi'(x).$$

Substituting into

$$\frac{\partial F}{\partial x} = M(x, y)$$

gives $\psi'(x) = 0$, which is easily integrated to obtain $\psi(x) = C_1$. Thus,

$$F(x, y) = x^3 y + x^2 \cos y - \frac{1}{2} y^2 + C_1,$$

and a one-parameter family of solutions is

$$x^3 y + x^2 \cos y - \frac{1}{2} y^2 = C.$$

The initial condition $y(0) = 2$ gives $C = -2$. Hence

$$x^2 \cos y + x^3 y - \frac{1}{2} y^2 = -2$$

is the implicit solution that satisfies the given initial condition. The solution curves can be plotted in Matlab, Maple, or Mathematica using the code below as shown in Figure 1.5.

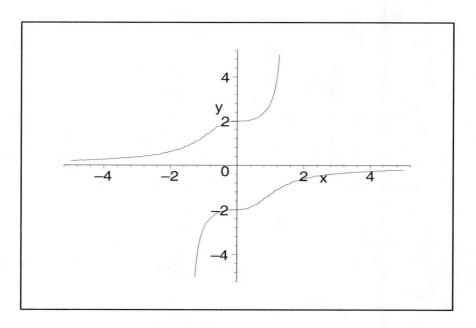

FIGURE 1.5:   Implicit plot for Example 6. The upper curve is the solution curve because it passes through the initial condition.

Computer Code 1.8:   **Single implicit plot,** $f(x,y) = C$

Matlab, Maple, Mathematica

Matlab

```
>>  [X,Y]=meshgrid(-4:.05:4,-4:.05:4);
>>  Z=X.^2.*cos(Y)+X.^3.*Y-Y.^2/2;
>>  contour(X,Y,Z,[-2 -2]); %Syntax for only ONE level curve
```

Maple

```
> eq1:=x^2*cos(y)+x^3*y-y^2/2=-2;
> with(plots):
> implicitplot(eq1,x=-4..4,y=-4..4,numpoints=1000);
```

Mathematica

```
<< Graphics`ImplicitPlot`
```
$eq1[x\_] = x^2 Cos[y] + x^3 y - \frac{y^2}{2}$ (*fractions entered from palette*)
```
ImplicitPlot[eq1[x]==-2, {x, -4, 4}, {y, -4, 4}]
```

Note that although both curves in Figure 1.5 satisfy the implicit equation, only one of these curves passes through the given initial condition and thus is the correct solution.

1.4.1   Solution by Grouping

There is a much slicker method for solving exact differential equations and it is known as the *method of grouping*. For better or worse, it requires a "working knowledge" of differentials and a certain amount of ingenuity. We again consider Example 5, this time in its differential form:

$$(3x^2 + 4xy)\,dx + (2x^2 + 2y)\,dy = 0.$$

We rewrite it in the form

$$3x^2\,dx + (4xy\,dx + 2x^2\,dy) + 2y\,dy = 0$$

which is

$$d(x^3) + d(2x^2 y) + d(y^2) = d(C).$$

That is,

$$d(x^3 + 2x^2 y + y^2) = d(C)$$

so that

$$x^3 + 2x^2 y + y^2 = C.$$

Clearly, this procedure is much quicker if we can find the appropriate grouping. Let's try this method one more time by again considering Example 6. We group the terms as

$$(2x \cos y \, dx - x^2 \sin y \, dy) + (3x^2 y \, dx + x^3 \, dy) - y \, dy = 0.$$

Thus, we have

$$d(x^2 \cos y) + d(x^3 y) - d \left(\frac{y^2}{2} \right) = d(C)$$

and so

$$x^2 \cos y + x^3 y - \frac{1}{2} y^2 = C$$

is a one-parameter family of solutions.

Important Note: If we use the method of grouping, we still need to check that the equation is exact for our first step.

Problems

Solve problems 1–11 by the methods of this section. Use the computer to plot the implicit or explicit solution. Use the given initial condition if appropriate or, if one is not given, plot the solutions for three different initial conditions.

1. $2xy^3 + (1 + 3x^2 y^2) \frac{dy}{dx} = 0$

2. $(2xy + 1) + (x^2 + 4y) \frac{dy}{dx} = 0$ with $y(0) = 1$

3. $(y \sec^2 x + \sec x \tan x) + (\tan x + 2y) \frac{dy}{dx} = 0$ with $y(0) = 1$

4. $(2y \sin x \cos x + y^2 \sin x) + (\sin^2 x - 2y \cos x) \frac{dy}{dx} = 0$, with $y(0) = 3$

5. $2xy + (x^2 - y^2) \frac{dy}{dx} = 0$

6. $(2 - 9xy^2)x + (4y^2 - 6x^3)y \frac{dy}{dx} = 0$ with $y(1) = 1$

7. $e^{-y} - (2y + xe^{-y}) \frac{dy}{dx} = 0$ with $y(1) = 3$

8. $(1 + y^2 \sin 2x) - y \cos 2x \frac{dy}{dx} = 0$

9. $3x^2 (1 + \ln y) = (2y - \frac{x^3}{y}) \frac{dy}{dx}$

10. $(2 + \frac{y}{x^2}) \, dx + (y - \frac{1}{x}) \, dy = 0$

11. $(\frac{x}{\sin y} + 2) + \frac{(x^2 + 1) \cos y}{\cos 2y - 1} \frac{dy}{dx} = 0$

12. Let x represent the units of labor and y represent the units of capital. If $f(x, y)$ measures the number of units produced, a differential equation satisfied by a level curve of it is

$$ax^{a-1} y^{1-a} + (1-a)x^a y^{-a} \frac{dy}{dx} = 0.$$

Solve this equation as (i) a separable equation and (ii) an exact equation. In doing so, obtain the well-known *Cobb-Douglas production function* $f(x, y) = Cx^a y^{1-a}$.

13. Determine the constant A such that the equation is exact, and solve the resulting exact equation:

    a. $(x^2 + 3xy) + (Ax^2 + 4y)\frac{dy}{dx} = 0$

    b. $\left(\frac{Ay}{x^3} + \frac{y}{x^2}\right) + \left(\frac{1}{x^2} - \frac{1}{x}\right)\frac{dy}{dx} = 0$

14. By following the proof of Theorem 1.4.1, show that an equivalent formulation of $F(x, y)$ is given by

$$F(x, y) = \int N(x, y)\frac{dy}{dx} + \int \left(M(x, y) - \frac{\partial}{\partial x}\int N(x, y)\frac{dy}{dx}\right).$$

Although this could easily be obtained by rearranging the previously obtained expression for F (1.17), do *not* simply rearrange terms.

15. Determine the most general function $N(x, y)$ so that

$$(x^3 + xy^2) + N(x, y)\frac{dy}{dx} = 0$$

is exact.

16. Determine the most general function $M(x, y)$ so that

$$M(x, y) + (2ye^x + y^2 e^{3x})\frac{dy}{dx} = 0$$

is exact.

17. Show that the homogeneous equation

$$(Ax + By) + (Cx + Ey)\frac{dy}{dx} = 0,$$

where A, B, C, and E are constants, is exact if and only if $B = C$.

18. Show that the homogeneous equation

$$(Ax^2 + Bxy + Cy^2) + (Ex^2 + Fxy + Gy^2)\frac{dy}{dx} = 0,$$

where A, B, C, E, F, and G are constants, is exact if and only if $B = 2E$ and $F = 2C$.

1.5 Linear Equations

Linear first-order differential equations are perhaps the most commonly arising class of differential equations in applications. A linear differential equation is defined as follows:

DEFINITION 1.6   *A first-order ordinary differential equation is linear in the dependent variable y and the independent variable x if it can be written as*

$$\frac{dy}{dx} + P(x)y = Q(x). \tag{1.22}$$

More generally, we often see equations of the form

$$a_0(x)y' + a_1(x)y = b(x)$$

but, provided $a_0(x) \neq 0$ for all x, we can always divide by $a_0(x)$ and define $P(x) = a_1/a_0$ and $Q(x) = b/a_0$ to obtain an equation of the form of (1.22).

In our work to follow, specifically in Chapters 3 and 4 we will refer to an equation of this form as a "linear nonhomogeneous equation."[2] In the case when $Q(x) = 0$, we refer to the equation as "homogeneous," but we caution the reader to be careful with the word "homogeneous" as this is not the same homogeneous that we have encountered previously. It is an unfortunate fact that mathematicians often use the same term for different mathematical notions; it is only by context that we can be sure what is being described.

Linear equations are not exact, but can be transformed into exact equations. To do so, we will multiply through by an appropriate *integrating factor*, thus transforming the equation into a form we can solve. Once this is done, we will see that a convenient formula results that will be valid for solving any linear first-order differential equation. We begin the first derivation of an integrating factor with an example.

Example 1: The first-order differential equation

$$x\frac{dy}{dx} + (x+1)y = x^3$$

is linear since it can be written

$$\frac{dy}{dx} + \left(1 + \frac{1}{x}\right)y = x^2.$$

Comparing with (1.22), we see that

$$P(x) = 1 + \frac{1}{x} \quad \text{and} \quad Q(x) = x^2.$$

---

[2]In the exercises, we give an alternative derivation of the solution by using variation of parameters. This derivation gives a preview of how to solve higher-order linear nonhomogeneous equations.

Now if we consider (1.22) in differential form, i.e.,

$$dy = (Q(x) - P(x)y) \, dx,$$

we find

$$(P(x)y - Q(x)) \, dx + dy = 0.$$

In this form, we have

$$M(x,y) = P(x)y - Q(x) \quad \text{and} \quad N(x,y) = 1$$

so that

$$\frac{\partial M}{\partial y} = P(x) \quad \text{and} \quad \frac{\partial N}{\partial x} = 0.$$

Hence, the equation is not exact, unless $P(x) = 0$, in which case we have

$$\frac{dy}{dx} = Q(x),$$

which is easily solved.

Question: Is there some way to obtain an exact equation?

The following definition answers this question.

DEFINITION 1.7 *If the differential equation*

$$M(x,y) \, dx + N(x,y) \, dy = 0 \qquad (1.23)$$

is not exact in a domain D but the differential equation

$$\mu(x,y)M(x,y) \, dx + \mu(x,y)N(x,y) \, dy = 0 \qquad (1.24)$$

is exact in D, then $\mu(x,y)$ is called an integrating factor of the differential equation (1.23).

Example 2: The differential equation

$$(3y + 4xy^2) \, dx + (2x + 3x^2y) \, dy = 0 \qquad (1.25)$$

is not exact since

$$\frac{\partial M}{\partial y} = 3 + 8xy \neq 2 + 6xy = \frac{\partial N}{\partial x}.$$

If we let

$$\mu(x,y) = x^2 y,$$

we can use (1.24) to rewrite (1.25) as

$$(x^2 y)(3y + 4xy^2) \, dx + (x^2 y)(2x + 3x^2 y) \, dy = 0.$$

Expanding gives

$$M = 3x^2 y^2 + 4x^3 y^3 \quad \text{and} \quad N = 2x^3 y + 3x^4 y^2.$$

Then

$$\frac{\partial}{\partial y}(3x^2 y^2 + 4x^3 y^3) = 6x^2 y + 12x^3 y^2 = \frac{\partial}{\partial x}(2x^3 y + 3x^4 y^2).$$

Thus the new equation is exact and hence $\mu(x, y) = x^2 y$ is an integrating factor.

Question: How do we find such a function?

Let's first look at this in the case where the integrating factor only depends on x. Suppose such a $\mu(x)$ existed. If we multiply the equation

$$(P(x)y - Q(x)) \, dx + dy = 0$$

by this function (currently unknown) we have

$$(\mu(x)P(x)y - \mu(x)Q(x)) \, dx + \mu(x) \, dy = 0.$$

Now, $\mu(x)$ is an integrating factor if and only if this equation is exact, that is,

$$\frac{\partial}{\partial y}\left(\mu(x)P(x)y - \mu(x)Q(x)\right) = \frac{\partial}{\partial x}\mu(x).$$

Since $\frac{\partial}{\partial y}(\mu(x)Q(x)) = 0$,

$$\mu(x)P(x) = \frac{d\mu}{dx}. \tag{1.26}$$

In this equation, $P(x)$ is known, whereas $\mu(x)$ (the integrating factor) is unknown. We can find $\mu(x)$ because equation (1.26) is separable. Thus

$$\frac{d\mu}{\mu} = P(x) \, dx.$$

Integrating gives

$$\mu(x) = e^{\int P(x)dx}. \tag{1.27}$$

Since (1.27) is an integrating factor, we have

$$e^{\int P(x)dx}\frac{dy}{dx} + e^{\int P(x)dx}P(x)y = Q(x)e^{\int P(x)dx},$$

which is the <u>same</u> as

$$\frac{d}{dx}\left(e^{\int P(x)dx}y\right) = Q(x)e^{\int P(x)dx}.$$

So

$$e^{\int P(x)dx}y = \int Q(x)e^{\int P(x)dx}dx + C,$$

which gives

$$y = e^{-\int P(x)\,dx}\left(\int Q(x)e^{\int P(x)dx}\,dx + C\right) \tag{1.28}$$

as the solution of the differential equation (1.22). Note that we have explicitly written the constant of integration even though the integral has not yet been evaluated. Depending upon your situation, one can memorize the formula (1.28) for the solution of a first-order linear equation; however, it is just as easy (if not out right preferable) to simply apply the method of solution each time.

We can summarize the method of solving a linear equation as follows:

1. Write the linear equation in the form of equation (1.22).
2. Calculate the integrating factor $e^{\int P(x)\,dx}$.
3. Evaluate the integral $\int Q(x)e^{\int P(x)dx}dx$ and then multiply this result by $e^{-\int P(x)\,dx}$.
4. The general solution to (1.22) is

$$y = Ce^{-\int P(x)\,dx} + e^{-\int P(x)\,dx}\int Q(x)e^{\int P(x)dx}\,dx.$$

In the event that we are given an initial condition $y(x_0) = y_0$, we would apply it at the time of integration, going from x_0 to a final (general) value x. If we let $\bar{p}(x) = \int P(x)dx$, then the general formula becomes

$$y = Ce^{-\bar{p}(x)} + e^{-\bar{p}(x)}\int Q(x)e^{\bar{p}(x)}\,dx,$$

and applying the initial condition gives us the solution

$$y = y_0 e^{\bar{p}(x_0)-\bar{p}(x)} + e^{-\bar{p}(x)}\int_{x_0}^{x} Q(t)e^{\bar{p}(t)}\,dt, \tag{1.29}$$

where the last limit of integration has changed to a dummy variable t.

Example 3: Solve

$$\frac{dy}{dx} + 2xy = 3x.$$

This equation is linear with $P(x) = 2x$ and $Q(x) = 3x$ so that an integrating factor is

$$e^{\int P(x)dx} = \exp\left(\int 2x dx\right)$$

$$= e^{x^2}.$$

We need to calculate

$$\int Q(x)e^{\int P(x)dx}dx = \int 3x(e^{x^2})$$

$$= \frac{3}{2}e^{x^2}.$$

Now we apply the formula for the solution as given in (1.28):

$$y = e^{-x^2}\left(\frac{3}{2}e^{x^2} + C\right),$$

which simplifies to

$$y = \frac{3}{2} + Ce^{-x^2}.$$

We can easily check that this is a solution of the original differential equation.

Example 4: Solve

$$\frac{dy}{dx} + \left(\frac{2x+1}{x}\right)y = e^{-2x}.$$

This is linear with

$$P(x) = \frac{2x+1}{x} \quad \text{and} \quad Q(x) = e^{-2x}$$

so that an integrating factor is

$$e^{\int P(x)dx} = e^{\int \frac{2x+1}{x} dx}$$

$$= e^{(2x+\ln|x|)}$$

$$= |x|e^{2x}.$$

We note that integrating factors are not unique. For instance, dropping the absolute value to obtain xe^{2x} gives another integrating factor of the differential equation. Thus, multiplying the original equation by this expression gives

$$xe^{2x}\frac{dy}{dx} + e^{2x}(2x+1)y = x.$$

If we had multiplied by $-xe^{2x}$, we would have obtained the *same* equation. This equation can be simplified to give

$$\frac{d}{dx}(xe^{2x}y) = x.$$

Integrating this equation gives

$$xe^{2x}y = \frac{1}{2}x^2 + C,$$

which becomes

$$y = \frac{1}{2}xe^{-2x} + \frac{C}{x}e^{-2x}.$$

These last few steps could have been avoided by using (1.28).

Example 5: Solve

$$(x^2 + 1)\frac{dy}{dx} + 4xy = x$$

with the initial condition $y(2) = 1$. Graphically compare your answer with the approximate solution obtained using Matlab, Maple, or Mathematica to numerically solve the original differential equation.

Rewriting this equation gives

$$\frac{dy}{dx} + \left(\frac{4x}{x^2 + 1}\right)y = \frac{x}{x^2 + 1},$$

hence

$$P(x) = \frac{4x}{x^2 + 1} \quad \text{and} \quad Q(x) = \frac{x}{x^2 + 1}$$

so that an integrating factor is

$$e^{\int P(x)dx} = e^{\int \frac{4x}{x^2+1}\,dx}$$

$$= e^{\ln(x^2+1)^2}$$

$$= (x^2 + 1)^2.$$

Once we have our integrating factor, we can use the solution as given in (1.28), first noting that

$$e^{-\int P(x)dx} = e^{-\ln(x^2+1)^2}$$

$$= (x^2 + 1)^{-2}. \tag{1.30}$$

Then

$$y = \frac{1}{(x^2+1)^2} \left(\int x(x^2+1) \, dx \right)$$

$$= \frac{1}{(x^2+1)^2} \left(\frac{1}{4}x^4 + \frac{1}{2}x^2 + C \right).$$

Now the initial condition, $y(2) = 1$, gives $C = 19$ and thus

$$y = \frac{\frac{1}{4}x^4 + \frac{1}{2}x^2 + 19}{(x^2+1)^2}$$

is the solution we seek.

We now compare the explicit solution we just obtained with the numerical approximation. The Matlab syntax is just as described earlier in the text. The Maple syntax introduced earlier could also be used here but we instead opt for a different plotting package for the benefit of the reader.

Computer Code 1.9:   Numerically solving a differential equation with built-in commands and plotting this solution

Matlab, Maple, Mathematica

Matlab
As is typical, we first create an m-file and then name it `Example5.m`, which contains the function that we want to approximate numerically:

Example5.m

```
                              Matlab
function f=Example(xn,yn)
%
% The original ode, rewritten as y'=f(x,y)
% is dy/dx=(x-4*x.*y)./(x.^2+1);
%
f= (xn-4*xn.*yn)./(xn.^2+1);
```

Then in the command window, we would type

```
                          Matlab
≫  [x1,y1]=ode45('Example4',[2,5],1);
≫  plot(x1,y1,'r')
≫  hold on
≫  x2=2:.01:5;
≫  y2=(x2.^4/4+x2.^2/2+19)./(x2.^2+1).^2;
≫  plot(x2,y2,'b:')
≫  xlabel('x');ylabel('y');
≫  title('solid is approx soln, dotted is explicit soln');
≫  hold off
```

Maple

With Maple, we need to remember to load the relevant packages (in this case, plot and DEtools) at the beginning.

```
                          Maple
> with(plots):
> with(DEtools):
> eq1:=(x.^2+1)*diff(y(x),x)+4*x*y(x)=x;
> soln1:=dsolve({eq1,y(2)=1},y(x),numeric);
> eq2:=odeplot(soln1,[x,y(x)],2..5,labels=[x,y]):
> eq3:=(x^4/4+x^2/2+19)/(x^2+1)^2;
> eq4:=plot(eq3,x=2..5,color=blue,linestyle=2):
> display([eq2,eq4],title="solid is approx soln, dotted
    is explicit soln");
```

Mathematica

With Mathematica, we need to remember to load the relevant packages at the beginning.

```
                       Mathematica
<< Graphics`Colors`
de[x_]=(x^2+1) y'[x] + 4 x y[x]
NDSolve[{de[x]==x,y[2]==1},y,{x,2,5}]
p1=Plot[Evaluate[y[x]/.%],{x,2,5},PlotStyle→Red,
  AxesLabel→{"x","y"}]
```
$$\text{sol1}[\text{x\_}]=\frac{\frac{x^4}{4} + \frac{x^2}{2} + 19}{(x^2 + 1)^2}; \quad (*\text{entered from palette}*)$$
```
p2=Plot[sol1[x],{x,2,5},PlotStyle→Blue,AxesLabel→{"x","y"}]
Show[p1,p2]
```

We note that with a stepsize of $h = 0.01$, the curves are nearly indistinguishable. This is one way to "check" your answers if you do not have the solution available to you.

Now that we know the techniques of solving linear equations, we consider some applications. Earlier we considered Newton's law of cooling. In that formulation, we assumed that the temperature of the surroundings was constant. This is not always realistic, as in some settings the temperature of the surroundings varies. For example, determining the temperature inside a building over a span of a 24-hour day is complicated because the outside temperature varies. If we assume that the building has no heating or air conditioning, the differential equation that needs to be solved to find the temperature $u(t)$ at time t inside the building is

$$\frac{du}{dt} = k(C(t) - u(t)), \tag{1.31}$$

where $C(t)$ is a function that describes the outside temperature and $k > 0$ is a constant that depends on the insulation of the building. Note that (1.31) is a linear equation. According to this equation, if $C(t) > u(t)$, then

$$\frac{du}{dt} > 0,$$

which implies that $u(t)$ increases, and if $C(t) < u(t)$, then

$$\frac{du}{dt} < 0,$$

so that $u(t)$ decreases.

Example 6: Suppose that during the month of April in Pomona, California, the outside temperature in degrees Fahrenheit is given by

$$C(t) = 70 - 10\cos\left(\frac{\pi t}{12}\right)$$

for $0 \le t \le 24$. Determine the temperature in a building that has an initial temperature of $60°F$ if $k = 1/4$. See Figure 1.6.

We see that the average temperature (i.e., the average of $C(t)$) is $70°F$ because

$$\int_0^{24} \cos\left(\frac{\pi t}{12}\right) dt = 0.$$

The initial-value problem that we must solve is

$$\frac{du}{dt} = k\left(70 - 10\cos\left(\frac{\pi t}{12}\right) - u\right)$$

Outside Temperature

FIGURE 1.6:   Outside temperature over 24 hours.

with initial condition $u(0) = 60$. The differential equation can be rewritten as

$$\frac{du}{dt} + ku = k\left(70 - 10\cos\left(\frac{\pi t}{12}\right)\right),$$

which is a linear equation and is thus solvable. This gives (check it!)

$$u(t) = \frac{10}{9 + \pi^2}\left(63 + 7\pi^2 - 9\cos\left(\frac{\pi t}{12}\right) - 3\pi\sin\left(\frac{\pi t}{12}\right)\right) + C_1 e^{-t/4}.$$

We then apply the initial condition $u(0) = 60$ to determine the arbitrary constant C_1 and obtain the solution

$$u(t) = \frac{10}{9 + \pi^2}\left(63 + 7\pi^2 - 9\cos\left(\frac{\pi t}{12}\right) - 3\pi\sin\left(\frac{\pi t}{12}\right)\right) - \frac{10\pi^2}{9 + \pi^2}e^{-t/4}.$$

A graph of this solution is shown in Figure 1.7. The graph shows that the temperature reaches its maximum of about $77°$F near $t = 15.5$, which is about 3:30 p.m.

Sometimes an equation may not immediately appear to be linear. Consider the differential equation

$$y^2\,dx + (3xy - 1)\,dy = 0.$$

This equation is <u>not</u> linear in y. We can also check that it is <u>not</u> homogeneous or exact. What do we do? Look harder. If we consider y as the independent variable and x as the dependent variable, we can write

$$\frac{dx}{dy} = \frac{1 - 3xy}{y^2},$$

Inside Temperature

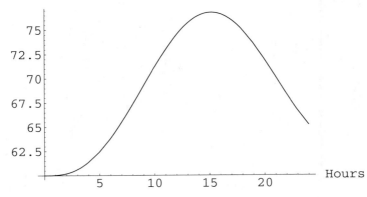

FIGURE 1.7:   Inside temperature over 24 hours.

which is

$$\frac{dx}{dy} + \frac{3x}{y} = \frac{1}{y^2},$$

and we see that it is in the form

$$\frac{dx}{dy} + P(y)\,x = Q(y),$$

which is linear in x, so that this equation can be solved using the theory we have just developed.

Hence, an integrating factor is

$$e^{\int P(y)dy} = e^{\int \frac{3}{y}\,dy}$$

$$= e^{\ln |y|^3}$$

$$= y^3.$$

We also have $\exp\left(-\int P(y)dy\right) = 1/y^3$. Then our solution is

$$x = \frac{1}{y^3}\left(\int \frac{1}{y^2}(y^3)dy\right) + \frac{C}{y^3}$$

$$= \frac{1}{y^3}\left(\frac{y^2}{2}\right) + \frac{C}{y^3}.$$

This becomes

$$x = \frac{1}{2y} + \frac{C}{y^3}$$

which is defined for all $y \neq 0$.

1.5.1 Special Integrating Factors

We saw how multiplying by an appropriate integrating factor converted a linear equation into an exact equation, which we could then solve. Multiplying by an appropriate integrating factor is a technique that will work in other situations as well.

We have seen that if the equation

$$M(x, y)\, dx + N(x, y)\, dy = 0$$

is not exact and if $\mu(x, y)$ is an integrating factor, then the differential equation

$$\mu(x, y)M(x, y)\, dx + \mu(x, y)N(x, y)\, dy = 0$$

is exact. Using the criterion for exactness, we must have

$$\frac{\partial}{\partial y}(\mu(x, y)M(x, y)) = \frac{\partial}{\partial x}(\mu(x, y)N(x, y)).$$

To simplify notation, we will write M, N instead of $M(x, y), N(x, y)$ when taking the partial derivatives, even though both M and N are functions of x and y. The criterion for exactness can then be written

$$\frac{\partial \mu}{\partial y}\, M(x, y) + \mu(x, y)\, \frac{\partial M}{\partial y} = \frac{\partial \mu}{\partial x}\, N(x, y) + \mu(x, y)\, \frac{\partial N}{\partial x}.$$

Rearranging gives

$$\frac{\partial \mu}{\partial y}\, M(x, y) - \frac{\partial \mu}{\partial x}\, N(x, y) = \mu(x, y)\, \frac{\partial N}{\partial x} - \mu(x, y)\, \frac{\partial M}{\partial y}. \qquad (1.32)$$

Thus $\mu(x, y)$ is an integrating factor if and only if it is a solution of the partial differential equation (1.32). We will not consider the solution of this partial differential equation. We will instead consider (1.32) in the case where μ only depends on x, i.e., $\mu(x, y) = \mu(x)$. (We can also consider the case when $\mu(x, y) = \mu(y)$ and the analogous formulation is left as one of the exercises.) In this situation, (1.32) reduces to

$$-\mu'(x)\, N(x, y) = \mu(x)\, \frac{\partial N}{\partial x} - \mu(x)\, \frac{\partial M}{\partial y}.$$

That is,

$$\frac{1}{\mu}\frac{d\mu}{dx} = \frac{1}{N(x, y)}\left(\frac{\partial M}{\partial y} - \frac{\partial N}{\partial x}\right). \qquad (1.33)$$

If the right-hand side of (1.33) involves two dependent variables, we run into trouble. If, however, it depends only upon x, then equation (1.33) is separable, in which case we obtain

$$\mu(x) = \exp\left[\int \frac{1}{N(x, y)}\left(\frac{\partial M}{\partial y} - \frac{\partial N}{\partial x}\right) dx\right]$$

as an integrating factor.

Example 7: Solve the differential equation

$$(2x^2 + y)\, dx + (x^2 y - x)\, dy = 0.$$

In this equation,

$$M(x, y) = 2x^2 + y \quad \text{and} \quad N(x, y) = x^2 y - x$$

so that

$$\frac{\partial M}{\partial y} = 1 \neq 2xy - 1 = \frac{\partial N}{\partial x}$$

and the equation is not exact. It can also be shown (try it!) that the differential equation is not separable, homogeneous, or linear. Now

$$\frac{1}{N(x, y)} \left(\frac{\partial M}{\partial y} - \frac{\partial N}{\partial x} \right) = \frac{1}{x^2 y - x} (1 - (2xy - 1))$$

$$= \frac{-2}{x}$$

depends only upon x. Thus,

$$\mu(x) = \exp\left(- \int \frac{2}{x}\, dx \right)$$

$$= e^{-2 \ln |x|}$$

$$= \frac{1}{x^2}$$

is an integrating factor. If we multiply the equation through by this factor we have

$$\left(2 + \frac{y}{x^2} \right) dx + \left(y - \frac{1}{x} \right) dy = 0.$$

Now this equation <u>is</u> exact since

$$\frac{\partial M}{\partial y} = \frac{1}{x^2} = \frac{\partial N}{\partial x}.$$

We can thus solve this differential equation using the exact method to obtain

$$2x + \frac{y^2}{2} - \frac{y}{x} = C.$$

1.5.2 Bernoulli Equation

We will now consider a class of differential equations that can be reduced to linear equations by an appropriate transformation. These equations are called *Bernoulli* equations and often arise in applications.

DEFINITION 1.8 *A first-order differential equation of the form*

$$\frac{dy}{dx} + P(x)\,y = Q(x)\,y^n \qquad n \in \mathbb{R} \tag{1.34}$$

is called a Bernoulli differential equation.

Note that when $n = 0$ or $n = 1$, the Bernoulli equation is actually a linear equation and can be solved as such. When $n \neq 0$ or 1, then we must consider an additional method.

THEOREM 1.5.1 *Suppose $n \neq 0$ or 1, then the transformation*

$$v = y^{1-n}$$

reduces the Bernoulli equation (1.34) to

$$\frac{dv}{dx} + (1-n)P(x)\,v = (1-n)Q(x), \tag{1.35}$$

which is a linear equation in v.

Proof: Multiply the Bernoulli equation by y^{-n} and thus obtain

$$y^{-n}\frac{dy}{dx} + P(x)\,y^{1-n} = Q(x). \tag{1.36}$$

Now let $v = y^{1-n}$ so that

$$\frac{dv}{dx} = (1-n)y^{-n}\frac{dy}{dx}.$$

Hence, equation (1.36) becomes

$$\frac{1}{1-n}\frac{dv}{dx} + P(x)\,v = Q(x),$$

that is,

$$\frac{dv}{dx} + (1-n)P(x)\,v = (1-n)Q(x).$$

Letting

$$P_1(x) = (1-n)P(x) \quad \text{and} \quad Q_1(x) = (1-n)Q(x)$$

gives

$$\frac{dv}{dx} + P_1(x)\, v = Q_1(x),$$

a linear differential equation in v. $\square$

Example 8: Solve the differential equation

$$\frac{dy}{dx} + y = xy^3.$$

This is a Bernoulli equation with $n = 3$. We thus let $v = y^{1-3} = y^{-2}$, so that

$$\frac{dv}{dx} = -2y^{-3}\frac{dy}{dx}.$$

Using (1.35) we obtain

$$\frac{dv}{dx} - 2v = -2x. \tag{1.37}$$

This is a linear differential equation with integrating factor

$$\exp\left(\int P(x)\,dx\right) = \exp\left(\int -2\,dx\right) = e^{-2x}.$$

We also calculate $\exp\left(-\int P(x)\,dx\right) = e^{2x}$. Thus the solution of (1.37) can be written

$$v = e^{2x}\left(\int -2xe^{-2x}\,dx\right).$$

Integrating by parts gives

$$v = e^{2x}\left(xe^{-2x} + \frac{1}{2}e^{-2x}\right) + Ce^{2x}.$$

Simplifying gives

$$v = x + \frac{1}{2} + Ce^{2x}.$$

But our original problem was in the variable y. We know $v = y^{-2}$ and thus the solution is

$$\frac{1}{y^2} = x + \frac{1}{2} + Ce^{2x}$$

which can be written as

$$y = \pm\left(\frac{1}{x + \frac{1}{2} + Ce^{2x}}\right)^{1/2}.$$

This solution is defined as long as the denominator is not equal to zero.

Problems

Solve the linear equations in problems 1–15 by considering y as a function of x, that is, $y = y(x)$.

1. $\frac{dy}{dx} + \frac{1}{x}y = x$

2. $\frac{dy}{dx} - \frac{2x}{1+x^2}y = x^2$

3. $(2x + 1)y' = 4x + 2y$

4. $y' + y \tan x = \sec x, \quad y(\pi) = 1$

5. $dy = \left(2x + \frac{xy}{x^2-1}\right) dx$

6. $\frac{dy}{dx} - y = 4e^x$ with initial condition $y(0) = 4$

7. $y' + xy = 2x$

8. $y' + \frac{1}{x}y = e^x$

9. $y' + 2y = -3x$

10. $\frac{dy}{dx} + \frac{y}{x} = \frac{\cos x}{x}$ with initial condition $y(\frac{\pi}{2}) = \frac{4}{\pi}$, $x > 0$

11. $y' - 3x^2 y = x^2$

12. $y' - 2xy = x^3$

13. $y' + y = \cos x$

14. $y' + y = e^x$

15. $\frac{dy}{dx} + 2xy = 1$ with the initial condition $y(2) = 1$

Solve the linear equations in problems 16–18 by considering x as a function of y, that is, $x = x(y)$.

16. $(x + y^2)dy = ydx$

17. $(2e^y - x)y' = 1$

18. $(\sin 2y + x \cot y)y' = 1$

Problems 19–21 address aspects of another derivation of the solution to a linear equation.

19. Recall that a linear equation is called *homogeneous* if $Q(x) = 0$, i.e., if it can be written as

$$\frac{dy}{dx} + P(x)\,y = 0.$$

a. Show that $y = 0$ is a solution (called the *trivial* solution).
b. Show that if $y = y_1(x)$ is a solution and k is a constant, then $y = ky_1(x)$ is also a solution.
c. Show that if $y = y_1(x)$ and $y = y_2(x)$ are solutions, then $y = y_1(x) + y_2(x)$ is a solution.

20. a. If $y = y_1(x)$ satisfies the homogeneous linear equation $\dfrac{dy}{dx} + P(x)\,y = 0$ and $y = y_2(x)$ satisfies the nonhomogeneous linear equation $\dfrac{dy}{dx} + P(x)\,y = r(x)$, show that $y = y_1(x) + y_2(x)$ is a solution to the nonhomogeneous linear equation

$$\frac{dy}{dx} + P(x)\,y = r(x).$$

   b. Show that if $y = y_1(x)$ is a solution of $\dfrac{dy}{dx} + P(x)\,y = r(x)$, and $y = y_2(x)$ is a solution of $\dfrac{dy}{dx} + P(x)\,y = q(x)$, then $y = y_1(x) + y_2(x)$ is a solution of

$$\frac{dy}{dx} + P(x)\,y = q(x) + r(x).$$

   c. Use the results obtained in parts **a** and **b** to solve

$$\frac{dy}{dx} + 2y = e^{-x} + \cos x.$$

21. a. Solve the linear homogeneous equation

$$\frac{dy}{dx} + P(x)\,y = 0$$

   by using separation of variables to obtain y_c, the *complementary* solution.[3]
   b. With the solution to part **a** written as Ay_c, we now seek to obtain a solution to

$$\frac{dy}{dx} + P(x)\,y = Q(x) \qquad (1.38)$$

   by assuming that the solution in part **a** can be modified as $A(x)y_c$ for some function $A(x)$ that must determined.[4] In order for $A(x)y_c$ to be a solution, it needs to satisfy the differential equation. Substitute the assumed solution into (1.38) to obtain

$$A'(x)y_c = Q(x) \qquad (1.39)$$

   after simplification. Thus, $A(x)y_c$ is a solution to (1.38) when $A'(x)y_c = Q(x)$.
   c. Substitute your answer from part **a** into (1.39) and integrate to obtain

$$A(x) = \int Q(x)e^{\int P(x)dx}\, dx. \qquad (1.40)$$

---

[3]This solution is sometimes called the *homogeneous* solution and is denoted y_h. The terms are used interchangeably.

[4]This is known as the *method of variation of parameters* and arises in the solution of nth order nonhomogeneous equations (Section 4.7) as well as the solution of systems of nonhomogeneous first-order equations (Section 6.3).

d. Denote the solution to (1.38) as $y_p = y_c A(x)$ where the subscript "p" stands for *particular* and y_p is known as the *particular solution*. Write the general solution as the sum of the complementary and particular solutions:

$$y = y_c + y_p \qquad (1.41)$$

and verify that this is the same solution obtained in the text via exact equations.

22. A pond that initially contains $500,000$ gal of unpolluted water has an outlet that releases $10,000$ gal of water per day. A stream flows into the pond at $12,000$ gal/day containing water with a concentration of 2 g/gal of a pollutant. Find a differential equation that models this process and determine what the concentration of pollutant will be after 10 days.

23. When wading in a river or stream, you may notice that microorganisms like algae are frequently found on rocks. Similarly, if you have a swimming pool, you may notice that in the absence of maintaining appropriate levels of chlorine and algaecides, small patches of algae take over the pool surface, sometimes overnight. Underwater surfaces are attractive environments for microorganisms because water removes waste and provides a continuous supply of nutrients. On the other hand, the organisms must spread over the surface without being washed away. If conditions become unfavorable, they must be able to free themselves from the surface and recolonize on a new surface.

The rate at which cells accumulate on a surface is proportional to the rate of growth of the cells and the rate at which the cells attach to the surface. An equation describing this situation is given by

$$\frac{dN(t)}{dt} = r(N(t) + A),$$

where $N(t)$ represents the cell density, r the growth rate, A the attachment rate, and t time.

a. If the attachment rate, A, is constant, solve

$$\frac{dN(t)}{dt} = r(N(t) + A)$$

with the initial condition $N(0) = 0$.

b. If $A = 3$ in a particular colony of cells, use the following table to find the growth rate at the end of each hour.

t	$N(t)$
1	3
2	9
3	21
4	45

Using this growth rate, estimate the algae population size at the end of 24 hours and 36 hours.

24. A 100-L beaker contains 10 kg of salt. Five liters of water are added to the beaker per minute and the overflow, after perfect mixing, is transferred into another 100-L beaker which initially contained pure water. The liquid in the second beaker is also perfectly mixed. When will the amount of salt in the second beaker reach its maximum and what is the value of this maximum?

25. Find the solution of the equation $y' \sin 2x = 2(y + \cos x)$ that remains bounded as $x \to \pi/2$.

26. Show that the equation

$$\frac{dx}{dt} + x = f(t)$$

has a unique solution bounded for $-\infty < t < +\infty$, where $|f(t)| \leq M$. Find the solution. Show further that this solution is "periodic" if the function $f(t)$ is periodic.

27. Suppose $a(t) > 0$, and $f(t) \to 0$ for $t \to \infty$. Show that every solution of the equation

$$\frac{dx}{dt} + a(t)x = f(t)$$

approaches 0 for $t \to \infty$.

28. In the same equation suppose that $a(t) > 0$, and let $x_0(t)$ be the solution for which the initial condition $x(0) = b$ is satisfied. Show that for every positive $\varepsilon > 0$ there is a $\delta > 0$, such that if we perturb the function $f(t)$ and the number b by a quantity less than δ, then the solution $x(t), t > 0$, is perturbed by less than ε. The word *perturbed* is understood in the following sense: $f(t)$ is replaced by $f_1(t)$ and b is replaced by b_1 where

$$|f_1(t) - f(t)| < \varepsilon, \quad |b_1 - b| < \delta.$$

This property of the solution $x(t)$ is called *stability for persistent disturbances*.

29. Find an appropriate integrating factor and use it to solve the following equations:
a. Solve $(3x^2 + y) \, dx + (x^2 y - x) \, dy = 0$
b. Solve $(x^4 - x + y) \, dx - x \, dy = 0$

30. Show that if $(\partial N/\partial x - \partial M/\partial y)/(xM - yN)$ depends only on the product xy, that is,

$$\frac{\frac{\partial N}{\partial x} - \frac{\partial M}{\partial y}}{xM - yN} = H(xy),$$

then the equation

$$M(x, y) \, dx + N(x, y) \, dy = 0$$

has an integrating factor of the form $\mu(xy)$. Find the general formula for $\mu(xy)$.

31. We derived a formula for an integrating factor if $\mu(x, y) = \mu(x)$. If $\mu(x, y) = \mu(y)$, derive the integrating factor formula

$$\mu(y) = \exp\left[\int \frac{1}{M(x, y)}\left(\frac{\partial N}{\partial x} - \frac{\partial M}{\partial y}\right) dy\right]. \qquad (1.42)$$

Solve problems 32–40 by finding an integrating factor of suitable form. Use the computer to plot the implicit or explicit solution for three different initial conditions.

32. $(x^2 + y^2 + x)dx + ydy = 0$
33. $xdx = (xdy + ydx)\sqrt{1 + x^2}$
34. $y(x + y)dx + (xy + 1)dy = 0$
35. $y(y^2 + 1)dx + x(y^2 - x + 1)dy = 0$
36. $ydx - xdy = 2x^3 \sin x \; dx$
37. $ydx + (e^x - 1)dy = 0$
38. $(x^2 - y^2 + y)dx + x(2y - 1)dy = 0$
39. $(2x^2y^2 + y)dx + (x^3y - x)dy = 0$
40. $(x^2 + 1)(2xdx + \cos y \; dy) = 2x \sin y \; dx$

Solve the Bernoulli equations in problems 41–50 by considering y as a function of x, that is, $y = y(x)$.

41. $y' + y = xy^2$
42. $y' - xy = y^3$
43. $y' = y^4 \cos x + y \tan x$
44. $y' + 3y = y^4$
45. $xy^2y' = x^2 + y^3$
46. $y' + 2xy = 4y$
47. $(x + 1)(y' + y^2) = -y$
48. $xy' - 2x^2\sqrt{y} = 4y$
49. $xy' + 2y + x^5y^3e^x = 0$
50. $xydy = (y^2 + x)dx$

51. Use the Bernoulli method of solution to solve the logistic equation

$$\frac{dN}{dt} = rN\left(1 - \frac{N}{K}\right).$$

1.6   Chapter 1: Additional Problems and Projects

ADDITIONAL PROBLEMS

In problems 1–7, determine whether the statement is true or false. If it is true, give reasons for your answer. If it is false, give a counterexample or other explanation of why it is false.

1. The equation $y'' + xy' - y = x^2$ is a linear ordinary differential equation that is considered an initial-value problem.

2. The equation $y^{(4)} - y^2 = \sin x$ is a nonlinear ordinary differential equation.

3. An implicit solution to $y' = f(x, y)$ can be written in the form $y(x) = f(x)$.

4. With an appropriate substitution, any exact equation can be put in the form of a linear equation.

5. A solution to the differential equation $y' = f(x, y)$ must be defined for all x.

6. An equation of the form $M(x, y) + N(x, y)y' = 0$ is considered exact if

$$\frac{\partial M(x, y)}{\partial y} = \frac{\partial N(x, y)}{\partial x}.$$

7. Every equation of the form $y' = f(y)$ is separable.

8. Verify that $y(x) = 1 + Ce^{x^4}$ is a solution to $y' = 4x^3(y - 1)$ on $(-\infty, \infty)$ for any constant C.

9. Verify that $y(x) = Ce^x - x^2 - 2x - 2$ is a solution to $y' = x^2 + y$ on $(-\infty, \infty)$ for any constant C.

10. Verify that $y^2(x) = \dfrac{1}{C - 2x}$ is an implicit solution to $y' = y^3$. For what C-values and x-values is it a solution?

11. Verify that $y(x) = \dfrac{3x - 1 + 3C}{x + C}$ is a solution to $y' = (y - 3)^2$. For what C-values and x-values is it a solution?

Solve each of the following separable differential equations. If an initial condition is given, also plot the solution in Matlab, Maple, or Mathematica.

12. $2x^2yy' + y^2 = 2$

13. $y' = 3\sqrt[3]{y^2}$ $y(2) = 0$

14. $y' - xy^2 = 2xy$

15. $e^{-x}(1 + \frac{dx}{dt}) = 1$

16. $xy' + y = y^2$ $y(1) = 0.5$

Solve each of the following homogeneous differential equations.

17. $xy' = y - xe^{y/x}$

18. $xy' - y = (x + y)\ln\frac{x+y}{x}$

19. $xy' = y\cos(\ln(\frac{y}{x}))$

20. $(y + \sqrt{xy})dx = xdy$

21. $xy' = \sqrt{x^2 - y^2} + y$

22. $(2x - 4y)dx + (x + y)dy = 0$

Solve each of the following exact differential equations.

23. $\frac{y}{x} + (y^3 + \ln x)\frac{dy}{dx} = 0$

24. $\frac{3x^2 + y^2}{y^2} - \frac{2x^3 + 5y}{y^3}\frac{dy}{dx} = 0$

25. $2x(1 + \sqrt{x^2 - y}) - \sqrt{x^2 - y}\frac{dy}{dx} = 0$

Solve each of the following differential equations by first finding an integrating factor of a suitable form.

26. $xy^2(xy' + y) = 1$

27. $y^2dx - (xy + x^3)dy = 0$

28. $(y - \frac{1}{x})dx + \frac{dy}{y} = 0$

29. $(x^2 + 3\ln y)ydx = xdy$

30. $y^2dx + (xy + \tan xy)dy = 0$

Solve the following linear or Bernoulli equations.

31. $xy' + y = x^4$

32. $y' - y = \sin x$

33. $y' + 4xy = 5x^3$

34. $xy' + y = xy^4$

35. $y' - y = xy^3$

36. $2y' + xy = 4xy^3$

Solve each of the following problems using one of the methods of this chapter.

37. *One Big Ol' Pot of Soup* Mike, a professor of mathematics and a part-time evening cook at a local diner, prepares a big pot of soup late at night, just before closing time. He does this so that there would be plenty of soup to feed customers the next day. Being food safety cautious, he knows that refrigeration is essential to preserve the soup

overnight; however, the soup is too hot to be put directly into the fridge when it is ready. (The soup had just boiled at 100°C, and the fridge is not powerful enough to accommodate a big pot of soup if it was any warmer than 20°C.) Mike is resourceful (as he is a student of M.M. Rao) and discovered that by cooling the pot in a sink full of cold water (kept running, so that its temperature was roughly constant at 5°C) and stirring occasionally, he could bring the temperature of the soup to 60°C in 10 min. How long before closing time should the soup be ready so that Mike could put it in the fridge?

38. A cup of very hot coffee from a fast food restaurant is at 120°F when it is served. If left on the table in a room with ambient temperature of 70°F, it is found to have cooled to 115°F in 1 min. If burns can result from spilling coffee at a temperature greater than 100°F, how many minutes will the coffee have to sit before the management is safe from lawsuits?

39. A 30-gal tank initially has 15 gal of saltwater containing 6 lb of salt. Saltwater containing 1 lb of salt per gallon is pumped into the top of the tank at the rate of 2 gal/min, while a well-mixed solution leaves the bottom of the tank at a rate of 1 gal/min. Determine the amount of salt in the tank at time t. How long does it take for the tank to fill? What is the amount of salt in the tank when it is full?

40. A 100-gal tank initially contains 100 gal of sugar-water at a concentration of 0.25 lb of sugar per gallon. Suppose sugar is added to the tank at a rate of p lb/min, sugar-water is removed at a rate of 1 gal/min, and the water in the tank is kept well mixed. Determine the concentration of sugar at time t. What value of p should be chosen so that, when 5 gal of sugar solution is left in the tank, the concentration is 0.5 lb of sugar per gallon?

41. A 20% nitric acid solution flows at a constant rate of 6 L/min into a 200-L tank of 0.5% nitric acid solution. If the well-stirred mixture leaves the tank at the rate of 8 gal/min, find the volume of nitric acid in the tank after t minutes. When will the percentage of nitric acid in the tank reach 10%?

42. Find a curve such that the point of intersection of an arbitrary tangent with the x-axis has an abscissa half as great as the abscissa of the point of tangency.

43. A boat is held by a cable that is wound around a post, the end being held by a dog. What is the braking force in the cable if it is wound around the post three times, the coefficient of friction between cable and post is 1/3, and the dog exerts a force of 10 kg on the free end of the cable?

PROJECTS FOR CHAPTER 1

Project 1: Particles in the Atmosphere

Under normal atmospheric conditions, the density of soot particles, $N(t)$, satisfies the differential equation

$$\frac{dN}{dt} = -k_c N^2(t) + k_d N(t)$$

where k_c (the *coagulation constant*) is a constant that relates how well particles stick together; and k_d (the *dissociation constant*) is a constant that relates how well particles fall apart. Both of these constants depend on temperature, pressure, particle size, and other external forces.

a. Solve the differential equation using separation of variables to obtain the solution

$$N(t) = \frac{e^{k_d t}}{\left(\frac{k_c}{k_d}\right) e^{k_d t} + C},$$

where c is an arbitrary constant. (You will need to use the technique of partial fractions.)

b. Find the value of the constant C that makes $N(t_0) = N_0$. The following table lists typical values of k_c and k_d.

k_c	k_d
163	5
125	26
95	57
49	85
300	26

Values of coagulation and dissociation constants for Project 1.

For *each* pair of values in the table, sketch the graph (use a graphing calculator or computer) of $N(t)$ if $N(0) = 0.01$, 0.05, 0.1, 0.5, 0.75, 1, 1.5, and 2. You will have five graphs with eight solution curves on each graph (corresponding to the eight initial values). Regardless of the initial condition, what do you notice in each case? Do pollution levels seem more sensitive to k_c or k_d?

c. Show that if $k_d > 0$, then

$$\lim_{t \to \infty} N(t) = \frac{k_d}{k_c}.$$

Why is the assumption $k_d > 0$ reasonable? For each pair in the table, calculate k_d/k_c. Which situation results in the highest pollution levels? How could the situation be changed?

Project 2: Insights into Graphing

In describing the solution of a differential equation, we use the term *general solution* for a family of solutions of the differential equation. These general solutions contain arbitrary constants. The term *particular solution* is used to discuss solutions that are free of arbitrary constants, usually as a result of requiring that the solution satisfy some initial condition. Sometimes a differential equation has a particular solution that cannot be obtained by selecting a specific value for the arbitrary constant in the general solution.

a. Using calculus show that the line

$$y = mx + \frac{a}{m}$$

is tangent to the parabola

$$y^2 = 4ax$$

for all values of m.

b. Show that the slope of the tangent to the parabola is given by

$$\frac{dy}{dx} = \frac{2a}{\sqrt{4ax}}$$

so that at $x = \frac{a}{m^2}$, the slope is m.

c. From this analysis, note that at the point of tangency P the tangent line and the parabola have the same direction. Thus, they have a common value of dy/dx, as well as of x and y. Show that at the tangency point P,

$$m = \frac{dy}{dx}$$

and that the tangent satisfies the equation

$$y = \frac{dy}{dx}x + \frac{a}{\frac{dy}{dx}}.$$

d. Show that the differential equation

$$y = \frac{dy}{dx}x + \frac{a}{\frac{dy}{dx}}$$

also holds for the parabola at P, where x, y, and dy/dx are the same as for the tangent. But, since P may be any point on the parabola, show that the equation of the parabola

$$y^2 = 4ax$$

must be a solution to the differential equation. This solution is a *singular solution* to the differential equation as it contains no arbitrary constants.

e. Solve the differential equation

$$y = \frac{dy}{dx}x + \frac{a}{\frac{dy}{dx}}.$$

Why is the solution different than that obtained in part d? Does this violate the uniqueness theorem?

Chapter 2

Geometrical and Numerical Methods for First-Order Equations

We have studied methods for solving particular types of differential equations, but as we have seen, there are many equations for which these methods do not apply. In this chapter we will investigate some useful geometrical and computer-assisted methods for analyzing the solution of a differential equation.

2.1 Direction Fields—the Geometry of Differential Equations

It should come as no surprise that most differential equations that one encounters cannot be solved. For example, even when an equation is separable, we have seen that some of the resulting integrals may not have antiderivatives in terms of elementary functions. Even when we exhaust all possible methods for analytically solving ordinary differential equations, we will encounter just as many equations where those methods *don't* apply.

All is not lost, however, as graphical and numerical methods have become commonplace in analyzing the behavior of differential equations. We will now turn to graphical and numerical methods for solving the first-order differential equation

$$\frac{dy}{dx} = f(x, y). \tag{2.1}$$

The numerical methods of solution that we cover in this section apply in an analogous way to higher-order equations and systems of equations.

We will begin our study of the geometry of differential equations by again considering (2.1) and assuming for now that solutions exist and are unique in some rectangle $\{(x, y) | a < x < b,\ c < y < d\}$. Suppose that $y_1(x)$ is a known solution to (2.1) on this rectangle. What does this mean? Recall that a solution gives us an identity when substituted into the differential equation. Thus,

$$\frac{dy_1(x)}{dx} = f(x, y_1(x)) \tag{2.2}$$

81

on this rectangle. Let us take a closer look at (2.2). The left-hand side of this equation is simply the derivative of the solution and this derivative is given by the right-hand side. But it really didn't take much work to evaluate the right-hand side. In fact, given any pair of values in the rectangle, say (x_0, y_0), we could find the slope of the solution that passes through that point in the x-y plane by simply evaluating $f(x_0, y_0)$. And we could do this for *each* point in this rectangle. Thus, we can calculate the slope of any solution curve without actually knowing an explicit solution.

The implications of this last statement are profound: given an initial condition, we can trace out solution curves without having to know the analytical solution. We can choose a grid of points in the x-y plane and evaluate the slope of the solution at each of these points. We draw a short line segment in the plane at that point with the calculated slope. The collection of these line segments is called the *slope field* or *direction field*. In many examples, the slope of the solution may be the same along a given curve. If we let (x, y) be the points on this curve, then

$$f(x, y) = k.$$

Any member of the family of curves that satisfies this equation is called an *isocline*, which means a curve along which the inclination of the tangents is the same. The use of isoclines is extremely useful if we are sketching a direction field by hand. We will illustrate their use in the first example and leave it to the reader to check their use in the others. Let us consider a simple example.

Example 1: Draw the direction field for the equation

$$\frac{dy}{dx} = y$$

and sketch the solution curve in this direction field that passes through the point $(0, 1)$.

We draw a representation of the direction field by selecting a grid of points and drawing a short tangent line at each point. The slope of each tangent line is determined by evaluating the right-hand side of our differential equation at the given point. This is very time consuming but is easily done on the computer. Equipped with the computer-generated direction field, we should be able to check that each point (x, y) in the plane has the slope on the graph, as determined by substituting particular values into the right-hand side of the given differential equation. In particular, we observe that the isoclines are given by $y = k$, i.e., by horizontal lines at the y-value of k. For example, all the line segments along $y = 1$ have a slope of 1 and all the line segments along $y = -2$ have a slope of -2. A direction field that only depends on y and not x is called *autonomous* and will be discussed in detail in later sections. The direction field for $\frac{dy}{dx} = y$ is shown in Figure 2.1a.

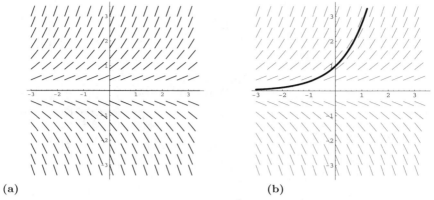

(a)   (b)

FIGURE 2.1: The direction field for $\frac{dy}{dx} = y$ is shown in **(a)**. The graph of $y(x)$ through the point $(0,1)$ on the direction field for $\frac{dy}{dx} = y$ is shown in **(b)**.

Using the condition that $y(x)$ passes through the point $(0,1)$, we can sketch the solution of $y(x)$ on the direction field by following the line segments in the direction field of Figure 2.1a. The sketch of the graph of $y(x)$ is shown in Figure 2.1b.

Notice how the solution passes tangentially through the direction field. That is, each line segment is tangent to the solution in the direction field. Now, of course, we could have easily calculated the solution to

$$\frac{dy}{dx} = y, \quad y(0) = 1$$

as $y(x) = e^x$ and we should observe that this *is* the curve plotted. Let's try this method again for an equation for which we didn't know the solution.

Example 2: Draw the direction field for the equation

$$\frac{dy}{dx} = x^2 + y^2$$

and sketch several solution curves that pass through the direction field.

The direction field for $\frac{dy}{dx} = x^2 + y^2$ is shown in Figure 2.2a. The reader should check a few pairs of points on the graph to make sure the direction field is correct. E.g., check the pairs $(3,0)$ and $(0,\frac{1}{2})$. One could also observe that the isoclines are given by $x^2 + y^2 = k$ which are circles of radius $\sqrt{k}$, which indicate that the line segments have the same slope along any circle in the x-y plane. On this direction field, we can plot several graphs of $y(x)$, shown in Figure 2.2b.

We note two points that should be obvious:

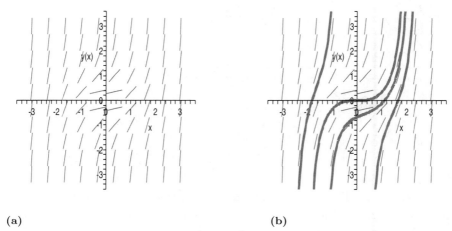

(a)   (b)

FIGURE 2.2:   The direction field for $\frac{dy}{dx} = x^2 + y^2$ is shown in (**a**). Graphs of several solutions $y(x)$ on the direction field for $\frac{dy}{dx} = x^2 + y^2$ are shown in (**b**).

1. The finer the mesh of the grid for the representation of the slope field, the better the approximate solution curve which we are able to draw. This is the same as saying the tangent line is a good approximation to a curve close to the point of tangency. And the more points we have the better we can sketch this approximation.

2. Drawing a direction field by hand is tedious—it is best to use a computer. It is, however, *essential* that we check a few points of the direction field (generated by the computer) by hand. As much as we would like computers to always give us the answers we want, this will never be the case.

2.1.1   Direction Fields in Matlab, Maple, and Mathematica

We will now consider plotting direction fields using software. This is a very convenient method for obtaining useful information about a differential equation.

Matlab
As we did when we plotted implicit solutions, we first need to create a `meshgrid`; that is, we need to create a set of pairs (x, y). We will then evaluate the slope of the solution through each pair of points. Let's consider Example 2 from this section.

Computer Code 2.1: **Direction field for first-order equation**
$y' = f(x, y)$

Matlab, Maple, Mathematica

```
                            Matlab
>>  [X,Y]=meshgrid(-3:.6:3,-3:.6:3); %We use CAPITAL letters
>>  DY=X.^2+Y.^2;      %DY is rhs of original equation
>>  DX=ones(size(DY));
>>  DW=sqrt(DX.^2+DY.^2);
>>  quiver(X,Y,DX./DW,DY./DW,.5,'.'); %plots direction field
>>  xlabel('x');
>>  ylabel('y');
```

Each line segment has an associated length ($\sqrt{1 + (dy/dx)^2}$ and Matlab automatically scales all of the vectors so they fit in the window. It can sometimes be difficult to see the smaller line segments and we fixed this by making the line segments have the same length of 1 (by dividing by DW). We also could have tried to adjust either the step size in meshgrid or by scaling all the vectors by a factor S, given in the example below with $S = 2, .5$:

```
                            Matlab
>>  quiver(X,Y,DX,DY,'.');     %unscaled direction field
>>  quiver(X,Y,DX,DY,2,'.');   %scaled by 2
>>  quiver(X,Y,DX,DY,.5,'.');  %scaled by .5
```

Note that the scaling factor, .5 in the above line, may be changed as necessary to make the length of the line segments "nice." Also, quiver normally draws the line segments with arrows, but since we want to think of the solution going in *both* directions, we simply plot the segment without the arrow. The '.' in the last entry of the quiver commands tells Matlab not to put an arrow. Maple, on the other hand, automatically scales the line segments so that each is of the same length. It is easier to see this way but it is also deceptive in telling how fast a solution is changing.

```
                            Maple
>  with(DEtools):  #Loads the needed diff eqns package
>  eq1:= diff(y(x),x)=x^2+y(x)^2;
>  dfieldplot(eq1,y(x),x=-3..3,y=-3..3,arrows=line,
     dirgrid=[10,10]);
```

As with Matlab, the default in Maple is to plot with the arrows. We force it to plot with line segments with the command `arrows=line` given at the end of the last input. The command `dirgrid` adjusts the spacing of the line segments with `dirgrid=[20,20]` being the default option.

| Mathematica |

```
<<Graphics`PlotField`;
    (*Loads vector field graphics package*)
eq1[x_]= x² + y²  (*entered from palette*)
PlotVectorField[{1,eq1[x]},{x,-3,3},{y,-3,3},
    ScaleFunction→(1&), Axes→Automatic, HeadLength→ 0]
```

As in the other packages, the default in Mathematica is also to plot with arrows; the plot option HeadLength→ 0 tells Mathematica to suppress the arrowheads. In each of these plots, we could sketch a solution by hand that passes through a given initial condition. This is worth trying a few times, as it helps clarify the relationship between the direction field and the solutions superimposed on the field. There are some problems in the exercises that will require the reader to sketch the solution by hand.

After one becomes comfortable with plotting direction fields and some corresponding solution curves, we will use our computer software packages to sketch the solution. For now, we will use the built-in routines of the programs and try to be as careful as possible in trusting the output from these "black boxes." Later in this chapter, we will examine numerical methods for calculating these solutions and will thus be able to gain significant insight into the inner workings of these built-in routines.

Calculating and Superimposing the Numerical Solutions

Example 3: We will now use Matlab, Maple, or Mathematica to calculate the numerical solution to

$$\frac{dy}{dx} = x^2 + y^2$$

and superimpose solution curves on the direction field that pass through the initial conditions $(-2.4, -3), (-1.6, -2.5), (-1, -3), (1, -3)$.

This particular example gives an illustration of why blindly plugging in code sometimes fails to give a good numerical approximation. Using the direction field previously obtained, we can sketch solutions passing through these conditions and we see that in all cases, the solution path grows very quickly and the slope gets extremely steep for large y-values. In Maple, we can handle this situation by either restricting the y-values of the numerical solution via the **range** option or by shrinking the error introduced in each step via the **relerr** option. In Mathematica the same idea applies—we specify the range of the y-values. In Matlab, we keep the error introduced at each step below a given value with '`AbsTol`' in the `odeset` function. In all of the cases, we will still get a warning but will be able to obtain plots. As just mentioned, the warning is due to the program having problems, numerically solving the differential equation beyond a certain x-value due to the very large slopes.

As with all the code in this book, there is rarely only one way to obtain a correct answer. The reader is urged to explore the help menus and some of the options available, including all the options for plotting different types of lines, symbols, etc. We point out in advance that we can smooth curves with numpoints or refine in Maple or with refine in Matlab.

We again stress how important it is to use technology carefully. In the current example, the solutions to $y' = x^2 + y^2$ quickly go to infinity; see Figure 2.2. This will present a problem in the three programs if we blindly tried to find the numerical solution over, say, the interval $[-3, 3]$. We may have warnings in each program and/or may need to carefully choose final x-values.

Computer Code 2.2: Obtaining numerical solutions to $y' = f(x, y)$ when solution $y(x)$ goes to ∞ for finite x; superimposing these solutions onto direction field

---

Matlab, Maple, Mathematica

Maple
If we again start with nothing stored in memory, we would enter

```
                              Maple
> with(plots):  with(DEtools):
> eq1:=diff(y(x),x)=x^2+y(x)^2;
> IC1:=y(-2.4)=-3;
> soln1:= dsolve({eq1,IC1},y(x),numeric,range=-10..10);
> IC2:=y(-1.6)=-2.5;
> soln2:= dsolve({eq1,IC2},y(x),numeric,range=-10..10);
> IC3:=y(-1)=-3;
> soln3:= dsolve({eq1,IC3},y(x),numeric,relerr=1e-12);
> IC4:=y(1)=-3;
> soln4:= dsolve({eq1,IC4},y(x),numeric,relerr=1e-12);
> eq2:=odeplot(soln1,[x,y(x)],numpoints=200,thickness=2):
> eq3:=odeplot(soln2,[x,y(x)],refine=2,thickness=2):
> eq4:=odeplot(soln3,[x,y(x)],numpoints=200,thickness=2):
> eq5:=odeplot(soln4,[x,y(x)],numpoints=200,thickness=2):
> eq6:=dfieldplot(eq1,y(x),x=-3..3,y=-3..3,arrows=line):
> display([eq2,eq3,eq4,eq5,eq6]);
```

Mathematica

In Mathematica, the command for obtaining a numerical solution is ND-Solve. This command finds a numerical approximation to a function that is equal to its first derivative at each point x between a and b, and that has the initial condition $y = b_1$ when x is a_1. NDSolve returns a rule to replace y by an object known as an InterpolatingFunction. An InterpolatingFunction object can be evaluated just like any other Mathematica function. This approximate function reproduces the values given at the data points and it gives approximate values in-between. For instance, the following commands will give a numerical solution to $y'(x) = x^2 + y^2$ with the initial condition $(-2.4, -3)$.

Mathematica

sol1 = NDSolve[$\{y'[x] == x^2 + y[x]^2, y[-2.4] == -3\}, y, \{x, -10, 10\}$]

Evaluating this command produces the result

$$\{\{y \rightarrow \text{InterpolatingFunction}[\{\{-2.67878, -1.03357\}\}, \text{``} <> \text{''}]\}\}$$

which is Mathematica's form for the numerical solution. We can use an InterpolatingFunction object like any other function that evaluates to a number. For example, we can check the initial condition using the command

$$\text{y}[-2.4]/.\text{sol1}$$

and see that it gives the result $\{-3\}$ as expected. Assuming that there is nothing stored in memory, the following are the commands used to obtain the numerical solutions superimposed upon the direction field.

```
                        Mathematica
<<Graphics'PlotField';
    (*Loads vector field graphics package*)
eq1[x_]=x² + y[x]²
sol1 = NDSolve[{y'[x] == eq1[x], y[−2.4] == −3}, y, {x, −10, 10}];
sol2 = NDSolve[{y'[x] == eq1[x], y[−1.6] == −2.5}, y, {x, −10, 10}];
sol3 = NDSolve[{y'[x] == eq1[x], y[−1] == −3}, y, {x, −10, 10}];
sol4 = NDSolve[{y'[x] == eq1[x], y[1] == −3}, y, {x, −10, 10}];
p1 = Plot[Evaluate[y[x]/. sol1], {x, −3, −1.05}, PlotRange → {−3, 3},
    PlotStyle → {Thickness[.01]}];
p2 = Plot[Evaluate[y[x]/. sol2], {x, −3, 3}, PlotRange → {−3, 3},
    PlotStyle → {Thickness[.01]}];
p3 = Plot[Evaluate[y[x]/. sol3], {x, −1, 2}, PlotRange → {−3, 3},
    PlotStyle → {Thickness[.01]}];
p4 = Plot[Evaluate[y[x] /. sol4], {x, 0.5, 2.5}, PlotRange → {−3, 3},
    PlotStyle → {Thickness[.01]}];
p5 = PlotVectorField[{1, x² + y²}, {x, −3, 3}, {y, −3, 3},
    ScaleFunction → (1&), Axes → Automatic, HeadLength → 0]
Show[p1, p2, p3, p4, p5];
```

Note that in each of the Plot commands there are specific values for the values of x. Try experimenting with some values here. You will discover that the behavior of the solutions necessitates appropriately chosen values for x. Also, note that the semicolons at the end of each NDSolve command suppress the displaying of the InterpolatingFunction object.

Matlab
For the Matlab code, we first need to create an m-file that defines the function for which we want to find the numerical solution. Remember that anything following a % is ignored but is put here for good programming technique.

Example.m

```
                        Matlab
function f=Example(xn,yn)
%
% The original ode is dy/dx=x^2+y^2
%
f= xn.^2+yn.^2;
```

The Matlab command to compute the numerical solution is ode45. To use it, we need to specify an initial x-value, a final x-value, and an initial condition. Let these be given as x0, xf, y0, respectively. In the command window, we can then type:

```
                              Matlab
>>  x0=-2.4;xf=-1;y0=-3;
>>  options=odeset('refine',10,'AbsTol',1e-20);
>>  [x,y]=ode45('Example',[x0,xf],y0,options);
>>  x0=-1.6;xf=2;y0=-2.5;
>>  [x1,y1]=ode45('Example',[x0,xf],y0,options);
>>  x0=-1;xf=2;y0=-3;
>>  [x2,y2]=ode45('Example',[x0,xf],y0,options);
>>  x0=1;xf=2.5;y0=-3;
>>  [x3,y3]=ode45('Example',[x0,xf],y0,options);
```

To plot the solution, we simply type `plot(x,y)` in the command window. If we wanted to see the solution plotted along with the direction field, we could use the commands:

```
                              Matlab
>>  [X,Y]=meshgrid(-3:.3:3,-3:.3:3);
>>  DY=X.^2+Y.^2;
>>  DX=ones(size(DY));
>>  DW=sqrt(DX.^2+DY.^2);
>>  quiver(X,Y,DX./DW,DY./DW,.5,'.');
>>  hold on
>>  plot(x,y)
>>  plot(x1,y1)
>>  plot(x2,y2)
>>  plot(x3,y3)
>>  axis([-3  3  -3  3])
>>  hold off
```

Dfield in Matlab

For a first-order system, a very user-friendly software supplement called **dfield** exists for Matlab and it is freely available for educational use.[1] There are two or three programs that you will need to download (depending on your version of Matlab) and install in your working directory. The program `dfield` (like its two-dimensional counterpart `pplane` that we will encounter in Section 6.4.1) is much easier to implement than either of the above methods for plotting direction fields and superimposing numerical solutions. We give a brief introduction here.

---

[1]The `dfield` Matlab supplement, written by John C. Polking, may be found at <http://math.rice.edu/~dfield/>.

Once you have placed the relevant programs in your working directory, type
`dfield6`. A new window should pop up, see the figure in Computer Code 2.3.
(Note that this is for Matlab 6.x. Download the appropriate files for other
versions of Matlab, e.g., `dfield7` is for Matlab 7.)

The dependent variable, by default, is $x(t)$ but we can easily change these in
the window so that everything will be in terms of $y(x)$. We again consider the
equations of the previous example and enter them in the `dfield` window as

---

Matlab dfield window

y'=x^2+y^2

---

and we must also specify that the independent variable is `x` (as opposed to
the default variable `t`).

For other problems, we may have parameters or expressions to enter. For
this problem, we also need to modify the minimum and maximum values of
x and y. To be consistent with the previous examples, we set the minimum
x-value to be -3, the maximum x-value to be 3, the minimum y-value to
be -3, and the maximum y-value to be 3. We click the *Proceed* button on
the window that we have been typing to generate the direction field. To plot
solutions, we simply click on the direction field at a desired initial condition.
Repeat a few times. The reader should experiment with the *Options* on the
display window and see other options for plotting.

Computer Code 2.3: **Pop-up window for `dfield` program,
used to obtain numerical solutions to first-order equations
(Note: this was done with Matlab 6.5 and hence the program
is `dfield6`.)**

Matlab only

Problems

Without solving the equations and without using the computer, match each differential equation in problems 1–4, and 5–8 with the graph of its direction field shown here.

1. $\dfrac{dy}{dx} = \dfrac{x^3 + 1}{y^3 + 1}$

2. $\dfrac{dy}{dx} = (x^3 + 1)(y^3 + 1)$

3. $\dfrac{dy}{dx} = \dfrac{y^3 + 1}{x^3 + 1}$

4. $\dfrac{dy}{dx} = \dfrac{1}{(x^3 + 1)(y^3 + 1)}$

(a)   (b)

(c)   (d)

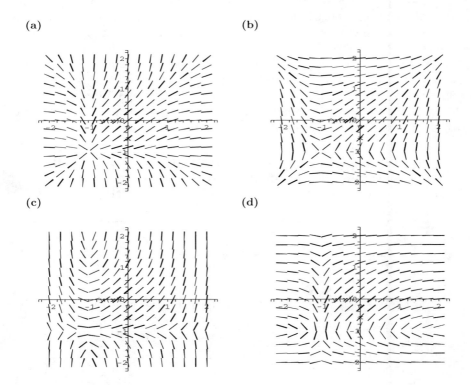

FIGURE 2.3:   Direction fields for problems 1–4.

5. $\dfrac{dy}{dx} = y(x^2 - y)$

6. $\dfrac{dy}{dx} = x(x^2 - y)$

7. $\dfrac{dy}{dx} = y^2(y - x^2)$

8. $\dfrac{dy}{dx} = \dfrac{x}{y - x^2}$

(a)   (b)

(c)   (d)

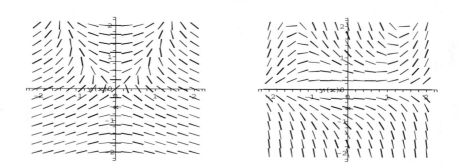

FIGURE 2.4:   Direction fields for problems 5–8.

Sketch the direction field for each of the first-order differential equations given in problems 9–25. If possible, draw some isoclines to confirm your sketch. (Get clarification from your instructor as to whether you should obtain these direction fields by hand or from the computer.) Verify that the direction fields you obtain in problems 9–12 are autonomous by observing that the slopes of the line segments are constant along fixed y-values. For each problem, use your sketch to draw various solution curves. Then draw the solution curve that passes through the given initial condition.

9. $\dfrac{dy}{dx} = y^4,$ $(1,1)$

10. $\dfrac{dy}{dx} = \cos y,$ $(0,0)$

11. $\dfrac{dy}{dx} = \sin y,$ $(\pi, 2)$

12. $\dfrac{dy}{dx} = e^{-y},$ $(0,1)$

13. $\dfrac{dy}{dx} = x^4,$ $(1,1)$

14. $\dfrac{dy}{dx} = \cos x,$ $(0,0)$

15. $\dfrac{dy}{dx} = \sin x,$ $(\pi, 2)$

16. $\dfrac{dy}{dx} = e^{-x},$ $(0,1)$

17. $\dfrac{dy}{dx} = xy,$ $(1,1)$

18. $\dfrac{dy}{dx} = x + y,$ $(0,0)$

19. $\dfrac{dy}{dx} = (x^2 + 1)(y + 1),$ $(0,1)$

20. $\dfrac{dy}{dx} = e^{x^2},$ $(0,1)$

21. $\dfrac{dy}{dx} = \dfrac{x - 1}{y - 1},$ $(0,0)$

22. $\dfrac{dy}{dx} = \dfrac{x^2 - 1}{y^2 + 1},$ $(0,0)$

23. $\dfrac{dy}{dx} = y(y^2 - 2),$ $(0,1)$

24. $\dfrac{dy}{dx} = xy(x^2 + 2),$ $(0,1)$

25. $\dfrac{dy}{dx} = \dfrac{x^3(y^2 + 1)}{y^2 + x^2},$ $(0,0)$

2.2   Existence and Uniqueness for First-Order Equations

We have now seen some approaches for gaining insight into the solution of an ordinary differential equation. Before going further we address the issue of when we can expect solutions to even exist or, if they exist, to be unique. This will become essential as we further develop numerical solution methods.

We begin with an example based on two direction fields that look very similar.

Example 1: Consider the direction fields for the differential equations

$$y' = 3y^{4/3} \text{ and } y' = 2y^{2/3}$$

(given in Figure 2.5), together with the initial condition $(\frac{-3}{2}, -1)$.

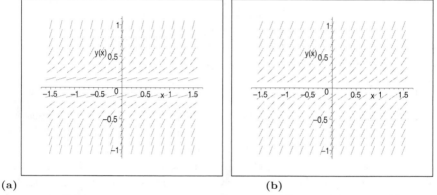

(a)   (b)

FIGURE 2.5: Direction fields for (a) $y' = 3y^{4/3}$ and (b) $y' = 2y^{2/3}$.

The reader should attempt to sketch the corresponding solution curve. It is not obvious that one of the solution curves passes through $y = 0$ while the other does not. The problem with passing through $y = 0$ is that $y = 0$ is itself a constant solution for both of the differential equations. The point where the two solutions intersect is a point where the solutions are *not unique*. In other words, if we start at the IC where they intersect, we cannot be sure which solution curve we should follow. We can see this more explicitly by solving the two equations, which easily can be done since both are separable equations. The differential equation

$$y' = 3y^{4/3} \text{ with IC } y\left(\frac{-3}{2}\right) = -1 \text{ has solution } y = \frac{-1}{\left(x + \frac{5}{2}\right)^3}$$

while

$$y' = 2y^{2/3} \text{ with IC } y\left(\frac{-3}{2}\right) = -1 \text{ has solution } y = \left(\frac{2}{3}x\right)^3.$$

The first of these approaches zero as x gets large and the second passes through $(0,0)$; see Figure 2.6.

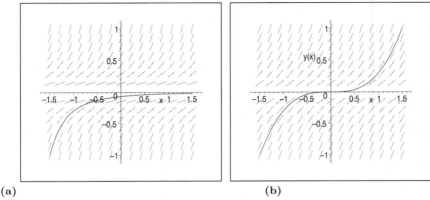

(a)   (b)

FIGURE 2.6: Direction fields for (a) $y' = 3y^{4/3}$ with IC $y\left(\frac{-3}{2}\right) = -1$ and (b) $y' = 2y^{2/3}$ with IC $y\left(\frac{-3}{2}\right) = -1$.

Thus, it appears that we do not have any problems with uniqueness of solutions in $y' = 3y^{4/3}$, but we do have a problem with uniqueness of solutions in $y' = 2y^{2/3}$ or at least we do at the point (0,0).

There are two things we need to be aware of at this point. The first is the need for a way of checking whether we can expect solutions to exist and be unique at a given initial condition. The other involves how to best calculate a numerical approximation to the solution of a differential equation—we trusted the computer to accurately sketch a solution curve in the previous section but how can we be sure that it will not overlook the slight differences between two direction fields? We will save our discussion of the latter until the next chapter and will only address the former for now. In order to understand the types of initial value problems that yield a *unique* solution the following result is required.

THEOREM 2.2.1 Existence and Uniqueness *Consider the initial-value problem*

$$y' = f(x, y) \quad \text{with} \quad y(x_0) = y_0.$$

If f and $\partial f / \partial y$ are continuous functions on the rectangular region

$$R : a < x < b, \quad c < y < d$$

containing the point (x_0, y_0), then there exists an interval

$$|x - x_0| < h$$

centered at x_0 on which there exists one and only one solution to the differential equation that satisfies the initial condition.

The proof of this theorem will be omitted, as it will take us off course. If you are interested, it can be found in C. Corduneanu's *Principles of Differential and Integral Equations* [10].

Remark: If the condition that $\partial f / \partial y$ is continuous on the rectangular region R containing the point (x_0, y_0) is not included in the assumptions of the theorem, then we say that at *least* one solution exists. We call this more "relaxed" result an *existence theorem* because the uniqueness of the solution is not guaranteed.

Example 2: Solve the differential equation

$$\frac{dy}{dx} = \frac{x}{y}$$

with the initial condition $y(0) = 0$. Does this result contradict the Existence and Uniqueness theorem?

The equation is separable and thus easily can be solved. Rewriting the equation

$$y \, dy = x \, dx,$$

integration yields the family of solutions

$$y^2 - x^2 = C.$$

Application of the initial condition gives us $C = 0$, so that

$$y^2 - x^2 = 0.$$

Solving for y gives two solutions

$$y = x \quad \text{and} \quad y = -x.$$

Does this result contradict the Existence and Uniqueness theorem? Although more than one solution satisfies this initial-value problem, the Existence and Uniqueness theorem is not contradicted because the function x/y is <u>not</u> continuous at the point $(0,0)$. The requirements of the theorem are not met at the place where uniqueness is violated.

Example 3: Verify that the initial-value problem

$$\frac{dy}{dx} = y$$

with the initial condition $y(0) = 1$ has a unique solution.

In this problem,

$$f(x,y) = y, \, x_0 = 0, \quad \text{and} \quad y_0 = 1.$$

Hence, both f and $\partial f/\partial y$ are continuous on all rectangular regions containing the point

$$(x_0, y_0) = (0, 1).$$

By the Existence and Uniqueness Theorem, there exists a unique solution to the differential equation that satisfies the initial condition $y(0) = 1$. We verify this by solving the initial-value problem.

The equation is separable and equivalent to

$$\frac{dy}{y} = dx.$$

A general solution is given by

$$y = ce^x$$

and with the initial condition $y(0) = 1$ gives

$$y = e^x.$$

The Existence and Uniqueness theorem gives sufficient, but not necessary, conditions for the existence of a unique solution of an initial-value problem. That is, there may be differential equations that do *not* satisfy the theorem but have solutions that are all unique.

Example 4: Consider the initial-value problem

$$\frac{dy}{dx} = \frac{x}{y^{2/3}}$$

with $y(0) = 0$. The Existence and Uniqueness theorem does not guarantee the existence of a solution because $xy^{-2/3}$ is discontinuous at $(0,0)$. The equation is separable and has solution

$$\frac{3}{5}y^{\frac{5}{3}} = \frac{1}{2}x^2 + c$$

so that

$$y = \left(\frac{5}{6}x^2 + c\right)^{3/5}.$$

Using the initial condition gives the *unique* solution

$$y = \left(\frac{5}{6}\right)^{\frac{3}{5}} x^{\frac{6}{5}}.$$

The theorem only provides sufficient conditions for a unique solution to exist. It does <u>not</u> say a unique solution won't exist if the conditions are <u>not</u> satisfied.

Example 5. Consider the initial-value problem

$$x \frac{dy}{dx} = y$$

with $y(0) = 0$. We can rewrite this as

$$\frac{dy}{dx} = \frac{y}{x}.$$

We see that $f(x, y) = y/x$ and thus $\partial f/\partial y = 1/x$. The Existence and Uniqueness theorem does not guarantee the existence of a solution because $f(x, y)$ is not continuous at the point $(0, 0)$.

The equation is separable and we can calculate the solution as

$$y = Cx.$$

Now, the initial condition $y(0) = 0$ gives the identity $0 = 0$, which means that $y = Cx$ is a solution to the differential equation for <u>any</u> value of C. Thus, there are infinitely many solutions to the initial-value problem; see Figure 2.7. Hence, the solution of the initial value problem is *not unique*.

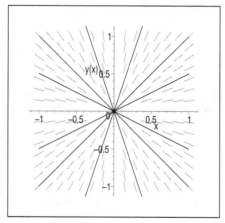

FIGURE 2.7: Direction fields for $\frac{dy}{dx} = \frac{y}{x}$. Note that all the solutions shown here satisfy the initial condition $y(0) = 0$.

Problems

1. Does the Existence and Uniqueness theorem guarantee a unique solution to the following initial-value problems on some interval? Explain.

a. $\dfrac{dy}{dx} + x^2 = y^2,$ with $y(0) = 0$

b. $\dfrac{dy}{dx} + x^2 = y^{-2},$ with $y(0) = 0$

c. $\dfrac{dy}{dx} = y + \dfrac{1}{1 - x},$ with $y(1) = 0$

2. Determine the region(s) of the x-y plane where solutions (i) exist and also where solutions (ii) exist and are unique, according to the Existence and Uniqueness theorem:

a. $\dfrac{dy}{dx} = 3x(y + 2)^{2/3}$

b. $\dfrac{dy}{dx} = (x - y)^{1/5}$

c. $\dfrac{dy}{dx} = x^2 y^{-1}$

d. $\dfrac{dy}{dx} = (x + y)^{-2}$

3. Determine the region(s) of the x-y plane where solutions (i) exist and also where solutions (ii) exist and are unique, according to the Existence and Uniqueness theorem:

a. $\dfrac{dy}{dx} = 2xy^{2/3}$

b. $\dfrac{dy}{dx} = 2x^{2/3}y$

c. $\dfrac{dy}{dx} = xy^{-2/3}$

d. $\dfrac{dy}{dx} = xy^{2/3}$

For each of the above parts, solve the differential equation subject to the initial condition $y(0) = 0$ and determine whether the solution you calculate crosses the solution $y = 0$.

4. Consider the initial-value problem

$$\frac{dy}{dx} = 2y^{1/2}, \quad y(1) = 3,$$

where $y \geq 0$. Use the Existence and Uniqueness theorem to determine if solutions will exist and be unique. Then solve the initial-value problem to obtain an analytic solution. Use this analytic solution to determine whether solution(s) passing through $y(1) = 3$ will be unique.

5. Consider the initial-value problem

$$\frac{dy}{dx} = 5(y - 2)^{3/5}, \quad y(0) = 2.$$

Use the Existence and Uniqueness Theorem to determine if solutions will exist and be unique. Then solve the initial value problem to obtain an analytic solution. Use this analytic solution to determine whether solution(s) passing through $y(0) = 2$ will be unique.

6. Consider the initial-value problem

$$\frac{dy}{dx} = x^{1/3}(1 - y^2), \quad y(0) = 1.$$

Use the Existence and Uniqueness theorem to determine if solutions will exist and be unique. Then solve the initial-value problem to obtain an analytic solution. Use this analytic solution to determine whether solution(s) passing through $y(0) = 1$ will be unique.

7. Consider the initial-value problem

$$\frac{dy}{dx} = 3(y - 1)^{1/3}, \quad y(0) = 1.$$

Use the Existence and Uniqueness theorem to determine if solutions will exist and be unique. Then solve the initial-value problem to obtain an analytic solution. Use this analytic solution to determine whether solution(s) passing through $y(0) = 1$ will be unique.

8. Consider the initial-value problem

$$\frac{dy}{dx} = \frac{x}{y^2}, \quad y(1) = 0.$$

Use the Existence and Uniqueness theorem to determine if solutions will exist and be unique. Then solve the initial value problem to obtain an analytic solution. Use this analytic solution to determine whether solution(s) passing through $y(0) = 1$ will be unique.

9. Consider the initial-value problem

$$\frac{dy}{dx} = \frac{\sin x}{(y - 1)^2}, \quad y(0) = 1.$$

Use the Existence and Uniqueness theorem to determine if solutions will exist and be unique. Then solve the initial value problem to obtain an analytic solution. Use this analytic solution to determine whether solution(s) passing through $y(0) = 1$ will be unique.

10. According to the Existence and Uniqueness theorem, the initial-value problem

$$\frac{dy}{dx} = |y|$$

with $y(1) = 0$ has a solution. Must this solution be unique? Solve this equation by considering two cases, $y \geq 0$ and $y < 0$, and using separation of variables. Is the solution unique?

2.3   First-Order Autonomous Equations—Geometrical Insight

We will now consider a method of geometrical analysis of differential equations to solve the first-order equation

$$\frac{dy}{dx} = f(y), \tag{2.3}$$

where $y = y(x)$ and x is *not* explicit on the right-hand side of the equation. We develop an important method of analyzing a differential equation to determine its behavior without solving it. *It is recommended that the reader be very familiar with the material in Appendix B.*

We first note that any first-order autonomous equation is separable and can thus be rewritten as

$$\int \frac{dy}{f(y)} = \int dx.$$

The integral on the left, however, is often not a simple integral. By this, we simply mean that we won't always be able to find an anti-derivative of it. Sometimes no such anti-derivative exists and other times we simply need "new" functions in order to write an anti-derivative. Our reason for needing to evaluate the integral was our desire to find the explicit (or implicit) solution, thereby allowing us to plot solutions in the x-y plane. For autonomous equations, however, we will gain insight into our solution in a qualitative manner. We will still be able to sketch solutions in the x-y plane but will lose the ability to answer a question such as "given the initial condition of $y(0) = 1$, what is the value of the solution at $x = 3$?" Our "qualitative description" means that we will be able to describe the long-term behavior of the solutions even though we will not write an explicit or implicit formula for the solution; that is, we will be able to answer the question "given the initial condition of $y(0) = 1$, what is the value of the solution as $x \to \infty$?"

2.3.1   The Phase Line

Our goal is to be able to predict the solution curve of the differential equation given in (2.3). We will gain an understanding of the behavior of the solution by observing that the expression $\frac{dy}{dx}$ is the rate of change of y with respect to x; whenever this expression is zero, y is constant and is called an *equilibrium* solution[2] denoted y^*; whenever it is positive, y is increasing; and

---

[2]We note that some books refer to equilibrium solutions as *critical points* because they satisfy the calculus definition $\frac{dy}{dx} = 0$. We will plot solutions in the y-$\frac{dy}{dx}$ plane and a "critical point" of the graph in this plane is *not* the same as the equilibrium solutions we are focusing on here. Thus, we will refer to y-values that satisfy $\frac{dy}{dx} = 0$ as equilibria.

whenever it is negative, y is decreasing. Thus for (2.3) we can say whether the solution y is increasing or decreasing simply based on the sign of $f(y)$.

Example 1: Consider the exponential growth differential equation

$$\frac{dy}{dx} = ry. \qquad (2.4)$$

The equation is separable with solution $y = Ce^{rx}$, where C is determined by the initial condition. We could plot these solutions but instead we want to use our "qualitative approach." To do so, we observe that the right-hand side is simply a straight line whose slope depends on r. There are two qualitatively different cases: $r > 0$ and $r < 0$. The horizontal axis is the y-axis and the vertical axis represents $\frac{dy}{dx}$; see Figure 2.8. When $r > 0$, the derivative is always positive, implying that the solution y is increasing. This is indicated by an arrow pointing to the right on the y-axis, with right being the direction of increase in y. In the other case, the arrow points to the left because the derivative is always negative. The y-axis with the equilibria and corresponding arrows drawn on it is called the *phase line* and contains the essential information necessary to describe the long-term behavior of the solution, i.e., what is $\lim_{x \to \infty} y(x)$ for a given initial condition $y(x_0) = y_0$. More can be said although it does not show in the arrows. In the first diagram, the derivative function is growing, which indicates not only that y is growing, but it is growing more and more rapidly.

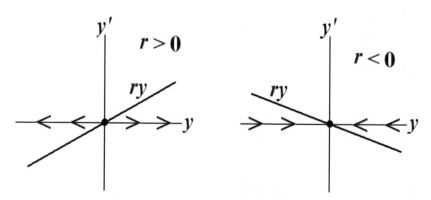

FIGURE 2.8: Phase line diagrams for equation (2.4).

We said that an equilibrium solution, y^*, corresponds to $\frac{dy}{dx} = 0$ and we thus have one equilibrium at $y^* = 0$. If our initial condition is *on* the equilibrium solution, e.g., $y(0) = 0$, then we will always remain at that equilibrium solution. For a general equilibrium, y^*, we state this as

$$\lim_{x \to \infty} y(x) = y^* \text{ for the initial condition } y(x_0) = y^*.$$

Note that y still depends on x (even though there is no x that is seen in the picture!) and the arrows indicate what happens to y as x increases.

It is natural for us to ask what happens to the solution of a differential equation as it proceeds from an initial condition close to an equilibrium solution. We will still consider the differential equation of Example 1 to gain insight into this. We will first examine the direction field for the two distinct cases, $r > 0$ and $r < 0$; see Figure 2.9. The unique feature about the direction field of an autonomous equation is that the equilibrium solution is represented by a horizontal line at the value y^*. Solutions starting near this horizontal line will either approach it or diverge away from it.

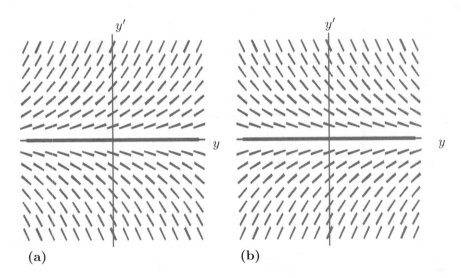

FIGURE 2.9:   The y-y' direction field for $\frac{dy}{dx} = ry$. In **(a)**, we have $r > 0$ and in **(b)** we have $r < 0$.

If we consider (a) corresponding to $r > 0$, we see that if we start with an initial value $y_0 = 0$, we will always stay at $y = 0$. If we start with an initial y-value $y_0 > 0$, our solutions will go to ∞, while if we start with an initial y-value $y_0 < 0$, our solutions will go to $-\infty$. Thus, starting on anywhere other than the equilibrium solution will yield a solution that diverges away from $y^* = 0$. We summarize the behavior near the equilibrium solution with a convenient definition. For convenience of notation, we will denote the derivative as y' instead of as $\frac{dy}{dx}$.

DEFINITION 2.1 *Consider the autonomous differential equation $y' = f(y)$, where $y = y(x)$. Assume that y^* is an equilibrium solution. Then*
1. *y^* is said to be **stable** if all solutions with an open set of initial conditions close by it always remain close by as $x \longrightarrow \infty$.*
2. *y^* is said to be **asymptotically stable** if all solutions with an open set of initial conditions close by it approach y^* as $x \longrightarrow \infty$.*
3. *y^* is said to be **unstable** if it is not stable.*

Note that every equilibrium solution that is asymptotically stable is also stable. We should also note that the equilibrium solution always corresponds to horizontal line segments in the direction field. Thus, a quick inspection of a direction field can tell us whether or not it corresponds to an autonomous equation.

Relating this definition to our previous example, we see that the equilibrium solution $y^* = 0$ is unstable if $r > 0$ because solutions with an initial condition close by y^* do not remain close by. When $r < 0$, however, solutions starting close by y^* approach it and thus $y^* = 0$ is asymptotically stable.

Our definition of stability can be put in terms of the traditional ϵ-δ definitions used in calculus. Specifically, y^* is stable if, for each $\epsilon > 0$, there exists a $\delta > 0$ such that

$$|y(x) - y^*| < \epsilon \text{ whenever } |y_0 - y^*| < \delta \text{ for all } x > 0,$$

where y_0 is the initial y-value.

Remark: In the case of a first-order autonomous equation, we can have an equilibrium solution that is *half-stable*. In this case, it is stable from one side but unstable from the other. A simple example is $y' = y^2$ where initial conditions to the left of $y = 0$ approach $y = 0$, whereas initial conditions to the right of $y = 0$ go to ∞.

What was the purpose of using the phase line if we ultimately plotted the direction field and concluded information about the stability of y^*? We actually did not need the direction field to reach the conclusions that we did—everything was already contained in the phase line. Specifically, the arrows on the line were drawn because they described how the solution was changing. In the case of $r > 0$, any solution starting with an initial condition to the right of $y^* = 0$ increased (as x increased) and thus moved away from y^*, whereas any solution starting with an initial condition to the left of $y^* = 0$ decreased (as x increased) and thus moved away from y^*. We can think of the arrows as pushing the solution in a given direction. On both sides of $y^* = 0$, the arrows are pushing the solution away and thus we classify y^* as unstable. For $r < 0$, it is the opposite situation and we see that the arrows are pointing toward $y^* = 0$ from both sides, thus indicating that solutions approach y^* as x increases. In this case, we consider $y^* = 0$ an asymptotically stable equilibrium solution. We can also sketch the solutions in the x-y plane without knowledge of the direction field. The information contained in the phase line is sufficient to sketch qualitatively accurate solutions.

By examining the phase line, we have the ability to explain what happens to the solution as $x \longrightarrow \infty$. As mentioned earlier, this description is a *qualitative* one—we cannot answer the question "given that $y(1) = -2$, what is $y(8)$?" Giving a *quantitative* answer to this question would require having the explicit or implicit solution, which may or may not be an easy solution to obtain. Using the phase line, we can also still give a plot of the solutions in the x-y plane but we again lose our ability to describe the value of the solution at a particular x-value. The usefulness of the phase line is in its ability to *quickly* give us qualitative information about *any* solution as $x \to \infty$.

Example 2: Consider the autonomous equation

$$y' = y^2(1 - y^2). \qquad (2.5)$$

Find the equilibria. Draw the corresponding phase line diagram and use it to determine the stability of the equilibrium solutions. Use the information contained in the phase line to sketch solutions in the x-y plane. Then conclude the long-term behavior for *any* initial condition.

To determine the equilibria, we set $y' = 0$ and solve for y. This gives us equilibria of $y = 0, -1, 1$, where $y = 0$ is a double root. The highest power and corresponding coefficient is $-y^4$ and thus we can use the information from Appendix B to quickly sketch the curve in the y-y' plane (see Figure 2.10). To draw the arrows on the phase line, we simply need to observe where the derivative y' is positive vs. negative. Between the equilibria $y^* = -1$ and $y^* = 0$ as well as between $y^* = 0$ and $y^* = 1$, we see that $y' > 0$ because the curve is above the horizontal axis. Thus we draw arrows going to the right to indicate an increase in the solution value. To the left of $y^* = -1$ and to the right of $y^* = 1$, we see that $y' < 0$ because the curve is below the horizontal axis and thus we draw arrows going to the left.

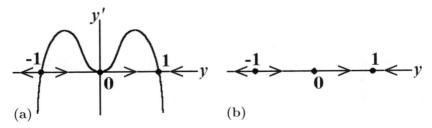

FIGURE 2.10: View of (2.5) in the (a) y-y' plane and (b) phase line.

Examining the phase line from the y-y' plane (see Figure 2.10), we see that the arrows on both sides of $y^* = -1$ are pointing away from it and it is thus unstable. The arrows on both sides of $y^* = 1$ are pointing toward it and it is thus asymptotically stable. For $y^* = 0$, we observe that arrows to the left of it point toward it but arrows to the right of it point away from it. Thus $y^* = 0$ is half-stable.

To sketch solutions in the x-y plane (see Figure 2.11), we first draw the equilibrium solutions as horizontal lines. From this we see that solutions will always be unique and thus *we will never be able to cross these equilibrium solutions.* If we begin with the initial condition $y_0 < -1$, the arrows on the phase line indicate that solutions will go to $-\infty$ as $x \to \infty$. We thus draw curves in the x-y plane that go away from the line $y = -1$. If the initial condition satisfies $-1 < y_0 < 0$, the arrows on the phase line indicate that we will approach the equilibrium solution $y^* = 0$. We thus draw curves in the x-y plane that go away from $y = -1$ and toward $y = 0$. Because the value of y' is small near the equilibria, the slope of the corresponding solution curve in the x-y plane will be small. Thus as $x \to -\infty$, the solution will have $y = -1$ as a horizontal asymptote, whereas when $x \to \infty$ the solution will have $y = 0$ as its horizontal asymptote. We can follow similar reasoning to draw solutions that begin with initial condition $0 < y_0 < 1$ and $y > 1$.

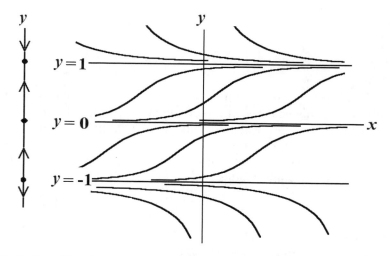

FIGURE 2.11: Sketch of some solutions of (2.5), based exclusively on phase line information; phase line (on left) from Figure 2.10 is now drawn vertically for easier comparison.

Equipped with the above information, we can now answer the final question of this example. Mathematically, we want to know $\lim_{x \to \infty} y(x)$ for any given $y(x_0) = y_0$. This would be a very challenging task if we had the explicit solution in front of us. With the phase line, however, we simply need to describe what happens to solutions with initial conditions between equilibrium solutions.

(i) if $y_0 < -1$, then $y(x) \to -\infty$ as $x \to \infty$

(ii) if $-1 < y_0 < 0$, then $y(x) \to 0$ as $x \to \infty$

(iii) if $0 < y_0 < 1$, then $y(x) \to 1$ as $x \to \infty$

(iv) if $y_0 > 1$, then $y(x) \to 1$ as $x \to \infty$

It is this type of "qualitative description" and knowledge of the "long-term behavior" that we are able to gain from considering the phase line instead of the explicit solution. Depending on the situation you may be considering, one of these approaches may be more desirable than the other.

2.3.2   Stability of Equilibrium via Taylor Series

Up to this point, our description of the equilibrium solutions has been based on insight gained from the phase line. Although this has been useful, it is often desirable to have an analytical method for determining stability of equilibria. This will be completely independent of the graphical interpretation we just learned but we will see that a combination of both methods will often be useful in analyzing a given problem.

Recall that the Taylor series expansion of a function $f(y)$ near y^* is given by

$$f(y) = f(y^*) + f'(y^*)(y - y^*) + \frac{1}{2!}f''(y^*)(y - y^*)^2 + \frac{1}{3!}f^{(3)}(y^*)(y - y^*)^3 + \cdots .$$

$$(2.6)$$

Taylor series are often useful in approximating the value of a function near a point y^* by only considering a finite number of terms of the Taylor series. In some situations, we are able to gain very good approximations with only the first two or three terms in the expansion.

We now expand the right-hand side of our autonomous equation in a Taylor series about the equilibrium solution, y^*:

$$y' = f(y^*) + f'(y^*)(y - y^*) + \cdots .$$

How has this helped us? We first observe that $f(y^*) = 0$ because y^* is an equilibrium solution. Then we observe that for y-values very close to y^*, the terms $(y - y^*)^2, (y - y^*)^3$, etc. are small compared to $f'(y^*)(y - y^*)$. Thus, close by the equilibrium y^*, we see that

$$y' \approx f'(y^*)(y - y^*)$$

$$(2.7)$$

is a good approximation. But this differential equation is separable and we know the solution is an exponential function. Moreover, the behavior of the solution will depend only on the sign of $f'(y)$. We can summarize this result in a useful theorem.

THEOREM 2.3.1 *Consider the autonomous differential equation $y' = f(y)$, with equilibrium solution y^*. If f has a Taylor expansion about y^*, then*
(a) y^* is stable when $f'(y^*) < 0$ and
(b) y^* is unstable when $f'(y^*) > 0$.

Note 1: In (a), the condition $f'(y^*) < 0$ says that the graph in the y-y' plane *decreases* through the horizontal axis, whereas in (b), the condition $f'(y^*) > 0$ says that the graph in the y-y' plane *increases* through the horizontal axis.

Note 2: The stability of y^* cannot be determined by this theorem in the case when $f'(y^*) = 0$. In this situation, the stability of y^* is determined by the higher-order terms of the Taylor expansion. We would need to use other means, e.g., the phase line, in order to conclude anything about the stability of the equilibrium.

Example 3: Find the equilibrium solutions of the autonomous equation $y' = y^2(1 - y^2)$. Use Theorem 2.3.1 to determine their stability.

We see the equilibria are $y = -1, 0, 1$. Our function is $f(y) = y^2(1 - y^2) = y^2 - y^4$ and the derivative is $f'(y) = 2y - 4y^3$. Applying the previous theorem shows that

(i) $f'(-1) = 2(-1) - 4(-1)^3 = 2 > 0 \implies y^* = -1$ is unstable

(ii) $f'(1) = 2(1) - 4(1)^3 = -2 < 0 \implies y^* = 1$ is stable

(iii) $f'(0) = 2(0) - 4(0)^3 = 0 \implies$ stability of $y^* = 0$ is inconclusive

In order to determine the stability of $y^* = 0$, we could use the graphical method of Example 2 to show that it is half-stable.

2.3.3 Bifurcations of Equilibria

Often we have a situation where the stability of a given equilibrium solution depends on the value of a *parameter*. We saw this in the first example when we considered $y' = ry$. In this case, r was the parameter and we saw that $y^* = 0$ was stable when $r < 0$, whereas $y^* = 0$ was unstable when $r > 0$. Thus changing the r-value changes the stability of the equilibrium solution. When the reason for the change in stability is due to the presence of another equilibrium solution, we can state the following definition:

DEFINITION 2.2 *Consider the equation $y' = f(r, y)$, where f is an infinitely differentiable function and r is a parameter. A bifurcation is a qualitative change in the number or stability of equilibrium solutions for this equation.*

There are three bifurcations that we will consider: transcritical, saddle-node, and pitchfork. We will give examples of the first two and will leave the third as an exercise.

Example 4: (Transcritical Bifurcation) Consider the equation

$$y' = ry - y^2, \tag{2.8}$$

where $r \in \mathbb{R}$. The equilibria are $y^* = 0, r$. We have three cases to consider: $r < 0$, $r = 0$, and $r > 0$. We could check the stability analytically using the Taylor series method previously discussed. However, it is more instructive to see the change in the stability of the solutions with the phase line. We first note that $y^* = 0$ is always an equilibrium solution. When $r < 0$, there is another equilibrium solution at $y^* = r$, lying to the left of $y^* = 0$. Drawing arrows on the phase line, we see that $y^* = r$ is unstable and $y^* = 0$ is stable. As r increases, the non-zero equilibrium approaches $y^* = 0$. At the bifurcation value of $r = 0$, the equilibrium solution $y^* = 0$ is a double root and is half-stable. As r becomes positive, the non-zero equilibrium solution is now to the right of $y^* = 0$. Drawing arrows on the phase line, we see that $y^* = 0$ is now unstable and $y^* = r$ is stable.

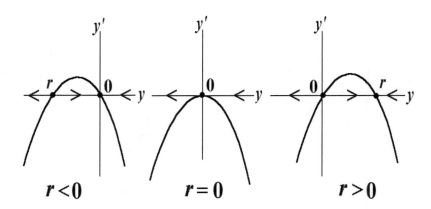

FIGURE 2.12: Phase line view of a transcritical bifurcation. The bifurcation happens **at** the value $r = 0$ but we determine the type of bifurcation by observing what happens before (for $r < 0$) and after (for $r > 0$).

This bifurcation is known as a *transcritical bifurcation*. It is usually easiest to think of in terms of the parameter r. When r is negative, we have two equilibrium solutions; as r continues to increase, the solutions approach each other and "crash" into each other when $r = 0$, switching stability as they do so; for

$r > 0$, the solution $y^* = 0$ is now the unstable one and $y^* = r$ is the stable one.

Example 5: (Saddlenode Bifurcation) Consider the equation

$$y' = r - y^2, \tag{2.9}$$

where $r \in \mathbb{R}$. The equilibria are $y^* = \pm\sqrt{r}$, which exist when $r \geq 0$. We again have three cases to consider: $r < 0$, $r = 0$, and $r > 0$. When $r < 0$, there are no equilibrium solutions. Drawing arrows on the phase line, we see that any initial condition gives a solution that diverges to $-\infty$. When $r = 0$, an equilibrium solution appears at $y^* = 0$ and is half-stable. Solutions beginning with an initial condition $y_0 < 0$ go to $-\infty$, whereas an initial condition $y_0 > 0$ gives a solution that approaches 0. When $r > 0$, there are now two equilibria. Drawing arrows on the phase line shows that the equilibria at $y^* = -\sqrt{r}$ is unstable and $y^* = \sqrt{r}$ are stable; see Figure 2.13.

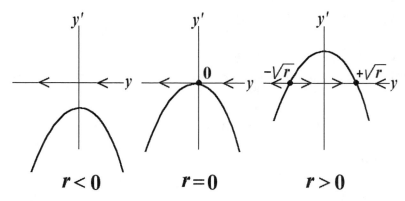

FIGURE 2.13: Phase line view of a saddlenode bifurcation. The bifurcation happens **at** the value $r = 0$ but we determine the type of bifurcation by observing what happens before (for $r < 0$) and after (for $r > 0$).

This bifurcation is known as a *saddlenode bifurcation* and we again think of it in terms of the parameter r. When r is negative, we have no equilibrium solutions; as r continues to increase, an equilibrium solution appears on the phase line, seemingly out of nowhere and is half-stable. For $r > 0$, this equilibrium then breaks into two parts, an unstable and a stable equilibrium.

Problems

For problems 1–22, (i) sketch the differential equation in the y-y' plane. (ii) Draw the phase line diagram, clearly stating the location and stability of the equilibria. (iii) State the long-term behavior for all initial conditions. (iv)

Assuming $y = y(x)$, sketch the corresponding graph in the traditional x-y plane. Although plotting the curves using the computer is always an option, the reader is highly encouraged to do the graphs below by hand and use the computer if desired only to check the work.

1. $y' = 2y + 3$
2. $y' = y^2 + 4y + 4$
3. $y' = y^2 - y - 6$
4. $y' = y(y + 2)(y - 3)$
5. $y' = (y - 2)^3(y^2 - 9)$
6. $y' = \sin y, \quad -2\pi < y < 2\pi$
7. $y' = e^{-y^2/2} - e^{-2}$
8. $y' = -y^2$
9. $y' = y^2(2 - y)$
10. $y' = (y - 1)^2(y - 2)^3(1 + y)$
11. $y' = \cos y + 1, \quad -2\pi < y < 2\pi$
12. $y' = y^3(2 + y)(5 + y)^7(3 - y)(y - 2)^4$

For the next two problems, assume that $v = v(t)$. The equations were discussed in detail in Section 1.3 as they are used to model the motion of a free-falling object. If $v(t)$ represents velocity, interpret the stability of the resulting equilibria. Note that (i) will be done in the v-v' plane and (iv) will be sketched in the t-v plane.

13. $v' = g - \frac{k}{m}v$
14. $v' = g - \frac{k}{m}v^2$

For the next eight problems, assume that $x = x(t)$. Note that (i) will be done in the x-x' plane and (iv) will be sketched in the t-x plane.

15. $x' = x^2(2 - x)(x + 3)^2$
16. $x' = (2 - x)^3(x^2 + 4)$
17. $x' = (2 - x)^3(x^2 + 4)^2$
18. $x' = -x^2(4 - x)(9 - x^2)$
19. $x' = x^5(1 - x)(1 - x^3)$
20. $x' = x(x - 3)(1 + x^3)(1 - x^2)^2$
21. $x' = x^2(1 - 2x)^3(x^2 - 1)$
22. $x' = x^3(x^2 + 5)(x - 4)^2(x + 5)$

Use the analytic approach of Theorem 2.3.1 to determine the stability of the equilibria in problems 23–27.

23. $y' = 1 - y^2$
24. $y' = y^2 - 1$

25. $y' = \sin y, \quad -2\pi < y < 2\pi$

26. $y' = y^3 + 1$

27. $y' = -y^3$

28. Consider the autonomous equation $y' = f(y)$ and suppose it is known that $f'(y^*) = 0, f''(y^*) = 0, f^{(3)}(y^*) < 0$ for the equilibrium solution y^*. Can we conclude anything about the stability of this solution?

29. A pitchfork bifurcation occurs when two equilibria appear out of one and change the stability of the existing equilibria as they appear; three equilibria are present after the bifurcation has occurred. Draw the phase line view of the following two bifurcations (before, at, and after the bifurcation):
    a. $y' = ry - y^3$ (supercritical pitchfork bifurcation)
    b. $y' = ry + y^3$ (subcritical pitchfork bifurcation)
    For both types of pitchfork bifurcations, you will need to consider the three different cases $r < 0$, $r = 0$, and $r > 0$.

2.4 Population Modeling: An Application of Autonomous Equations

The autonomous equations of the previous section have application in the area of population modeling. We can derive some important equations by thinking of the *per capita* rate of change of a population. We will let $x(t)$ denote the given population. The rate of change is simply the derivative with respect to time and the per capita rate of change is then given by

$$\frac{1}{x}\frac{dx}{dt}.$$

We have considered the simplest model of population growth, where we assume the per capita rate of change is constant and is equal to some parameter r:

$$\frac{1}{x}\frac{dx}{dt} = r.$$

Rearranging this gives the familiar exponential growth differential equation that we have seen before: $x' = rx$. The assumption that the rate of change is constant is not usually realistic for large populations. Why? Think of bacteria growing in a petri dish with finite space and nutrients with which to grow. When the population is small, the exponential growth model is acceptable, but when limited space and resources become an issue, the assumptions of this model break down. A more accurate model arises if we assume a linear decrease in the per capita rate of change of the population. This basically says that the more bacteria there are, the less quickly the overall population will grow due to limited resources. If there are too many bacteria, the overall population level may not change or may even decrease. The equation can be written

$$\frac{1}{x}\frac{dx}{dt} = r - k_1 x,$$

which we will rewrite as

$$\frac{dx}{dt} = rx\left(1 - \frac{x}{K}\right),$$

where $K = r/k_1$. This is known as the *logistic* differential equation. For small populations, we note that the logistic equation is approximated well by the exponential model. To see this, observe that if x is much smaller than K, then $\frac{x}{K}$ is very small and the $rx(1 - \frac{x}{K})$ term is approximately rx. Thus for small x, the differential equation is approximately $\frac{dx}{dt} = rx$. For the logistic equation, we can conclude that if $x(t)$ is small relative to K, then $x(t)$ exhibits nearly exponential growth. When $x(t)$ is near K, it levels out to be nearly constant. The behavior in-between is harder to predict, but it would seem reasonable that there is a smooth transition from exponential growth to no growth. If x is greater than K, the results are similar except the derivative is negative, so $x(t)$ starts off decaying exponentially until it nears K where it levels out. These behaviors are illustrated in Figure 2.14 and we sketched these solution curves without explicitly calculating the solution!

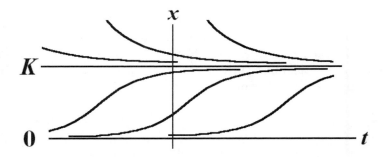

FIGURE 2.14:   Logistic curves for different initial values.

The constant r gives the intrinsic rate of growth of the population and the constant K is called the *carrying capacity* since, if x represents a population, the population rises or decays until it reaches this level.

So far, we have given verbal descriptions of the logistic equation. We could also analyze it by solving it using separation of variables or by using the graphical or numerical methods in the previous section. Choosing the latter approach, we observe that $f(x) = rx(1 - x/K)$ is a parabola, opening downward with zeros at $x = 0$ and $x = K$. (See Appendix B for a brief review of graphing factored polynomials.) With the parabola drawn in the y-y' plane, the phase line arrows can be determined by inspection. Whenever the graph is above the x-axis, the derivative is positive and indicates that x is increasing and thus the arrow points right. Analogously, when the graph is below the x-axis, x is decreasing and thus the arrows point to the left. Figure 2.15 shows a phase line diagram for $\frac{dx}{dt} = rx(1 - x/K)$.

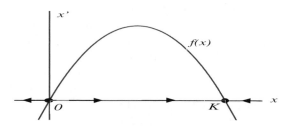

FIGURE 2.15: Phase line diagrams for the logistic model.

Since linear and exponential models are solved readily, one does not need to resort to the phase line to analyze the behavior of their solutions. However, the use of phase line analysis in these cases gives a quick illustration of the behavior of solutions in the long run. With the quadratic or logistic model, the value of the phase line analysis method becomes clear. Namely, it is a conceptual way to identify equilibria and test their stability. With the logistic model, this happens at $x = 0$ and $x = K$. Looking at the arrows on the phase plane, we observe that if x is perturbed from the equilibria at K, the growth rates as indicated by the arrows tend to push x back toward K. Thus, $x = K$ is a stable equilibrium (or stable steady-state solution). There is another steady-state at $x = 0$; observe that a slight perturbation from zero causes x to move away from zero. (For population models, we are interested only in positive x, but this statement is true for negative x as well.) Thus $x = 0$, is an unstable equilibrium (or an unstable steady-state solution).

If we wanted a qualitative plot in the t-x plane, we could still construct from the situation sketched in the x-x' plane: if we start with an initial condition x_0 in the interval $(0, K)$, then the solution $x(t)$ will approach K as $t \to \infty$; if we start with an initial condition $x_0 > K$, then $x(t)$ will again approach K as $t \to \infty$; if the initial condition satisfies $x_0 < 0$, then $x(t)$ will approach $-\infty$ as $t \to \infty$. As mentioned earlier, we can also observe that the larger the magnitude of $f(x)$, the larger the change in the solution. Thus, the solution changes most rapidly at the value $K/2$. The change is very small near the equilibria.

Again considering the logistic model, we could have solved it as a separable equation using the technique of partial fractions in an intermediate step. The solution can be understood by looking at it in the right way. But the qualitative analysis provided here gives us a way to quickly describe the long-term behavior of the solution for any initial x-value (i.e., any initial condition). In particular, we can readily see that *any* initial population between 0 and K will grow. In terms of modeling, this is actually a weakness in the logistic model. It suggests that a non-zero population, no matter how small, will result in growth. There are several resolutions to this problem. One is to put restrictions on the range of model validity. One might say that this model holds only for populations above the size of, say, 100. On the other hand, one might want a model that tends toward zero if the population gets too small. One such model is an *Allee effect model.*

An Allee effect model is an improvement over the logistic model in that if a population gets too small the growth rate becomes negative. For small population or large population, we thus have the per capita growth rate being negative and we can write

$$\frac{1}{x}\frac{dx}{dt} = r(a - x)(x - b),$$

where $b > a > 0$. This is used to handle a number of complex processes that occur with small populations, such as difficulties finding mates or inbreeding effects, without complicating an otherwise simple model. The Allee effect model is also the next natural step in our progression for it involves $\frac{dx}{dt}$ with a cubic equation.

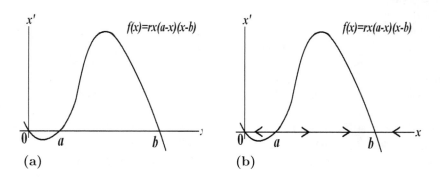

FIGURE 2.16:   (a) Cubic from Allee effect model.   (b) Phase line diagram for Allee effect model.

An Allee effect model is a cubic whose graph is given in Figure 2.16. The most useful forms to see this cubic equation are in the form $\frac{dx}{dt} = rx(a - x)(x - b)$, $r > 0$, $b > a > 0$, which is useful if the equilibria $x = a$, $x = b$ are known, and in the form $\frac{dx}{dt} = x(r - a(x - b)^2)$ if one is interested in the form xr (where the constants r, a, b would need to be chosen appropriately to give the desired cubic shape). A typical Allee effect growth curve and phase line are presented in Figure 2.16. There are three equilibria: one corresponding to the carrying capacity, one corresponding to zero population, and one at the threshold between "large" and "small" population sizes. With this model, both the zero population and the carrying capacity are stable equilibria, which generally corresponds to intuition. The "large-small" threshold point is an unstable equilibrium. Populations that start out above this value tend toward the carrying capacity; populations below this point tend to extinction.

We again note that the differential equation for the Allee effect model is separable and the resulting integral would involve partial fractions. Our qualitative phase line analysis was much quicker. We have now worked through analysis of three specific cases: linear, quadratic, and cubic. From this foundation, generalizations to other functions are obvious. A very nice discussion of population models with polynomial growth rates can be found in the technical report by Bewernick et al. [6].

Problems

1. Consider the equation
$$x' = x^2(1 - x),$$
   where $x \geq 0$ represents a population.
   a. Determine the equilibria and their stability. Do the stability results differ from the logistic equation?
   b. Compare the growth rate of the two models for small x.

2. Consider the Allee model
$$\frac{1}{x}\frac{dx}{dt} = r(a - x)(x - b),$$
   with $r = 2, a = 1, b = 6$. Determine the equilibria and their stability.

3. Consider the Allee model
$$x' = rx(a - x)(x - b),$$
   with $r = 1, a = 2, b = 10$. Determine the equilibria and their stability.

4. Consider the equation
$$x' = x^2(1 - x)(x - 4),$$
   where $x \geq 0$ represents a population.
   a. Determine the equilibria and their stability. Do the stability results differ from an equation with the Allee effect?
   b. Compare the growth rate of the two models for small x.

5. When is the exponential population model appropriate? When is the logistic population model appropriate? When is an Allee model appropriate? Discuss the benefits of each of these models and their drawbacks.

6. Suppose that a certain harmful bacteria, once introduced into the body, always persists. The body's immune system is usually able to keep the levels of the bacteria low unless the level of bacteria introduced is initially too large. We thus consider the equation

$$x' = x(x - a)(x - 5),$$

where $x \geq 0$ represents the population of this harmful bacteria and $0 < a < 5$ is a parameter.

a. Determine the equilibria and their stability.

b. Describe what happens to the bacteria for any initial condition. (You will have two cases to consider.)

c. Give a biological interpretation of the parameter a. In particular, determine whether a healthy person is likely to have a larger or smaller a-value than an unhealthy person. Are there other factors that may change the a-value?

7. Suppose a certain harmful bacteria is governed by the equation

$$x' = x(10 - x)[(x - 5)^2 - a],$$

where $x \geq 0$ represents the population of this harmful bacteria and $a < 25$ is a parameter (a can also be negative). Suppose that a bacteria level $x > 8$ represents a lethal level (and the person thus dies).

a. Determine the equilibria and their stability when $a > 0$.

b. Determine the equilibria and their stability when $a = 0$.

c. Determine the equilibria and their stability when $a < 0$.

d. What type of bifurcation did the system undergo?

e. What is the biological significance of going from $a > 0$ to $a < 0$? Does the patient survive this change in the parameter?

f. Give possible biological factors that may influence the value of the parameter a.

8. Write a differential equation describing a population of bacteria that has each of the following characteristics:

(i) Below a level $x = 1$, the bacteria will die off.

(ii) If the initial level of bacteria satisfies $1 < x_0 < 10$, the body is able to keep the bacteria at the "safe" level of $x = 6$.

(iii) If the initial level of bacteria satisfies $x > 10$, then the number of bacteria grows without bound.

9. Write a differential equation describing a population of bacteria that has each of the following characteristics:

(i) Below a level $x = 1/2$, the bacteria will die off.

(ii) If the initial level of bacteria satisfies $1 < x_0 < 8$, the body is able to keep the bacteria at the "safe" level of $x = 2$.

(iii) If the initial level of bacteria satisfies $x > 8$, then the number of bacteria grows without bound.

10. Suppose that the per capita growth rate (x'/x) of a certain population is described by the polynomial

$$x(2 - x)^2(x - 4).$$

   a. Write the corresponding differential equation.

   b. Determine the equilibria and their stability.

11. Suppose that the per capita growth rate (x'/x) of a certain population is described by the polynomial

$$r(x - a)(x - 1), \quad r, a > 0.$$

   a. Write the corresponding differential equation.

   b. Determine the equilibria and their stability. You will have three cases to consider $(a < 1, a = 1, a > 1)$.

   c. Give a biological interpretation for what happens to the population levels for each of your answers in part b.

2.5   Numerical Approximation with the Euler Method

We will now consider Euler's method, the first of two methods of numerically solving a differential equation of the form

$$\frac{dy}{dx} = f(x, y).$$

In examining direction fields, we observed that solutions must pass through the field tangentially. A natural goal is to want the computer to somehow sketch the solution curves. There are numerous methods for solving differential equations, some explicit and some implicit. We can choose methods that will give us as much accuracy as we want; however, the greater the accuracy desired, the more the computational work involved. As mentioned, we highlight two explicit methods here and leave an implicit method for the exercises. As we will soon see, Euler's method is a simple-to-understand method and turns out to be very unreliable. It does, however, give us insight into the second method we will consider—the Runge-Kutta method, which is widely used in practical applications due to its balance of computational work vs. accuracy.

Extremely important note: At this point we caution the reader about the potential pitfalls of computing numerical solutions. There is not a single method that will always work. Some methods have a wider applicability than others; some are easier to work with than others; all of the methods will run into difficulties for specialized problems. The reader should always view a graphical and/or numerical output with a bit of skepticism. Even if no other approach seems to work, the reader should attempt some type of calculation or analysis to make sure the solution behaves as it is expected. The best advice we can offer is **don't blindly trust your computer output.** You should always be able to justify why the computer output is a believable answer.

The Method

The first method is essentially an algorithm that formalizes the method of solution we used to draw a direction field. The method is an old one, essentially due to Euler.

Here is how it works. Pick an initial value, say $y(0) = a$. Compute the equation of the tangent line at this point, namely,

$$y_0(x) = a + f(0, a)x.$$

This approximate solution is good for very small changes. Euler's method is sensitive and tends to diverge away from a good approximation. We specify the *step size, h*, which tells the algorithm how far to go before recomputing a tangent line. In essence we are choosing a mesh size that is uniform on the x-axis but not the y-axis. The next equation of a tangent line is

$$y_1(x) = y_0(h) + f(h, y_0(h))h.$$

The process is continued until the last desired x-value is reached. The smaller the h-value, the smaller we expect the error to be *and* the more calculations we require to reach the specified value of x. Usually when Euler's method is used the equations of the lines are not written; rather the series of points that the lines connect is given. The coordinates of each point $x_{i+1} = x_i + h = x_0 + ih$ are calculated by the Euler method formula:

$$y_{i+1} = y_i + f(x_i, y_i)h, \quad i = 0, 1, 2, \ldots \tag{2.10}$$

Note that the value y_{i+1} is an approximation to the true solution. We obviously want the two to be as close together as possible. To generate these approximate solutions, we apply the following algorithm:

1. Specify (x_0, y_0), h

2. Divide the interval along the x-axis by the step size h to obtain the total number of steps n and a sequence of x-values: $x_i = x_0 + ih$, with $i = 0, 1, 2, 3, \ldots, n$

3. For each x_i, calculate the next approximate solution value $y_{i+1} = y_i + f(x_i, y_i)h$

Euler's method is easy to implement on a spreadsheet, calculator, or with any programming language geared toward mathematical computations.

Example 1: Use Euler's method with $h = 0.1$ to approximate the solution of

$$\frac{dy}{dx} = xy$$

with $y(0) = 1$ on $0 \leq x \leq 1$. Determine the exact solution and compare the results.

Since $\frac{dy}{dx} = xy$ we have $f(x, y) = xy$. The initial condition $y(0) = 1$ says that $x_0 = 0$ and $y_0 = 1$ so with $h = 0.1$ we have

$$x_1 = x_0 + h = 0.1, \quad x_2 = x_0 + 2h = 0.2, \quad \cdots, \quad x_{10} = x_0 + 10h = 1.0.$$

Euler's formula (2.10) applied to this example becomes

$$y_{i+1} = y_i + (0.1)(x_i)(y_i).$$

We calculate y_1 as

$$y_1 = y_0 + (.1)(x_0)(y_0) = 1 + (.1)(0)(1) = 1.$$

We then use x_1 and the newly found y_1 to calculate y_2:

$$y_2 = y_1 + (.1)(x_1)(y_1) = 1 + (.1)(.1)(1) = 1.01.$$

Since we are solving the equation on the interval $0 \leq x \leq 1$, we need to carry out these calculations until $x_i = 1$, that is, eight additional times. The results are summarized in the table in Figure 2.17.

x_i	y_i	x_i	y_i
0.0	1.0	0.6	1.15873
0.1	1.0	0.7	1.22825
0.2	1.01	0.8	1.31423
0.3	1.0302	0.9	1.41937
0.4	1.06111	1.0	1.54711
0.5	1.10355		

FIGURE 2.17: Euler's method with $h = 0.1$ for the equation $\frac{dy}{dx} = xy$.

The closed form solution to $\frac{dy}{dx} = xy$ with the condition $y(0) = 1$ is found, using separation of variables, to be

$$y(x) = e^{x^2/2}.$$

(Give it a try!) The table in Figure 2.18 compares the values using Euler's method and the exact solution. We see that initially the approximation is good, but its accuracy declines as x increases. A smaller step size h would increase the accuracy of the approximation but, as mentioned earlier, requires more steps.

x_i	Euler y_i	Exact $y(x_i)$	Error $|y_i - y(x_i)|$
0.0	1.0	1.0	0
0.1	1.0	1.0050	0.005
0.2	1.01	1.0202	0.0102
0.3	1.0302	1.0460	0.0158
0.4	1.06111	1.08329	0.02218
0.5	1.10355	1.13315	0.02960
0.6	1.15873	1.19722	0.03849
0.7	1.22825	1.27762	0.04937
0.8	1.31423	1.37713	0.06290
0.9	1.41937	1.49930	0.07993
1.0	1.54711	1.64872	0.10161

FIGURE 2.18:   Euler's method with $h = 0.1$ compared with the exact solution for the equation $\frac{dy}{dx} = xy$.

Figure 2.19 shows the graph of $y(x) = e^{x^2/2}$ plotted together with the ten points from Euler's method from Figure 2.18.

There are many methods that are considered superior for accuracy to Euler's method, but Euler's method has the advantage of simplicity, intuition, and clarity. Regarding the accuracy, the table in Figure 2.18 compared the approximate answer via Euler's method vs. the exact method. In general, we can place a bound on this difference, known as the *local truncation error*, by saying that

$$|y_i - y(x_i)| \leq Ch,$$

where C is a constant that depends on the function and the specified interval. See, for example, Burden and Faires [9]. It is also an important method to understand because when $h = 1$, it treats differential equations as difference equations. (Differential equations depend on a continuous independent variable, whereas difference equations depend on a discrete independent variable. As an example, difference equations can be used to model an insect population in which the adults lay eggs and die before the young hatch—the independent variable is the generation of the population.) In particular, we can use a numerical differential equation solver to give numerical solutions to difference equations provided we specify that Euler's method be used and that a step size $h = 1$ is always used. See, for example, Martelli [25].

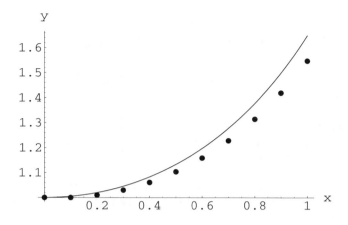

FIGURE 2.19: Plot of $y(x) = e^{x^2/2}$ with points obtained via Euler's method.

2.5.1 Implementation in Matlab, Maple, and Mathematica

The beauty of this method is that it can be implemented in a computer program. We have already given the built-in routines for the programs and this begins to give us insight into how those routines might work.
Maple
The Maple code is relatively straightforward. As we will see, we must specify exactly which method we wish to use, as well as the step size, etc. For the example of this section, we could type the following:

Computer Code 2.4: **Numerically solving a first-order equation with Euler's method; viewing the output and superimposing this approximate solution onto direction field**

Matlab, Maple, Mathematica

```
                          Maple
> eq1:=diff(y(x),x)=x*y(x);
> IC:=y(0)=1;
> xvalues:=array([0,.1,.2,.3,.4,.5,.6,.7,.8,.9,1.0]);
> soln1:= dsolve({eq1,IC},y(x),numeric,method=
    classical[foreuler],output=xvalues,stepsize=.1);
```

If we wanted to see the numerical solution at the x-values .25 and 1.1, we could instead *not* have specified `xvalues`:

```
                              Maple
> soln1:= dsolve({eq1,IC},y(x),numeric,method=
   classical[foreuler],stepsize=.1);
#the output should say soln2=proc(x_classical) ...   end proc
> soln1(.25);
> soln1(1.1);
> with(plots):
# Note we don't need to calculate the values again:
> odeplot(soln1,[x,y(x)],view=[-3..3,0..4],labels=[x,y],
   numpoints=500);
```

As shown earlier, we could also superimpose our numerical solution onto its direction field with the commands:

```
                              Maple
> with(DEtools):
> eq2:=odeplot(soln1,[x,y(x)],labels=[x,y],numpoints=500,
   thickness=2):
> eq3:=dfieldplot(eq1,y(x),x=0..1,y=1..1.6,arrows=line):
> display([eq2,eq3]);
```

Mathematica

```
                          Mathematica
de[x_]=x y[x]
solution=NDSolve[{y'[x]==de[x],y[0]==1},y,{x,0,1.2},
   StartingStepSize→ .1,Method→ {FixedStep,Method→
   ExplicitEuler}]
y[x_] = y[x]/.solution[[1]]
y[.25]
y[1.1]
<<Graphics'PlotField'
p1=PlotVectorField[{1,x y},{x,0,1},{y,1,1.6},
   ScaleFunction→ (1&),Axes→ Automatic,HeadLength→ 0,
   AxesLabel→ {"x","y"}]
p2=Plot[y[x],{x,0,1},PlotStyle→ {Thickness[0.01]},
   AxesLabel→ {"x","y"}]
Show[p1,p2]
```

Matlab

The Matlab code requires the use of m-files (Matlab's name for a text file that contains code to be executed), which are explained in Appendix A. We first create this m-file, called `Example.m`, which contains the function that we want to approximate numerically. The syntax is the *same* as for the m-file that we used for Matlab's built-in `ode45` routine.

Example.m

```
                          Matlab
function f=Example(xn,yn)
%
% The original ode is dy/dx=x*y
%
f= xn.*yn;
```

Then we need to create another m-file, `Euler1.m`, which is the implementation of Euler's method. It requires the use of the m-file `Example.m` which we just created.

Euler1.m

```
                          Matlab
function [xout,yout]=Euler1(fname,xvals,y0,h)

%fname = the function f for the equation
%xvals =vector that contains initial x0 and final xf
%y0 =initial y
%h =stepsize
x0=xvals(1); xf=xvals(2);

xn=x0; yn=y0;
xout=xn; yout=yn;
steps=(xf-x0)/h;
for j=1:steps
    fn=feval(fname,xn,yn);
    xn=xn+h;
    yn=yn+h*fn;
    xout=[xout; xn];
    yout=[yout; yn];
end
%end of function Euler1.m
```

We implement `Euler1.m` and view its output by typing the following commands in the command window:

```
                              Matlab
>>  [x,y]=Euler1('Example',[0,1],1,.1);
>>  [x y]
```

To plot the solution along with the vector field, we could then type

```
                              Matlab
>>  [X,Y]=meshgrid(0:.2:1,1:.1:1.6);%We use CAPITAL letters
>>  DY=X.*Y;
>>  DX=ones(size(DY));
>>  DW=sqrt(DX.^2+DY.^2);
>>  quiver(X,Y,DX./DW,DY./DW,.5,'.'); %plots direction field
>>  hold on
>>  plot(x,y)
>>  hold off
```

Problems

In problems 1–5, solve the given differential equation by hand using Euler's method with $h = 0.1$ to find approximate values for $y(x_1)$, $y(x_2)$, $y(x_3)$, and $y(x_4)$. Then compare the results with the given explicit solution at x_1, x_2, x_3, and x_4. It will be easiest if you display the results as we did in the table in Figure 2.18.

1. $dy/dx = x^3$, $y(1) = 1$; explicit solution: $y = \frac{1}{4}(x^4 + 3)$

2. $dy/dx = x^4 y$, $y(1) = 1$; explicit solution: $y = e^{(x^5-1)/5}$

3. $dy/dx = -y^2 \cos x$, $y(0) = 1$; explicit solution: $y = \dfrac{1}{1 + \sin x}$

4. $dy/dx = \frac{\sin x}{y^3}$, $y(\pi) = 2$; explicit solution: $y = (12 - 4\cos x)^{1/4}$

5. $dy/dx = ye^{-x}$, $y(0) = 1$; explicit solution: $y = \exp(1 - e^{-x})$

   In problems 6–11, solve the given differential equation with your computer software package using Euler's method with $h = 0.1$ to find approximate values for $y(x_1)$, $y(x_2)$, $y(x_3)$, $y(x_4)$, $y(x_5)$, $y(x_6)$, $y(x_7)$, and $y(x_8)$. Then compare the results with the given explicit solution at x_1, x_2, x_3, x_4, x_5, x_6, x_7, and x_8. It will be easiest if you display the results as we did in Table 2.18.

6. $dy/dx = xy^2$, $y(0) = 1$; explicit solution: $y = \dfrac{2}{2 - x^2}$

7. $dy/dx = y + \cos x$, $y(0) = 0$; explicit solution: $y = \frac{1}{2}(\sin x - \cos x + e^x)$

8. $dy/dx = y + \sin x$, $y(0) = 2$; explicit solution: $y = \frac{-1}{2}(\cos x + \sin x + 5e^x)$

9. $dy/dx = e^{-y}$, $y(0) = 2$; explicit solution: $y = \ln(x + e^2)$

10. $dy/dx = x + y$, $y(0) = 0$; explicit solution: $y = -x - 1 + e^x$

11. $dy/dx = (x+1)(y^2+1)$, $y(0) = 0$; explicit solution: $y = \tan(\frac{1}{2}x^2 + x)$

12. Use Euler's method to numerically solve the following two logistic differential equations for time t between 0 and 1. Assume that $x(0) = 100$. Solve both using the following h values: 1, 0.5, 0.25, and 0.1. Compare the solutions of the first equation for the different h values. Explain what is happening as h gets small. Repeat the comparison for the second equation. Compare the solutions of the two equations to each other for the different h values. Explain why two differential equations that are identical except for the parameter values exhibit different behaviors while h is in the process of "getting small."

   a. $\dfrac{dx}{dt} = 1.7x \left(1 - \dfrac{x}{1000}\right)$

   b. $\dfrac{dx}{dt} = 2.7x \left(1 - \dfrac{x}{1000}\right)$

13. Consider the initial value problem

$$\frac{dy}{dx} = f(x, y), \ a < x < b, \ y(x_0) = y_0,$$

with true solution $y(x)$. Derive the formula used in Euler's method at the ith step in the following manner: assume the approximate and exact solutions agree at $(x_i, y(x_i))$ and Taylor expand the exact solution at the next calculated x-value $y(x_i + h)$. Ignore terms of $O(h^2)$.

2.6 Numerical Approximation with the Runge-Kutta Method

We have just employed Euler's method for solving a first-order differential equation. There are many other methods. Two that are frequently encountered are second-order and fourth-order Runge-Kutta methods. The method was named after the German mathematicians Carl Runge (1856–1927) and Wilhelm Kutta (1867–1944), who developed the theory long before the advent of modern computers. These work in manners similar to Euler's method except that instead of using first-order approximations (lines), they use second-order (parabolas) and fourth-order (quartics) curves, although they do it in ways that avoid the computation of higher-order derivatives. The Runge-Kutta methods can also be interpreted as multiple applications of Euler's method at *intermediate* values, that is, between x_i and $x_i + h$. It is this interpretation that we will use for graphical purposes.

The algorithms are more complicated than Euler's but, nevertheless, are easily enough programmed. We will consider the Runge-Kutta fourth-order algorithm which is by far the method most commonly used by scientists and engineers in their respective fields. The idea is to use a weighted average of the slopes of field segments at four points "near" the current point (x_i, y_i). The location of each slope evaluation is determined by an application of Euler's method.

In order to figure out a good formula for giving weights to the function value at the four points we assume we have computed y_i, which is the approximation to $y(x_i)$. In order to compute the next approximate solution value y_{i+1}, we observe that the fundamental theorem of calculus gives

$$y(x_{i+1}) - y(x_i) = \int_{x_i}^{x_{i+1}} y'(x)dx = \int_{x_i}^{x_i+h} y'(x)dx. \tag{2.11}$$

We then use Simpson's rule to obtain an approximation to this integral:

$$y(x_{i+1}) - y(x_i) \approx \frac{h}{6}\left[y'(x_i) + 4y'\left(x_i + \frac{h}{2}\right) + y'(x_{i+1})\right].$$

We said that we want to use a weighted average of the slopes at four points "near" the current point. We obtain a fourth point by performing *two* evaluations at the middle point $x_i + \frac{h}{2}$. Thus our approximation is rewritten

$$y(x_{i+1}) - y(x_i) \approx \frac{h}{6}\left[y'(x_i) + 2y'\left(x_i + \frac{h}{2}\right) + 2y'\left(x_i + \frac{h}{2}\right) + y'(x_{i+1})\right] \tag{2.12}$$

and we will update the slope at the successive evaluations.

In order to see the evaluations at the four points, we give a graphical interpretation of the Runge-Kutta fourth-order method and then state the formula for calculating the next approximate value, y_{i+1}. See Figure 2.20 for the graphical interpretation. We begin at the point (x_i, y_i) and calculate the coordinates of the next point of the approximate solution (x_{i+1}, y_{i+1}).

1. First evaluation $k_1 = f(x_i, y_i)$
We calculate the slope of the solution that passes through our current value (x_i, y_i) by evaluating $f(x_i, y_i)$. Call this slope k_1. Euler's method tells us to continue at this slope for one step size and the Euler approximation of the next value is $(x_i + h, y_i + hk_1)$. The Runge-Kutta method takes *half* of this increase in the y coordinate, which is easily seen to be $hk_1/2$, and goes to the x-value *halfway* between x_i and the next step $x_i + h$.

2. Second evaluation $k_2 = f(x_i + \frac{h}{2}, y_i + \frac{hk_1}{2})$
We now perform our first of two "mid-value" calculations. Our "mid-value" x-value is the halfway point $x_i + \frac{h}{2}$ and the y-value is $y_i + \frac{hk_1}{2}$; we evaluate the slope of the solution passing through this point, which is $f(x_i + \frac{h}{2}, y_i + \frac{hk_1}{2})$.

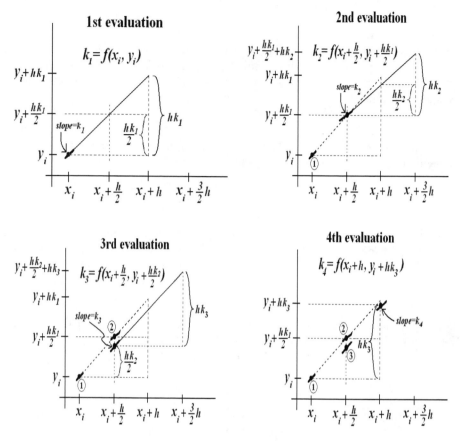

FIGURE 2.20: One step of fourth-order Runge-Kutta. Given the point (x_i, y_i), the next y-value is calculated by $y_{i+1} = y_i + h(k_1 + 2k_2 + 2k_3 + k_4)/6$.

For simplicity, call this value k_2. If we followed this slope for one step size (to the x-value $x_i + \frac{3h}{2}$) we would be at the y-value $y_i + \frac{hk_1}{2} + hk_2$. We again take *half* of this increase, which is seen to be $hk_2/2$. We go back to the "mid-value" x-value and a different y-value, $y_i + \frac{hk_2}{2}$.

3. Third evaluation $k_3 = f(x_i + \frac{h}{2}, y_i + \frac{hk_2}{2})$
We now perform our second of two "mid-value" calculations. Our "mid-value" x-value is still the halfway point $x_i + \frac{h}{2}$ and the y-value is now $y_i + \frac{hk_2}{2}$; we again evaluate the slope of the solution passing through this point, which is $f(x_i + \frac{h}{2}, y_i + \frac{hk_2}{2})$. For simplicity, call this value k_3. If we followed this slope for one step size (to the x-value $x_i + \frac{3h}{2}$) we would be at the y-value $y_i + \frac{hk_2}{2} + hk_3$. We note the increase, which is seen to be hk_3. We now go to the x-value $x_i + h$

and the y-value $y_i + hk_3$, which takes into account this last estimated increase.

4. Fourth evaluation $k_4 = f(x_i + h, y_i + hk_3)$
We now do our final evaluation at this point by calculating the slope of the solution which passes through the value $(x_i + h, y_i + hk_3)$. Call this slope k_4. Note that the (x, y) value where we evaluate is *not* the original one predicted by Euler's method.

We thus have four slopes close by our current (x_i, y_i) pair: one at x_i, one at $x_i + h$, and two at $x_i + \frac{h}{2}$. The Runge-Kutta fourth-order takes a weighted average of these slopes to calculate the next point:

$$y_{i+1} = y_i + \frac{h}{6}(k_1 + 2k_2 + 2k_3 + k_4).$$

As with the Euler method, we can put a bound on the error between the exact solution value $y(x_i)$ and the approximate value calculated by the Runge-Kutta method. The local truncation error, which is simply this difference, satisfies

$$|y_i - y(x_i)| \le Mh^4$$

where M is a constant that depends on the function and the specified interval. For a more detailed explanation, see Burden and Faires [9].
    We summarize the above in the following algorithm:

1. Specify (x_0, y_0), h

2. Divide the interval along the x-axis by the step size h to obtain the total number of steps n and a sequence of x-values: $x_i = x_0 + ih$, $i = 0, 1, 2, 3, \ldots, n$

3. For each x_i, calculate $k_{j,i}$, $j = 1, 2, 3, 4$ (where the additional subscript will be used to denote the calculation from the ith value), defined by
    a. $k_{1,i} = f(x_i, y_i)$
    b. $k_{2,i} = f(x_i + h/2, y_i + (hk_{1,i})/2)$
    c. $k_{3,i} = f(x_i + h/2, y_i + (hk_{2,i})/2)$
    d. $k_{4,i} = f(x_i + h, y_i + hk_{3,i})$

4. Calculate $y_{i+1} = y_i + h(k_{1,i} + 2k_{2,i} + 2k_{3,i} + k_{4,i})/6$, which is the next y-value of the approximate solution, as calculated by the fourth-order Runge-Kutta method.

Example 1: We will now use the Runge-Kutta algorithm with $h = 0.1$ to approximate the solution of

$$\frac{dy}{dx} = xy$$

with $y(0) = 1$ on $0 \leq x \leq 1$. This is the same equation we approximated using Euler's method.

The initial condition $y(0) = 1$ gives $x_0 = 0$ and $y_0 = 1$, so with $h = 0.1$ we have

$$k_{1,0} = f(0, 1) = 0,$$

$$k_{2,0} = f\left(0 + 0.05, 1 + (0.1)\left(\frac{0}{2}\right)\right) = 0.05,$$

$$k_{3,0} = f\left(0 + 0.05, 1 + (0.1)\left(\frac{0.05}{2}\right)\right) = 0.0501,$$

$$k_{4,0} = f(0 + 0.1, 1 + (0.1)(0.0501)) = 0.1005,$$

and thus the next value is

$$y_1 = 1 + \left(\frac{0.1}{6}\right)(0 + 2(0.05) + 2(0.0501) + 0.1005) = 1.005012.$$

The remaining calculations are performed similarly and are summarized in the table in Figure 2.21 along with the true solution. Note the close agreement of the Runge-Kutta approximate value of the solution with the true value of y_n.

x_i	Runge-Kutta y_i	True $y(x_i)$	Error $|y_i - y(x_i)|$
0.0	1.0	1.0	0.0
0.1	1.0050125	1.0050125	0.0000000
0.2	1.0202013	1.0202013	0.0000000
0.3	1.0460279	1.0460279	0.0000000
0.4	1.0832871	1.0832872	0.0000000
0.5	1.1331485	1.1331484	0.0000001
0.6	1.1972174	1.1972173	0.0000001
0.7	1.2776213	1.2776213	0.0000000
0.8	1.3771278	1.3771277	0.0000001
0.9	1.4993025	1.4993024	0.0000001
1.0	1.6487213	1.6487210	0.0000003

FIGURE 2.21: Runge-Kutta's method with $h = 0.1$ compared with the analytic solution $(e^{x^2/2})$ for the equation $\frac{dy}{dx} = xy$.

It is often better to have closed-form solutions as they may provide theoretical insights, and there are no problems with errors associated with approximate methods. Unfortunately most of the differential equations one encounters cannot be solved. Numerical solvers allow us to proceed with our analysis, but are limited in that one must explicitly express all parameters as numbers and give all necessary initial conditions.

2.6.1   Implementation in Matlab, Maple, and Mathematica

As with Euler's method, the fourth-order Runge-Kutta method can be implemented in a computer program. Many of the built-in routines in a computer software package will use some variation of this routine (usually utilizing a variable step size), although other routines may be chosen in specific situations. *The Matlab code given below should be saved as it can be used for numerically solving any first-order system in question by simply changing the m-file where the equation is entered.*

Maple
The Maple code is relatively straightforward. As we will see, we must specify exactly which method we wish to use, as well as the step size, etc. For the example of this section, we could type the following:

Computer Code 2.5:   **Numerically solving a first-order equation with fourth-order Runge-Kutta method; viewing the output and superimposing this approximate solution onto direction field**

Matlab, Maple, Mathematica

```
                          Maple
> with(plots):
> eq1:=diff(y(x),x)=x*y(x);
> IC:=y(0)=1;
> xvalues:=array([0,.1,.2,.3,.4,.5,.6,.7,.8,.9,1.0]);
> soln1:= dsolve({eq1,IC},y(x),numeric,method=classical[rk4],
  output=xvalues,stepsize=.1);
```

The only difference between this code and that of Euler's method is that we now say `method=classical[rk4]`—everything else remains the same. If we wanted to see the numerical solution at the x-values .25 and 1.1, we could instead *not* have specified `xvalues`:

```
                          Maple
> soln1:= dsolve({eq1,IC},y(x),numeric,method=classical[rk4],
  stepsize=.1);
#the output should say soln2=proc(x_classical) ...  end proc
> soln1(.25);
> soln1(1.1);
```

To plot the solution, we don't need to calculate the values again—we simply type

```
                          Maple
> odeplot(soln1,[x,y(x)],view=[-3..3,0..4],labels=[x,y],
  numpoints=500);
```

As shown earlier, we could also superimpose our numerical solution onto its direction field with the commands:

```
                          Maple
> with(DEtools):
> eq2:=odeplot(soln1,[x,y(x)],labels=[x,y],numpoints=500,
  thickness=2):
> eq3:=dfieldplot(eq1,y(x),x=0..1,y=1..1.6,arrows=line):
> display([eq2,eq3]);
```

Mathematica

```
                       Mathematica
de[x_]=x y[x]
solution=NDSolve[{y'[x]==de[x],y[0]==1},y,{x,0,1.2},
  StartingStepSize→ .1,Method→ {FixedStep,Method→
  ExplicitRungeKutta}]
y[x_] = y[x]/.solution[[1]]
y[.25]
y[1.1]
<<Graphics'PlotField'
p1=PlotVectorField[{1,x y},{x,0,1},{y,1,1.6},
  ScaleFunction→ (1&),Axes→ Automatic,HeadLength→ 0,
  AxesLabel→ {"x","y"}]
p2=Plot[y[x],{x,0,1},PlotStyle→ {Thickness[0.01]},
  AxesLabel→ {"x","y"}]
Show[p1,p2]
```

Matlab

The Matlab code again requires the use of m-files. We use the m-file `Example.m`, which was created in the previous section and contains the function that we want to approximate numerically:

Example.m

```
                              Matlab
function f=Example(xn,yn)
%
% The original ode is dy/dx=x*y
%
f= xn.*yn;
```

The implementation of the fourth-order fixed-step Runge-Kutta also requires an m-file. The m-file `RK4.m` gives this implementation. Remember that comments following a % are ignored by Matlab but are often instructive for the programmer in remembering the purpose of a given line. We consider both x and y to be scalars in the code below but y could actually be a vector, as we will see in the chapter on higher-order equations. In Matlab, if `y` is a vector then `y'` is the transpose of this vector. In other words, if `y` is a column vector, then `y'` is a row vector and vice versa. So that our syntax used is Matlab's built-in `ode45`, we point out two minor modifications, both of which involve the use of the 'prime' in Matlab. The first (`yn=yn'`) occurs just before the 'for-loop,' while the other occurs near the end of the loop (`yout=[yout; ynext']`). These lines have no effect if we have only one first-order equation but will be needed for more equations or higher-order systems to be considered later. And again, it is simply so that a call to `RK4.m` and Matlab's `ode45` are only different because the former requires a step size whereas the latter does not.

RK4.m

```
                          Matlab
function [xout,yout]=RK4(fname,xvals,y0,h)

%fname = the function f for the equation
% Note:  fname does NOT need to match file name--leave as is!
%xvals =vector with initial x0 and final xf
%y0 =initial y
%h =stepsize
%Note that function RK4step is contained in RK4.m!!
%Function RK4.m ends with "% end of m-file RK4.m"

x0=xvals(1); xf=xvals(2);
xn=x0; yn=y0;
xout=xn; yout=yn;
steps=(xf-x0)/h;
yn=yn';%Note the 'prime' after yn;
    %Needed for same syntax as ode45 in higher dimensions

for j=1:steps
    k1 = feval(fname,xn,yn);
    k2 = feval(fname,xn+(h/2),yn+(h*k1/2));
    k3 = feval(fname,xn+(h/2),yn+(h*k2/2));
    k4 = feval(fname,xn+h,yn+h*k3);
    ynext = yn +h*(k1 + 2*k2 + 2*k3 + k4)/6;
    xnext = xn+h;
    xout=[xout; xnext];
    yout=[yout; ynext']; %Note the 'prime' after ynext;
    %Needed for same syntax as ode45 in higher dimensions
    xn=xnext; %xnext is the new xn value
    yn=ynext; %ynext is the new yn value
end
% end of m-file RK4.m
```

In the command window, we would type

```
                          Matlab
>>  [x,y]=RK4('Example',[0,1],1,.1);
>>  [x y]
```

to show the values obtained in the text. To plot the solution along with the vector field, we could then type

```
                          Matlab
≫   [X,Y]=meshgrid(0:.2:1,1:.1:1.6);%We use CAPITAL letters
≫   DY=X.*Y;
≫   DX=ones(size(DY));
≫   DW=sqrt(DX.^2+DY.^2);
≫   quiver(X,Y,DX./DW,DY./DW,.5,'.');  %plots direction field
≫   hold on
≫   plot(x,y)
≫   hold off
```

Once this program is entered, it can be used over and over again—only the m-file `Example.m` that contains the right-hand side of the differential equation needs to be changed. This program is the same as the one used for higher-order equations (e.g., Section 3.5), as well as the variable step size routine `ode45` (discussed briefly in Section 2.1.1). The variable step size methods will allow us to obtain more accurate numerical answers because they take into account the changing direction field. In general terms, a larger step size is taken when the direction field doesn't change much and a smaller step size is taken when the direction field is changing more rapidly. We will not go into the details of variable step size methods here but will simply refer the interested reader to some of the references at the end of the book.

We finish this section with an example that compares the accuracy of the fixed step size method for different step sizes.

Example 2: Graph the numerical solution for

$$\frac{dy}{dx} = \sin(x^2)$$

with $y(0) = 1$ on $-7 \leq x \leq 7$. Use the fourth-order Runge-Kutta method and compare the graphs for the step sizes of $h = 1.0, 0.5, 0.1$.

Computer Code 2.6: **Comparing numerical solutions obtained with different step sizes using fourth-order Runge-Kutta method; plotting solutions with different line styles, colors**

Matlab, Maple, Mathematica

Maple
We again need to remember to load the relevant packages before using them.

```
                        Maple
> with(plots):
> with(DEtools):
> eq1:=diff(y(x),x)=sin(x^2);
> IC:=y(0)=1;
> soln1:= dsolve({eq1,IC},y(x),numeric,method=classical[rk4],
    stepsize=1);
> soln2:= dsolve({eq1,IC},y(x),numeric,method=classical[rk4],
    stepsize=.5);
> soln3:= dsolve({eq1,IC},y(x),numeric,method=classical[rk4],
    stepsize=.1);
> eq1a:=odeplot(soln1,[x,y(x)],view=[-7..7,-0.5..2.5],
    labels=[x,y],numpoints=500,color=red,linestyle=2):
> eq2a:=odeplot(soln2,[x,y(x)],view=[-7..7,-0.5..2.5],
    labels=[x,y],numpoints=500,color=blue,linestyle=3):
> eq3a:=odeplot(soln3,[x,y(x)],view=[-7..7,-0.5..2.5],
    labels=[x,y],numpoints=500, color=green, linestyle=4):
> display([eq1a,eq2a,eq3a],title="dy/dx=sin(x^2); dotted is
    h=1.0, dashed is h=0.5, dot-dash is h=0.1");
```

Mathematica

```
                     Mathematica
de[x_]=Sin[x^2]
solution1=NDSolve[{y'[x]==de[x],y[0]==1},y,{x,0,1.2},
   StartingStepSize→ 1,Method→ {FixedStep,Method→
   ExplicitRungeKutta}]
y1[x_] = y[x]/.solution1[[1]]
solution2=NDSolve[{y'[x]==de[x],y[0]==1},y,{x,0,1.2},
   StartingStepSize→ .5,Method→ {FixedStep,Method→
   ExplicitRungeKutta}]
y2[x_] = y[x]/.solution2[[1]]
solution3=NDSolve[{y'[x]==de[x],y[0]==1},y,{x,0,1.2},
   StartingStepSize→ .1,Method→ {FixedStep,Method→
   ExplicitRungeKutta}]
y3[x_] = y[x]/.solution3[[1]]
p1=Plot[y1[x],{x,0,1},PlotStyle→ {Thickness[0.01],
   Dashing[{.02}]},AxesLabel→ {"x","y"}]
p2=Plot[y2[x],{x,0,1},PlotStyle→ {Thickness[0.01],
   Dashing[{.05}]},AxesLabel→ {"x","y"}]
p3=Plot[y3[x],{x,0,1},PlotStyle→ {Thickness[0.01],
   Dashing[{.01,.05,.05,.05}]},AxesLabel→ {"x","y"}]
Show[p1,p2,p3]
```

Matlab
We first create an m-file, call it `Example3.m`, which contains the function that we want to approximate numerically:

Example3.m

```
                          Matlab
function f=Example(xn,yn)
%
% The original ode is dy/dx=sin(x.^2)
%
f= sin(xn.^2);
```

Then in the command window, we would type

```
                          Matlab
>>  [x1,y1]=RK4('Example3',[0,1],1,1.0);
>>  [x2,y2]=RK4('Example3',[0,1],1,0.5);
>>  [x3,y3]=RK4('Example3',[0,1],1,0.1);
>>  plot(x1,y1,'r:');
>>  hold on
>>  plot(x2,y2,'b-')
>>  plot(x3,y3,'g-.')
>>  axis[-7,7,-0.5,2.5]
>>  xlabel('x');ylabel('y');
>>  title('dy/dx=sin(x^2); dotted is h=1.0, dashed is h=0.5,
     dot-dash is h=0.1');
>>  hold off
```

Problems

In problems 1–5, solve by hand using the fourth-order Runge-Kutta method with $h = 0.1$ to find approximate values for $y(x_1)$, and $y(x_2)$. Then compare the results with the given explicit solution at x_1 and x_2. It will be easiest if you display the results as we did in Table 2.21.

1. $dy/dx = x^3$, $y(1) = 1$; explicit solution: $y = \frac{1}{4}(x^4 + 3)$

2. $dy/dx = x^4 y$, $y(1) = 1$; explicit solution: $y = e^{(x^5-1)/5}$

3. $dy/dx = -y^2 \cos x$, $y(0) = 1$; explicit solution: $y = \dfrac{1}{1 + \sin x}$

4. $dy/dx = \frac{\sin x}{y^3}$, $y(\pi) = 2$; explicit solution: $y = (12 - 4\cos x)^{1/4}$

5. $dy/dx = ye^{-x}$, $y(0) = 1$; explicit solution: $y = \exp(1 - e^{-x})$

In problems 6–11, solve with your computer software package using the fourth-order Runge-Kutta method with $h = 0.1$ to find approximate values for $y(x_1), y(x_2), y(x_3), y(x_4)$, $y(x_5), y(x_6), y(x_7)$, and $y(x_8)$. Then compare the results with the given explicit solution at $x_1, x_2, x_3, x_4, x_5, x_6, x_7$, and x_8. It will be easiest if you display the results as we did in the table in Figure 2.21.

6. $dy/dx = xy^2$, $y(0) = 1$; explicit solution: $y = \dfrac{2}{2 - x^2}$

7. $dy/dx = y + \cos x$, $y(0) = 0$; explicit solution: $y = \frac{1}{2}(\sin x - \cos x + e^x)$

8. $dy/dx = y + \sin x$, $y(0) = 2$; explicit solution: $y = \frac{-1}{2}(\cos x + \sin x + 5e^x)$

9. $dy/dx = e^{-y}$, $y(0) = 2$; explicit solution: $y = \ln(x + e^2)$

10. $dy/dx = x + y$, $y(0) = 0$; explicit solution: $y = -x - 1 + e^x$

11. $dy/dx = (x + 1)(y^2 + 1)$, $y(0) = 0$; explicit solution: $y = \tan(\frac{1}{2}x^2 + x)$

In problems 12–16, use the fourth-order Runge-Kutta method with $h = 0.01$ and plot the numerical solution. Choose a viewing window that will allow you to see the behavior of the solution. If your solution appears to diverge rapidly or illustrate strange behavior, try reducing your step size to $h = .001$.

12. $dy/dx = e^{-x^2}$, $y(0) = 1$

13. $dy/dx = x^3 y - x^2 y^2$, $y(-1) = 1$

14. $dy/dx = x^3 e^y + 3x^2 \sin y$, $y(.5) = 1$

15. $dy/dx = |1 - x^2|y + x^3$, $y(1) = 1$

16. $dy/dx = y\sqrt{x^2 + y^2 + 1} + \cos(xy)$, $y(0) = 1$

17. Use the fourth-order Runge-Kutta method to numerically solve the following two logistic differential equations for time t between 0 and 1. Assume that $x(0) = 100$. Solve both using the following h values: 1, 0.5, 0.25, and 0.1. Compare the solutions of the first equation for the different h-values. Explain what is happening as h gets small. Repeat the comparison for the second equation. Compare the solutions of the two equations to each other for the different h-values. Explain why two differential equations that are identical except for the parameter values exhibit different behaviors while h is in the process of "getting small."

a. $\dfrac{dx}{dt} = 1.5x \left(1 - \dfrac{x}{1000}\right)$

b. $\dfrac{dx}{dt} = 2.5x \left(1 - \dfrac{x}{1000}\right)$

2.7   An Introduction to Autonomous Second-Order Equations (optional)

We will now consider a special class of second-order differential equations, useful in applications, that can be solved by methods we know. With first-order equations, we considered autonomous equations to be those that did not have the independent variable appearing explicitly in the problem. The definition here is similar.

DEFINITION 2.3   *A second-order differential equation of the form*

$$y''(t) = h(y, y') \tag{2.13}$$

is called autonomous when h does not depend explicitly on the independent variable t.

Equation (2.13) can be interpreted as a differential equation whose solutions provide the position $y(t)$ of a body that moves according to Newton's law for a special kind of forcing function h.

One, perhaps unexpected, way to solve (2.13) is to consider the velocity as a function of the position. That is, we let

$$v(y) = \frac{dy}{dt}$$

so that v is a function of y, rather than t. We also note that this determines the acceleration as

$$\begin{aligned}
\frac{d^2 y}{dt^2} &= \frac{d}{dt}\frac{dy}{dt} \\
&= \frac{d}{dt} v(y) \\
&= \frac{dv}{dy}\frac{dy}{dt} \\
&= v\frac{dv}{dy}.
\end{aligned}$$

Thus, the task of solving the second-order equation (2.13) is reduced to solving the first-order equation

$$v\frac{dv}{dy} = h(y, v).$$

Solutions can be plotted in the y-v plane and are called *orbits*. If we can solve this transformed equation for $v(y)$, the remaining separable differential equation

$$v(y) = \frac{dy}{dt}$$

can be solved to obtain $y(t)$.

Example 1: A simple application of the preceding method is given by the motion of a frictionless spring whose deflection y satisfies the initial value problem

$$m \frac{d^2y}{dt^2} + ky = 0$$

for a positive mass m and a positive spring constant k, with prescribed initial values $y(0) = y_0$ and $y'(0) = v_0$ for the position and velocity. (We will give a formal study of harmonic motion later.) Letting

$$v(y) = \frac{dy}{dt},$$

the equation becomes

$$mv \frac{dv}{dy} = -ky$$

which is separable and exact (check it!), so that

$$mv\,dv + ky\,dy = 0.$$

Integrating gives

$$\frac{m}{2} v^2 + \frac{k}{2} y^2 = c.$$

The initial conditions $y(0) = y_0$ and $y'(0) = v_0$ give

$$c = \frac{m}{2} v_0^2 + \frac{k}{2} y_0^2.$$

This gives

$$\frac{m}{2} v^2 + \frac{k}{2} y^2 = \frac{m}{2} v_0^2 + \frac{k}{2} y_0^2,$$

which can be interpreted as saying that the sum of the kinetic and potential energies must remain constant, i.e., the total energy must remain constant. Solving this expression for $v(y) = dy/dt$ gives

$$\frac{dy}{dt} = v(y) = \pm \sqrt{\frac{k}{m}} \sqrt{\alpha^2 - y^2}, \qquad (2.14)$$

where

$$\alpha^2 = \frac{m}{2} v_0^2 + \frac{k}{2} y_0^2$$

is a constant determined by the initial conditions. The sign of v is uniquely determined by the prescribed initial velocity v_0. It now remains for us to solve (2.14) for $y(t)$. But (2.14) is separable and thus

$$\frac{dy}{\sqrt{\alpha^2 - y^2}} = \pm \sqrt{\frac{k}{m}}\, dt,$$

so that

$$\int \frac{dy}{\sqrt{\alpha^2 - y^2}} = \int \pm\sqrt{\frac{k}{m}}\, dt$$

which is

$$\sin^{-1}\left(\frac{y}{\alpha}\right) = \pm\sqrt{\frac{k}{m}}\, t + \phi.$$

Here ϕ is some constant. Thus, the deflection $y(t)$ of the spring is given by

$$y(t) = \alpha \sin\left(\pm\sqrt{\frac{k}{m}}\, t + \phi\right)$$

$$= (\pm\alpha\cos\phi)\sin\left(\sqrt{\frac{k}{m}}\, t\right) + (\alpha\sin\phi)\cos\left(\sqrt{\frac{k}{m}}\, t\right)$$

where this last expression follows from the trigonometric identity $\sin(u \pm v) = \sin(u)\cos(v) \pm \cos(u)\sin(v)$. Letting

$$C_1 = \pm\alpha\cos\phi \quad \text{and} \quad C_2 = \alpha\sin\phi$$

gives

$$y(t) = C_1 \sin\left(\sqrt{\frac{k}{m}}\, t\right) + C_2 \cos\left(\sqrt{\frac{k}{m}}\, t\right).$$

Now using the initial conditions $y(0) = y_0$ and $y'(0) = v_0$ we have

$$y(t) = y_0 \sin\left(\sqrt{\frac{k}{m}}\, t\right) + v_0 \cos\left(\sqrt{\frac{k}{m}}\, t\right),$$

which is the familiar equation for *simple harmonic motion*.

Problems

1. Show that the autonomous equation

$$yy'' = (y')^2$$

   has solution

$$y = ce^{kx}$$

   for constants c and k.

2. In a first physics course, students derive the equation of motion for a frictionless simple pendulum as

$$m\theta'' + g\sin\theta = 0 \tag{2.15}$$

where θ is the angle that the pendulum makes with the vertical; however, the next step is to assume the angle is small and use the small angle approximation $(\sin\theta \approx \theta + \cdots)$ to rewrite this equation as

$$m\theta'' + g\theta = 0,$$

which is conveniently the equation for simple harmonic motion. This approximation obviously fails if θ becomes too large. Let's revisit (2.15) and consider this as an autonomous equation. Let $v = d\theta/dt$ and use the methods of this section to derive a total energy formula. Use this expression to verify the plot of some of the orbits in Figure 2.22. Alternatively, plot this total energy formula to obtain the orbits in Figure 2.22. Interpret the three qualitatively different orbits, keeping in mind that the pendulum is allowed to whirl over the top.

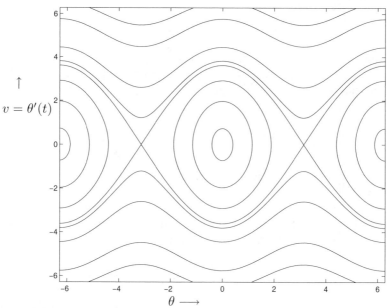

FIGURE 2.22:   Orbits for simple pendulum with $m = 3$, $g = 9.8$.

3. Suppose the position of a body of mass m that is shot up from the surface of the Earth satisfies the inverse square law

$$m\,x''(t) = -mg\frac{R^2}{x^2(t)},$$

where $x(t)$ is the distance from the Earth's center at time t, R is the

Earth's radius, and the initial conditions are

$$x(0) = R > 0 \quad \text{and} \quad x'(0) = v_0 > 0.$$

If $v_0^2 \geq 2gR$, show that the body will never return to Earth. The solution defines the *escape velocity* for the body. (Hint: Determine the expression for $v(x)$ and its sign.)

4. The nonlinear equation

$$\frac{d^2y}{dx^2} = \sqrt{1 + \left(\frac{dy}{dx}\right)^2}$$

describes the position y of a suspension cable, which either supports a bridge or hangs under its own weight. The equation is of the form (2.13), but it is also of the form

$$y''(x) = f(x, y'),$$

where $y = y(x)$ and the right-hand side f is independent of y.
a. Let $u(x) = dy/dx$ and derive a first-order equation in terms of only u, x, that is of the form:

$$\frac{du}{dx} = f(x, u).$$

b. Solve this equation to obtain

$$x - x_0 = \ln(u + \sqrt{1 + u^2})$$

for some constant x_0.
c. Exponentiate the above equation and the negative of it and combine appropriately to obtain the expression

$$u = \frac{1}{2}(e^{x-x_0} + e^{-(x-x_0)}).$$

d. Using the original substitution $u(x) = dy/dx$, obtain the equation

$$y(x) = \cosh(x - x_0) + c,$$

which is the (familiar) equation for the *catenary*, where x_0 is the location of the zero of y' (i.e., the minimum of y) and c is 0 if that minimum value is 1.

2.8 Chapter 2: Additional Problems and Projects

ADDITIONAL PROBLEMS

In problems 1–6, determine whether the statement is true or false. If it is true, give reasons for your answer. If it is false, give a counterexample or other explanation of why it is false.

1. The Existence and Uniqueness theorem allows us to state that the solution to $y' = x^2 y^{1/3}$ passing through the initial condition $y(1) = 0$ will not be unique.

2. The Runge-Kutta method for approximating the solution to an ODE is superior to Euler's method because it uses $1/2$ the step size of Euler's method to calculate the approximate solution.

3. The Euler and Runge-Kutta methods are both used for numerically approximating the solution to a differential equation. Euler's method is superior to the Runge-Kutta method because the step size is smaller and thus the approximations are better at each step.

4. The Existence and Uniqueness theorem allows us to conclude that solutions to $y' = \tan y$ will always exist and be unique.

5. The Allee model of population growth is often considered superior to the Logistic model because it models the need for a critical population level in order for the species to survive.

6. Autonomous equations can be written as $\frac{dx}{dt} = f(x)$ and are useful because we are able to obtain long-term behavior of the solutions quickly.

7. Determine the region(s) of the x-y plane where solutions (i) exist and also where solutions (ii) exist and are unique, according to the Existence and Uniqueness theorem:

   a. $\dfrac{dy}{dx} = (xy - 1)^{1/3}$

   b. $\dfrac{dy}{dx} = \sqrt{xy}$

   c. $\dfrac{dy}{dx} = y \tan x$

   d. $\dfrac{dy}{dx} = \sec y$

8. Determine the region(s) of the x-y plane where solutions (i) exist and also where solutions (ii) exist and are unique, according to the Existence and Uniqueness theorem:

   a. $\dfrac{dy}{dx} = 2x - y^{2/3}$

   b. $\dfrac{dy}{dx} = 2x^{2/3} - y$

c. $\dfrac{dy}{dx} = \sin x - \tan y$

d. $\dfrac{dy}{dx} = e^{y/(x+1)}$

9. In the direction field given below, sketch the solution to $y' = y - x^2$ that passes through the initial condition $y(-1) = 2$.

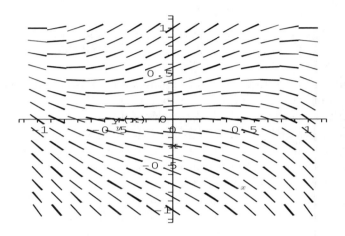

10. In the direction field given below, sketch the solution to $y' = y^2 - x^2$ that passes through the initial condition $y(-1) = 2$.

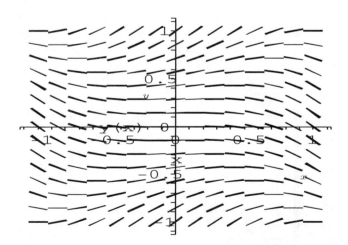

11. Consider the autonomous equation $y' = (3 + y)(2 - y)$.
    a. Find the equilibria.
    b. Draw a graph in the y-y' plane and indicate the stability of these points by drawing the appropriate arrows on the phase line.
    c. Explicitly state the stability of the equilibrium solutions.
    d. Sketch solutions in the x-y plane.

12. The following equation has been proposed to model the level of a certain harmful bacteria in the body:

$$x' = x(x - a)(x - 4),$$

where $0 < a < 4$, $x = x(t), x \geq 0$.
    a. Draw the phase line diagram and determine the stability of the equilibrium solutions.
    b. Describe the long-term behavior of the solution for any initial condition.
    c. Sketch solutions in the t-x plane.

13. Consider the autonomous ODE $y' = (1 - y)(2 + y)$.
    a. Find the equilibria.
    b. Draw a graph in the y-y' plane and indicate the stability of these points by drawing the appropriate arrows on the phase line.
    c. Explicitly state the stability of the equilibrium solutions.
    d. Sketch solutions in the x-y plane.

14. Match the two given direction fields with their respective equations.
    a. $y' = y(x^2 - 1)$, b. $y' = y \sin \pi x$, c. $y' = y(1 - x^2)$, d. $y' = y(y^2 - 1)$

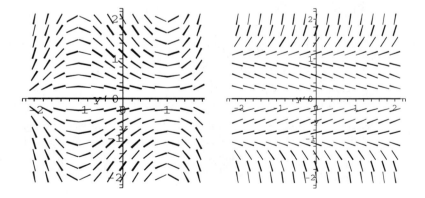

15. Match the two given direction fields with their respective equations.
   a. $y' = y^2(xy-1)$, b. $y' = x^2(xy^2-1)$, c. $y' = xy(1-x^2)$, d. $y' = y(xy-1)$

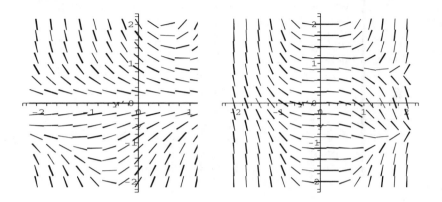

Use Matlab, Maple, or Mathematics to solve each of problems 16–25. Plot the numerical solution using the fourth-order Runge-Kutta method with $h = 0.1$. Compare this to the solution obtained using Euler's method with $h = 0.1$.

16. $xy' + x^2 + xy - y = 0$, $y(0) = 1$

17. $2xy' + y^2 = 1$, $y(1) = \pi$

18. $(2xy^2 - y)^2 dx + xdy = 0$, $y(\pi) = 1$

19. $(xy' + y)^2 = x^2 y'$, $y(1) = 2$

20. $y - y' = y^2 + xy'$, $y(2) = 0$

21. $(x + 2y^3)y' = y$, $y(-1) = 1$

22. $(y')^3 - y'e^{2x} = 0$, $y(-\pi) = 1$

23. $x^2 y' = y(x + y)$, $y(\sqrt{2}) = 1$

24. $(1 - x^2)dy + xydx = 0$, $y(0) = 5$

25. $(y')^2 + 2(x - 1)y' - 2y = 0$, $y(0) = -3$

Solve each of problems 26–43 analytically. If it is required by your instructor, use the computer to plot the implicit or explicit solution for three different initial conditions.

26. $x^2 y' - 2xy = 3y$

27. $y = (xy' + 2y)^2$

28. $y' + (3x - 6)y' = 3y$

29. $2x^3 yy' + 3x^2 y^2 + 7 = 0$

30. $xy' = e^y + 2y'$
31. $x^2(y')^2 + y^2 = 2x(2 - yy')$
32. $dy + (xy - xy^3)dx = 0$
33. $x(x - 1)y' + 2xy = 1$
34. $(1 - x^2)y' - 2xy^2 = xy$
35. $y' + y = xy^3$
36. $(xy^2 - x)dx + (y + xy)dy = 0$
37. $(\sin x + y)dy + (y \cos x - x^2)dx = 0$
38. $yy' + y^2 \cot x = \cos x$
39. $(e^y + 2xy)dx + (e^y + x)x dy = 0$
40. $x(x + 1)(y' - 1) = y$
41. $y' + x\sqrt[3]{y} = 3y$
42. $y' = \frac{x}{y}e^{2x} + y$
43. $xy' = 2\sqrt{y}\cos x - 2y$

We have seen a variety of different methods for solving first-order equations ranging from standard analytical techniques to numerical approaches. There are numerous other types of first-order equations not covered here, many of which can be solved with the help of your computer programs. Use dsolve *(in Matlab's symbolic package)*, dsolve *(in Maple)*, or NDSolve *(in Mathematica) to solve problems 44–65 analytically. State as many different characteristics about the equation as you can and use your software package to identify the type of equation. If it is required by your instructor, use the computer to plot the implicit or explicit solution for three different initial conditions.*

44. $y + y' \ln^2 y = (x + 2\ln y)y'$
45. $x + yy' - y^2(1 + (y')^2)$
46. $y' = \frac{1}{x - y^2}$
47. $x - \frac{y}{y'} = \frac{2}{y}$
48. $2(y')^3 - 3(y')^2 + x = y$
49. $(x + y)^2 y' = 1$
50. $\frac{dx}{x} = (\frac{1}{y} - 2x)dy$
51. $2(x - y^2)dy = ydx$
52. $2x^2 y' = y^2(2xy' - y)$
53. $\frac{y - xy'}{x + yy'} = 2$
54. $xy(xy' - y)^2 + 2y' = 0$
55. $3(y')^3 - xy' + 1 = 0$
56. $x(y')^2 = y - y'$

57. $y(y - xy') = \sqrt{x^4 + y^4}$

58. $xy' + y = \ln y'$

59. $x^2(dy - dx) = (x + y)y\,dx$

60. $(x \cos y + \sin 2y)y' = 1$

61. $(y')^2 - yy' + e^x = 0$

62. $(xy' - y)^3 = (y')^3 - 1$

63. $(4xy - 3)y' + y^2 = 1$

64. $y'\sqrt{x} = \sqrt{y - x} + \sqrt{x}$

65. $3(y')^4 = y' + y$

PROJECTS FOR CHAPTER 2

Project 1: Spruce Budworm

We consider the famous spruce budworm model which was used to describe the dynamics of this population on a forest in Canada. Periodically, there were large outbreaks of the budworm that devastated the forest. The forest would recover and then a similar outbreak would occur. A simple first-order autonomous ODE was proposed to explain the behavior [24],[27],[34]:

$$\frac{dx}{dt} = rx \left(1 - \frac{x}{k}\right) - \frac{x^2}{1 + x^2}. \tag{2.16}$$

All of the variables and parameters are *dimensionless*, that is, they don't have units associated with them and were obtained by dividing by appropriate factors, see the references. We think of x as the population level of the budworm, r as the growth rate of the budworm, and k as the carrying capacity of the budworm.

a. Show that the system has between one and three biologically meaningful equilibria.

b. Determine the stability of the equilibria. Your answer will have a dependence on the values of the parameters r and k.

c. Find the equation, in terms of the parameters r and k, that describes where the system changes from one to three equilibria.

d. Give a biological interpretation for the system with one equilbrium point vs. three equilibria.

Project 2: Multistep Methods of Numerical Approximation.

Many numerical methods for solving differential equations exist. In this project, we introduce the student to a *multistep* method that takes into account previous (and sometimes future!) values of the solution in order to calculate the next solution. Multitstep methods of all orders exist but we focus on second-order methods for simplicity of formulation [9]. Students will need to have some programming knowledge in one of Matlab, Maple, or Mathematica in order to make progress on this project.

We consider the differential equation

$$\frac{dy}{dx} = xy, \quad y(0) = 1$$

discussed in Sections 2.5 and 2.6.

Explicit Method: Consider the formula

$$y_{i+1} = y_i + \frac{h}{2}\left(3f(x_i, y_i) - f(x_{i-1}, y_{i-1})\right), \quad i = 1, 2, \cdots, N-1, \quad (2.17)$$

which is used to approximate the solution of the initial value problem, $y' = f(x, y)$, $y(x_0) = y_0$. The local truncation error is given by $\frac{5}{12}y^{(3)}(\xi_i)h^2$, for some $\xi_i \in (x_{i-2}, x_{i+1})$.

1. Using Matlab, Maple, or Mathematica, implement the above formula, known as the *second-order Adams-Bashforth formula*, an explicit method, to approximate the solution at $x = 1$. Use a step size of $h = .1$ and use the approximate value of $y(.1)$ from Runge-Kutta with $h = .1$ for your value of y_1.

2. Compare the error with that of Euler's method (a first-order method) and the Runge-Kutta method (a fourth-order method).

Implicit Method: Now consider the formula

$$y_{i+1} = y_i + \frac{h}{12}\left(5f(x_{i+1}, y_{i+1}) + 8f(x_i, y_i) - f(x_{i-1}, y_{i-1})\right), \quad (2.18)$$

$i = 1, 2, \cdots, N-1$, which is again used to approximate the solution of the initial value problem, $y' = f(x, y)$, $y(x_0) = y_0$. The local truncation error is given by $\frac{-1}{24}y^{(4)}(\xi_i)h^2$, for some $\xi_i \in (x_{i-1}, x_{i+1})$. This is known as the *second-order Adams-Moulton formula*, an implicit method. Note here that in order to approximate the solution at y_{i+1}, we need to know the value of $f(x_{i+1}, y_{i+1})$! How can this be done?

1. Substitute the original differential equation for $f(x_{i+1}, y_{i+1})$ in the Adams-Moulton formula. Solve for y_{i+1}.

2. Using Matlab, Maple, or Mathematica, implement the resulting formula of part 1 with $h = .1$ again using the Runge-Kutta approximation for y_1.

3. Compare the error with that of Euler's method (a first-order method) and the Runge-Kutta method (a fourth-order method).

Predictor-Corrector Method: This method combines the explicit and implicit methods used above. Instead of solving the Adams-Moulton (AM) formula for y_{i+1}, we use the approximate result for y_{i+1} obtained from the Adams-Bashforth (AB) formula. The result of substituting the AB result into the AM result usually gives a more accurate answer than the original AB answer and is almost always quicker than solving the AM formula by hand, as done above. For the instructions below, again use $h = .1$ and the Runge-Kutta approximation for y_1.

1. Using Matlab, Maple, or Mathematica, implement as follows:

   a. Perform one step of AB.

   b. Substitute this value into the AM formula.

   c. Perform one step of AM.

   d. If desired, resubstitute this last AM result into the AM expression again.

   e. Compare the results with the AB, AM, and exact solutions.

This method is called a *predictor-corrector method* because the AB method predicts the function value at x_{i+1} and the AM method corrects this with a higher-order approximation.

Chapter 3

Elements of Higher-Order Linear Equations

Following our work with first-order equations, it is natural for us to consider higher-order equations. The first methods of solving differential equations of second or higher order with constant coefficients were due to Euler. D'Alembert dealt with the case when the auxiliary equation had equal roots. Some of the symbolic methods of finding the particular solution were not given until about 100 years later by Lobatto (1837) and Boole (1859).

3.1   Some Terminology

We begin our study with a definition.

DEFINITION 3.1   *A linear differential equation of order n in the dependent variable y and the independent variable x is an equation that is in, or can be expressed in, the form*

$$a_0(x)\frac{d^n y}{dx^n} + a_1(x)\frac{d^{n-1} y}{dx^{n-1}} + \ldots + a_{n-1}(x)\frac{dy}{dx} + a_n(x)\,y = F(x) \qquad (3.1)$$

where a_0 is not identically zero.

We shall assume throughout that $a_0, a_1, \ldots, a_n$ and F are continuous real functions on a real interval $a \leq x \leq b$ and that

$$a_0(x) \neq 0 \text{ for any } x \text{ on } a \leq x \leq b.$$

The right-hand side $F(x)$ is called the *nonhomogeneous term*. If F is identically zero, then

$$a_0(x)\frac{d^n y}{dx^n} + a_1(x)\frac{d^{n-1} y}{dx^{n-1}} + \ldots + a_{n-1}(x)\frac{dy}{dx} + a_n(x)\,y = 0 \qquad (3.2)$$

is called a *homogeneous* equation. It is sometimes called the *reduced equation* of the given nonhomogeneous equation.

Example 1: The equation

$$\frac{d^2y}{dx^2} + 7x\,\frac{dy}{dx} + x^5 y = e^x$$

is a linear nonhomogeneous differential equation of the second order.

Example 2: The equation

$$\frac{d^3y}{dx^3} + x\,\frac{d^2y}{dx^2} + 3x\,\frac{dy}{dx} + y = 0$$

is a linear homogeneous differential equation of the third order.

3.1.1   Existence and Uniqueness of Solutions

Our goal is to consider methods of solving these higher-order differential equations. In particular, we will be interested in second-order linear equations as these equations occur often in physics and engineering applications and give rise to the special functions we will consider. Before we pursue methods of solution, it is best to give an existence theorem for initial-value problems associated with a linear ordinary differential equation of order n.

THEOREM 3.1.1 Existence and Uniqueness *Consider the linear ordinary differential equation of order n (3.1) where $a_0, a_1, \ldots, a_n$ and F are continuous real functions on a real interval $a \leq x \leq b$ and*

$$a_0(x) \neq 0 \text{ for any } x \text{ on } a \leq x \leq b.$$

If x_0 is any point of the interval $a \leq x \leq b$ and $c_0, c_1, \ldots, c_{n-1}$ are n arbitrary constants, then there exists a unique solution f of (3.1) such that

$$f(x_0) = c_0, \ f'(x_0) = c_1, \ \ldots, \ f^{(n-1)}(x_0) = c_{n-1},$$

and this solution is defined over the entire interval $a \leq x \leq b$.

What this theorem says is that for the differential equation (3.1) there is precisely one solution that passes through the value c_0 at x_0 and whose k^{th} derivative takes on the value c_k for each $k = 1, 2, \ldots, n - 1$ at $x = x_0$. Further, the theorem asserts that this unique solution is defined for $a \leq x \leq b$.

Example 3: For the initial value problem

$$\frac{d^2y}{dx^2} + 7x\,\frac{dy}{dx} + x^5 y = e^x \ \text{ with } \ y(2) = 1 \ \text{ and } \ y'(2) = 5,$$

there exists a unique solution since the coefficients $1, 7x, x^5$ as well as the nonhomogenous term e^x are all continuous on $(-\infty, \infty)$.

Example 4: Determine if $y = c_1 + c_2 x^2$ gives rise to a unique solution for $xy'' - y' = 0$, $y(0) = 0, y'(0) = 1$ on $(-\infty, \infty)$.

We can easily verify that this is a solution. However, if we simply apply the IC, we obtain the two equations

$$0 = y(0) = c_1 + c_2 \cdot 0^2$$
$$1 = y'(0) = 2c_2 \cdot 0,$$

the second of which gives us the inconsistent statement $1 = 0$. What happened? A look at the differential equation shows that $a_0(x) = 0$ when $x = 0$, which is the x-value of the initial condition. Thus Theorem 3.1.1 does not apply for the initial condition we were given.

The theorem has an important corollary in the homogeneous case.

COROLLARY 3.1.1 *Let f be the solution of the nth order homogeneous linear differential equation (3.2) such that*

$$f(x_0) = 0, \ f'(x_0) = 0, \ \ldots, \ f^{(n-1)}(x_0) = 0$$

where x_0 is a point of the interval $a \leq x \leq b$ in which the coefficients $a_0, a_1, \ldots, a_n$ are all continuous and $a_0(x) \neq 0$. Then $f(x) = 0$ for all x on $a \leq x \leq b$.

Example 5: The third-order linear homogeneous ordinary differential equation

$$\frac{d^3 y}{dx^3} + x \frac{d^2 y}{dx^2} + 3x \frac{dy}{dx} + y = 0 \ \text{ with } \ f(8) = f'(8) = f''(8) = 0$$

has the unique (trivial) solution $f(x) = 0$ for all x.

We have now talked about existence and uniqueness of solutions. Our goal will be to obtain the most general possible solution. We will come up with various ways to obtain solutions of the given differential equation but we need to know how to find the minimum number of solutions that can be combined to give every possible solution. This is absolutely essential to understand. Finding this minimum possible list of solutions that can generate all other solutions can be thought of (in a very rough sense) as analogous to finding the prime factors of a given integer. This minimum set is known as a *fundamental set of solutions* and satisfies the property that each element is a solution and each "contributes something new" to the set.

3.1.2 The Need for Sufficiently Different Solutions

We are now going to consider a case where we can easily obtain a solution and use a second example as motivation for studying linear independence and other related concepts from linear algebra. Most of the equations we will consider in this chapter have y as the dependent variable and x as the independent variable. The simplest homogeneous linear equation we have considered thus far is first order, with constant coefficients. It is of the form

$$a_0 \frac{dy}{dx} + a_1 y = 0.$$

What is the form of the solution? In this simple case, we observe that the equation is separable and the solution will be an exponential function $e^{(-a_1/a_0)x}$. If we consider the second-order homogeneous constant coefficient linear equation

$$a_0 \frac{d^2y}{dx^2} + a_1 \frac{dy}{dx} + a_2 y = 0,$$

we ask if it is still possible to have a solution of the form $y = e^{rx}$, since it worked in the first-order case. To find the value of r that will make it a solution we substitute into the differential equation and find values for r that will satisfy the resulting condition. Let's try this with a specific example.

Example 6: Consider the equation

$$\frac{d^2y}{dx^2} + 3\frac{dy}{dx} + 2y = 0. \tag{3.3}$$

If we substitute $y = e^{rx}$, we have

$$r^2 e^{rx} + 3re^{rx} + 2e^{rx} = 0, \quad \text{which gives} \quad e^{rx}(r^2 + 3r + 2) = 0.$$

Because e^{rx} is never zero, we consider only the terms in parentheses. This equation then gives the conditions for the r-values that make e^{rx} a solution:

$$r^2 + 3r + 2 = 0,$$

which factors into $(r+2)(r+1) = 0$. This holds if $r = -1$ or $r = -2$. Thus, two solutions to the differential equation are e^{-x} and e^{-2x}.

It is not a coincidence that we found *two* different solutions to this *second-order* equation. We observe that for this situation

$$C_1 e^{-x} + C_2 e^{-2x} \tag{3.4}$$

is also a solution, as the reader should check! It turns out that *any* solution to the differential equation (3.3) must be of the form (3.4). Will things always be so nice? Let's try another example to see.

Example 7: Consider the equation

$$\frac{d^2y}{dx^2} + 2\frac{dy}{dx} + y = 0. \tag{3.5}$$

If we substitute $y = e^{rx}$, we see that it will be a solution whenever

$$r^2 + 2r + 1 = 0,$$

which factors into $(r + 1)^2 = 0$. This holds if $r = -1$. A solution of (3.5) can thus be written as $y = e^{-x}$. Is this the only solution? The answer is a resounding "no!" and we observe by substitution that xe^{-x} is also a solution. The details of why we might guess xe^{-x} as a second solution are left for discussion in Section 3.3.

What went wrong in this last example? And how do we deal with such situations? To give satisfactory answers to these questions, we first need some basic concepts from linear algebra, including an understanding of linear independence and linear combination.

Problems

1. Determine whether $y''' + y' = 0$ is guaranteed to have a unique solution passing through $y(0) = 0, y'(0) = 1, y''(0) = 0$. Then try to find constants c_1, c_2, c_3 such that $y = c_1 + c_2 \cos x + c_3 \sin x$ satisfies the IVP.

2. Determine whether $xy'' - 2xy + 2y = 0$ is guaranteed to have a unique solution passing through $y(0) = 0, y'(0) = 0$. Then try to find constants c_1, c_2 such that $y = c_1 x + c_2 x^2$ satisfies the IVP.

3. Determine whether $xy'' - xy - 3y = 0$ is guaranteed to have a unique solution passing through $y(0) = 0, y'(0) = 0$. Then try to find constants c_1, c_2 such that $y = c_1 x^3 + \frac{c_2}{x}$ satisfies the IVP.

4. Determine whether $x^2 y'' - 6xy + 12y = 2x^2$ is guaranteed to have a unique solution passing through $y(0) = 0, y'(0) = 0$. Then try to find constants c_1, c_2 such that $y = c_1 x^3 + c_2 x^4 + x^2$ satisfies the IVP.

5. Determine whether $y'' - 4y = 0$ is guaranteed to have a unique solution passing through $y(0) = 4, y'(0) = 0$. Then try to find constants c_1, c_2 such that $y = c_1 e^{2x} + c_2 e^{-2x}$ satisfies the IVP.

6. Find all values of (x, y) where Theorem 3.1.1 guarantees the existence of a unique solution.
   a. $y'' + 3xy' + y = \sin x$
   b. $y''' + 3(\tan x)y' + xy = x$
   c. $xy'' + 3x^2 y = \sin x$
   d. $y'' + 5y' + 6y = \frac{\cos x}{x}$

7. Find all values of (x, y) where Theorem 3.1.1 guarantees the existence of a unique solution.
   a. $y^{(4)} + 4y'' + 4x^2 y = e^x$
   b. $y'' + 3xy' + xy = \ln x$
   c. $y'' + 6y' - 16y = e^{-x^2} - 1$
   d. $(x - 1)^2 y'' + 5(x^2 - 1)y' + 6y = 1$

3.2   Essential Topics from Linear Algebra

Our discussion of topics from linear algebra will deal with both general functions as well as solutions of differential equations. The interplay between linear algebra and differential equations is a rich one and we encourage the reader to embrace it. In the ensuing discussion, we will not assume that the reader has had a course in linear algebra. The reader may find it useful to read Appendix C.1–C.2 for an introduction into the subject of linear algebra.

3.2.1   Linear Combinations of Solutions

Now we will borrow some terminology from linear algebra.

DEFINITION   3.2   *If* $f_1, f_2, \ldots, f_m$ *are* m *given functions and* $c_1, c_2, \ldots, c_m$ *are* m *constants, then*

$$c_1 f_1 + c_2 f_2 + \ldots + c_m f_m$$

is called a linear combination of $f_1, f_2, \ldots, f_m$.

Our goal is to tie this in with our study of differential equations and this is done with the following theorem:

THEOREM 3.2.1 *Let* $f_1, f_2, \ldots, f_m$ *be* **any** m **solutions** *of the homogeneous linear differential equation (3.2), then*

$$c_1 f_1 + c_2 f_2 + \ldots + c_m f_m$$

is also a **solution** *of (3.2) where* $c_1, c_2, \ldots, c_m$ *are arbitrary constants.*

Thus, this theorem gives the useful fact that **any linear combination of solutions to (3.2) is a solution to (3.2)**.

Example 1: Consider the differential equation

$$\frac{d^2 y}{dx^2} + y = 0.$$

We note that $y_1(x) = \sin x$ and $y_2(x) = \cos x$ are solutions. (Check it!) Thus,

$$y(x) = c_1 \sin x + c_2 \cos x$$

is also a solution for any constants c_1 and c_2. (Check this, too!)

3.2.2 Linear Independence

We have found that solutions exist and are unique. We also know that any linear combination of solutions of a homogeneous linear differential equation is still a solution. The question is: When we have solutions to (3.2), how do we know when we have the most general solution? To answer this, we need to introduce some additional concepts from linear algebra.

DEFINITION 3.3   *The n functions $f_1, f_2, \ldots, f_n$ are linearly dependent on $a \le x \le b$ if there exists constants $c_1, c_2, \ldots, c_n$, not all zero, such that*

$$c_1 f_1(x) + c_2 f_2(x) + \ldots + c_n f_n(x) = 0$$

for all x such that $a \le x \le b$. We say that the n functions $f_1, f_2, \ldots, f_n$ are linearly independent on $a \le x \le b$ if they are not linearly dependent there. That is, the functions $f_1, f_2, \ldots, f_n$ are linearly independent on $a \le x \le b$ if

$$c_1 f_1(x) + c_2 f_2(x) + \ldots + c_n f_n(x) = 0$$

for all x such that $a \le x \le b$ implies $c_1 = c_2 = \ldots = c_n = 0$.

Example 2: x and $-3x$ are linearly dependent on $0 \le x \le 1$ since there are constants c_1 and c_2, not both zero, such that

$$c_1 x + c_2(-3x) = 0$$

for all x on $0 \le x \le 1$. Just take $c_1 = 3$ and $c_2 = 1$.

Example 3: The functions e^x and $2e^x$ are linearly dependent, since

$$-2 \cdot (e^x) + 1 \cdot (2e^x) = 0.$$

The constants are $c_1 = -2$ and $c_2 = 1$.

Example 4: We show that x and x^2 are linearly independent on $-1 \le x \le 1$. To see this, we show that

$$c_1 x + c_2 x^2 = 0 \tag{3.6}$$

for all x on $0 \le x \le 1$ implies both $c_1 = 0$ and $c_2 = 0$.

Since the equation hold for all x, it must hold when $x = 1$. This gives

$$c_1(1) + c_2(1)^2 = 0.$$

It must also hold when $x = -1$ and this gives

$$c_1(-1) + c_2(-1)^2 = 0.$$

Solving the second equation gives

$$c_1 = c_2$$

and substitution into the first gives

$$2c_2 = 0,$$

which only holds when $c_2 = 0$. Thus we also have $c_1 = 0$. Hence x and x^2 are linearly independent on $0 \leq x \leq 1$.

   In the previous example, we note that we could have chosen any number of x-values *except* $x = 0$. This is because both functions are zero at $x = 0$. Determining whether a set of functions is linearly independent can sometimes become cumbersome by using the definition if there are many functions to consider. Sometimes it may be useful to use our computer software packages to determine the values of the coefficients and we now illustrate this.

Example 5:   Determine whether $1, x, x^2, x^3$ are linearly independent on the real line.
We need to see if

$$c_1 \cdot 1 + c_2 x + c_3 x^2 + c_4 x^3 = 0 \quad \text{for all} \quad x \tag{3.7}$$

can only happen when $c_1 = c_2 = c_3 = c_4 = 0$. Because (3.7) holds for all x by assumption, it must hold when $x = 1, -1, 2, 3$, for example. Substituting these four x-values into (3.7) gives the four equations

$$c_1 + c_2 + c_3 + c_4 = 0$$
$$c_1 - c_2 + c_3 - c_4 = 0$$
$$c_1 + 2c_2 + 4c_3 + 8c_4 = 0$$
$$c_1 + 3c_2 + 9c_3 + 27c_4 = 0.$$

We want to see if there are non-zero c-values that will make this set of equations true. This simply means we need to solve the equations simultaneously and our computer packages can do this quickly.

Computer Code 3.1:   **Solving a system of equations**

Matlab, Maple, Mathematica

Matlab
In Matlab, we could either enter the coefficients of the c_i and solve as a system of equations or utilize the `solve` command. Both are presented here:

Matlab

```
>>  A=[1,1,1,1; 1,-1,1,-1; 1,2,4,8; 1,3,9,27]
>>  b=[0;0;0;0] %this is the right hand side of each equation
>>  A\b
>>  %Alternatively we could use the following long line
>>  %which requires Matlab's Symbolic Math Toolbox
>>  [c1,c2,c3,c4]=solve('c1+c2+c3+c4=0','c1-c2+c3-c4=0',
    'c1+2*c2+4*c3+8*c4=0','c1+3*c2+9*c3+27*c4=0')
```

The result of the first approach gives a vector with entries c_1, c_2, c_3, c_4 in that order, while the second approach gives labeled answers.

Maple

In Maple, we could also enter the coefficients of the c_i but we instead choose to use the `solve` command to find the values of c_i:

Maple

```
>  eq1:=c[1]+c[2]*x+c[3]*x^2+c[4]*x^3;
>  eq2a:=subs(x=1,eq1);
>  eq2b:=subs(x=-1,diff(eq1,x));
>  eq2c:=subs(x=2,diff(eq1,x$2));
>  eq2d:=subs(x=3,diff(eq1,x$3));
>  eq3:=solve({eq2a,eq2b,eq2c,eq2d},{c[1],c[2],c[3],c[4]});
```

The result comes out with the c_i-values labeled but we need to be careful that they may not be given in the natural order.

Mathematica

Mathematica

```
Solve[{c1+c2+c3+c4==0, c1-c2+c3-c4==0, c1+2c2+4c3+8c4==0,
   c1+3c2+9c3+27c4==0}, {c1,c2,c3,c4}]
```

Regardless of the program, we see that the only way for (3.7) to be true is when $c_1 = c_2 = c_3 = c_4 = 0$. This shows that the functions are linearly independent for all x-values.

With this additional concept of linear independence, we can give the following theorem.

THEOREM 3.2.2 *The nth order homogeneous linear differential equation (3.2) always possesses n solutions that are linearly independent. Further, if $f_1, f_2, \ldots, f_n$ are n linearly independent solutions of (3.2), then every solution f of (3.2) can be expressed as a linear combination*

$$c_1 f_1(x) + c_2 f_2(x) + \ldots + c_n f_n(x) \tag{3.8}$$

of these n linearly independent solutions by proper choice of the constants $c_1, c_2, \ldots, c_n$. The expression (3.8) is called the **general solution** *of (3.2) and is defined on (a, b), the interval on which solutions exist and are unique.*

We thus have that the solutions $f_1, \ldots, f_n$ can be combined to give us any solution we desire. Could another set of n functions also work or is this set unique? Actually, any set of functions that satisfies the following three conditions will work and we call the set a *fundamental set of solutions* of (3.2).

Three Conditions of a Fundamental Set of Solutions
1. The number of functions (elements) in this set must be the same as the order of the ODE.
2. Each function in this set must be a solution to the ODE.
3. The functions must be linearly independent.

We again note that a fundamental set of solutions is *not* unique. In order to obtain a fundamental set, we may need to add or remove functions from a given set. Once we have a fundamental set of solutions, we can construct all possible solutions from it.

What remains now is to determine whether or not the n solutions of (3.2) are linearly independent. This is necessary in order to determine a fundamental set of solutions.

The following definition is for n general functions and does *not* assume the functions are solutions of a differential equation.

DEFINITION 3.4   *Let $f_1, f_2, \ldots, f_n$ be n real functions each of which has an $(n-1)^{st}$ derivative on the interval $a \leq x \leq b$. The determinant*

$$W(x) = W(f_1, f_2, \ldots, f_n)(x) = \begin{vmatrix} f_1(x) & f_2(x) & \ldots & f_n(x) \\ f_1'(x) & f_2'(x) & \ldots & f_n'(x) \\ \vdots & \vdots & \ddots & \vdots \\ f_1^{(n-1)}(x) & f_2^{(n-1)}(x) & \ldots & f_n^{(n-1)}(x) \end{vmatrix},$$

in which primes denote derivatives, is called the Wronskian of these n functions.

The determinant is a useful concept in linear algebra. The formulas for calculating it are straightforward but very computationally expensive (i.e., it takes a long time to do it). A general formula for the determinant exists but we will just state it for the "2 × 2" and "3 × 3" cases:

$$\begin{vmatrix} a & b \\ c & d \end{vmatrix} = ad - bc \tag{3.9}$$

and

$$\begin{vmatrix} a_1 & a_2 & a_3 \\ b_1 & b_2 & b_3 \\ c_1 & c_2 & c_3 \end{vmatrix} = a_1(b_2c_3 - b_3c_2) - a_2(b_1c_3 - b_3c_1) + a_3(b_1c_2 - b_2c_1). \tag{3.10}$$

We refer the reader to Appendix C for a more thorough introduction to determinants.

From the formula for determinant, we observe that

$$W(x) = W(f_1, f_2, \ldots, f_n)(x)$$

is a real function defined on $a \leq x \leq b$.

THEOREM 3.2.3 *Let $f_1, f_2, \ldots, f_n$ be defined as in Definition 3.4.*
1. If $W(x_0) \neq 0$ for some $x_0 \in (a, b)$, then it follows that $f_1, f_2, \ldots, f_n$ are linearly independent on (a, b).
2. If $f_1, f_2, \ldots, f_n$ are linearly dependent on (a, b), then $W(x) = 0$, for all $x \in (a, b)$

We note that this theorem *does not* say that "if $W(x) = 0$ for all $x \in (a, b)$, then the $f_i(x)$ are linearly dependent."[1] This last statement will only be true over some restricted domain; see problem 22. *To show linear dependence of functions, we must use the definition.* However, if the functions under consideration are solutions to a differential equation, then linear dependence is much easier to determine, as this next theorem shows.

THEOREM 3.2.4 *Let $f_1, f_2, \ldots, f_n$ be defined as in Definition 3.4. Suppose that the f_i are each a solution of a given linear homogenous differential equation of order n. Then exactly one of the following statements is true:*
1. $W(x) \neq 0$, for all $x \in (a, b)$
2. $W(x) = 0$, for all $x \in (a, b)$
Moreover, $W(x) \neq 0$ for all $x \in (a, b)$ if and only if the $\{f_i\}$ are linearly independent on (a, b). Similarly, $W(x) = 0$ for all $x \in (a, b)$ if and only if the $\{f_i\}$ are linearly dependent on (a, b).

---

[1]We also note that part 2 of this theorem is just the *contrapositive* of part 1 and hence equivalent to it. But we list both for emphasis.

	Solutions of same ODE	General Functions
Linearly Independent	by definition *or* show $W(x) \neq 0$ for *some* x	by definition *or* show $W(x) \neq 0$ for *some* x
Linearly Dependent	by definition *or* show $W(x) = 0$ for *some* x	by definition *only*

FIGURE 3.1: Checking linear independence/dependence for solutions vs. functions. Note that the entire column under **Solutions of same ODE** could also read "for **all** x" since the two are equivalent by Theorem 3.2.4.

Note that this theorem tells us that if we have solutions of the same differential equation, we only need to check the Wronskian at *one* point in order to determine whether the set is linearly dependent. The previous theorem thus gives us an "easy" check for both linear independence and linear dependence in this case. We show its usefulness with an example.

Example 6: We have seen that $\sin x$ and $\cos x$ are solutions of $\dfrac{d^2y}{dx^2} + y = 0$.

Given our discussion, we should ask if the two functions are linearly independent. To check this, we calculate the Wronskian:

$$W(\sin x, \cos x) = \begin{vmatrix} \sin x & \cos x \\ \cos x & -\sin x \end{vmatrix}$$

$$= -\sin^2 x - \cos^2 x = -(\sin^2 x + \cos^2 x)$$

$$= -1 \neq 0 \text{ for all real } x.$$

Thus, $\sin x$ and $\cos x$ are linearly independent and so are a fundamental set of solutions to the differential equation and

$$y(x) = c_1 \sin x + c_2 \cos x$$

is the general solution.

As mentioned previously, $W(x)$ can show the linear independence of functions, even if they are *not* solutions of a differential equation. The table in Figure 3.1 gives a summary for checking linear independence (or dependence) of a set of functions.

Example 7: The functions e^x, e^{-x}, and e^{2x} are linearly independent since if we calculate the Wronskian $W(x)$ we have

$$W(e^x, e^{-x}, e^{2x}) = \begin{vmatrix} e^x & e^{-x} & e^{2x} \\ e^x & -e^{-x} & 2e^{2x} \\ e^x & e^{-x} & 4e^{2x} \end{vmatrix},$$

which is expanded as

$$W(e^x, e^{-x}, e^{2x}) = e^x \begin{vmatrix} -e^{-x} & 2e^{2x} \\ e^{-x} & 4e^{2x} \end{vmatrix} - e^{-x} \begin{vmatrix} e^x & 2e^{2x} \\ e^x & 4e^{2x} \end{vmatrix} + e^{2x} \begin{vmatrix} e^x & -e^{-x} \\ e^x & e^{-x} \end{vmatrix}$$

$$= e^x(-4e^x - 2e^x) - e^{-x}(4e^{3x} - 2e^{3x}) + e^{2x}(1 + 1)$$

$$= -6e^{2x} \neq 0 \text{ for all real } x.$$

We could have also tried $x = 0$ (for example) to see that the Wronskian was non-zero. The only drawback for "guessing" values if we want to show linear independence in the case when we *do not* have solutions is that an answer of zero at a specific x-value tells us nothing.

We can also use our computer software at this point to help us out. Although Maple and Mathematica have symbolic solving capabilities, some versions of Matlab *do not* have symbolic solving capabilities and we thus will only show how to use them to calculate the determinant of a system with numbers (and not general functions). It is possible to use the programs to calculate the Wronskian symbolically, for example. We encourage the interested reader to explore the symbolic capabilities using the help menus and reference material therein.

Computer Code 3.2: **Calculating the Wronskian of three functions**

Matlab, Maple, Mathematica

```
                        ┌─────────┐
                        │ Matlab  │
                        └─────────┘
>>  x=0
>>  A=[exp(x),exp(-x),exp(2*x); exp(x),-exp(-x),2*exp(2*x);
    exp(x),exp(-x),4*exp(2*x)] %calculated by hand
    %everything above should be entered on one line
>>  det(A)
%with symbolic math toolbox, the following is easier
>>  clear x
>>  syms x
>>  f1=exp(x)
>>  f2=exp(-x)
>>  f3=exp(2*x)
>>  diff([f1,f2,f3],x) %2nd row of wronskian
>>  A=[f1,f2,f3; diff([f1,f2,f3],x);diff([f1,f2,f3],x,2)]
>>  Wronsk=det(A)
>>  x=0
>>  eval(Wronsk)
```

Maple

```
> with(linalg):
> f1:=exp(x);
> f2:=exp(-x);
> f3:=exp(2*x);
> A:=matrix(3,3,[[f1,f2,f3],diff([f1,f2,f3],x),
    diff([f1,f2,f3],x$2)]);
> Wronsk:=det(A);
> simplify(subs(x=0,Wronsk));
> Wronsk2:=wronskian([f1,f2,f3],x); #built-in command
> simplify(subs(x=0,Wronsk2));
```

Mathematica

```
f1[x_]=eˣ; (*entered from palette, as usual*)
f2[x_]=e⁻ˣ;
f3[x_]=e²ˣ;
A={{f1[x],f2[x],f3[x]},{f1'[x],f2'[x],f3'[x]},
    {f1''[x],f2''[x],f3''[x]}};
A//MatrixForm
Wronsk[x_]=Det[A]
Wronsk[0]
```

It would be simple to try other values of x for which we wanted to substitute. This would be useful as the following example shows the potential problem of guessing values of x that may give a non-zero value for the determinant.

Example 8: Determine whether the functions $\sin 2x$ and $\sin x$ are linearly dependent or independent by using the Wronskian.
We can calculate the Wronskian as

$$W(x) = \begin{vmatrix} \sin 2x & \sin x \\ 2\cos 2x & \cos x \end{vmatrix}. \tag{3.11}$$

We can easily check that $W(0) = 0$, which doesn't help. $W(\pi) = 0$ as well. We might be tempted to stop guessing and switch to a check by definition, but "wise" guessing shows that $W(\pi/2) = 2$. Thus the two functions are linearly independent.

While it may be easier to substitute an x-value into the matrix and then calculate the determinant, this is not recommended when you have access to one of the programs. As can be seen in this example, it is just as easy to calculate $W(x)$, especially if we use Matlab, Maple, or Mathematica.

We note that the problem that arose here is that the functions are both zero at the same point. We simply need to evaluate the functions at places where they (i) don't intersect tangentially or (ii) aren't zero at the same point. Neither of these problems occurs with solutions of linear, homogeneous second-order differential equations because the uniqueness theorem disallows any intersections of solutions. And again we point out that it is often useful to have our computer software packages available to aid us in trying to understand the course material.

3.2.3   Solution Spaces, Subspaces, and Bases

The concept of linear independence is only the "tip of the iceberg" in the subject of linear algebra. We mentioned that the interplay between differential equations and linear algebra is rich and we will now further explore some of the connections between the two. Although this section may be particularly slow reading for the student without formal training in linear algebra, we encourage all readers to study the concepts introduced here because it will help in the overall understanding of the *differential equation* course material.

To recap from the previous section, we know how to check whether a given set of solutions is a fundamental set of solutions. We said there were three main points: (1) the number of functions (elements) in this set must be the same as the order of the ODE; (2) each function in this set must be a solution to the ODE; (3) the functions must be linearly independent. As mentioned before, we would like to be able to obtain the most general possible solution of a given differential equation. Our fundamental set gives us a set whose elements we combine to form all possible solutions. We can think of these elements of this set as the fundamental building blocks needed to generate solutions. Understanding the characteristics of these building blocks and the possible things that they can generate is worth examining a bit more. This special set is called a *basis* and we will give a formal definition after introducing a few more concepts.

In our study of differential equations, we know that the elements in a fundamental set are not unique. For example, we could easily check that

$$\{\sin x, \cos x\} \quad \text{and} \quad \{4\sin x, 2\sin x + \sqrt{2}\cos x\}$$

are *both* fundamental sets of solutions for the equation

$$y''(x) + y(x) = 0.$$

We also know from Theorem 3.2.1 that any linear combination of solutions is still a solution. Many of these traits that we have been discussing fit together in an abstract setting and introduce us to the concept of a *solution space* and, more generally, to that of a *vector space*. We call a set V a vector space if its elements satisfy the same properties of addition and multiplication by scalars that geometric vectors do. More precisely, we have the following definition.

DEFINITION 3.5   *Let V be a set of elements (called vectors); this set, together with vector addition $+$ and scalar multiplication, is termed a vector space if for c_1, c_2 scalars and $\mathbf{u}, \mathbf{v}, \mathbf{w}$ in V the following axioms hold:*
(i) $\mathbf{u} + \mathbf{v} \in V$
(ii) $c_1\mathbf{v} \in V$
(iii) $\mathbf{u} + \mathbf{v} = \mathbf{v} + \mathbf{u}$
(iv) $\mathbf{u} + (\mathbf{v} + \mathbf{w}) = (\mathbf{u} + \mathbf{v}) + \mathbf{w}$
(v) There is a zero vector, $\mathbf{0}$, *such that* $\mathbf{0} + \mathbf{u} = \mathbf{u}$
(vi) Every vector has an additive inverse $-\mathbf{u}$ *such that* $\mathbf{u} + (-\mathbf{u}) = \mathbf{0}$
(vii) $c_1(\mathbf{u} + \mathbf{v}) = c_1\mathbf{u} + c_1\mathbf{v}$
(viii) $(c_1 + c_2)\mathbf{u} = c_1\mathbf{u} + c_2\mathbf{u}$
(ix) $c_1(c_2\mathbf{u}) = (c_1 c_2)\mathbf{u}$
(x) $1\mathbf{u} = \mathbf{u}$

Example 9: Consider all the solutions of a given nth order linear homogeneous differential equation. Together they form a vector space, specifically called the *solution space* of the given differential equation. We can easily check this by letting a, b be any real numbers and y_1, y_2, y_3 be solutions. Then applications of Theorem 3.2.1, which says that linear combinations of solutions are still solutions for *any* scalars, show that each of the above axioms are true. We leave it to the reader to show this. Thus the space of all solutions is a vector space.

For differential equations, we have seen that elements of the solution space (a vector space) of a given differential equation are solutions to this differential equation. The concept of vector space is one that is taken for granted in many courses that incorporate math and we'll illustrate the importance of knowing the vector space to which you are referring. In physics and engineering, we often consider a "vector" to be something with magnitude and direction. An example is

$$\mathbf{v}_1 = 2\hat{\imath} - 3\hat{\jmath}.$$

If you are told to sketch this vector, where would you draw it? Many of you would probably answer "in the x-y plane" and would go 2 units to the right and 3 units down from the origin to draw it. Now sketch the vector

$$\mathbf{v}_2 = 2\hat{\imath} - 3\hat{\jmath} + 0\hat{k}.$$

Is there any difference? Yes! The second vector, $\mathbf{v}_2$, really lives in the x-y-z space even though its z-component is 0. Mathematically, this is a significant distinction. In multivariable calculus, vectors are often written using the notation $\hat{\mathbf{i}}, \hat{\mathbf{j}}, \hat{\mathbf{k}}$. We will instead write vectors in a different form:

$$\mathbf{v}_1 = 2\hat{\mathbf{i}} - 3\hat{\mathbf{j}} = \begin{pmatrix} 2 \\ -3 \end{pmatrix} \quad \text{and} \quad \mathbf{v}_2 = 2\hat{\mathbf{i}} - 3\hat{\mathbf{j}} + 0\hat{\mathbf{k}} = \begin{pmatrix} 2 \\ -3 \\ 0 \end{pmatrix}.$$

Vector $\mathbf{v}_1$ lives in $\mathbb{R}^2$ and elements in this space must have exactly two components both of which are real numbers (as opposed to complex numbers). On the other hand, vector $\mathbf{v}_2$ lives in $\mathbb{R}^3$ and elements in this space must have exactly three components, all of which are real numbers. In order to answer questions about these kinds of vectors, we need to specify the vector space to which we are referring.

As a familiar example, consider all vectors of the form $\begin{pmatrix} x \\ y \end{pmatrix}$. With the normal componentwise addition and subtraction of vectors, we can easily check that all the axioms of a vector space are satisfied. The vector form represents the general form of a vector that lives in $\mathbb{R}^2$, which is simply the familiar x-y plane.

There are many examples of vector spaces but we will focus our efforts on understanding the specific example of a solution space for a given differential equation. We stated that all possible solutions of a linear homogeneous differential equation form the solution space for that differential equation. We have also described a fundamental set of solutions for a given nth order linear homogeneous differential equation as being the elements (solutions) necessary to generate all possible solutions. This brings us to two more important topics, each of which is satisfied by this fundamental set of solutions.

DEFINITION 3.6 *A set of elements of a vector space is said to* **span** *the vector space if any vector in the space can be written as a linear combination of the elements in the spanning set.*

It should be believable that if we don't have a set that spans the vector space, then we can possibly add more vectors so that it does span. In terms of differential equations, the following example illustrates this idea.

Example 10: Consider the differential equation $y''(x) + y(x) = 0$. We can easily check that

$$\{\sin x, \sqrt{3} \sin x\}$$

are solutions but that this set *cannot* generate all possible solutions. For example, we could never write $2\cos x$ as a linear combination of these two elements (solutions). If we also include $2\cos x$, the resulting set

$$\{\sin x, \sqrt{3} \sin x, 2\cos x\}$$

does span the space. That is, all solutions of $y''(x) + y(x) = 0$ can be found by taking linear combinations of the solutions in $\{\sin x, \sqrt{3}\sin x, 2\cos x\}$.

At this point, we hope you are wondering why we don't just take the fundamental set of solutions as our spanning set. The short answer is *we could* and *we will!* Our ultimate goal is to find a set of vectors that spans the given space and is also *linearly independent*. Having a set with these two characteristics gives us the smallest possible set that can generate every element in the space.

DEFINITION 3.7   *A* **basis** *of a vector space is any collection of elements that spans the vector space and is also linearly independent.*

Example 11: Consider an nth order linear homogeneous differential equation. Any fundamental set of solutions for this equation is a basis for the solution space of this differential equation. Why? (i) We required the elements in the fundamental set to be solutions, so that they are in the solution space. (ii) We required the elements (solutions) in the fundamental set to be linearly independent. That n linearly independent solutions necessarily span the space was stated in Theorem 3.2.2. We are using the fundamental set as a concrete illustration of a basis for a vector space (the solution space).

Finding a spanning set sometimes requires adding more vectors to the given set, thus increasing the number of elements. Finding a linearly independent set sometimes requires discarding vectors that can be written as linear combinations of vectors in the given set, thus decreasing the number of elements. Finding a basis for a vector space is finding the balance between a set that has enough elements to span but not so many that there is any overlap or redundancy. We can state a nice theorem that relates these two concepts.

THEOREM 3.2.5 *Let V be a vector space with a basis consisting of n vectors. Any set that spans V is either a basis of V (and thus has n vectors) or can be reduced to a basis of V by removing vectors. Any set that is linearly independent in V is either a basis of V (and thus has n vectors) or can be extended to a basis of V by inserting vectors.*

We note that for this theorem, we do *not* say that we can add or remove any vectors we wish. In the case of a spanning set, it is necessary to only remove the ones that don't contribute anything new, while in the case of the linearly independent set we can only add vectors that are not already included.

The number of elements in the basis also tells us the size of the vector space. This may seem trivial in the case of a solution space—we know that for an nth order linear homogeneous differential equation, any fundamental set of solutions, i.e., any basis, must have n elements. The size of a general vector space is often a very useful piece of information.

> **DEFINITION 3.8**   *Let* $\{\mathbf{u}_1, \mathbf{u}_2, \cdots, \mathbf{u}_n\}$ *be a basis for a given vector space. The* **dimension** *of this vector space is equal to the number of elements in the basis, i.e., n in this case.*

Example 12: Consider the Cartesian plane, $\mathbb{R}^2$. The set

$$\left\{ v_1 = \begin{pmatrix} 1 \\ 0 \end{pmatrix}, \quad v_2 = \begin{pmatrix} 0 \\ 1 \end{pmatrix} \right\}$$

is a basis for the space. Why? If we consider any vector $\begin{pmatrix} a \\ b \end{pmatrix}$, we can write this as a linear combination of the two basis vectors, $a v_1 + b v_2$. Thus the set spans the space. It should also be clear that $c_1 v_1 + c_2 v_2 = 0$ only when $c_1 = c_2 = 0$ and thus the two vectors are linearly independent.[2] This particular basis is known as the *standard basis*.

We finish this section with one additional concept, that of a *subspace*. In Section 3.3, we will see that if we know one non-zero solution to an nth order linear homogeneous differential equation, then we can reduce the equation to an $(n-1)$st order linear homogeneous equation. In terms of a basis of the solution space, we are saying that if we remove one of the solutions from the solution space of the original differential equation, then the remaining solutions will form a solution space of a different differential equation (of order one less than the original differential equation). Removing elements from a basis yet still having a basis (but of a *different* space) gives rise to this notion of a subspace.

> **DEFINITION 3.9**   *Let W be a subset of a vector space V. We call W a* **subspace** *if the following three properties hold:*
> *(i) W contains the zero vector.*
> *(ii) W is closed under addition; that is, if $\mathbf{u}, \mathbf{v}$ are in W then so is $\mathbf{u} + \mathbf{v}$.*
> *(iii) W is closed under scalar multiplication; that is, if c is a scalar and $\mathbf{u}$ a vector W then so is $c\mathbf{u}$.*

Example 13: Consider the equation $y''' - y'' + y' - y = 0$. It is straightforward to check that a fundamental set of solutions is given by $\{\cos x, \sin x, e^x\}$. A subspace of this solution space can be created by taking $\{\cos x, \sin x\}$ as the basis for it. We know this particular subspace is also the solution space for the

---

[2]As mentioned earlier, a first step is to check that the vectors under consideration live in the proper space. Checking that you have a basis for a vector space is analogous to checking the conditions for a fundamental set of solutions. You need the following: (1) the same number of vectors as the size of the space, (2) the vectors need to be in the proper space, and (3) the vectors must be linearly independent.

equation $y'' + y = 0$. Any solution in the subspace remains in the subspace. In order to leave the subspace, we must make a linear combination with an element *not* already in the subspace, in this case with e^x.

Example 14: Consider the two-dimensional Cartesian plane $\mathbb{R}^2$ and its standard basis. The y-axis is a subspace of $\mathbb{R}^2$ because any linear combination of vectors on this axis necessarily remains on this axis. In fact, any line through the origin is a subspace of $\mathbb{R}^2$.

Problems

1. Show that if y_1 and y_2 are solutions of the first-order differential equation

$$y' + p(x)y = 0$$

   then y_1 and y_2 are linearly dependent.
2. Show that x and $2x$ are linearly dependent on $[0, 1]$.
3. Show that e^{2x} and xe^{2x} are linearly independent for all x.
4. Show that $\sin x$ and x are linearly independent for all x.
5. Show that e^x and $x + 1$ are linearly independent for all x.

   In problems 6–21, determine whether the following sets of functions are linearly independent for all x where the functions are defined.

6. $\{e^x, e^{5+x}\}$
7. $\{\sin 2x, \cos 2x\}$
8. $\{x^3 - 4, x, 3x\}$
9. $\{x^3 - 4x, x, 2x^3\}$
10. $\{\sqrt{x} - 4x, x, 2\sqrt{x}\}$
11. $\{2, \tan^2 x, \sec^2 x\}$
12. $\{x, 2x - 2, x + 3\}$
13. $\{x^3 - 3, x^3 - 3x, x - 1\}$
14. $\{e^x, e^{-x}, 1\}$
15. $\{x^3, 1 - x\}$
16. $\{1 + x^2, x, x^2\}$
17. $\{x^2, x + 2, x^2 - \frac{x}{2} - 1\}$
18. $\{x, e^{2x+3}, e^{2x-1}\}$
19. $\{x, e^{2x+2}, e^{2x-5}\}$
20. $\{\sin x, \cos 2x, e^x\}$
21. $\{\sec^2 x, \cos^2 x, \tan^2 x, \sin^2 x\}$

22. Consider the functions $f_1 = x$, $f_2 = |x|$.
    a. Show that $\{f_1, f_2\}$ is linearly dependent on $[0, 1]$.
    b. Show that $\{f_1, f_2\}$ is linearly dependent on $[-1, 0]$.
    c. Show that $\{f_1, f_2\}$ is linearly *independent* on $[-1, 1]$.
    d. Show that $W(x, |x|) = 0$ for all x.
    Thus we have found a set of functions, $\{x, |x|\}$, whose Wronskian is always zero even though the set is linearly independent.

23. Repeat a–d in problem 22 for the functions $f_1 = x^3$, $f_2 = |x|^3$.

24. Verify that $y = c_1 x + c_2 x \ln x$, $0 < x < \infty$ is a two-parameter family of solutions to $x^2 y'' - xy' + y = 0$. Determine whether this solution can give rise to a unique solution that passes through $y(1) = 3$, $y'(1) = -1$. Does this violate the existence and uniqueness theorem?

25. Verify that $y = c_1 + c_2 x^2$ is a two-parameter family of solutions to $xy'' - y' = 0$ on $(0, \infty)$. Determine whether this solution gives rise to a unique solution that passes through $y(0) = 0$, $y'(0) = 1$. Does this violate the existence and uniqueness theorem?

26. Verify that $y = c_1 x^2 + c_2 x$ is a two-parameter family of solutions to $x^2 y'' - 2xy' + 2y = 0$ on $(-\infty, \infty)$. Determine whether this solution gives rise to a unique solution that passes through $y(0) = 3$, $y'(0) = 1$. Does this violate the existence and uniqueness theorem?

In problems 27–36, show that the given set of functions forms a fundamental set of solutions to the differential equation on the specified interval. (In the language of Section 3.2.3, this question could equivalently state "show that the given set of functions forms a basis for the solution space of the given differential equation on the specified interval.")

27. $\{3 + x, 2 - 5x\}$, $y'' = 0$ for all x
28. $\{4, x - 1\}$, $y'' = 0$ for all x
29. $\{e^{-2x}, e^{3x}\}$, $y''(x) - y'(x) - 6y(x) = 0$ for all x
30. $\{e^{2x}, e^{5x}\}$, $y''(x) - 7y'(x) + 10y(x) = 0$ for all x
31. $\{\sin 2x, \cos 2x\}$, $y''(x) + 4y(x) = 0$ for all x
32. $\{e^x, xe^x\}$, $y''(x) - 2y'(x) + y(x) = 0$, on $(-\infty, \infty)$
33. $\{e^{-3x}, e^{x/2}\}$, $2y''(x) + 5y'(x) - 3y(x) = 0$, on $(-\infty, \infty)$
34. $\{1, \cos x, \sin x\}$, $y''' + y' = 0$ on $(-\infty, \infty)$
35. $\{1, e^{-x}, e^{-2x}\}$, $y''' + 3y'' + 2y' = 0$ on $(-\infty, \infty)$
36. $\{1, \cos x, \sin x, x \cos x, x \sin x\}$, $y^{(5)} + 2y''' + y' = 0$ on $(-\infty, \infty)$

In problems 37–54, determine whether the given set forms a fundamental set of solutions to the differential equation (or equivalently, forms a basis for the solution space of the given differential equation) on the specified interval. Clearly state your reasons for your answers.

37. $\{x^2 + 4, 5 - x\}$, $y'' + y = 0$ for all x

38. $\{1, e^{3x}\}$, $y'' - 3y' = 0$ for all x

39. $\{x^3, x^{-1}\}$, $x^2 y'' + 6xy' + 12y = 0$ on $(0, \infty)$

40. $\{\sin x - 5 \cos x, 3 \sin x\}$, $y'' + y = 0$, on $(-\infty, \infty)$

41. $\{\sin 5x + \cos 5x, \cos 5x - \sin 5x\}$, $y'' + 25y = 0$, on $(-\infty, \infty)$

42. $\{\sin 2x, \cos 3x\}$, $y'' + 4y = 0$, on $(-\infty, \infty)$

43. $\{\sin x, e^x\}$, $y''(x) - y(x) = 0$ on $(-\infty, \infty)$

44. $\{x^3, x^4\}$, $x^2 y'' - 6xy' + 12y = 0$ on $(0, \infty)$

45. $\{x^2, x, e^{-x}\}$, $y^{(4)}(x) - y'''(x) = 0$ on $(-\infty, \infty)$

46. $\{e^{-3x}, xe^{-3x}\}$, $y'' - 6y' + 9y = 0$ on $(-\infty, \infty)$

47. $\{e^{-x}, xe^{-x}\}$, $y'' + 2y' + y = 0$ on $(-\infty, \infty)$

48. $\{e^{2x}, xe^{2x}\}$, $y'' - 4y' + 4y = 0$ on $(-\infty, \infty)$

49. $\{x, 3, e^x\}$, $y''' - y'' = 0$ on $(-\infty, \infty)$

50. $\left\{ \dfrac{1}{\sqrt{x}}, \dfrac{1}{x}, \dfrac{\sqrt{x}+2}{x} \right\}$, $2x^2 y'' + 5xy' + y = 0$ on $(0, \infty)$

51. $\{e^{4x}, e^{2x}\}$, $y'' - 6y' + 8y = 0$ on $(-\infty, \infty)$

52. $\{e^{-2x} \sin x, e^{-2x} \cos x\}$, $y'' + 4y' + 5y = 0$, on $(0, \infty)$

53. $\{3, e^{2x}, e^{2x} + 2\}$, $y''' - 4y' = 0$ on $(-\infty, \infty)$

54. $\{e^{-x}, e^x, xe^{-x}, xe^x\}$, $y^{(4)} - 2y'' + y = 0$ on $(-\infty, \infty)$

In problems 55–62, show that the given set of vectors forms a basis for the specified vector space.

55. $\left\{ \begin{pmatrix} 0 \\ -1 \end{pmatrix}, \begin{pmatrix} 1 \\ 1 \end{pmatrix} \right\}$, for $\mathbb{R}^2$

56. $\left\{ \begin{pmatrix} 2 \\ -1 \end{pmatrix}, \begin{pmatrix} -1 \\ 1 \end{pmatrix} \right\}$, for $\mathbb{R}^2$

57. $\left\{ \begin{pmatrix} \sqrt{2} \\ 0 \end{pmatrix}, \begin{pmatrix} 0 \\ \sqrt{5} \end{pmatrix} \right\}$, for $\mathbb{R}^2$

58. $\left\{ \begin{pmatrix} 10 \\ 1 \end{pmatrix}, \begin{pmatrix} 3 \\ 2 \end{pmatrix} \right\}$, for $\mathbb{R}^2$

59. $\left\{ \begin{pmatrix} 1 \\ 1 \\ 1 \end{pmatrix}, \begin{pmatrix} 1 \\ 1 \\ 0 \end{pmatrix}, \begin{pmatrix} 1 \\ 0 \\ 0 \end{pmatrix} \right\}$, for $\mathbb{R}^3$

60. $\left\{ \begin{pmatrix} 2 \\ 1 \\ -1 \end{pmatrix}, \begin{pmatrix} 1 \\ 0 \\ 1 \end{pmatrix}, \begin{pmatrix} 0 \\ -2 \\ 0 \end{pmatrix} \right\}$, for $\mathbb{R}^3$

61. $\left\{ \begin{pmatrix} 1 \\ 0 \\ 0 \\ 2 \end{pmatrix}, \begin{pmatrix} 0 \\ 1 \\ 0 \\ -1 \end{pmatrix}, \begin{pmatrix} 3 \\ 2 \\ 0 \\ 1 \end{pmatrix}, \begin{pmatrix} 1 \\ 0 \\ 2 \\ 0 \end{pmatrix} \right\}$, for $\mathbb{R}^4$

62. $\left\{ \begin{pmatrix} 1 \\ 0 \\ 0 \\ 0 \end{pmatrix}, \begin{pmatrix} -1 \\ 1 \\ 0 \\ 0 \end{pmatrix}, \begin{pmatrix} 1 \\ -1 \\ 1 \\ 0 \end{pmatrix}, \begin{pmatrix} -1 \\ 0 \\ -2 \\ 2 \end{pmatrix} \right\}$, for $\mathbb{R}^4$

In problems 63–73, determine whether the given set of vectors forms a basis for the specified vector space.

63. $\left\{ \begin{pmatrix} 1 \\ -1 \end{pmatrix}, \begin{pmatrix} 1 \\ 2 \end{pmatrix} \right\}$, for $\mathbb{R}^2$

64. $\left\{ \begin{pmatrix} 1 \\ -1 \end{pmatrix}, \begin{pmatrix} -1 \\ 1 \end{pmatrix} \right\}$, for $\mathbb{R}^2$

65. $\left\{ \begin{pmatrix} 3 \\ 1 \end{pmatrix}, \begin{pmatrix} 0 \\ 1 \end{pmatrix}, \begin{pmatrix} -1 \\ 1 \end{pmatrix} \right\}$, for $\mathbb{R}^2$

66. $\left\{ \begin{pmatrix} 4 \\ 5 \end{pmatrix}, \begin{pmatrix} 7 \\ 8 \end{pmatrix} \right\}$, for $\mathbb{R}^2$

67. $\left\{ \begin{pmatrix} 1 \\ 0 \end{pmatrix}, \begin{pmatrix} 0 \\ 1 \end{pmatrix}, \begin{pmatrix} 2 \\ -5 \end{pmatrix}, \right\}$ for $\mathbb{R}^2$

68. $\left\{ \begin{pmatrix} 3 \\ -1 \end{pmatrix}, \begin{pmatrix} -1 \\ 2 \end{pmatrix}, \begin{pmatrix} -1 \\ 1 \end{pmatrix} \right\}$, for $\mathbb{R}^3$

69. $\left\{ \begin{pmatrix} 3 \\ -1 \\ 0 \end{pmatrix}, \begin{pmatrix} -1 \\ 0 \\ 1 \end{pmatrix}, \begin{pmatrix} 0 \\ 1 \\ 1 \end{pmatrix} \right\}$, for $\mathbb{R}^4$

70. $\left\{ \begin{pmatrix} 2 \\ -3 \\ 0 \\ 2 \end{pmatrix}, \begin{pmatrix} -1 \\ 1 \\ 0 \\ 5 \end{pmatrix}, \begin{pmatrix} 0 \\ 0 \\ 1 \\ 1 \end{pmatrix} \right\}$, for $\mathbb{R}^4$

71. $\left\{ \begin{pmatrix} 3 \\ 0 \\ 0 \end{pmatrix}, \begin{pmatrix} -1 \\ 0 \\ 1 \end{pmatrix}, \begin{pmatrix} 0 \\ 1 \\ 1 \end{pmatrix} \right\}$, for $\mathbb{R}^3$

72. $\left\{ \begin{pmatrix} -1 \\ 1 \\ 0 \end{pmatrix}, \begin{pmatrix} 1 \\ -1 \\ 1 \end{pmatrix}, \begin{pmatrix} 0 \\ 0 \\ 1 \end{pmatrix} \right\}$, for $\mathbb{R}^3$

73. $\left\{ \begin{pmatrix} 2 \\ 1 \\ 3 \end{pmatrix}, \begin{pmatrix} -1 \\ 0 \\ 0 \end{pmatrix}, \begin{pmatrix} 1 \\ 1 \\ 1 \end{pmatrix} \right\}$, for $\mathbb{R}^3$

74. Show that $\mathbb{R}^2$ is a vector space.

75. Show that $\mathbb{R}^3$ is a vector space.

76. Show that the set of all 2×2 matrices with real entries, $\mathbb{R}^{2\times 2}$, is a vector space.

77. Show that

$$\begin{pmatrix} 1 & 0 \\ 0 & 0 \end{pmatrix}, \quad \begin{pmatrix} 0 & 1 \\ 0 & 0 \end{pmatrix}, \quad \begin{pmatrix} 0 & 0 \\ 1 & 0 \end{pmatrix}, \quad \begin{pmatrix} 0 & 0 \\ 0 & 1 \end{pmatrix}$$

is a basis for $\mathbb{R}^{2 \times 2}$ (see previous question).

78. Show that the plane $z = 0$ (i.e., the x-y plane) is a subspace of $\mathbb{R}^3$.

79. Show that the plane $x = 0$ (i.e., the y-z plane) is a subspace of $\mathbb{R}^3$.

80. Show that any plane passing through the origin in $\mathbb{R}^3$ is a subspace of $\mathbb{R}^3$.

3.3   Reduction of Order—The Case $n = 2$

In this section we develop a method of simplifying an nth order differential equation if we already know a solution. The following theorem is useful.

THEOREM 3.3.1 *Let $f(x)$ be a solution of the nth order homogeneous linear differential equation (3.2), with $f(x) \neq 0$ for all $x \in (a, b)$. Then the transformation*

$$y = f(x)v$$

reduces (3.2) to an $(n - 1)$st order homogeneous linear differential equation in the dependent variable

$$w = \frac{dv}{dx}.$$

This result is very useful, for if we know one nontrivial solution then we can reduce the order of the differential equation.

We will consider this theorem for the case with $n = 2$. Suppose f is a known, nontrivial solution of the second-order homogeneous linear equation

$$a_0(x)\frac{d^2 y}{dx^2} + a_1(x)\frac{dy}{dx} + a_2(x)y = 0. \qquad (3.12)$$

Let $y = f(x)v$ where f is the known solution and v is a function of x to be determined.

Taking derivatives and simplifying gives

$$\frac{dy}{dx} = f(x)\frac{dv}{dx} + f'(x)v$$

and

$$\frac{d^2 y}{dx^2} = f(x)\frac{d^2 v}{dx^2} + 2f'(x)\frac{dv}{dx} + f''(x)v.$$

Substituting the derivatives into the differential equation (3.12) gives

$$a_0(x)\left(f(x)\frac{d^2 v}{dx^2} + 2f'(x)\frac{dv}{dx} + f''(x)v\right)$$

$$+ a_1(x)\left(f(x)\frac{dv}{dx} + f'(x)v\right) + a_2(x)f(x)v = 0.$$

Rearranging the terms then yields

$$a_0(x)f(x)\frac{d^2v}{dx^2} + (2a_0(x)f'(x) + a_1(x)f(x))\frac{dv}{dx}$$
$$+ (a_0(x)f''(x) + a_1(x)f'(x) + a_2(x)f(x))\,v = 0. \quad (3.13)$$

But, we have assumed that $f(x)$ is a solution of

$$a_0(x)\frac{d^2y}{dx^2} + a_1(x)\frac{dy}{dx} + a_2(x)y = 0,$$

that is,

$$a_0(x)f''(x) + a_1(x)f'(x) + a_2(x)f(x) = 0$$

so that the coefficient of v is zero. So the equation (3.13) becomes

$$a_0(x)f(x)\frac{d^2v}{dx^2} + (2a_0(x)f'(x) + a_1(x)f(x))\frac{dv}{dx} = 0.$$

We now make the substitution

$$w = \frac{dv}{dx} \quad (3.14)$$

and obtain the first-order homogeneous linear differential equation

$$a_0(x)f(x)w'(x) + (2a_0(x)f'(x) + a_1(x)f(x))\,w(x) = 0.$$

This equation is separable and you can show that

$$w(x) = c\exp\left[-\frac{\int \frac{a_1(x)}{a_0(x)}dx}{f^2(x)}\right]$$

is the solution. By (3.14), we have $\frac{dv}{dx} = w$ and thus integrating gives us

$$v = \int c\exp\left[-\frac{\int \frac{a_1(x)}{a_0(x)}dx}{f^2(x)}\right]dx.$$

We also have $y = f(x)v$ and thus

$$y = f(x)\int c\exp\left[-\frac{\int \frac{a_1(x)}{a_0(x)}dx}{f^2(x)}\right]dx \quad (3.15)$$

is a solution of the differential equation (3.12).

Recall that our goal is to find two linearly independent solutions to (3.12). We assumed that $f(x)$ is a solution and we showed that (3.15) is a solution. If we let

$$g(x) = f(x)\int c\exp\left[-\frac{\int \frac{a_1(x)}{a_0(x)}dx}{f^2(x)}\right]dx,$$

we can show that the solutions $f(x)$ and $g(x)$ are linearly independent by considering their Wronskian:

$$\begin{vmatrix} f(x) & g(x) \\ f'(x) & g'(x) \end{vmatrix} = fg' - gf',$$

where

$$g'(x) = f(x)c\exp\left[-\frac{\int \frac{a_1(x)}{a_0(x)}\,dx}{f^2(x)}\right] + f'(x)\int c\exp\left[-\frac{\int \frac{a_1(x)}{a_0(x)}\,dx}{f^2(x)}\right]\,dx.$$

Substituting in for g and g' gives

$$fg' - gf'$$

$$= f(x)\left(f(x)c\exp\left[-\frac{\int \frac{a_1(x)}{a_0(x)}\,dx}{f^2(x)}\right] + f'(x)\int c\exp\left[-\frac{\int \frac{a_1(x)}{a_0(x)}\,dx}{f^2(x)}\right]\,dx\right)$$

$$-f'(x)\left(f(x)\int c\exp\left[-\frac{\int \frac{a_1(x)}{a_0(x)}\,dx}{f^2(x)}\right]\,dx\right)$$

$$= f^2(x)\int c\exp\left[-\frac{\int \frac{a_1(x)}{a_0(x)}\,dx}{f^2(x)}\right]\,dx \tag{3.16}$$

$$\neq 0. \tag{3.17}$$

Thus $f(x)$ and $g(x)$ are linearly independent and the linear combination

$$c_1 f(x) + c_2 g(x) = c_1 f(x) + c_2 f(x)\int c\exp\left[-\frac{\int \frac{a_1(x)}{a_0(x)}\,dx}{f^2(x)}\right]\,dx \tag{3.18}$$

$$= c_1 f(x) + f(x)\int \tilde{c}_2 \exp\left[-\frac{\int \frac{a_1(x)}{a_0(x)}\,dx}{f^2(x)}\right]\,dx \tag{3.19}$$

is the general solution of (3.12) where $\tilde{c}_2 = c_2 \cdot c$.

Example 1: Given that $y = e^{2x}$ is a solution of

$$\frac{d^2 y}{dx^2} + \frac{dy}{dx} - 6y = 0,$$

find a solution that is linearly independent of $y = e^{2x}$ by reducing the order. Let $y = ve^{2x}$; this gives

$$\frac{dy}{dx} = e^{2x}\frac{dv}{dx} + 2e^{2x}v \quad \text{and} \quad \frac{d^2 y}{dx^2} = e^{2x}\frac{d^2 v}{dx^2} + 4e^{2x}\frac{dv}{dx} + 4e^{2x}v. \tag{3.20}$$

Substituting these derivatives gives

$$\left(e^{2x}\frac{d^2v}{dx^2} + 4e^{2x}\frac{dv}{dx} + 4e^{2x}v\right) + \left(e^{2x}\frac{dv}{dx} + 2e^{2x}v\right) - 6\left(e^{2x}v\right) = 0.$$

This simplifies to

$$e^{2x}\frac{d^2v}{dx^2} + 5e^{2x}\frac{dv}{dx} = 0.$$

Now let $w = \frac{dv}{dx}$ so that

$$e^{2x}\frac{dw}{dx} + 5e^{2x}w = 0.$$

This last equation is separable with solution $w = ce^{-5x}$.

We now choose a specific c-value to obtain the particular solution. For example, choosing $c = -5$ gives

$$\frac{dv}{dx} = -5e^{-5x}.$$

Integrating gives $v(x) = e^{-5x}$. Thus,

$$y = e^{2x}v = e^{2x}e^{-5x} = e^{-3x}$$

and the general solution is

$$y = c_1e^{2x} + c_2e^{-3x}.$$

We note that choosing $c = -5$ in the previous example only made our integral "nice." If we chose, for example, $c = 1$, we still would have obtained an equally valid solution since we ultimately multiplied the second solution by a constant c_2. Reducing the order of the equation works on any linear homogeneous equation, even if the coefficients are not constant as we will see in the next example.

Example 2: Given that $y = x$ is a solution of

$$(x^2 + 1)\frac{d^2y}{dx^2} - 2x\frac{dy}{dx} + 2y = 0,$$

find a solution that is linearly independent of $y = x$ by reducing the order.

We see that $y = x$ is indeed a solution, since $y' = 1$ and $y'' = 0$ so that substitution gives

$$(x^2 + 1)(0) - 2x(1) + 2x = 0.$$

We now let $y = vx$ so that

$$\frac{dy}{dx} = x\frac{dv}{dx} + v \quad \text{and} \quad \frac{d^2y}{dx^2} = x\frac{d^2v}{dx^2} + 2\frac{dv}{dx}. \tag{3.21}$$

Substituting these derivatives gives

$$(x^2 + 1)\left(x\frac{d^2v}{dx^2} + 2\frac{dv}{dx}\right) - 2x\left(x\frac{dv}{dx} + v\right) + 2xv = 0,$$

which implies

$$x(x^2 + 1)\frac{d^2v}{dx^2} + 2\frac{dv}{dx} = 0.$$

Now let $w = \frac{dv}{dx}$ so that

$$x(x^2 + 1)\frac{dw}{dx} + 2w = 0.$$

This equation is separable. After we separate the variables we have

$$\frac{1}{w}\,dw = \frac{-2}{x(x^2 + 1)}\,dx = \left(\frac{2x}{x^2 + 1} - \frac{2}{x}\right)dx,$$

where the last equality results from the partial fraction expansion. We then integrate and solve for w to obtain

$$w = \frac{c(x^2 + 1)}{x^2}.$$

We again choose a specific c-value to obtain the particular solution, say $c = 1$, and obtain

$$\frac{dv}{dx} = \frac{(x^2 + 1)}{x^2} = 1 + \frac{1}{x^2}.$$

Integrating gives $v(x) = x - \frac{1}{x}$. Thus,

$$y = xv = x\left(x - \frac{1}{x}\right) = x^2 - 1.$$

So the general solution is

$$y = c_1 x + c_2(x^2 - 1).$$

Problems

In problems 1–21, use the given solution to reduce the order of the differential equation. Use the methods of Chapters 1, 2, and Section 3.1.2 to solve the reduced equation.

1. $\frac{d^2y}{dx^2} - 2\frac{dy}{dx} + y = 0$, solution $y = e^x$

2. $4\frac{d^2y}{dx^2} + 4\frac{dy}{dx} + y = 0$, solution $y = e^{-x/2}$

3. $\dfrac{d^2y}{dx^2} + 25y = 0$, solution $y = \cos 5x$

4. $\dfrac{d^2y}{dx^2} - 9y = 0$, solution $y = e^{3x}$

5. $\dfrac{d^2y}{dx^2} - 6\dfrac{dy}{dx} + 9y = 0$, solution $y = e^{3x}$

6. $y'' - 8y' + 16y = 0$, solution $y = e^{4x}$

7. $2y'' + 5y' - 3y = 0$, solution $y = e^{x/2}$

8. $xy'' - y' = 0$, solution $y = x^2$

9. $xy'' + y' = 0$, solution $y = \ln x$

10. $x^2y'' - xy' - 3y = 0$, solution $y = x^3$

11. $x^2y'' - xy' + y = 0$, solution $y = x\ln x$

12. $2x^2y'' + 5xy' + y = 0$, solution $y = \dfrac{1}{\sqrt{x}}$

13. $(x+1)^2\dfrac{d^2y}{dx^2} - 3(x+1)\dfrac{dy}{dx} + 3y = 0$, solution $y = x + 1$

14. $(2x+1)\dfrac{d^2y}{dx^2} - 4(x+1)\dfrac{dy}{dx} + 4y = 0$, solution $y = e^{2x}$

15. $(x^2 - 1)y'' - 2xy' + 2y = 0$, solution $y = x^2 + 1$

16. $(x^2 + 1)y'' + (2 - x^2)y' - (2 + x)y = 0$, solution $y = \dfrac{1}{x}$

17. $\dfrac{d^3y}{dx^3} - 2\dfrac{d^2y}{dx^2} - \dfrac{dy}{dx} + 2y = 0$, solution $y = e^{2x}$

18. $\dfrac{d^3y}{dx^3} - 4\dfrac{d^2y}{dx^2} + 5\dfrac{dy}{dx} - 2y = 0$, solution $y = e^x$

19. $y''' - 5y'' + 8y' - 4y = 0$, solution $y = xe^{2x}$

20. $y''' + 5y'' = 0$, solution $y = x$

21. $y''' - 3y'' = 0$, solution $y = 2$

22. Show that the separable differential equation

$$a_0(x)f(x)w'(x) + (2a_0(x)f'(x) + a_1(x)f(x))\,w(x) = 0$$

has solution

$$w(x) = c\exp\left[-\frac{\int \frac{a_1(x)}{a_0(x)}\,dx}{f^2(x)}\right].$$

23. Show that if $f(x)$ is a solution to the differential equation

$$a_0(x)\frac{d^2y}{dx^2} + a_1(x)\frac{dy}{dx} + a_2(x)y = 0$$

and $g(x)$ is another solution given by

$$g(x) = f(x) \int c \exp \left[-\frac{\int \frac{a_1(x)}{a_0(x)} dx}{f^2(x)} \right] dx,$$

then $\{f(x), g(x)\}$ forms a fundamental solution set.

24. Consider the hyperbolic functions

$$\cosh x = \frac{e^x + e^{-x}}{2} \quad \text{and} \quad \sinh x = \frac{e^x - e^{-x}}{2}.$$

Show that

a. $\dfrac{d}{dx} \cosh x = \sinh x.$

b. $\dfrac{d}{dx} \sinh x = \cosh x.$

c. $\cosh^2 x - \sinh^2 x = 1.$

d. $\cosh x$ and $\sinh x$ are linearly independent functions.

e. A general solution of $y'' - y = 0$ is

$$y = c_1 \cosh x + c_2 \sinh x.$$

3.4   Operator Notation

We will now introduce some notation to make our subsequent work a bit easier. The nth derivative of a function $y(x)$ is given in *operator notation* by

$$D^n y = \frac{d^n y}{dx^n}.$$

Using this notation, the left-hand side of the nth order linear homogeneous differential equation (3.2)

$$a_0(x) \frac{d^n y}{dx^n} + a_1(x) \frac{d^{n-1} y}{dx^{n-1}} + \ldots + a_{n-1}(x) \frac{dy}{dx} + a_n(x) y = 0$$

can be expressed as

$$a_0(x) \frac{d^n y}{dx^n} + a_1(x) \frac{d^{n-1} y}{dx^{n-1}} + \ldots + a_{n-1}(x) \frac{dy}{dx} + a_n(x) y$$

$$= a_0(x) D^n y + a_1(x) D^{n-1} y + \ldots + a_{n-1}(x) Dy + a_n(x) y$$

$$= \left(a_0(x) D^n + a_1(x) D^{n-1} + \ldots + a_{n-1}(x) D + a_n(x) \right) y.$$

Thus, we can write (3.2) as

$$\left(a_0(x)\, D^n + a_1(x)\, D^{n-1} + \ldots + a_{n-1}(x)\, D + a_n(x)\right) y = 0,$$

and in fact we can write the nth order linear nonhomogeneous equation (3.1) as

$$\left(a_0(x)\, D^n + a_1(x)\, D^{n-1} + \ldots + a_{n-1}(x)\, D + a_n(x)\right) y = F(x).$$

With this in mind, the following definition is natural.

DEFINITION 3.10   *The expression*

$$P(D) = a_0(x)\, D^n + a_1(x)\, D^{n-1} + \ldots + a_{n-1}(x)\, D + a_n(x),$$

where $a_0(x), a_1(x), \ldots, a_n(x)$ *are (possibly constant) real-valued functions with* $a_0(x) \neq 0$*, is called an* n*th order linear differential operator.*

With this notation, we can write an nth order linear homogeneous differential equation compactly as

$$P(D)y = 0$$

and an nth order linear nonhomogeneous equation as

$$P(D)y = F(x).$$

Example 1: Write the differential equation $2y''' - 3y'' + 5y' - y = 2x$ in operator form.

The operator is given as

$$P(D) = 2D^3 - 3D^2 + 5D - 1$$

so that the equation is

$$P(D)y = 2x.$$

Example 2: Write the differential equation $2xy'' - (3x^2 + 1)y' + e^x y = \cos x$ in operator form.

The operator is given as

$$P(D) = 2xD^2 - (3x^2 + 1)D + e^x$$

so that the equation is

$$P(D)y = \cos x.$$

Before we proceed with our further study of nonhomogeneous equations, it will be of future benefit if we obtain some results for computing with differential operators. As we almost always will be dealing with differential operators with *constant coefficients*, we state a few important results with these.

Let y_1 and y_2 both be n times continuously differentiable functions and let $P(D)$ be an nth order linear differential operator with constant coefficients. We have

$$P(D)(y_1 + y_2) = \left(a_0\, D^n + a_1\, D^{n-1} + \ldots + a_{n-1}\, D + a_n\right)(y_1 + y_2)$$

$$= a_0\, D^n(y_1 + y_2) + a_1\, D^{n-1}(y_1 + y_2) + \ldots$$

$$+ a_{n-1}\, D(y_1 + y_2) + a_n\,(y_1 + y_2)$$

$$= a_0\, D^n y_1 + a_1\, D^{n-1} y_1 + \ldots + a_{n-1}\, Dy_1 + a_n\, y_1$$

$$+ a_0\, D^n y_2 + a_1\, D^{n-1} y_2 + \ldots + a_{n-1}\, Dy_2 + a_n\, y_2$$

$$= P(D)y_1 + P(D)y_2.$$

Thus, we have the *linearity property*

$$P(D)(y_1 + y_2) = P(D)y_1 + P(D)y_2. \tag{3.22}$$

It can also be shown, but with a bit more effort, that if $P_1(D)$ and $P_2(D)$ are two linear differential operators with constant coefficients, then

$$P_1(D)P_2(D) = P_2(D)P_1(D) \tag{3.23}$$

so that constant-coefficient linear differential operators *commute*. It is essential to realize, however, that applying an operator to a function is not really a multiplication, although it seems like it at times.

Example 3: Consider $P_1(D) = xD, P_2(D) = D, y = 3x$. Then

$$P_2(D)P_1(D)(y) = (D)(xD)(3x) = (D)(xD(3x)) = (D)(x(3)) = (D)(3x) = 3$$

but

$$P_1(D)P_2(D)(y) = (xD)(D)(3x) = (xD)(D(3x)) = (xD)(3) = x(0) = 0.$$

We again stress that **the order can be exchanged only in the case of differential operators with constant coefficients.** To see this in general, we write

$P_1(D)P_2(D)$

$$= \left(a_0\,D^n + a_1\,D^{n-1} + \ldots + a_{n-1}\,D + a_n\right)\left(b_0\,D^n + b_1\,D^{n-1} + \ldots + b_n\right)$$

$$= (a_0b_0)D^{2n} + (a_1b_0 + a_0b_1)D^{2n-1} + (a_2b_0 + a_1b_1 + a_0b_2)D^{2n-2} + \cdots$$

$$+ (a_0b_n + a_1b_{n-1} + a_2b_{n-2} + \cdots + a_{n-2}b_2 + a_{n-1}b_1 + a_nb_0)D^n + \cdots$$

$$+ (a_{n-2}b_n + a_{n-1}b_{n-1} + a_nb_{n-2})D^2 + (a_{n-1}b_n + a_nb_{n-1})D + a_nb_n$$

$$= (b_0a_0)D^{2n} + (b_1a_0 + b_0a_1)D^{2n-1} + (b_2a_0 + b_1a_1 + b_0a_2)D^{2n-2} + \cdots$$

$$+ (b_0a_n + b_1a_{n-1} + b_2a_{n-2} + \cdots + b_{n-2}a_2 + b_{n-1}a_1 + b_na_0)D^n + \cdots$$

$$+ (b_{n-2}a_n + b_{n-1}a_{n-1} + b_na_{n-2})D^2 + (b_{n-1}a_n + b_na_{n-1})D + b_na_n$$

$$= \left(b_0\,D^n + b_1\,D^{n-1} + \ldots + b_n\right)\left(a_0\,D^n + a_1\,D^{n-1} + \ldots + a_{n-1}\,D + a_n\right)$$

$$= P_2(D)P_1(D). \tag{3.24}$$

It is left as an exercise to use sigma notation to show this property in a more elegant fashion. With the property $P_1(D)P_2(D) = P_2(D)P_1(D)$, we see that we can always factor a constant-coefficient linear differential operator into powers of first-order terms, just as we would a polynomial, *as long as we allow the use of complex numbers*. Allowing for complex roots will be essential in finding the solutions of constant-coefficient homogeneous linear differential equations. The properties of the differential operator will also be essential when studying the Annihilator method for solving certain constant coefficient nonhomogeneous differential equations; see Section 4.6.

Example 4: The linear differential operator

$$P(D) = D^4 + 5D^3 + 6D^2$$

can be written as

$$P(D) = D^2(D^2 + 5D + 6) = D^2(D + 2)(D + 3).$$

Note that it could also be written as $P(D) = (D+2)(D+3)D^2$, for example, since the factors commute.

Example 5: The linear differential operator

$$P(D) = D^2 + 9$$

can be written as
$$P(D) = (D + 3i)(D - 3i).$$

Example 6: The linear differential operator
$$P(D) = D^2 + 2D + 2$$

has roots $-1 \pm i$ and thus can be written as
$$P(D) = (D - (-1 + i))(D - (-1 - i)) = (D + 1 - i)(D + 1 + i).$$

In practice we will usually not need to factor the operator completely into linear factors but it is conceptually important to realize that we, in fact, are able to do so.

Example 7: Apply $D^2 + 3$ to the function
$$f(x) = x^4 - \cos x.$$

We apply the operator to this function to obtain
$$(D^2 + 3)(f(x)) = D^2 f(x) + 3f(x) = f''(x) + 3f(x).$$

We thus need to calculate the second derivative of this function. Taking two derivatives gives
$$f'(x) = 4x^3 + \sin x, \quad f''(x) = 12x^2 + \cos x.$$

We thus obtain
$$\begin{aligned} f''(x) + 3f(x) &= (12x^2 + \cos x) + 3(x^4 - \cos x) \\ &= 3x^4 + 12x^2 - 2\cos x. \end{aligned} \tag{3.25}$$

The operator notation is useful for writing differential equations in compact form and we will see that it will be especially useful in the next section.

Problems

In problems 1–6, apply the given differential operator $P(D)$ to the functions and simplify as much as possible.

1. $P(D) = D - 5$
   a. $f_1(x) = 3x + 7$,   b. $f_2(x) = \cos x$,   c. $f_3(x) = e^{5x}$
2. $P(D) = D + 1$
   a. $f_1(x) = e^{-x}$,   b. $f_2(x) = 4 + \sin x$,   c. $f_3(x) = e^{-2x}$
3. $P(D) = D^2 - 1$
   a. $f_1(x) = 4e^{-x} + e^{2x}$,   b. $f_2(x) = x^3 + 4$,   c. $f_3(x) = e^x - 5e^{-x}$

4. $P(D) = D^2 + 1$
   a. $f_1(x) = x^3 + 2x$, b. $f_2(x) = \cos x + \sin x$, c. $f_3(x) = \sin 2x$
5. $P(D) = D^2 - 4D + 5$
   a. $f_1(x) = e^{2x} \cos x$, b. $f_2(x) = 4x^2 + e^x$, c. $f_3(x) = x^3 - \cos 2x$
6. $P(D) = D^2 + 2D + 1$
   a. $f_1(x) = e^x$, b. $f_2(x) = xe^{-x}$, c. $f_3(x) = \sin x$

In problems 7–14, calculate $P(D)Q(D)(y)$ and $Q(D)P(D)(y)$. Compare the results.

7. $P(D) = D^2 - D, Q(D) = D^3 + 2D, y = x^6 - 2x^2$
8. $P(D) = D - 1, Q(D) = D^2 + D - 3, y = x^2 + \sin x$
9. $P(D) = D, Q(D) = D + 3x, y = 2x + 1$
10. $P(D) = D, Q(D) = D + 3x, y = \sin x$
11. $P(D) = D^2 + xD + 2, Q(D) = D + 1, y = e^x$
12. $P(D) = D^2 + xD + 2, Q(D) = D + 1, y = x^3 + 5x$
13. $P(D) = D, Q(D) = D + x, y = e^{-x^2}$
14. $P(D) = D, Q(D) = D + x, y = 1 + \cos x$

15. Use the commutative property of the constant coefficient differential operator to quickly evaluate
$$D^3(D - 2)(D + 3)(x^2 - 2x + 7)$$

16. Use the linearity property of the differential operator to quickly evaluate
$$D^3(x^2 - 5 + e^{2x})$$

17. Factor the operators into linear factors and/or irreducible quadratic factors
    a. $D^2 - D - 6$, b. $D^2 - 11D + 12$, c. $D^4 + 2D^2 + 1$
18. Factor the operators into linear factors and/or irreducible quadratic factors
    a. $D^4 - 2D^2 + 1$, b. $D^3 - D^2 + 4D$, c. $D^3 - 5D^2 + 4D$
19. Factor the operators into linear factors and/or irreducible quadratic factors
    a. $D^3 - 1$, b. $D^3 + 8$, c. $D^3 - 2D^2 + 1$
20. Rewrite the following differential equations using operator notation and factor completely.
    a. $y'' + 10y' + 16y = 0$, b. $y^{(5)} + 4y''' + 4y' = 0$
21. Rewrite the following differential equations using operator notation and factor completely.
    a. $y''' + 27y = 0$, b. $y'' + 3y' - 4y = 0$

22. Rewrite the following differential equations using operator notation and factor completely.

    a. $y''' - 8y = 0$,   b. $y''' + 6y'' + 9y' = 0$

23. Use sigma notation to show that the constant-coefficient differential operator commutes. That is, let

$$P_1(D) = \sum_{i=0}^{n} a_i D^{n-i} \quad \text{and} \quad P_2(D) = \sum_{i=0}^{n} b_i D^{n-i}$$

and show that $P_1(D)P_2(D) = P_2(D)P_1(D)$.

3.5   Numerical Considerations for nth Order Equations

Although we will learn how to solve nth order homogeneous equations that have constant coefficients in Section 4.1, we will often encounter situations where the coefficients are not constant. In addition, sometimes it will be difficult to exactly find the roots of the characteristic polynomial that arises from our differential equation. In both of these situations, we still need to know the behavior of the solution and it is often helpful to calculate this solution numerically. The method we learn in this section also applies to *nonlinear* equations and this will be useful in later chapters (and real life!). The methods that we learned in Sections 2.5 and 2.6 have analogous formulations in higher dimensions, with the Runge-Kutta method still being the much preferred method. We will not go through the details of this method again, but will refer the interested reader to many of the texts given in the References, e.g., Burden and Faires [9]. Instead, we will show how to use Maple and Matlab to calculate solutions of an nth order equation. For the Maple code, if we want to plot solutions in the x-y plane we only need to change the equation that is input. For the Matlab code, we will also need to change the m-file containing the equation. (This will be necessary for both Maple and Matlab when we want to plot solutions in the y-y' *phase plane*. But first we will need to reduce the nth order equation to a system of n first-order equations.)

For both implementations we will consider two different differential equations:

Example 1: $2xy''(x) + x^2 y'(x) + 3x^3 y(x) = 0$, $y(1) = 2, y'(1) = 0$.

Example 2: $y^{(4)}(x) + x^2 y'(x) + y(x) = \cos(x)$, $y(0) = 1, y'(0) = 0, y''(0) = 0, y'''(0) = 1$.

In both cases, we will want to plot the solution from the initial x-value until $x = 5$ using a step size of $h = 0.05$. We will also give the approximate solution at the value $x = 5$. We note that dsolve is used to numerically calculate the solution.

Computer Code 3.3: **Numerically solving a higher order differential equation with fourth-order Runge-Kutta method and built-in commands; plotting this solution**

Maple, Mathematica

```
                            Maple
> with(plots):  #loads the plots package
> eq1:=2*x*diff(y(x),x$2)+x^2*diff(y(x),x)+3*x^3*y(x)=0;
> IC1:=y(1)=2, D(y)(1)=0; # specifies the initial condition
> soln1:= dsolve({eq1,IC1},y(x),numeric,
  method=classical[rk4],stepsize=.05);
> soln1(5.0); # this is needed only because we wanted y(5)
> odeplot(soln1,[x,y(x)],0..6,numpoints=300);
> eq2:=diff(y(x),x$4)+ x^2*diff(y(x),x)+y(x)=cos(x);
> IC2:=y(0)=1, D(y)(0)=0, D(D(y))(0)=0,D(D(D(y)))(0)=1;
> soln2:= dsolve({eq2,IC2},y(x),numeric,
  method=classical[rk4],stepsize=.05);
> soln2(5.0); # this is needed only because we wanted y(5)
> odeplot(soln2,[x,y(x)],0..6,numpoints=300);
```

The command odeplot utilizes the numerical solution that we just calculated. To just obtain the numerical plot without concern for any specific function values or method of numerical solution, we could have typed

---

```
                            Maple
> with(DEtools):  #loads the differential equations package
> eq1:=2*x*diff(y(x),x$2)+x^2*diff(y(x),x)+3*x^3*y(x)=0;
> IC1:=y(1)=2, D(y)(1)=0; # specifies the initial condition
> DEplot(eq1,y(x),x=1..5,[[IC1]],stepsize=.05);
> eq2:=diff(y(x),x$4)+ x^2*diff(y(x),x)+y(x)=cos(x);
> IC2:=y(0)=1, D(y)(0)=0, D(D(y))(0)=0,D(D(D(y)))(0)=1;
> DEplot(eq2,y(x),x=0..5,[[IC2]],stepsize=.05);
```

```
                    Mathematica
de[x_]=2x y''[x]+x^2 y'[x] + 3x^3 y[x]
solution1=NDSolve[{de[x]==0,y[1]==2, y'[1]==0},y,{x,0,6},
   StartingStepSize→ .05,Method→ {FixedStep,
   Method→ ExplicitRungeKutta}]
y1[x_] = y[x]/.solution1[[1]]
y1[5]
Plot[y1[x],{x,0,6},AxesLabel→ {"x","y"}]
de[x_]= y''''[x]+x^2 y'[x] + y[x]
solution2=NDSolve[{de[x]==Cos[x],y[0]==1, y'[0]==0,
y''[0]==0,y'''[0]==1},y,{x,0,6}, StartingStepSize→ .05,
   Method→ {FixedStep,Method→ ExplicitRungeKutta}]
y2[x_] = y[x]/.solution2[[1]]
y2[5]
Plot[y2[x],{x,0,6},AxesLabel→ {"x","y"}]
```

The Matlab code will require an additional step before we can use the relevant code and commands. This will help us better understand how the Runge-Kutta method works with a higher-order equation.

3.5.1   Converting an nth Order Equation to a System of n First-Order Equations

We said that we will skip the details of the Runge-Kutta method for numerically solving higher-order equations. But we will briefly discuss how the method is applied. For a first-order equation, Section 2.6 showed how we considered four different slopes of the direction field in our efforts to calculate one step. For a second-order equation, we will convert our differential equation to the form

$$u_1' = f(x, u_1, u_2)$$
$$u_2' = g(x, u_1, u_2). \qquad (3.26)$$

In doing so we then see that we can carry out the first-order Runge-Kutta calculation in *each* variable. Thus we now have *two* directions to worry about besides the independent x-direction. But the idea is exactly the same as before—we will still consider four different slopes of the direction field (now three-dimensional!) in order to take one step. The direction field, as you might imagine, gets rather ugly and it is hopeless to even attempt to draw it for a third- or higher-order equation.

The obvious question is "how do we go from a second-order equation to something of the form of (3.26)?" Let us consider the equation

$$a_0(x)y'' + a_1(x)y' + a_2(x)y = F(x) \qquad (3.27)$$

as our second-order equation that we would like to convert. We need two variables, u_1 and u_2. We then set $u_1 = y$ and $u_2 = y'$. The left-hand sides of (3.26) are taken care of and we need to relate the right-hand side to (3.27). We have

$$u_1' = y'$$
$$u_2' = y''$$

but need to write the right-hand sides in terms of our new variables. The first equation is simple enough—y' is the same as u_2. What is y''? Well, we can solve the original equation (3.27) for y'' and then substitute. That is, we write

$$y'' = \frac{1}{a_0}\left(-a_1(x)y' - a_2(x)y + F(x)\right).$$

The right-hand side still needs to be put in terms of the new variables and we substitute $u_1 = y$ and $u_2 = y'$ to obtain

$$y'' = \frac{1}{a_0}\left(-a_1(x)u_2 - a_2(x)u_1 + F(x)\right).$$

Our system of two first-order equations is thus

$$u_1' = u_2$$
$$u_2' = \frac{1}{a_0}\left(-a_1(x)u_2 - a_2(x)u_1 + F(x)\right). \tag{3.28}$$

If we were given initial conditions, these could be converted to give $u_1(x_0) = y_0$, $u_2(x_0) = y_1$. The situation is analogous for a higher-order equation.

How to convert to a system of first-order equations
1. *Introduce the same number of variables as the order of the equation. Rename the 0th through $(n-1)$st derivatives of the function. For example, a fifth-order differential equation requires five new variables, u_1, u_2, u_3, u_4, u_5. And we set $u_1 = y$, $u_2 = y'$, $u_3 = y''$, $u_4 = y^{(3)}$, and $u_5 = y^{(4)}$.*

2. *Solve the original nth order equation for the highest derivative, $y^{(n)}$. For the fifth-order example, this means we solve for $y^{(5)}$.*

3. *Write the first derivative of each new variable as the left-hand side of an equation (you will have n equations). Then rewrite the corresponding right-hand sides in terms of the new variables. For our fifth-order example, the first four equations would be $u_1' = u_2$, $u_2' = u_3$, $u_3' = u_4$, $u_4' = u_5$. The left-hand side of the fifth equation is u_5' and its right-hand side is obtained by substituting the new variables into the expression obtained in Step **2** above.*

We can always convert an nth order equation to a system of first-order equations but we cannot always take a general first-order system and make it into an nth order differential equation. And although we illustrated this conversion with a linear equation, we can actually convert *any* nth order equation to a system of n first-order equations, provided we are able to solve for the highest derivative (i.e., perform Step **2.**). Before giving two examples of the above method, we mention that u_1 is the solution of the original nth order equation. This will be useful when we want to plot the solution. As a bonus, we also will have $n-1$ derivatives of this solution. For physical problems, for example, it may be useful to have both position and velocity as functions of the independent variable time.

Example 3: Reduce the equations of Examples 1 and 2 to systems of first-order equations.

For Example 1, we are given a second-order equation and thus we will obtain two first-order equations. We solve for the highest derivative to obtain

$$y'' = \frac{-x}{2}y' - \frac{3x^2}{2}y.$$

We then set $u_1 = y$, $u_2 = y'$. Our system of first-order equations is thus

$$u_1' = u_2$$
$$u_2' = \frac{-x}{2}u_2 - \frac{3x^2}{2}u_1. \tag{3.29}$$

The initial condition then becomes $u_1(1) = 2, u_2(1) = 0$.

For Example 2, we have a fourth-order equation and thus we will have four first-order equations. Although this system is not homogeneous, this will not affect the steps we take. We solve for the highest derivative to obtain

$$y^{(4)} = -x^2 y'(x) - y(x) + \cos(x).$$

We then set $u_1 = y$, $u_2 = y', u_3 = y''$, $u_4 = y'''$. Our system of first-order equations is thus

$$u_1' = u_2$$
$$u_2' = u_3$$
$$u_3' = u_4$$
$$u_4' = -x^2 u_2 - u_1 + \cos(x). \tag{3.30}$$

The initial condition then becomes $u_1(0) = 1, u_2(0) = 0, u_3(0) = 0, u_4(0) = 1$.

Matlab

To input these examples into Matlab, we simply need to enter these equations into our m-file. The syntax for both **ode45** and **RK4.m** requires us to only enter the right-hand sides of the equation and we must do so in vector format. The variable that contains the solution is **yn** and thus $u_1 = y(1), u_2 =$

$y(2), u_3 = y(3), u_4 = y(4)$. We create two different m-files. Remember that anything following a % is ignored by Matlab and is only entered for clarity of programming. Thus, Example1.m and Example2.m are each really only two lines long.

Computer Code 3.4: **Numerically solving a higher-order differential equation with fourth-order Runge-Kutta method and built-in commands; plotting this solution**

Matlab only

Example1.m

```
                          Matlab
function f=Example1(xn,yn)
%
%The original ode is 2*x*y''(x) +x^2*y'(x)+3*x^3*y(x)=0
%The system of first order equations is
%u1'=u2
%u2'=(-x/2)*u2-((3*x^2)/2)*u1
%
%We let yn(1)=u1, yn(2)=u2
%
f= [yn(2); (-xn/2)*yn(2)-((3*xn^2)/2)*yn(1)];
```

Example2.m

```
                          Matlab
function f=Example2(xn,yn)

%The original ode is y^(4)(x)+ x^2*y'(x)+y(x)=cos(x)
%The system of first order equations is
%u1'=u2
%u2'=u3
%u3'=u4
%u4'=-x^2*u2-u1+cos(x)
%
%We let yn(1)=u1, yn(2)=u2, yn(3)=u3, yn(4)=u4

f= [yn(2); yn(3); yn(4); -xn^2*yn(2)-yn(1)+cos(xn)];
```

Using either the Matlab's built-in ode45 or the fixed step implementation of the Runge-Kutta code from Section 2.6, we can enter the following commands in the command window to calculate the solutions:

```
                              Matlab
>>  x0=1; xf=5;
>>  y0=[2, 0];
>>  [x1,y1]=RK4('Example1',[x0,xf],y0,.05);
>>  subplot(211),plot(x1,y1(:,1))
>>  xlabel('x'); ylabel('y(1)');
>>  subplot(212),plot(x1,y1(:,2))
>>  xlabel('x'); ylabel('y(2)');
>>  t=length(x1);
>>  y1(t,:)  %shows the entries of y1 corresponding to the
            %last entry of x1
>>  figure
>>  x0=0; xf=5;
>>  y0=[1, 0, 0, 1];
>>  [x2,y2]=RK4('Example2',[x0,xf],y0,.05);
>>  plot(x2,y2)
>>  legend('y(1)','y(2)','y(3)','y(4)')
>>  xlabel('x'); ylabel('y(1), y(2)');
```

Note that if we wanted to use Matlab's `ode45`, we would simply type

```
                      Matlab
>>  [x1,y1]=ode45('Example1',[x0,xf],y0);
```

instead of the above two lines that use RK4.

Problems

There are three parts to each of these problems: (i) convert the equations in problems 1–17 to a system of first-order equations (do this part even if your software allows you to type in the equation without having to first convert it); (ii) use your software package and the given initial conditions to estimate the numerical solution at the value $x = 5.0$ for the step sizes $h = 1.0, 0.2, 0.01$ (this will require three different runs of the software code); (iii) then plot the solutions from the IC to $x = 5.0$ for the above step sizes and compare the solutions graphically (this will require three plots, i.e., one for each step size).

1. $y'' + y' + 2xy = x$, $y(-1) = 1, y'(-1) = 0$
2. $7y'' + 4y' - 3y = 0$, $y(0) = 0, y'(0) = 1$

3. $(x + 2)y'' + 3y = x, \ y(0) = 0, y'(0) = 4$

4. $y'' + x^2 y' + 12y = 0, \ y(0) = 0, y'(0) = 7$

5. $y'' + 4y' + 3\sin(y) = 1, \ y(0) = -1, y'(0) = \pi$

6. $(x^2 + 2)y'' + 3y^2 = e^x, \ y(0) = 1, y'(0) = 2$

7. $x^3 y'' + (\sin x)y' + y = 3 * x, \ y(1) = 1, y'(1) = \frac{3}{2}$

8. $y'' + 2y' + 10y = \sin x, \ y(\pi) = e^1, y'(\pi) = 1$

9. $y''' + xy''y' + y = e^x, \ y(1) = 1, y'(1) = 3, y''(1) = 1$

10. $x^2 y''' + xy''y' + \sin(y) = e^{-x}, y(1) = 1, y'(1) = 0, y''(1) = 1$

11. $8y''' + y'' = 0, \ y(0) = 1, y'(0) = 0, y''(0) = 2$

12. $3y''' + xy'' + xy' + y^3 = \ln x, \ y(1) = 0, y'(1) = 0, y''(1) = 1$

13. $2y''' - (y'')^2 + y' + xy = e^{-x^2}, \ y(0) = 1, y'(0) = 0, y''(0) = 0$

14. $y''' - 8\dfrac{1}{x^3}y = 0, \ y(1) = 1, y'(1) = 1, y''(1) = 0$

15. $y^{(4)} + 7y''' + 6y'' - 32y' + 32y = 0, \ y(1) = -1, y'(1) = 0, y''(1) = 0, y'''(1) = 1$

16. $(x+2)y^{(4)} - y''' + 6y'' + 3y' - 2y = 0, \ y(-1) = 3, y'(-1) = -1, y''(-1) = 0, y'''(-1) = 1$

17. $y^{(5)} - 9y' = 0, y(1) = 2, y'(1) = 0, y''(1) = 0, y'''(1) = 1, y^{(4)}(1) = 1$

3.6 Chapter 3: Additional Problems and Projects

ADDITIONAL PROBLEMS

In problems 1–5, determine whether the statement is true or false. If it is true, give reasons for your answer. If it is false, give a counterexample or other explanation of why it is false.

1. The Existence and Uniqueness theorem allows us to state that the solution to $xy'' - y = 0$ passing through the initial condition $y(0) = 0, y'(0) = 1$ will not be unique.

2. The Existence and Uniqueness theorem allows us to state that the solution to $y'' - y = \cosh x$ passing through the initial condition $y(0) = 0, y'(0) = 1$ will not be unique.

3. Reducing the order of a differential equation requires the knowledge of a solution.

4. All differential operators commute; that is, if $P(D)Q(D) = Q(D)P(D)$.

5. Only nth order *linear* differential equations can be rewritten as a system of first-order equations.

6. Determine whether $x^2y'' - 2y = 0$ is guaranteed to have a unique solution passing through $y(0) = 1, y'(0) = 1$. Then try to find constants c_1, c_2 such that $y = c_1x^2 + c_2\frac{1}{x}$ satisfies the IVP.

7. Determine whether $y'' - 2y' + y = \dfrac{e^x}{x}$ is guaranteed to have a unique solution passing through $y(1) = 0, y'(1) = 1$. Then try to find constants c_1, c_2 such that $y = c_1e^x + c_2xe^x + e^xx(\ln x - 1)$ (with $x > 0$) satisfies the IVP.

8. Find all values of (x, y) where Theorem 3.1.1 guarantees the existence of a unique solution passing through the given initial condition.
   a. $y'' + 3xy' + y = \frac{\sin x}{x-1}$
   b. $3y'' + 5x^2y' + y = \frac{4}{e^x}$
   c. $y'' + 7\frac{x^2}{y} = \sin x$
   d. $y'' - \ln|x|y' + 2y = \frac{\cos x}{x^2+1}$

In problems 9–15, determine whether the following sets of functions are linearly independent for all x (where the functions are defined).

9. $\{e^x, x+1\}$
10. $\{x^2, x^3\}$
11. $\{\sin x, \tan x\}$
12. $\{\ln x, x-1\}$
13. $\{\sqrt{x} - 4x, x, 2x\}$
14. $\{x^2 - 4x, 3x, x^2\}$
15. $\{2, \cos^2 x, \sin^2 x\}$

In problems 16–22, determine whether the given set forms a fundamental set of solutions to the differential equation (or equivalently, forms a basis for the solution space of the given differential equation) on the specified interval. Clearly state your reasons for your answers.

16. $\{x, 4+x\}$, $y'' + y' = 0$ for all x
17. $\{e^{-3x}, e^{3x}\}$, $y'' - 9y = 0$ for all x
18. $\{e^x, e^{-2x}\}$, $y'' - y' - 2y = 0$ for all x
19. $\{e^{-x}, e^{5x}\}$, $(D^2 - 6D + 5)(y) = 0$ for all x
20. $\{e^{2x}\sin 2x, e^{2x}\cos 2x\}$, $(D^2 - 4D + 8)(y) = 0$ for all x
21. $\{e^{3x}\sin x, e^{3x}\cos x\}$, $y'' - 6y' + 10y = 0$ for all x

22. $\{e^{2x}\sin x, e^{5x}\cos x, 3\}$, $y''' - 6y'' + 10y' = 0$ for all x

In problems 23–33, determine whether the given set of vectors forms a basis for the specified vector space.

23. $\left\{ \begin{pmatrix} 1 \\ -1 \end{pmatrix}, \begin{pmatrix} 3 \\ 4 \end{pmatrix} \right\}$, for $\mathbb{R}^2$

24. $\left\{ \begin{pmatrix} 2 \\ 1 \end{pmatrix}, \begin{pmatrix} 4 \\ -1 \end{pmatrix} \right\}$, for $\mathbb{R}^2$

25. $\left\{ \begin{pmatrix} 2 \\ 4 \end{pmatrix}, \begin{pmatrix} -4 \\ -8 \end{pmatrix} \right\}$, for $\mathbb{R}^2$

26. $\left\{ \begin{pmatrix} 5 \\ -2 \end{pmatrix}, \begin{pmatrix} -2 \\ -3 \end{pmatrix} \right\}$, for $\mathbb{R}^2$

27. $\left\{ \begin{pmatrix} 1 \\ 1 \\ 0 \end{pmatrix}, \begin{pmatrix} 2 \\ 0 \\ -1 \end{pmatrix}, \begin{pmatrix} 3 \\ 2 \\ 2 \end{pmatrix} \right\}$, for $\mathbb{R}^3$

28. $\left\{ \begin{pmatrix} 1 \\ 1 \\ 0 \end{pmatrix}, \begin{pmatrix} 2 \\ 0 \\ -1 \end{pmatrix}, \begin{pmatrix} 4 \\ 2 \\ -1 \end{pmatrix} \right\}$, for $\mathbb{R}^3$

29. $\left\{ \begin{pmatrix} 1 \\ 1 \\ 0 \end{pmatrix}, \begin{pmatrix} 3 \\ 0 \\ -1 \end{pmatrix}, \begin{pmatrix} 1 \\ -2 \\ -1 \end{pmatrix}, \begin{pmatrix} 1 \\ 3 \\ -1 \end{pmatrix} \right\}$, for $\mathbb{R}^3$

30. $\left\{ \begin{pmatrix} 1 \\ 1 \\ 0 \end{pmatrix}, \begin{pmatrix} 3 \\ 1 \\ -1 \end{pmatrix}, \begin{pmatrix} 1 \\ 0 \\ -1 \end{pmatrix}, \begin{pmatrix} 2 \\ 3 \\ 2 \end{pmatrix} \right\}$, for $\mathbb{R}^4$

31. $\left\{ \begin{pmatrix} 1 \\ 0 \\ -1 \\ 1 \end{pmatrix}, \begin{pmatrix} 0 \\ 0 \\ -1 \\ 1 \end{pmatrix}, \begin{pmatrix} 2 \\ 2 \\ 1 \\ 2 \end{pmatrix}, \begin{pmatrix} 1 \\ 0 \\ 0 \\ 0 \end{pmatrix} \right\}$, for $\mathbb{R}^4$

32. $\left\{ \begin{pmatrix} 1 \\ 1 \\ -1 \\ 1 \end{pmatrix}, \begin{pmatrix} 2 \\ 0 \\ -1 \\ -3 \end{pmatrix}, \begin{pmatrix} 2 \\ 2 \\ 1 \\ 2 \end{pmatrix}, \begin{pmatrix} 1 \\ 0 \\ 2 \\ 0 \end{pmatrix} \right\}$, for $\mathbb{R}^4$

33. $\left\{ \begin{pmatrix} 1 \\ 0 \\ -1 \\ 2 \end{pmatrix}, \begin{pmatrix} 0 \\ 0 \\ -1 \\ 1 \end{pmatrix}, \begin{pmatrix} 2 \\ 2 \\ 1 \\ 2 \end{pmatrix}, \begin{pmatrix} 1 \\ 0 \\ 0 \\ 0 \end{pmatrix} \right\}$, for $\mathbb{R}^4$

In Problems 34–38, use the given solution to reduce the order of the differential equation. Find a second solution.

34. $y'' + 4y' + 4y = 0$, solution $y = e^{-2x}$

35. $y'' - 4y' - 5y = 0$, solution $y = e^{-x}$

36. $y'' + 2y' - 35y = 0$, solution $y = e^{5x}$

37. $y'' + 10y' + 25y = 0$, solution $y = e^{-2x}$

38. $y'' + 10y' + 9y = 0$, solution $y = e^{-9x}$

39. Rewrite the following differential equations using operator notation and factor completely.
    a. $y'' + 9y' + 14y = 0$,   b. $y''' - 5y'' + 4y' = 0$

40. Rewrite the following differential equations using operator notation and factor completely.
    a. $y'' - y' - 12y = 0$,   b. $y''' - 16y' = 0$

41. Rewrite the following differential equations using operator notation and factor completely.
    a. $y'' + y' - 20y = 0$,   b. $y''' - 27y = 0$

In problems 42–46, (i) convert the equations to a system of first-order equations; (ii) use your software package and the given initial conditions to estimate the numerical solution at the value $x = 5.0$ for the step sizes $h = 0.1$; (iii) then plot the solutions from the IC to $x = 5.0$.

42. $y'' - 4y' + 3y = 0$, $y(0) = 0, y'(0) = 1$

43. $2y''' + y' = 0$, $y(0) = 1, y'(0) = 0, y''(0) = 2$

44. $x^2 y'' - 2y = 0$, $y(1) = 0, y'(1) = 0, y''(1) = 1$

45. $y'' + \sin y = 0$, $y(0) = 1, y'(0) = 1$

46. $y'' + (1 - y^2)y' + y = \cos x$, $y(-1) = 1, y'(-1) = 1$

PROJECT FOR CHAPTER 3

Project 1: A Subspace?

In Section 4.6, we will examine the annihilator method as a way of solving a nonhomogeneous ODE. In Example 13 of Section 3.2.3, we stated that $y''' - y'' - y = 0$ has fundamental set of solutions $\{\cos x, \sin x, e^x\}$. Using operator notation we can rewrite the ODE as

$$(D - 1)(D^2 + 1)(y) = 0$$

and we observed that $\{\cos x, \sin x\}$ is a subspace. Specifically, $\{\cos x, \sin x\}$ is the solution space of $(D^2 + 1)(y) = 0$.

We now want to examine the solution space of

$$(D^2 + 1)(y) = 3e^x. \tag{3.31}$$

Show that

$$\{\cos x, \sin x\}$$

is not a basis for the subspace of the solution space of (3.31). (Recall that
a basis of a solution space is the same as the fundamental set of solutions
for that solution space.) Show that $\frac{3}{2}e^x$ is a solution of equation (3.31).
Determine whether $\{e^x\}$ is the basis of some subspace of the solution space.
Then determine whether $\{\frac{3}{2}e^x + \cos x, \frac{3}{2}e^x + \sin x\}$ is a basis for the solution
space of (3.31). Now apply the operator $A(D) = D - 1$ to equation (3.31).
Show that $\{\cos x, \sin x, e^x\}$ forms a basis of the solution space of this new
equation.

Consider a general equation

$$P(D)(y) = f(x) \tag{3.32}$$

and suppose that $A(D)$ is an operator such that

$$A(D)P(D)(y) = 0. \tag{3.33}$$

What can you say about the solution spaces of equations (3.32) and (3.33)?
Do they relate to each other? Test your conjecture when $P(D) = D^2 - 4$,
$A(D) = D - 2$, and the basis for $A(D)P(D)(y) = 0$ is $\{e^{2x}, e^{-2x}, \frac{1}{5}e^x\}$.

Chapter 4

Techniques of Higher-Order Linear Equations

4.1 Homogeneous Equations with Constant Coefficients

Now we will consider how to actually find linearly independent solutions for

$$a_0 \frac{d^n y}{dx^n} + a_1 \frac{d^{n-1} y}{dx^{n-1}} + \ldots + a_{n-1} \frac{dy}{dx} + a_n\, y = 0, \tag{4.1}$$

a homogeneous linear equation with constant coefficients $a_0, a_1, \ldots, a_n$.

In the case of a second-order, constant-coefficient equation, we "guessed" a solution of the form e^{rx} (in Section 3.1.2) based on what happened in the case of the linear first-order homogeneous equation. For the nth order equation, we are looking for a function with the property

$$\frac{d^k}{dx^k}[f(x)] = cf(x) \text{ for all } x.$$

This is a property of the exponential function:

$$\frac{d^k}{dx^k} e^{rx} = r^k e^{rx}.$$

Thus we substitute $y = e^{rx}$ into (4.1) and evaluate the derivatives to obtain

$$a_0 r^n e^{rx} + a_1 r^{n-1} e^{rx} + \ldots + a_{n-1} r e^{rx} + a_n e^{rx} = 0,$$

which can be rewritten as

$$e^{rx}(a_0 r^n + a_1 r^{n-1} + \ldots + a_{n-1} r + a_n) = 0.$$

We again observe that $e^{rx} \neq 0$ so that our solutions are obtained from solving a polynomial in r:

$$a_0 r^n + a_1 r^{n-1} + \ldots + a_{n-1} r + a_n = 0.$$

This polynomial is known as the *characteristic equation* and its roots are known as *eigenvalues or characteristic values*. (Sometimes the equation is also referred to as the *auxiliary equation*.) Corresponding to each eigenvalue is the *eigenvector* $y = e^{rx}$, which is a solution to (4.1).

Now, recalling the *Fundamental Theorem of Algebra*, we know **an nth degree polynomial has n roots if we allow for the possibility of complex roots** (and we count repeated roots, too). There are three cases to consider based on these n roots. We may have roots that are real and distinct, roots that are repeated, or roots that are complex. In general, we may have a combination of all three. We will consider each of these cases.

4.1.1   Case 1: Distinct Real Roots

We begin with a theorem that characterizes this case.

THEOREM 4.1.1 *For the nth order homogeneous linear differential equation (4.1) with constant coefficients, if the characteristic equation has n distinct real roots $r_1, r_2, \ldots, r_n$, then $e^{r_1 x}, e^{r_2 x}, \cdots, e^{r_n x}$ are linearly independent solutions of (4.1). The general solution is given by*

$$y = c_1 e^{r_1 x} + c_2 e^{r_2 x} + \ldots + c_n e^{r_n x}$$

where $c_1, c_2, \ldots, c_n$ are arbitrary constants.

It should be clear that $e^{r_1 x}, e^{r_2 x}, \cdots, e^{r_n x}$ are solutions; see Figure 4.1. We can see they are linearly independent by calculating $W(0)$, that is, the Wronskian evaluated at $x = 0$. Since they are solutions, recall that $W(x)$ is either always 0 or never 0.

For instance, in the case of a third-order equation, we would have

$$W(0) = \begin{vmatrix} e^{r_1 x} & e^{r_2 x} & e^{r_3 x} \\ r_1 e^{r_1 x} & r_2 e^{r_2 x} & r_3 e^{r_3 x} \\ r_1^2 e^{r_1 x} & r_2^2 e^{r_2 x} & r_3^2 e^{r_3 x} \end{vmatrix}_{x=0} = \begin{vmatrix} 1 & 1 & 1 \\ r_1 & r_2 & r_3 \\ r_1^2 & r_2^2 & r_3^2 \end{vmatrix},$$

which can then be expanded and simplified as

$$W(0) = 1 \cdot (r_2 r_3^2 - r_2^2 r_3) - 1 \cdot (r_1 r_3^2 - r_1^2 r_3) + 1 \cdot (r_1 r_2^2 - r_1^2 r_2)$$

$$= r_2(r_3^2 - r_2 r_3 + r_1 r_2) - r_1(r_3^2 - r_1 r_3 + r_1 r_2)$$

$$= r_2(r_3^2 - r_2 r_3 + r_1 r_2) - r_1(r_3^2 - r_1 r_3 + r_1 r_2) - r_2 r_1 r_3 + r_1 r_2 r_3$$

$$= r_2(r_3^2 - r_2 r_3 + r_1 r_2 - r_1 r_3) - r_1(r_3^2 - r_1 r_3 + r_1 r_2 - r_2 r_3)$$

$$= r_2(r_3 - r_1)(r_3 - r_2) - r_1(r_3 - r_1)(r_3 - r_2)$$

$$= (r_2 - r_1)(r_3 - r_1)(r_3 - r_2), \tag{4.2}$$

which is never zero since the roots are distinct.

In general, we have for an nth order equation that

$$\prod_{i>j}(r_i - r_j)(-1)^{n+1},$$

where the symbol Π (capital Pi) denotes the product of all the factors that follow.

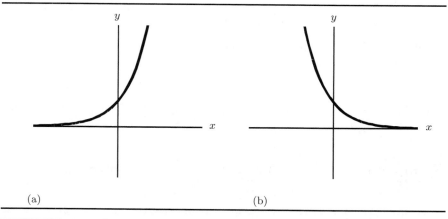

(a)   (b)

FIGURE 4.1: Basic shapes of solution curves for real, distinct roots: (a) e^{rx} for $r > 0$; (b) e^{rx} for $r < 0$.

Example 1: Consider the differential equation

$$\frac{d^2 y}{dx^2} - \frac{dy}{dx} - 6y = 0.$$

The characteristic equation is

$$r^2 - r - 6 = 0$$

which has the roots

$$r = -2 \quad \text{and} \quad r = 3.$$

Thus, e^{-2x} and e^{3x} are linearly independent solutions and the general solution is thus

$$y = c_1 e^{-2x} + c_2 e^{3x}.$$

Example 2: Consider the differential equation

$$\frac{d^3 y}{dx^3} - 4\frac{d^2 y}{dx^2} + \frac{dy}{dx} + 6y = 0.$$

The characteristic equation is

$$r^3 - 4r^2 + r + 6 = 0$$

for which we have some difficulty solving for r. In most cases, the characteristic equation will not factor nicely and numerical methods will perhaps be needed. It is also useful to employ a computer algebra system or recall some college algebra techniques.

In the present case we get lucky, of course, because $r = -1$ is a root. This means that $r + 1$ is a factor. The other factors can be found by division to be $r - 2$ and $r - 3$. That is

$$r^3 - 4r^2 + r + 6 = (r + 1)(r - 2)(r - 3).$$

Hence, e^{-x}, e^{2x}, e^{3x} are solutions. Thus

$$y = c_1 e^{-x} + c_2 e^{2x} + c_3 e^{3x}$$

is the general solution.

For this example, we could have also utilized the computer as follows:

Computer Code 4.1:   Finding all roots of a polynomial

Matlab, Maple, Mathematica

Matlab
```
>>  p=[1  -4  1  6] %these are the coeff of the poly
    %if a coefficient is 0, you must put 0 in its position
>>  roots(p)

>>  %If we have Symbolic Math Toolbox, then use
>>  solve('r^3 - 4*r^2 + r + 6 = 0')
```

Maple
```
>  eq1:=r^3 - 4*r^2 + r + 6 = 0
>  eq2:=solve(eq1,r);
```

Mathematica
```
Solve[r^3-4r^2+r+6 == 0,r]  (*polynomial entered from palette*)
```

4.1.2 Case 2: Repeated Real Roots

Many times we will have a root that is repeated and this requires a modification of the solution from case 1.

Example 3: Consider the differential equation

$$\frac{d^2y}{dx^2} - 6\frac{dy}{dx} + 9y = 0.$$

The characteristic equation is

$$r^2 - 6r + 9 = 0,$$

that is,

$$(r - 3)^2 = 0.$$

The roots are

$$r_1 = 3 \quad \text{and} \quad r_2 = 3 \text{ (double root!)}.$$

So we have the solution e^{3x} corresponding to $r_1 = 3$, and the solution e^{3x} corresponding to $r_2 = 3$. Clearly, e^{3x} is not linearly independent of e^{3x}. So we have a small problem.

Since we already know one solution is e^{3x}, we can reduce the order of the equation (see Section 3.3). Let

$$y = e^{3x}v.$$

Thus

$$\frac{dy}{dx} = e^{3x}\frac{dv}{dx} + 3e^{3x}v$$

and

$$\frac{d^2y}{dx^2} = e^{3x}\frac{d^2v}{dx^2} + 6e^{3x}\frac{dv}{dx} + 9e^{3x}v.$$

Substituting into the original differential equation and simplifying gives

$$e^{3x}\frac{d^2v}{dx^2} = 0.$$

Letting $w = \frac{dv}{dx}$, we have

$$e^{3x}\frac{dw}{dx} = 0.$$

This gives that $dw/dx = 0$ and thus $w = c$. We can let $c = 1$ so

$$v = x + c_0$$

and thus

$$v(x)e^{3x} = (x + c_0)e^{3x}$$

is a solution to the second-order equation. Now we know that $(x + c_0)e^{3x}$ and e^{3x} are linearly independent (check the Wronskian!), so taking $c_0 = 0$, we have corresponding to the double root 3, two linearly independent solutions

$$e^{3x} \quad \text{and} \quad xe^{3x}.$$

Thus, the general solution is

$$y = c_1 e^{3x} + c_2 x e^{3x}.$$

Generalizing this example, if a second-order homogeneous linear equation with constant coefficients has r as a double root to its characteristic equation, then e^{rx} and xe^{rx} are the corresponding linearly independent equations; see Figure 4.2. Extending this idea further, we have the following theorem.

THEOREM 4.1.2 *Consider the nth order homogeneous linear differential equation (4.1) with constant coefficients.*
1. *If the characteristic equation has the real root r occurring k times, then the part of the general solution corresponding to this k-fold repeated root is*

$$(c_1 + c_2 x + c_3 x^2 + \ldots + c_k x^{k-1})e^{rx}.$$

2. *If, further, the remaining roots of the characteristic equation are the distinct real numbers $r_{k+1}, r_{k+2}, \ldots, r_n$, then the general solution is*

$$y = (c_1 + c_2 x + c_3 x^2 + \ldots + c_k x^{k-1})e^{rx} + c_{k+1}e^{r_{k+1}x} + \ldots + c_n e^{r_n x}.$$

3. *If, however, any of the remaining roots are also repeated, then the parts of the general solution to (4.1) corresponding to each of these other repeated roots are expressions similar to that corresponding to r in part **1.***

Example 4: Solve the equation

$$\frac{d^3 y}{dx^3} - 4\frac{d^2 y}{dx^2} - 3\frac{dy}{dx} + 18y = 0.$$

The characteristic equation is

$$r^3 - 4r^2 - 3r + 18 = 0$$

which has roots $3, 3, -2$, (check it!). Theorems 4.1.1 and 4.1.2 tell us that three linearly independent solutions are e^{3x}, xe^{3x}, and e^{-2x}.

The general solution is thus

$$y = c_1 e^{3x} + c_2 x e^{3x} + c_3 e^{-2x}.$$

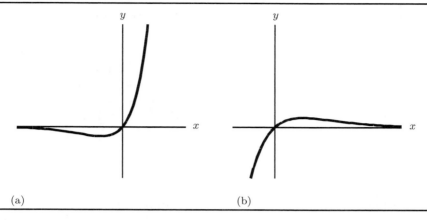

FIGURE 4.2: Basic shapes of solution curves for real, repeated roots: (a) xe^{rx} for $r > 0$; (b) xe^{rx} for $r < 0$. The figures here are to be combined with the corresponding part of Figure 4.1. For example, a second-order equation with $r < 0$ will be a linear combination of Figures 4.1b and 4.2b.

Example 5: Suppose a sixth order homogeneous linear differential equation with constant coefficients had the following roots of the characteristic equation:

$$-1, -1, 2, 3, 3, 3.$$

Then six linearly independent solutions are $e^{-x}, xe^{-x}, e^{2x}, e^{3x}, xe^{3x}, x^2 e^{3x}$ and the general solution to the differential equation is

$$y = c_1 e^{-x} + c_2 xe^{-x} + c_3 e^{2x} + c_4 e^{3x} + c_5 xe^{3x} + c_6 x^2 e^{3x}.$$

4.1.3 Case 3: Complex Roots (Non-Real)

Now suppose that the characteristic equation has $a + bi$, a complex number, as a root. (Here, a and b are real numbers, $b \neq 0$ and $i^2 = -1$.)

Because complex roots of a polynomial with real coefficients always come in conjugate pairs, we know that $a - bi$ is also a root. In obtaining the characteristic equation, we observed that e^{rx} solves the differential equation but *did not* require that r be real. Thus, $e^{(a+ib)x}$ and $e^{(a-ib)x}$ are linearly independent solutions and the corresponding part of the general solution is

$$k_1 e^{(a+bi)x} + k_2 e^{(a-bi)x} \tag{4.3}$$

where k_1 and k_2 are arbitrary (real) constants.

Note that the solutions defined by $e^{(a+bi)x}$ and $e^{(a-bi)x}$ are complex functions of the real variable x. We are interested in real linearly independent solutions.

A useful and interesting fact from the theory of complex variables is *Euler's formula*:

$$e^{i\theta} = \cos\theta + i\sin\theta \qquad (4.4)$$

which is valid for all θ. This expression can be obtained in many ways and we refer the interested reader to the exercises for two such excursions. Before we return to our discussion of complex eigenvalues, we give one remarkable application of Euler's formula; namely, letting $\theta = \pi$ we have

$$e^{i\pi} = \cos\pi + i\sin\pi.$$

That is, $e^{i\pi} = -1$ or

$$e^{i\pi} + 1 = 0.$$

This astounding expression relates the five most famous (and useful) constants $0, 1, e, \pi$, and i. Now, let's get back on track.

If we apply Euler's formula to equation (4.3), we have

$$k_1 e^{(a+bi)x} + k_2 e^{(a-bi)x} = e^{ax}\left(k_1 e^{ibx} + k_2 e^{-ibx}\right)$$

$$= e^{ax}\left(k_1(\cos bx + i\sin bx) + k_2(\cos bx - i\sin bx)\right)$$

$$= e^{ax}\left((k_1 + k_2)\cos bx + i(k_1 - k_2)\sin bx\right)$$

$$= e^{ax}\left(c_1\sin bx + c_2\cos bx\right),$$

where $c_1 = k_1 + k_2$ and $c_2 = i(k_1 - k_2)$. Thus, corresponding to the roots $a \pm bi$ is the solution

$$e^{ax}\left(c_1\sin bx + c_2\cos bx\right),$$

where $e^{ax}\sin bx$ and $e^{ax}\cos bx$ are two, linearly independent, real-valued solutions; see Figure 4.3. We have the following theorem.

THEOREM 4.1.3 *Consider the nth order homogeneous linear differential equation (4.1) with constant coefficients.*
1. *If the characteristic equation has conjugate complex roots $a + bi$ and $a - bi$, neither repeated, then $e^{ax}\sin bx$ and $e^{ax}\cos bx$ are linearly independent solutions. The corresponding part of the general solution may be written as*

$$e^{ax}\left(c_1\sin bx + c_2\cos bx\right).$$

2. *If, however, $a + bi$ and $a - bi$ are each roots of multiplicity k of the characteristic equation, then the corresponding part of the general solution may be written*

$$e^{ax}\left(c_1 + c_2 x + c_3 x^2 + \ldots + c_k x^{k-1}\right)\sin bx$$
$$+ e^{ax}\left(c_{k+1} + c_{k+2} x + c_{k+3} x^2 + \ldots + c_{2k} x^{k-1}\right)\cos bx. \qquad (4.5)$$

Example 6: Solve the equation $\dfrac{d^2y}{dx^2} + 9y = 0$.

The characteristic equation is

$$r^2 + 9 = 0$$

which has solution $r = \pm 3i$, so that $a = 0$ and $b = 3$. Then $\sin 3x$ and $\cos 3x$ are two linearly independent solutions and the general solution is

$$y = c_1 \sin 3x + c_2 \cos 3x.$$

Example 7: Solve the equation

$$(D^2 - 6D + 25)(y) = 0.$$

Try not to be confused by the use of the operator notation—it is actually easier to write the characteristic equation. The characteristic equation is

$$r^2 - 6r + 25 = 0$$

which has roots $r = 3 \pm 4i$. Two linearly independent solutions are $e^{3x} \sin 4x$ and $e^{3x} \cos 4x$ and thus

$$y = e^{3x}(c_1 \sin 4x + c_2 \cos 4x)$$

is the general solution.

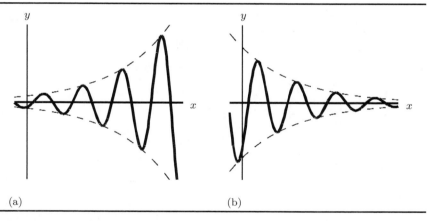

FIGURE 4.3: Basic shapes of solution curves for complex roots: (a) linear combination of $\{e^{ax}\cos(bx), e^{ax}\sin(bx)\}$ for $a > 0$; (b) linear combination of $\{e^{ax}\cos(bx), e^{ax}\sin(bx)\}$ for $a < 0$. Notice the *envelope* $\pm e^{ax}$ that bounds the solution in both parts.

Example 8: Solve the equation

$$\frac{d^4y}{dx^4} - 4\frac{d^3y}{dx^3} + 14\frac{d^2y}{dx^2} - 20\frac{dy}{dx} + 25y = 0.$$

The characteristic equation is

$$r^4 - 4r^3 + 14r^2 - 20r + 25 = 0.$$

This is nontrivial to solve by hand; however, in our respective computer programs we could type

Computer Code 4.2:   Finding all roots of a polynomial (complex roots occur here)

Matlab, Maple, Mathematica

```
                              Matlab
 >> p=[1  -4  14  -20  25] %these are the coeff of the poly
    %if a coefficient is 0, you must put 0 in its position
 >> roots(p)

 >> %If we have Symbolic Math Toolbox, then use
 >> solve('r^4 -4*r^3 + 14*r^2 - 20*r + 25 = 0')
```

```
                              Maple
 > eq1:=r^4 -4*r^3 + 14*r^2 - 20*r + 25 = 0;
 > eq2:=solve(eq1,r);
```

```
                          Mathematica
 Solve[r^4 - 4r^3 + 14r^2 - 20r + 25 == 0, r]  (*entered from palette*)
```

The roots are seen to be

$$1 + 2i, \ 1 - 2i, \ 1 + 2i, \ \text{and} \ 1 - 2i.$$

These are double complex roots, and linearly independent solutions are $e^x \sin 2x$, $xe^x \sin 2x$, $e^x \cos 2x$, and $xe^x \cos 2x$. The general solution is thus

$$y = e^x \left[(c_1 + c_2 x)\sin 2x + (c_3 + c_4 x)\cos 2x\right].$$

We have yet to consider any initial value problems, but these are very straightforward as can be seen in this next example.

Example 9: Solve the initial value problem

$$\frac{d^2y}{dx^2} - 6\frac{dy}{dx} + 25y = 0,$$

with the conditions $y(0) = -3$ and $y'(0) = -1$. Plot the solution on the interval $-1 \leq x \leq 1$.

In Example 8 above, we obtained the general solution as

$$y = e^{3x}(c_1 \sin 4x + c_2 \cos 4x).$$

We observe that $y(0) = -3$ implies that $c_2 = -3$ and $y'(0) = -1$ gives

$$4c_1 + 3c_2 = -1$$

so that $c_1 = 2$. Thus the solution is

$$y = e^{3x}(2 \sin 4x - 3 \cos 4x).$$

We give a brief reminder on plotting this solution.

Computer Code 4.3:   Basic plot with axes labeled, title inserted

Matlab, Maple, Mathematica

Matlab
```
>>  x=-1:.05:1;
>>  y=exp(3*x).*(2*sin(4*x)-3*cos(4*x));
>>  plot(x,y,'b:')
>>  xlabel('x');ylabel('y');
>>  title('Soln of y\prime\prime-6y\prime+25y=0');
```

Maple
```
> eq1:=exp(3*x)*(2*sin(4*x)-3*cos(4*x));
> plot(eq1,x=-1..1,color=blue,linestyle=2,labels=[x, y],
  title="Soln of y\"-6y'+25y=0");
```

Mathematica
```
f[x_] = e^{3x}(2Sin[4x] - 3Cos[4x])
Plot[f[x],{x,-1,1},PlotStyle→ Blue, AxesLabel→ {"x","y"}]
```

4.1.4   Second-Order Linear Homogeneous Equations with Constant Coefficients

The theory we have developed in our method of solving nth order linear homogeneous equations with constant coefficients relies upon finding the roots of the corresponding nth degree characteristic polynomial. Since polynomials of degree two can be explicitly solved, all solutions of the second-order linear homogeneous equation with constant coefficients

$$a_0 \frac{d^2y}{dx^2} + a_1 \frac{dy}{dx} + a_2\, y = 0 \qquad (4.6)$$

are obtained from the roots of the quadratic equation

$$a_0 r^2 + a_1 r + a_2 = 0.$$

These roots are easily obtained by the quadratic formula as

$$r_{1,2} = \frac{-a_1 \pm \sqrt{a_1^2 - 4a_0 a_2}}{2a_0} \qquad (4.7)$$

so that the general solution to (4.6) is

$$y(x) = \begin{cases} c_1 e^{r_1 x} + c_2 e^{r_2 x} & \text{if } a_1^2 > 4a_0 a_2 \\ (c_1 + c_2 x)e^{-a_1 x/2a_0} & \text{if } a_1^2 = 4a_0 a_2 \\ e^{-a_1 x/2a_0}\left[c_1 \cos(\omega x) + c_2 \sin(\omega x)\right] & \text{if } a_1^2 < 4a_0 a_2, \end{cases} \qquad (4.8)$$

where c_1 and c_2 are arbitrary constants and $\omega = \dfrac{1}{a_0}\sqrt{a_0 a_2 - \dfrac{a_1^2}{4}}$.

Example 10: Solve the equation

$$\frac{d^2y}{dx^2} + 8\frac{dy}{dx} + 15y = 0.$$

Since the characteristic equation is

$$r^2 + 8r + 15 = 0,$$

we see that it has two real roots, which are

$$r_1 = -3 \text{ and } r_2 = -5.$$

Thus the general solution is

$$y(x) = c_1 e^{-3x} + c_2 e^{-5x}.$$

Example 11: Solve the equation

$$\frac{d^2y}{dx^2} + 2\frac{dy}{dx} + 2y = 0.$$

Since the characteristic equation is

$$r^2 + 2r + 2 = 0,$$

we see that it has two complex roots. Applying equation (4.8) gives

$$\frac{1}{a_0}\sqrt{a_0 a_2 - \frac{a_1^2}{4}} = 1.$$

Thus the general solution is

$$y(x) = e^x \left[c_1 \cos x + c_2 \sin x \right].$$

Example 12: Solve the equation

$$\frac{d^2 y}{dx^2} + 6\frac{dy}{dx} + 9y = 0.$$

Since the characteristic equation is

$$r^2 + 6r + 9 = 0,$$

we see that it has a repeated root of $r = -3$. The solution is thus

$$y(x) = (c_1 + c_2 x)\left(e^{-3x}\right).$$

Problems

In problems 1–31, find a general solution of each equation. If an initial condition is given, use your computer software package to plot the explicit solution you obtain. Do not use the numerical solving methods of Section 3.5 to calculate and plot the solution.

1. $y'' + 8y' + 12y = 0$
2. $7y'' + 4y' - 3y = 0$
3. $8y''' + y'' = 0$
4. $y'' - 2y' = 0$
5. $y'' + y' - 12y = 0$, $y(0) = 0$, $y'(0) = 7$
6. $y'' + y' - 2y = 0$, $y(0) = 1$, $y'(0) = 1$
7. $y'' + 4y' + 4y = 0$, $y(0) = 0$, $y'(0) = 1$
8. $y'' + 4y' + 3y = 0$, $y(1) = 1$, $y'(1) = 2$
9. $4y'' + 4y' + y = 0$
10. $2y'' + 5y' + 2y = 0$, $y(0) = \pi$, $y'(0) = 1$
11. $(D^2 - 4D + 5)(y) = 0$
12. $(D^2 + 2D + 10)(y) = 0$, $y(-\pi) = 0$, $y'(-\pi) = 1$

13. $(D^2 + 1)^2(y) = 0$

14. $(D^2 + 4)(y) = 0$, $y(\pi) = 1$, $y'(\pi) = 1$

15. $y''' - 8y = 0$, $y(0) = 0$, $y'(0) = 1$, $y''(0) = 0$

16. $y^{(4)} - y = 0$

17. $(D^4 - 16)(y) = 0$

18. $(D^2 - 2D + 1)(y) = 0$

19. $y^{(4)} + 7y''' + 6y'' - 32y' - 32y = 0$

20. $D^3(D^2 - 6D + 9)(y) = 0$

21. $D^2(D - 1)^2(D^2 + 1)(y) = 0$

22. $y^{(5)} - 10y''' + 9y' = 0$

23. $y^{(4)} + 2y'' + y = 0$

24. $(D^2 + 4)^2(y) = 0$

25. $y''' - 3y'' + 3y' - y = 0$

26. $y''' - y'' - y' + y = 0$

27. $y^{(4)} - 5y'' + 4y = 0$

28. $D(D^2 + 4)(D^2 - 2D + 1)(y) = 0$

29. $y^{(5)} + 8y''' + 16y' = 0$

30. $y^{(4)} + 16y = 0$

31. $(D^4 + 64)(y) = 0$

32. Find a differential equation for which the characteristic equation has roots with corresponding multiplicities: $r_1 = -2$, $k_1 = 1$; $r_2 = 0$, $k_2 = 2$.

33. Find a differential equation for which the characteristic equation has roots with corresponding multiplicities $r_1 = 3i$, $k_1 = 2$; $r_{2,3} = 1 \pm i$, $k_2 = 1$.

34. Find a differential equation for which the characteristic equation has roots with corresponding multiplicities $r_1 = 0$, $k_1 = 4$; $r_{2,3} = 2 \pm 3i$, $k_2 = 3$.

35. Find a differential equation for which the characteristic equation has roots with corresponding multiplicities $r_1 = 2 \pm 3i$, $k_1 = 2$; $r_3 = -5$, $k_3 = 1$; $r_4 = 2$, $k_4 = 3$.

36. Show that a general solution of the differential equation

$$ay'' + 2by' + cy = 0$$

where

$$b^2 - ac > 0$$

can be written as

$$y = e^{-bx/a}\left[c_1 \cosh\left(\frac{x\sqrt{b^2 - ac}}{a}\right) + c_2 \sinh\left(\frac{x\sqrt{b^2 - ac}}{a}\right)\right].$$

37. In this problem, we consider a power series approach to obtaining Euler's formula.

    Hint: Use the Taylor series for e^x about the origin to "formally" replace x by $i\theta$. Recalling that $i = \sqrt{-1}$, simplify to obtain

    $$e^{i\theta} = 1 - \frac{\theta^2}{2!} + \frac{\theta^4}{4!} + \ldots + i\left(\theta - \frac{\theta^3}{3!} + \ldots\right) \qquad (4.9)$$

38. As another method of obtaining Euler's formula, consider

    $$z_1 = e^{i\theta} \text{ and } z_2 = \cos\theta + i\sin\theta.$$

    Show that these two expressions are both solutions of the complex-valued initial value problem

    $$\frac{dz}{d\theta} = iz \text{ with } z(0) = 1$$

    and give reasons why you can conclude that they must be identical.

4.2   A Mass on a Spring

We will now consider the motion of a mass attached to a spring by taking into account the resistance of the medium and the possible external forces acting on the mass. We note that although we have used $y(x)$ as our variable for much of the book, we switch to $x(t)$ for this section in order to stay consistent with the notation used in much of engineering and physics; see Figure 4.4. We need a fact from physics to begin our in-depth examination.

Hooke's law: It is experimentally observed that the magnitude of the force needed to produce a certain elongation of a spring is directly proportional to the amount of the elongation, provided the elongation is not too great. That is,

$$|F| = ks \qquad (4.10)$$

where $|F|$ is the magnitude of the force F, s is the amount of elongation, and k is a constant of proportionality, called the spring constant, which depends upon the characteristics of the spring.

Example 1: If a 30-lb weight stretches a spring 2 ft, then Hooke's law gives

$$30 = 2k$$

or

$$k = 15 \text{ lb/ft}.$$

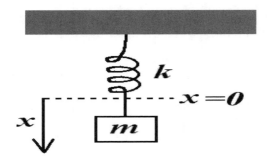

FIGURE 4.4:   Mass on a spring.

When a mass is hung upon a spring that has a spring constant k and produces elongation s, the force F of the mass upon the spring has magnitude ks. At the same time, the spring exerts a force upon the mass called the *restoring force*. This restoring force is equal in magnitude, but opposite in sign to F and hence is $-ks$.

Formulation of the problem: Suppose the spring has natural (unstretched) length L. The mass m is attached to the spring and the spring stretches to its equilibrium position, stretching the spring an amount ℓ. The stretched length is $L + \ell$.

   For convenience, we assume the origin of a coordinate system at this equilibrium position. We also assume the positive direction as down. Thus, the value x of this coordinate system is positive, zero, or negative, depending upon whether the mass is below, at, or above equilibrium, respectively.

Forces acting upon the mass: In this coordinate system, forces tending to pull the mass downward are positive, while those tending to pull it upward are negative. The forces are:

1. F_1, **the force of gravity.** This is given as

$$F_1 = mg,$$

where m is the mass and g is gravity. F_1 is positive since it acts downward.

2. F_2, **the restoring force of the spring.** Since $x + \ell$ is the total elongation of the spring, by Hooke's law, the magnitude of this force is $k(x + \ell)$. When the mass is <u>below</u> the end of the unstretched spring, the force acts upward and is thus <u>negative</u>. Since $x + \ell$ is positive, we have in this case

$$F_2 = -k(x + \ell).$$

Similarly, if the mass is above the end of the unstretched spring, the spring is acting downward and thus this force is <u>positive</u>. However, $x + \ell$ is <u>negative</u>, so that in this case we again have

$$F_2 = -k(x + \ell).$$

At the equilibrium point, the force of gravity is equal to the restoring force, so that

$$-mg = -k(0 + \ell)$$

or

$$mg = k\ell.$$

Hence, we have the equation

$$F_2 = -kx - mg.$$

3. F_3, the resisting force of the medium. This force is also known as the *damping force*. The magnitude of this force is not known exactly; however, it is known that for small velocities, it is approximately

$$|F_3| = b \left| \frac{dx}{dt} \right|$$

where $b > 0$ a constant, which is known as the *damping constant*. Note that when the mass is moving <u>downward</u>, F_3 acts in the <u>upward</u> direction so that $F_3 < 0$. Here, moving downward implies x increases so that $dx/dt > 0$. Thus, when moving downward

$$F_3 = -b\frac{dx}{dt}.$$

Similarly, when moving upward x decreases, so that $dx/dt < 0$. Thus, in this case, we again have

$$F_3 = -b\frac{dx}{dt}.$$

4. F_4, any external forces that act upon the mass. We will let the resultant of all such forces at time t be $F(t)$ and write

$$F_4 = F(t).$$

Now we can apply Newton's second law:

$$F = ma$$

where

$$F = F_1 + F_2 + F_3 + F_4,$$

the sum total of the forces involved. We thus have

$$m\frac{d^2x}{dt^2} = mg - kx - mg - b\frac{dx}{dt} + F(t)$$

or

$$m\frac{d^2x}{dt^2} + b\frac{dx}{dt} + kx = F(t). \tag{4.11}$$

This is the differential equation A for the motion of the mass on a spring. It is a nonhomogeneous second-order linear differential equation with constant coefficients.

If $b = 0$, the motion is called *undamped*; otherwise, it is called *damped*. If $F(t) = 0$ for all t, the motion is called *free*; otherwise, it is called *forced*.

4.2.1 Undamped Oscillations

In the case of free undamped motion, both $b = 0$ and $F(t) = 0$ for all t. This gives the differential equation

$$m\frac{d^2x}{dt^2} + kx = 0,$$

where $m > 0$ is the mass and $k > 0$ is the spring constant. Thus, if we divide through by m, we have

$$\frac{d^2x}{dt^2} + \frac{k}{m}x = 0.$$

Since $m, k > 0$, we let $\omega_0^2 = \frac{k}{m}$ and substitution yields

$$\frac{d^2x}{dt^2} + \omega_0^2 x = 0.$$

The corresponding characteristic equation is

$$r^2 + \omega_0^2 = 0$$

which has roots

$$r = \pm\omega_0 i.$$

The general solution is thus

$$x(t) = c_1 \sin\omega_0 t + c_2 \cos\omega_0 t.$$

Now we suppose that the mass was initially displaced a distance x_0 from the equilibrium with an initial velocity v_0. That is,

$$x(0) = x_0 \text{ and } x'(0) = v_0.$$

Using these initial conditions, we find that

$$c_1 = \frac{v_0}{\omega_0} \text{ and } c_2 = x_0$$

so that

$$x(t) = \frac{v_0}{\omega_0}\sin\omega_0 t + x_0\cos\omega_0 t. \tag{4.12}$$

Although equation (4.12) completely describes the motion of the spring at any given time, it is often easier to picture the solution curves if we write the solution in an alternate form (*amplitude-phase form*), as

$$x(t) = A\cos(\omega_0 t - \phi),$$

where A will be an *amplitude* and ϕ the *phase constant*. The quantity $(\omega_0 t - \phi)$ is called the *phase of the motion*.

To accomplish this we first observe that the maximum amplitude of (4.12) is given by

$$A = \sqrt{\left(\frac{v_0}{\omega_0}\right)^2 + x_0^2}. \tag{4.13}$$

Each of the two terms contributes to this A. To see this contribution, we consider the fractions

$$\frac{x_0}{A} \text{ and } \frac{v_0/\omega_0}{A}.$$

These quantities are both between 0 and 1 (Why?). They also satisfy the relation

$$\left(\frac{x_0}{A}\right)^2 + \left(\frac{v_0/\omega_0}{A}\right)^2 = 1,$$

which clues us in to set

$$\cos\phi = \frac{x_0}{A} \text{ and } \sin\phi = \frac{v_0/\omega_0}{A}$$

for some ϕ. Rewriting these two expressions and substituting in for x_0 and v_0/ω_0 gives

$$x(t) = A\cos\omega_0 t \cos\phi + A\sin\omega_0 t \sin\phi.$$

Recalling the trig identity

$$\cos(a \pm b) = \cos a \cos b \mp \sin a \sin b$$

gives us that

$$x(t) = A\cos(\omega_0 t - \phi)$$
$$= \sqrt{\left(\frac{v_0}{\omega_0}\right)^2 + x_0^2}\cos(\omega_0 t - \phi). \tag{4.14}$$

Note that we could have obtained a similar expression

$$x(t) = A\sin(\omega_0 t + \phi) \tag{4.15}$$

if we had set

$$\sin\phi = \frac{x_0}{A} \text{ and } \cos\phi = \frac{v_0/\omega_0}{A},$$

and used the trig identity $\sin(a \pm b) = \sin a \cos b \pm \cos a \sin b$. Both are equally correct.

This equation gives the displacement, x, of the mass from the equilibrium as a function of time t. The motion described by $x(t)$ is called *simple harmonic motion*. The constant A given by equation (4.13) is the amplitude of the motion and gives the maximum (positive) displacement. The motion is periodic, with the mass oscillating between $x = -A$ and $x = A$. We have $x = A$ if and only if

$$\sqrt{\frac{k}{m}}t - \phi = \pm 2n\pi.$$

So the maximum displacement occurs if and only if

$$t = \sqrt{\frac{m}{k}}(2n\pi + \phi) > 0.$$

The time interval between maxima is called the *period*, T. Thus

$$T = \frac{2\pi}{\sqrt{\frac{k}{m}}} = \frac{2\pi}{\omega_0}.$$

Just as the spring constant is inherent to each spring, the quantity

$$\omega_0 = \sqrt{\frac{k}{m}} \tag{4.16}$$

is inherent to the mass-spring system and is often called the *"natural" angular frequency*.

Example 2: An 8-lb weight is placed upon the lower end of a coil spring suspended from the ceiling. The weight comes to rest in its equilibrium position, thereby stretching the spring 6 in. The weight is then pulled down 3 in. below its equilibrium position and released at $t = 0$ with an initial velocity of 1 ft/sec, directed downward. Neglecting the resistance of the medium and assuming that no external forces are present, determine the amplitude, period and frequency of the resulting motion.

This is an example of free undamped motion. Since the 8-lb weight stretches the spring 6 in. $= 1/2$ ft, Hooke's law gives $8 = k(1/2)$ or

$$k = 16 \text{ lb/ft}.$$

Further, *mass = weight/gravity* so

$$m = \frac{8}{32} \text{ slugs}.$$

Thus, the differential equation describing free undamped motion becomes

$$\frac{8}{32}\frac{d^2x}{dt^2} + 16x = 0$$

or

$$\frac{d^2x}{dt^2} + 64x = 0.$$

Now since the weight was released downward with an initial velocity of 1 ft/sec, from a position 3 in. or 1/4 ft below equilibrium we have

$$x(0) = \frac{1}{4} \text{ and } x'(0) = 1.$$

The characteristic polynomial is $r^2 + 64 = 0$ so that $r = \pm 8i$. This gives

$$x(t) = c_1 \sin 8t + c_2 \cos 8t.$$

Using this with the initial conditions gives

$$c_1 = \frac{1}{8} \text{ and } c_2 = \frac{1}{4}.$$

Hence,

$$x(t) = \frac{1}{8} \sin 8t + \frac{1}{4} \cos 8t.$$

A graph of this displacement is shown in Figure 4.5.

To obtain amplitude-phase form we have

$$A = \sqrt{\left(\frac{1}{8}\right)^2 + \left(\frac{1}{4}\right)^2} = \frac{\sqrt{5}}{8}$$

as the amplitude. Thus

$$x(t) = \frac{\sqrt{5}}{8} \cos(8t + \phi).$$

The period is

$$\frac{2\pi}{8} = \frac{\pi}{4} \text{ sec}$$

and the frequency is $4/\pi$. We can find ϕ by solving

$$\cos \phi = \frac{x_0}{A} = \frac{2\sqrt{5}}{5} \text{ and } \sin \phi = \frac{v_0}{\omega_0 A} = \frac{\sqrt{5}}{5}$$

for ϕ. From these equations we find

$$\phi \approx 0.46 \text{ rad.}$$

spring displacement

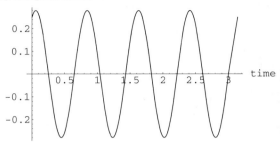

FIGURE 4.5:   Displacement of the spring in the example.

4.2.2   Damped Oscillations

When we allow for friction in our system, some very useful and important concepts arise. In the case of this damped motion, we again consider

$$m\frac{d^2x}{dt^2} + b\frac{dx}{dt} + kx = F(t), \tag{4.17}$$

now with $b \neq 0$ but still with $F(t) = 0$. We again have a homogeneous constant coefficient equation and thus we can solve it. Calculating the roots of the characteristic equation gives

$$r = \frac{-b \pm \sqrt{b^2 - 4mk}}{2m} \tag{4.18}$$

$$= \frac{-b}{2m} \pm \sqrt{\left(\frac{b}{2m}\right)^2 - \frac{k}{m}} \tag{4.19}$$

Recalling the three different cases that arose in our study of constant coefficient equations, we see that each case can be realized here depending on the value of $b^2 - 4mk$.

Case 1: $b^2 > 4mk$
Then the roots are real and distinct. If we denote the roots r_1 and r_2, the general solution can be written as

$$x(t) = c_1 e^{r_1 t} + c_2 e^{r_2 t}. \tag{4.20}$$

Note that $r_1 < 0$ *and* $r_2 < 0$ (check this!). Thus the solution will approach 0 as $t \to \infty$. Depending on the initial condition, it may cross the rest position at most one time but the spring motion will not oscillate as it dies out.[1] This case is called *overdamped* motion; see Figure 4.6.

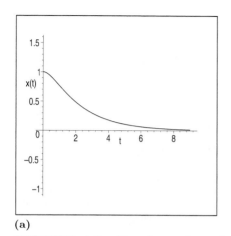

(a)

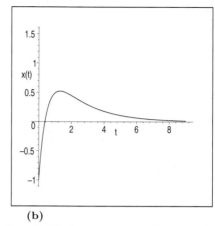

(b)

FIGURE 4.6: Overdamped motion: $x'' + (5/2)x' + x = 0$. The initial conditions are again $x(0) = 1$, $x'(0) = 0$ in **(a)** and $x(0) = -1$, $x'(0) = 4$ in **(b)**. Note that we have overshoot in **(b)** as the initial condition was chosen such that the mass passes through its rest position exactly one time.

Case 2: $b^2 = 4mk$
Then the roots are real and equal. If we denote the root r, the general solution can be written as

$$x(t) = c_1 e^{rt} + c_2 t e^{rt}. \tag{4.21}$$

Note that $r < 0$ and the solution again approaches 0 as $t \to \infty$. This last statement is obvious for the first term $c_1 e^{rt}$. The reader can check (for instance, using L'Hospital's rule) that the second term also approaches zero. Depending on the initial condition, the mass may cross the rest position at most one time before dying out without oscillation.[2] This case is called *critically damped* motion; see Figure 4.7. It also represents the motion that will die off to zero fastest.

Case 3: $b^2 < 4mk$
Then the roots are complex conjugates. If we denote the roots as $-\alpha \pm i\beta$ (that is, we set $\alpha = (b/2m)$ and $\beta = \sqrt{(k/m) - (b/2m)^2}$ and note that $\alpha > 0$

---

[1]You are asked to show in Exercise 21 that this condition is given by $|v_0| > |bx_0/(2m)|$, where v_0 is the initial velocity of the mass and x_0 is the initial displacement.
[2]As before, this condition is given by $|v_0| > |bx_0/(2m)|$, where v_0 is the initial velocity of the mass and x_0 is the initial displacement.

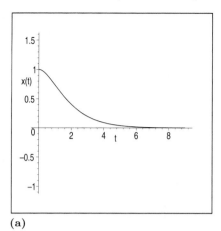
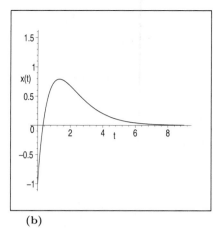

(a)   (b)

FIGURE 4.7:   Critically damped motion: $x'' + 2x' + x = 0$. The initial conditions are $x(0) = 1$, $x'(0) = 0$ in **(a)** and are $x(0) = -1$, $x'(0) = 4$ in **(b)**. Note that we again have overshoot in **(b)**. We also note that the motion dies off faster than in the overdamped case, even though both systems show **no oscillations**.

and $\beta > 0$), then the general solution can be written as

$$x(t) = c_1 e^{-\alpha t} \sin \beta t + c_2 e^{-\alpha t} \cos \beta t. \tag{4.22}$$

We again state that $\alpha > 0$ (and thus $-\alpha < 0$) and the solution again approaches 0 as $t \to \infty$. In this case, the spring will oscillate as its amplitude dies off to zero. This case is called *underdamped* motion; see Figure 4.8.

We can use the earlier trigonometric tricks to rewrite this general solution as[3]

$$x(t) = Ae^{-\alpha t} \cos(\beta t - \phi), \tag{4.23}$$

where

$$A = \sqrt{c_1^2 + c_2^2}, \quad \cos \phi = \frac{c_2}{A}, \quad \sin \phi = \frac{c_1}{A}. \tag{4.24}$$

The angular frequency of oscillation is now given by

$$\beta = \sqrt{\frac{k}{m} - \left(\frac{b}{2m}\right)^2}$$

$$= \sqrt{\omega_0^2 - \alpha^2} \tag{4.25}$$

and the amplitude of the oscillations is bounded by the exponentially decreasing functions $\pm Ae^{-\alpha t}$, which form an envelope that governs this decay.

---

[3]We again note that if we set $A = \sqrt{c_1^2 + c_2^2}$, $\cos \phi = \frac{c_1}{A}$, $\sin \phi = \frac{c_2}{A}$ we would obtain $x(t) = Ae^{-\alpha t} \sin(\beta t + \phi)$.

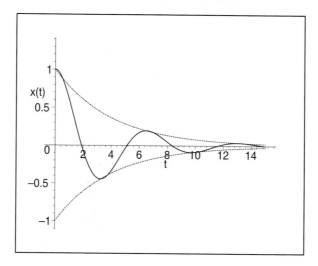

FIGURE 4.8: Underdamped motion: $x'' + (1/2)x' + x = 0$. The initial conditions are $x(0) = 1$, $x'(0) = 0$. The envelope $\pm e^{-t/4}$ is also drawn to show the decaying function that bounds the oscillations.

Example 3: The differential equation of motion of a damped spring is given by

$$3x''(t) + 5x'(t) + x(t) = 0.$$

Classify its motion as overdamped, critically damped, or underdamped.

To classify the motion, we simply need to know whether the roots of the characteristic equation are real and distinct, real and repeated, or complex. In other words, we need to know the sign of $b^2 - 4mk$. In this example, we have

$$b^2 - 4mk = 5^2 - 4 \cdot 3 \cdot 1 = 13.$$

Thus the motion is overdamped.

We stress that the case in which we end up always depends on the relationship between the mass, the spring constant, and the damping coefficient. In the case of no damping, we always end up in case 3. We also note that in each of the three cases, the motion of the spring dies out as $t \to \infty$. This motion is called the *transient* motion. We have not yet considered how to solve this problem when $F(t) \neq 0$, and will learn how to solve such problems in Section 4.4. The presence of a forcing function $F(t) \neq 0$ leads to an interference in the motion of the spring and will change the long-term or *steady-state* behavior of the spring. In the event that the frequency of the forcing is close to the natural frequency of the spring, a phenomenon called *resonance* occurs. In this situation the amplitude of the motion is amplified, sometimes dramatically.

Problems

1. A 12-lb weight is placed upon the lower end of a suspended coil spring. The weight comes to rest in its equilibrium position, stretching the spring 1.5 in. The weight is then pulled down 2 in. below its equilibrium position and released from rest at $t = 0$. Find the displacement of the weight as a function of time; determine the amplitude, period, and the frequency of the resulting motion. Assume there is no resistance of the medium and no external force.

2. A 16-lb weight is placed upon the lower end of a suspended coil spring. The weight comes to rest in its equilibrium position, stretching the spring 6 in. The weight is then pulled down 4 in. below its equilibrium position and released at $t = 0$ with an initial velocity downward of 2 ft/sec. Find the displacement of the weight as a function of time; determine the amplitude, period, and the frequency of the resulting motion. Assume there is no resistance of the medium and no external forces.

3. A 4-lb weight is placed upon the lower end of a suspended coil spring. The weight comes to rest in its equilibrium position, stretching the spring 6 in. At $t = 0$, the weight is set into motion with an initial velocity downward of 2 ft/sec. Find the displacement of the weight as a function of time; determine the amplitude, period, and the frequency of the resulting motion. Assume there is no resistance of the medium and no external force.

4. A weight of mass m is attached to one end of a spring, and the other end is fixed. The spring constant is k, i.e., if the weight is displaced from equilibrium a distance x, the spring exerts a force of kx N toward equilibrium. Moreover, if the weight moves with speed v, there is a resistance of nv N, where n is a constant. At time $t = 0$, the weight is placed in the equilibrium position and propelled with velocity v in the direction in which the spring acts. Find the motion of the weight in the two cases $n^2 < 4km$ and $n^2 > 4km$.

5. Suppose that m and k are given in the preceding problem. Determine n so that the weight reaches its equilibrium position in the shortest possible time (that is, so that the solution $x(t)$ approaches its equilibrium value as quickly as possible, $t > 0$).

In problems 6–13, we consider some mass-spring analogues.

6. Two pulleys are attached to a shaft; they have moments of inertia I_1 and I_2. To twist one of the pulleys an angle ϕ with respect to the other requires an elastic shaft-deforming torque of $K\phi$. Find the frequency of the torsional oscillations of the shaft.

7. A weight of mass m is attached to one end of an elastic rod. The other end moves so that its position at time t has coordinate $B \sin \omega t$. The elastic force exerted by the rod is proportional to the difference in

displacements of its ends. Neglecting the mass of the rod, and friction, find the amplitude A of the forced vibrations of the mass. Can the relation $A > B$ hold?

8. An electric circuit consists of a voltage source that supplies voltage V volts, a resistor of resistance R ohms, and an inductance of L henrys, together with a switch that is closed at time $t = 0$. Find the current as a function of the time.

9. Solve the preceding problem, replacing the inductance L by a capacitor of capacity C farads. The capacitor is uncharged when the switch is closed.

10. A resistor of resistance R ohms is connected to a capacitor of capacity C farads that has a charge q coulombs at time $t = 0$. The circuit is closed at $t = 0$. Find the current as a function of time for $t > 0$.

11. An inductor, resistor, and capacitor are connected in series. At time $t = 0$, the circuit is closed, the capacitor having a charge of q coulombs at that time. Find the current as a function of time, and the frequency of current change in case the current does actually change sign periodically.

12. A voltage source supplies voltage $E = V \sin \omega t$ in a circuit consisting of the voltage source, and a resistor and capacitor in series. Find the steady-state current in the circuit.

13. A voltage source supplies voltage $E = V \sin \omega t$ and is connected to a resistor, inductor, and capacitor in series. Find the steady-state current in the circuit. What frequency ω is needed to obtain the maximum possible current?

Problems 14–20 discuss topics from the "Damped Oscillations" subsection. For each problem, classify the motion of the mass on a spring as either underdamped, critically damped, or overdamped. Give analytical reasons for your conclusions. Then numerically solve the differential equation and plot its numerical solution using your computer code from Section 3.5 with IC $x(0) = 2$, $x'(0) = -1$ to confirm your results. Do not attempt to find the solution by hand. Assume $x = x(t)$, that is, x is a function of t.

14. $x'' + x' + x = 0$

15. $4x'' + \frac{1}{2}x' + 3x = 0$

16. $x'' + 4x' + 3x = 0$

17. $\pi x'' + 4x' + 2x = 0$

18. $x'' + 2x' + x = 0$

19. $3x'' + 7x' + x = 0$

20. $x'' + 6x' + 9x = 0$

21. In the text, we stated that in overdamped and critically damped motion, it is still possible for the mass to pass through its rest position

at most *one* time before coming to rest. Use the general solution of the damped mass-spring system to show that this condition is given by $|v_0| > |bx_0/(2m)|$, where v_0 is the initial velocity of the mass and x_0 is the initial displacement.

22. In a first physics course, the motion of a damped spring is usually given as

$$x(t) \approx c_1 e^{-bt/2m} \cos\left(\sqrt{\frac{k}{m}}\,t\right) + c_2 e^{-bt/2m} \sin\left(\sqrt{\frac{k}{m}}\,t\right), \qquad (4.26)$$

with the qualification that this approximation is valid if the damping constant b is small. Verify this approximation by using a Taylor series expansion about $b = 0$.

23. Consider the mass on a spring equation (4.17). If $m = 1, b = 2$, find the range of k-values that gives underdamped motion.

24. Consider the mass on a spring equation (4.17). If $m = 1, k = 3$, find the range of b-values that gives critically damped motion.

25. Consider the mass on a spring equation (4.17). If $b = 2, k = 1$, find the range of m-values that gives overdamped motion.

26. Consider the mass on a spring equation (4.17). If $b = 2, k = 2$, find the range of m-values that gives underdamped motion.

Problems 27–31 involve numerical explorations with the computer. They concern the generalized equation of motion for a mass on a spring given by

$$m(t)x''(t) + b(t)x'(t) + k(t)x(t) = F(t).$$

Here, we allow the possibility of changing mass, coefficient of friction, and spring constant, as well as a forcing function. Thus, we will not be able to use $b^2 - 4mk$ to determine the motion of the spring and we will instead use the computer to find the numerical solution.

27. Consider a container attached to a spring. Suppose the container is full of water that is evaporating in the hot Los Angeles sun. Suppose the mass function is given by $m(t) = 2e^{-t/10} + 1$. Let $b(t) = .2$, $k(t) = 1$, and $F(t) = 0$. Plot numerical solutions from $t = 0$ to $t = 50$ from the initial conditions $x(0) = 1$, $x'(0) = 0$.

28. Consider a block of constant mass $m = 2$ sliding back and forth on a sheet of ice. As the block slides, the ice melts and eventually exposes a rougher surface beneath. Assume the damping coefficient is given by $b(t) = \arctan(t - 20) + \pi/2$. Let $k(t) = 4$, $F(t) = 0$. Plot the motion of the spring over a large enough range for t so that you are able to see the results. Assume $x(0) = 1$, $x'(0) = 0$.

29. Springs typically lose some of their stiffness over time. Suppose a spring coefficient is given by $k(t) = 5e^{-t/25}$. Suppose also that $b(t) = 1$, $m(t) = 12$, $F(t) = 0$. Plot the motion of the spring over a large enough range for t so that you are able to see the results. Assume $x(0) = 1$, $x'(0) = 0$.

30. Consider the equation $4x'' + x' + 4x = \sin(\omega t)$. (i) Let $\omega = 10$. What happens to the motion for large t? Does it decay to zero? (ii) Now let $\omega = 1$. Again, what happens to the motion for large t? (iii) Numerically explore the behavior of the solution for various ω-values. Is there an ω value that gives the largest oscillations?

31. Consider the equation $4x'' + 0.1x' + 4x = \sin(\omega t)$. (i) Let $\omega = 10$. What happens to the motion for large t? Does it decay to zero? (ii) Now let $\omega = 1$. Again, what happens to the motion for large t? (iii) Numerically explore the behavior of the solution for various ω-values. Is there an ω value that gives the largest oscillations?

4.3   Cauchy-Euler (Equidimensional) Equation

Up to now, all of the second-order differential equations that we have considered were linear with constant coefficients. As we have seen, these equations occur in applications, but they are not the only type of second-order differential equations for which we can develop a method of obtaining an explicit solution. We will now consider the *Cauchy-Euler equation*[4] which is defined by the second-order differential equation

$$a_0 x^2 y'' + a_1 x y' + a_2 y = f(x), \tag{4.27}$$

where the coefficients a_i are constant.

Example 1: Electric Potential of a Charged Spherical Shell

Following Lomen and Lovelock [23], a differential equation that occurs in describing the electric potential of a charged spherical shell is given by

$$x^2 y'' + 2xy' - n(n+1)y = 0 \tag{4.28}$$

where x is the distance from the center of the spherical shell and y is the potential. Here n is a positive constant. This differential equation is of a form we have not previously considered: it is second order, but does not have constant coefficients. Comparing it with equation (4.27), we see that it is a homogeneous Cauchy-Euler equation.

To solve this equation, we will now consider a method of solution of (4.28) analogous to our work in Section 3.1.2; however, here we replace xy' and x^2y''. To do this, we will change variables from the independent variable x to a new

---

[4]This equation is also called an equidimensional equation, or Euler-Cauchy equation.

independent variable t, where

$$\frac{dy}{dt} = x\frac{dy}{dx}.$$

Applying the chain rule, we have that

$$\frac{dy}{dt} = \frac{dy}{dx}\frac{dx}{dt}$$

so that we can relate x and t by the differential equation

$$x'(t) = x(t).$$

This the differential equation which has solution

$$x(t) = ce^t$$

and thus suggests that we use the change of variables

$$x = e^t, \text{ if } x > 0 \quad \text{and } x = -e^t, \text{ if } x < 0. \tag{4.29}$$

In this example, $x > 0$ as distance is measured positively, so we take $x = e^t$ which gives $t = \ln x$. Differentiation with respect to x then gives

$$\frac{dt}{dx} = \frac{1}{x}.$$

Thus, using the chain rule,

$$\frac{dy}{dt} = \frac{dy}{dx}\frac{dx}{dt} = \frac{dy}{dx}e^t = x\frac{dy}{dx},$$

that is

$$x\frac{dy}{dx} = \frac{dy}{dt}. \tag{4.30}$$

Using this in (4.28) we can replace the $2xdy/dx$ term by $2dy/dt$. Now similarly, the x^2d^2y/dx^2 term can be replaced; if we differentiate (4.30) with respect to x, we have

$$x\frac{d^2y}{dx^2} + \frac{dy}{dx} = \frac{d}{dx}\frac{dy}{dt} = \frac{d^2y}{dt^2}\frac{dt}{dx}.$$

Multiplying by x gives

$$x^2\frac{d^2y}{dx^2} + x\frac{dy}{dx} = \frac{d^2y}{dt^2}$$

so that

$$x^2\frac{d^2y}{dx^2} = \frac{d^2y}{dt^2} - \frac{dy}{dt}. \tag{4.31}$$

Substituting (4.30) and (4.31) into (4.28), we have

$$\frac{d^2y}{dt^2} + \frac{dy}{dt} - n(n+1)y = 0. \tag{4.32}$$

This is a homogenous second-order linear differential equation with characteristic equation

$$r^2 + r - n(n+1) = 0$$

which has solution

$$r = n \text{ and } r = -n - 1.$$

Thus, the solution to (4.32) is

$$y(t) = c_1 e^{nt} + c_2 e^{-(n+1)t}.$$

Expressing this equation in the original variable x, we have the general solution to (4.28) as

$$y(x) = c_1 x^n + c_2 x^{-(n+1)}.$$

If we consider the potential inside the spherical shell, from symmetry considerations we expect the potential to be zero at the center $x = 0$, which we have if we choose $c_2 = 0$. If we are dealing with the potential outside the spherical shell, we expect the potential to go to 0 as $x \to \infty$, so we would choose $c_1 = 0$. Thus, for either situation we obtain a bounded solution of the differential equation.

Following the work in this example, we see that the homogeneous Cauchy-Euler equation

$$a_0 x^2 y'' + a_1 x y' + a_2 y = 0, \tag{4.33}$$

where the coefficients a_i are constant, has solutions of the form $y = x^r$. Substituting $y = x^r$ into (4.33) gives

$$a_0 x^2 (r(r-1)) x^{r-2} + a_1 x r x^{r-1} + a_2 x^r = 0,$$

so that simplifying we have

$$x^r (a_0 r^2 + (a_1 - a_0) r + a_2) = 0.$$

Noting that $x = 0$ yields the trivial solution, we have the characteristic equation

$$a_0 r^2 + (a_1 - a_0) r + a_2 = 0. \tag{4.34}$$

We again have three different cases to consider depending on the roots of this equation: if the roots are real and distinct $(r_1 \neq r_2)$, if the roots are real and repeated $(r_1 = r_2)$, and if the roots are complex $(r_1 = a + bi)$. However, we note here that we do not need to reinvent the wheel. We know what the solutions look like in each of these cases; the only difference here is that we need to apply the appropriate change of variables $x = \ln t$ as motivated in (4.29).

In the case that the roots are real and distinct, two linearly independent solutions are x^{r_1} and x^{r_2} and the general solution is

$$y(x) = c_1 x^{r_1} + c_2 x^{r_2}.$$

If the roots are real and repeated ($r_1 = r_2$), then two linearly independent solutions are x^{r_1} and $x^{r_1} \ln x$ and the general solution is

$$y(x) = c_1 x^{r_1} + x^{r_1} \ln x.$$

If the roots are complex again, non-real ($r_1 = a + bi$), then two real-valued linearly independent solutions are $x^a \sin(b \ln x)$ and $x^a \cos(b \ln x)$ so that the general solution is

$$y(x) = c_1 x^a \sin(b \ln x) + c_2 x^a \cos(b \ln x).$$

Example 2: Solve

$$2x^2 \frac{d^2 y}{dx^2} + 3x \frac{dy}{dx} - y = 0.$$

Changing variables from x to t where $x = e^t$ gives

$$2 \frac{d^2 y}{dt^2} + \frac{dy}{dt} - y = 0,$$

which has characteristic equation

$$2r^2 + r - 1 = 0.$$

The roots of this equation are $r_1 = 1/2$ and $r_2 = -1$ so that the general solution is

$$y = c_1 x^{\frac{1}{2}} + c_2 x^{-1}.$$

Example 3: Solve

$$x^2 \frac{d^2 y}{dx^2} + 2x \frac{dy}{dx} + 2y = 0$$

for $x > 0$.

Applying the change of variables $x = e^t$ gives

$$\frac{d^2 y}{dt^2} + \frac{dy}{dt} + 2y = 0,$$

which has characteristic equation

$$r^2 + r + 2 = 0.$$

The roots of this equation are

$$r = \frac{-1 \pm \sqrt{7} i}{2}$$

so that the general solution is

$$y(x) = c_1 \frac{\sin(\sqrt{7} \ln x)}{\sqrt{x}} + c_2 \frac{\cos(\sqrt{7} \ln x)}{\sqrt{x}}.$$

Problems

In problems 1–16, solve the homogeneous Cauchy-Euler equation by finding the general solution of the following differential equations, assuming $x > 0$.

1. $x^2 y'' + 4xy' + 2y = 0$
2. $x^2 y'' + 3xy' + y = 0$
3. $x^2 y'' + xy' + 4y = 0$
4. $x^2 y'' + 3xy' + 2y = 0$
5. $x^2 y'' + 11xy' + 21y = 0$
6. $2x^2 y'' + xy' - y = 0$
7. $2x^2 y'' - 3xy' + 3y = 0$
8. $x^2 y'' - 3xy' + 3y = 0$
9. $x^2 y'' - 3xy' + 4y = 0$
10. $x^2 y'' + 3xy' - 2y = 0$
11. $x^2 y'' + 5xy' - 3y = 0$
12. $3x^2 y'' + 3xy' + 9y = 0$
13. $7x^2 y'' + 5xy' + y = 0$
14. $x^2 y'' + 5xy' + 8y = 0$
15. $3x^2 y'' + 13xy' + 11y = 0$
16. $3x^2 y'' + 5xy' + y = 0$
17. Show that if $y_1(t)$ and $y_2(t)$ are linearly independent functions of t and satisfy a second-order linear differential equation with constant coefficients, then $Y_1(x) = y_1(\ln x)$ and $y_2(x) = y_2(\ln x)$ are linearly independent functions of x for $x \neq 0$.

4.4   Nonhomogeneous Equations

Up to this point we have been focusing on the solution of the homogeneous equation

$$a_0(x) \frac{d^n y}{dx^n} + a_1(x) \frac{d^{n-1} y}{dx^{n-1}} + \ldots + a_{n-1}(x) \frac{dy}{dx} + a_n(x) y = 0, \qquad (3.2)$$

often concentrating on the case in which (3.2) has constant coefficients. We will now begin discussing the nonhomogeneous equation first mentioned in Section 3.1:

$$a_0(x) \frac{d^n y}{dx^n} + a_1(x) \frac{d^{n-1} y}{dx^{n-1}} + \ldots + a_{n-1}(x) \frac{dy}{dx} + a_n(x) y = F(x). \qquad (3.1)$$

We will see how our work with homogeneous equations will play a role. In general it is not easy to solve this equation; sometimes we can get lucky, while other times we may only be able to guess a particular solution. Fortunately, many real world phenomena are accurately modeled with sinusoidal forcing functions and we will be able to solve these with the methods of this chapter.

THEOREM 4.4.1 *Let v be any solution of the nonhomogeneous nth order linear differential equation (3.1). Let u be any solution of the corresponding homogeneous equation (3.2); then $u + v$ is also a solution of the given nonhomogeneous equation.*

Example 1: Note that $y = x$ is a solution of the nonhomogeneous equation

$$\frac{d^2y}{dx^2} + y = x$$

and that $y = \sin x$ is a solution of the corresponding homogeneous equation

$$\frac{d^2y}{dx^2} + y = 0.$$

Thus, the sum $y = x + \sin x$ is a solution to

$$\frac{d^2y}{dx^2} + y = x.$$

This example motivates the following theorem.

THEOREM 4.4.2 *Let y_p be a solution of the nth order nonhomogeneous linear differential equation (3.1). Let*

$$y_c = c_1y_1 + c_2y_2 + \ldots + c_ny_n$$

be the general solution of the corresponding homogeneous equation (3.2). Then every solution ϕ of the nth order nonhomogeneous linear differential equation (3.1) can be expressed in the form

$$\phi = y_c + y_p.$$

The general solution of (3.2), y_c, is called the *complementary function* of (3.1) and any particular solution of (3.1) is called a *particular solution* of (3.1) and is denoted y_p. (Note that the general solution of (3.2) is sometimes called the *homogeneous solution* and is denoted y_h. Thus both y_c and y_h denote the same thing.) We can summarize this theorem as

general solution of nonhomogeneous

$$= \text{ general solution of homogeneous}$$

$$+ \text{ particular solution of nonhomogeneous}$$

Note that we still have not discussed exactly how to find a particular solution for a given differential equation—we are simply stating that once we do know a particular solution, we can write the general solution to the nonhomogeneous equation.

Example 2: Consider the differential equation

$$\frac{d^2y}{dx^2} + y = x.$$

The complementary function is the general solution

$$y_c = c_1 \sin x + c_2 \cos x$$

of the homogeneous equation. A particular solution is given by $y_p = x$ as mentioned in Example 1, so that the general solution of the nonhomogeneous equation is

$$y = c_1 \sin x + c_2 \cos x + x.$$

As mentioned before this example, we will show how to obtain this y_p in the next section, but for now we pursue further how a particular solution fits into the larger theory of solving nonhomogeneous linear differential equations.

Example 3: Consider the differential equation

$$y'' - 4y = \sin x. \tag{4.35}$$

A particular solution is given by

$$y_p = \frac{-1}{5} \sin x.$$

We can see that y_p is actually a solution since

$$y_p' = \frac{-1}{5} \cos x \quad \text{and} \quad y_p'' = \frac{1}{5} \sin x.$$

Substituting into the original differential equation gives

$$y_p'' - 4y = \frac{1}{5} \sin x - 4 \left(\frac{-1}{5} \sin x \right) = \sin x,$$

which shows that $y_p = \frac{-1}{5} \sin x$ is a particular solution.

Now it is straightforward to calculate the homogeneous solution

$$y_c = c_1 e^{-2x} + c_2 e^{2x}.$$

Thus the general solution is

$$y(x) = y_c(x) + y_p(x) = c_1 e^{-2x} + c_2 e^{2x} - \frac{1}{5} \sin x.$$

If instead the right-hand side of equation (4.35) was given by $A \sin x$, that is, if we considered

$$y'' - 4y = A \sin x \tag{4.36}$$

for some constant $A \neq 0$, then Ay_p is a particular solution to (4.36) where $y_p = \frac{-1}{5} \sin x$ was the particular solution obtained when $A = 1$.

4.4.1   Forced Mass on a Spring

Perhaps the most studied example of nonhomogeneous equations is that of a forced mass on a spring. We again consider the differential equation describing the motion of a mass on a spring and will let $x(t)$ denote the displacement:

$$m \frac{d^2 x}{dt^2} + b \frac{dx}{dt} + kx = F(t). \tag{4.37}$$

We have discussed the case when $F(t) = 0$ in Section 4.2.2 and will now consider the more general case when $F(t) \neq 0$. The physical interpretation of this type of situation is one in which the support of the spring, i.e., the object to which the spring is attached, is shaken or forced according to some known function. A typical assumption is that the forcing function can be written as

$$F(t) = F_0 \sin(\omega t), \tag{4.38}$$

where F_0 is the magnitude of the forcing function and ω is the frequency of this forcing. We note that this ω is completely independent of the inherent natural frequency, $\omega_0 = \sqrt{\frac{k}{m}}$, of the undamped mass-spring system. Many types of forced vibrations can be well approximated with such an assumption. This is also mathematically convenient because we can "guess" that the form of the particular solution should involve $\sin(\omega t)$ and maybe even $\cos(\omega t)$. What we really want to understand is how changing the frequency of the forcing oscillation, ω, can affect the motion of the attached mass.

Example 4: Consider a forced mass on a spring governed by the equation

$$x''(t) + \frac{1}{2} x'(t) + x(t) = \sin(2t). \tag{4.39}$$

Physically, we have that our mass and spring constant are both one but, more importantly, we see that the the the damping coefficient is small in comparison. Calculating $b^2 - 4mk$ shows that our motion is underdamped. How does the forcing frequency $\omega = 2$ affect things?

We can calculate the complementary solution as

$$x_c(t) = c_1 e^{-t/4} \sin\left(\frac{\sqrt{15}}{4} t \right) + c_2 e^{-t/4} \cos\left(\frac{\sqrt{15}}{4} t \right) \tag{4.40}$$

because we again have constant coefficients. A particular solution is given by

$$x_p(t) = \frac{-3}{10} \sin(2t) - \frac{1}{10} \cos(2t)$$

and the reader should check this. Our general solution is then $x_c(t) + x_p(t)$.

It is very worthwhile at this point to scrutinize these two parts of the solution. The solution of the homogeneous equation, $x_c(t)$ is exactly case 3 from the previous section describing the possible motions of a damped spring. We know from those results (or could easily calculate if we already forgot) that

$$\lim_{t \to \infty} x_c(t) = \lim_{t \to \infty} c_1 e^{-t/4} \sin\left(\frac{\sqrt{15}}{4}t\right) + c_2 e^{-t/4} \cos\left(\frac{\sqrt{15}}{4}t\right) = 0.$$

(This is another instance where writing the solution in the amplitude-phase form makes the evaluation of this limit easy.) As mentioned earlier, we call this motion *transient* because it dies off as $t \to \infty$; however, it is **not** the case that $x_p(t)$ dies off. Indeed,

$$\lim_{t \to \infty} x_p(t) \quad \text{does not exist}$$

because the sine and cosine continue to oscillate as $t \to \infty$. What does this mean? It simply (and importantly!) indicates that the motion of the mass does not stop. We can write bounds on the motion of these *steady-state* oscillations using our standard trigonometric identities:

$$B = \sqrt{\left(\frac{3}{10}\right)^2 + \left(\frac{1}{10}\right)^2} = \frac{1}{\sqrt{10}}.$$

See Figure 4.9a for a graph of the solution with the initial condition $x(0) = 1$, $x'(0) = 0$.

It may seem completely obvious that by forcing the spring support, we perpetuate the motion of the spring. What may seem strange is that, for a given mass-spring system, the frequency at which we choose to force the support may have a significant effect on the motion of the mass.

Example 5: Consider the same mass-spring system as in Example 4 but now we will use a different forcing frequency:

$$x''(t) + \frac{1}{2}x'(t) + x(t) = \sin\left(\sqrt{\frac{7}{8}}t\right). \tag{4.41}$$

The homogeneous solution is the same as before and we can check that

$$x_p(t) = \frac{-8\sqrt{14}}{15} \cos\left(\sqrt{\frac{7}{8}}t\right) + \frac{8}{15} \sin\left(\sqrt{\frac{7}{8}}t\right)$$

is a particular solution. The homogeneous (transient) solution again dies off and the particular (steady-state) solution determines the long-term behavior. We can again write bounds on the motion of these *steady-state* oscillations using our standard trigonometric identities:

$$B = \sqrt{\left(\frac{8\sqrt{14}}{15}\right)^2 + \left(\frac{8}{15}\right)^2} = \frac{8}{\sqrt{15}}.$$

For the *same* initial condition and mass-spring system, changing the frequency of the forcing function dramatically increased the amplitude of the steady-state oscillations; see Figure 4.9b.

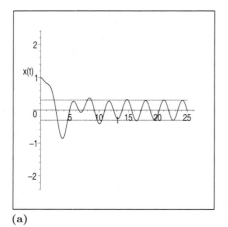

(a)

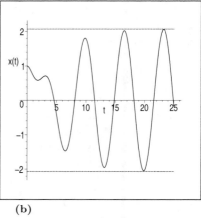

(b)

FIGURE 4.9: Forced damped motion with $m = 1$, $b = 1/2$, $k = 1$ for the initial condition $x(0) = 1$, $x'(0) = 0$. In **(a)**, the forcing function is $F(t) = \sin(2t)$; see Example 4. In **(b)**, the forcing function is $F(t) = \sin(\sqrt{7/8}\,t)$; see Example 5. Note the dramatic difference in the amplitude of the steady-state oscillations.

This phenomenon of periodically forcing an oscillating system in a way that excites or amplifies the motion is known as *resonance*. We will discuss resonance in more detail after we learn how to find particular solutions. But we mention two often-used examples. The first is that of a group of marching soldiers that will go out-of-step as they walk across a bridge. They do this because they don't want to introduce a periodic forcing that might put their own lives in danger! The second was a catastrophic event that actually was *not* due to resonance. In 1940, the Tacoma Narrows bridge collapsed due to strong winds that caused the bridge to sway with large oscillations.[5] Resonance was

---

[5]Try a web search using "Tacoma Narrow bridge" and you should be able to find video

originally used to explain its collapse even though it's clear that a bridge naturally has large damping (and a strong effect due to resonance requires small damping). It wasn't until the 1990s that it became better understood how *nonlinear* effects actually caused the collapse of the bridge [22], [20]. We will not go into any details of the mathematics but encourage the reader to use the numerical techniques of past sections to explore some of this research.

4.4.2 Overview

Examining the forced mass on a spring has given us sufficient motivation as to the importance of the study of nonhomogeneous equations. *Our goal in the next few sections will be to find ways to obtain a particular solution.* The two methods on which we will focus are the method of undetermined coefficients and variation of parameters—both are simply ways of accomplishing our goal.

Method of Undetermined Coefficients
This method will only work when the right-hand side of the equation, i.e., the forcing function $F(x)$, has the form of one of the solutions of a linear homogeneous equation with constant coefficients and when the left-hand side is linear with constant coefficients. For example, we will be able to solve

$$y^{(5)} + 2y'' + 7y = \sin(3x) + x^2 e^{2x} \cos(x) + e^{-2x}$$

but we will not be able to solve equations such as

$$y'' + xy = \sin(x) \quad \text{or} \quad y'' + y = \tan(x).$$

To use this method, we will assume a certain form of the particular solution, with coefficients to be found, and we will substitute this assumed form into the differential equation and find the coefficients.

We will initially focus on the mechanics of obtaining the particular solution from an assumed form. We will then examine two methods used to obtain the "assumed form" of the particular solution. The first will be *superposition* (or more simply thought of as tables, memorization, or copying the form of F) and the second will be the *annihilator method.*

Variation of Parameters
This method will work in two places where the method of undetermined coefficients fails—it works when the coefficients are non-constant regardless of the form of the forcing function $F(x)$. For example, it will work on all three of the examples given above:

$$y^{(5)} + 2y'' + 7y = \sin(3x) + x^2 e^{2x} \cos(x) + e^{-2x},$$
$$y'' + xy = \sin(x), \quad \text{and} \quad y'' + y = \tan(x).$$

---

footage, pictures, and additional information.

You may wonder why we should even learn the method of undetermined coefficients, it turns out that if we have a simple equation such as

$$y'' + y = x$$

the method of undetermined coefficients will be far less work than variation of parameters.

The main drawback of the variation of parameters, as we will see, is that it requires us to know a fundamental set of solutions before beginning. If our equation has constant coefficients, this will not be a problem. But if we have a non-constant coefficient equation, we must obtain the solution by whatever means necessary.

Problems

1. Given that a particular solution of $y'' - 5y' - 6y = 3e^x$ is $y_p = \dfrac{-3e^x}{10}$, find a particular solution to $y'' - 5y' - 6y = -5e^x$.

2. Given that a particular solution of $y'' + y' - 6y = \sin x$ is $y_p = \dfrac{-1}{50}\cos x - \dfrac{7}{50}\sin x$, find a particular solution to $y'' + y' - 6y = 3\cos x$.

3. Given that a particular solution of $y'' + 9y = \cos x$ is $y_p = \dfrac{1}{6}x\sin 3x + \dfrac{1}{18}x\cos 3x$, find a particular solution to $y'' + 9y = 12\cos x$.

4. Given that a particular solution of $y'' + y' = 2x$ is $y_p = x^2 - 2x$, find a particular solution to $y'' + y' = Ax$ for any constant $A \neq 0$.

5. Given that a particular solution of $y'' - 5y' + 6y = 1$ is $y = \frac{1}{6}$, a particular solution of $y'' - 5y' + 6y = x$ is $y = \frac{x}{6} + \frac{5}{36}$, and a particular solution of $y'' - 5y' + 6y = e^x$ is $y = \frac{e^x}{2}$, find a particular solution of

$$y'' - 5y' + 6y = 2 - 12x + 6e^x.$$

6. Given that a particular solution of $y'' + 4y' + 4y = x$ is $y = \frac{x-1}{4}$ and a particular solution of $y'' + 4y' + 4y = e^{-2x}$ is $y = \frac{x^2}{2}e^{-2x}$, find a particular solution of

$$y'' + 4y' + 4y = 8x - 3e^{-2x}.$$

For problems 7–11, use Matlab, Maple, or Mathematica to numerically solve the differential equation subject to the given forcing frequencies. In all cases, use the initial condition of $x(0) = 1$, $x'(0) = 0$ and numerically plot the solution from $t = 0$ to $t = 30$. From your graphs, (i) determine which of the three frequencies is the resonant frequency, (ii) approximate the steady-state amplitude of the solution.

7. $x'' + 2x' + 6x = \sin(\omega t)$, with $\omega = 1,\ 2,\ 3$

8. $x'' + 2x' + 3x = \sin(\omega t)$, with $\omega = 1,\ 2,\ 3$

9. $3x'' + x' + x = \sin(\omega t)$, with $\omega = \sqrt{1/18},\ \sqrt{5/18},\ \sqrt{1/2}$

10. $3x'' + 2x' + 4x = \sin(\omega t)$, with $\omega = \sqrt{5/9},\ \sqrt{7/9},\ \sqrt{10/9}$

11. $x'' + 2x' + 5x = \sin(\omega t)$, with $\omega = 1,\ \sqrt{2},\ \sqrt{3}$

4.5 Method of Undetermined Coefficients via Tables

In this section, we give the first of two approaches for using the method of undetermined coefficients. Our overall goal, as mentioned earlier, is to find a particular solution given some function $F(x)$. We will discuss how to obtain the general form of the particular solution for certain functions $F(x)$. The method of undetermined coefficients then takes this general form of the particular solution and uses the original differential to determine the coefficients of the assumed form of the particular solution.

We again consider the nonhomogeneous differential equation

$$a_0 \frac{d^n y}{dx^n} + a_1 \frac{d^{n-1} y}{dx^{n-1}} + \ldots + a_{n-1} \frac{dy}{dx} + a_n\, y = F(x), \qquad (4.42)$$

which is simply (3.1) with the specification that $a_0, a_1, \ldots, a_n$ are constants and $F(x)$, the nonhomogeneous term, is a general nonzero function of x.

Recall that the general solution may be written

$$y = y_c + y_p$$

where y_c is the complementary function, that is, the solution to the corresponding homogeneous equation; y_p is a particular solution. We will now consider the first of the two methods that fall under the category of the method of undetermined coefficients.

Example 1: Consider

$$\frac{d^2 y}{dx^2} - 2 \frac{dy}{dx} - 3y = 2e^{4x}.$$

We seek a particular solution, y_p, of this nonhomogeneous equation. The right side is the exponential $2e^{4x}$. We know that we need to find a function y_p that will be a solution. That is, when we substitute y_p into the left-hand side of the equation it should equal the right-hand side. Looking for a function that when added to its first and second derivatives gives $2e^{4x}$ suggests that the particular solution should also be an exponential of the form

$$y_p = Ae^{4x}.$$

Here A is a constant, which we call the "undetermined coefficient."

Thus

$$y_p' = 4Ae^{4x} \text{ and } y_p'' = 16Ae^{4x}$$

which gives

$$16Ae^{4x} - 2(4Ae^{4x}) - 3Ae^{4x} = 2e^{4x}.$$

This simplifies to

$$5Ae^{4x} = 2e^{4x}$$

which must satisfy the differential equation for all x, thus

$$5A = 2$$

so that $A = 2/5$. Thus, a particular solution is

$$y_p = \frac{2}{5}e^{4x}.$$

Example 2: Now consider the differential equation

$$\frac{d^2y}{dx^2} - 2\frac{dy}{dx} - 3y = 2e^{3x}.$$

This is the same equation we just considered, except that the right-hand side is now $2e^{3x}$. Lets proceed similarly and consider

$$y_p = Ae^{3x}.$$

Thus

$$y_p' = 3Ae^{3x} \text{ and } y_p'' = 9Ae^{3x}.$$

Substitution gives

$$9Ae^{3x} - 2(3Ae^{3x}) - 3Ae^{3x} = 2e^{3x}$$

which simplifies to

$$0 = 2e^{3x}.$$

But, this last equation <u>does not</u> hold for any real x, so that there is <u>no</u> particular solution of the form

$$y_p = Ae^{3x}.$$

What went wrong? Consider the reduced equation of the nonhomogeneous equation:

$$\frac{d^2y}{dx^2} - 2\frac{dy}{dx} - 3y = 0.$$

The corresponding characteristic equation

$$r^2 - 2r - 3 = 0$$

has roots 3 and -1, so that e^{3x} and e^{-x} are linearly independent solutions. This is why the method failed; the particular solution

$$y_p = Ae^{3x}$$

is already a solution to the homogeneous equation! That is, since Ae^{3x} satisfies the homogeneous equation, it reduces the left side to zero, not $2e^{3x}$.

How do you find a particular solution in this case?

Recall we showed (in the double root case) that e^{rx} and xe^{rx} are linearly independent, so we try a solution of the form

$$y_p = Axe^{3x}.$$

Thus,

$$y_p' = 3Axe^{3x} + Ae^{3x} \text{ and } y_p'' = 9Axe^{3x} + 6Ae^{3x}$$

so that upon substitution we have

$$9Axe^{3x} + 6Ae^{3x} - 2(3Axe^{3x} + Ae^{3x}) - 3Axe^{3x} = 2e^{3x}.$$

This simplifies to

$$4Ae^{3x} = 2e^{3x},$$

which must be valid for all x, so that $4A = 2$, which implies $A = 1/2$.
    A particular solution is

$$y_p = \frac{1}{2}xe^{3x}.$$

We need some terminology to help formalize this method of undetermined coefficients.

DEFINITION 4.1   *We shall call a function f a UC function if it is either*
1. *A function defined by one of the following:*
i) x^n, *where n is a nonnegative integer*
ii) e^{ax}, *where a is a nonzero constant*
iii) $\sin(bx + c)$ *and/or* $\cos(bx + c)$, *where b and c are constants, $b \neq 0$*
2. *A function defined as a finite product of two or more functions of these types.*

The method of undetermined coefficients applies when the nonhomogeneous function F in the differential equation is a finite linear combination of UC functions. So you can think of the UC functions as "building blocks" of F.
    Note here that for a given UC function f, each successive derivative of f is either itself a constant multiple of a UC function or else a linear combination of UC functions.

UC function	UC set
1. 7	$\{1\}$
2. $x^2 + 3$	$\{x^2, x, 1\}$
3. $2e^{3x}$	$\{e^{3x}\}$
4. $\sin(2x)$	$\{\sin(2x), \cos(2x)\}$
5. $5\cos(2x)$	$\{\sin(2x), \cos(2x)\}$
6. $\sin(2x) - 32\cos(2x)$	$\{\sin(2x), \cos(2x)\}$
7. $x^3 e^{-x}$	$\{x^3 e^{-x}, x^2 e^{-x}, xe^{-x}, e^{-x}\}$
8. $3e^{6x}\cos(x-1)$	$\{e^{6x}\sin(x-1), e^{6x}\cos(x-1)\}$
9. $x\sin(2x)$	$\{x\sin(2x), x\cos(2x), \sin(2x), \cos(2x)\}$
10. $x\sin(2x) + 3\cos(2x)$	$\{x\sin(2x), x\cos(2x), \sin(2x), \cos(2x)\}$
11. $x^2 e^{5x}\sin(3x)$	$\{x^2 e^{5x}\sin(3x), x^2 e^{5x}\cos(3x), xe^{5x}\sin(3x),$
	$xe^{5x}\cos(3x), e^{5x}\sin(3x), e^{5x}\cos(3x)\}$

FIGURE 4.10:   Examples of UC functions and the corresponding UC set. See the table in Figure 4.11 for the general list.

For a UC function f, we will define the *UC set* of f as the set of functions consisting of (i) f itself and (ii) all successive derivatives of f that are linearly independent of f.

The elements arising from (i) and (ii) above must be nonzero. They may also be multiplied by arbitrary nonzero constants without changing the essential "building blocks" contained in the UC set. Thus a UC set of the form $\{3x, 5\}$ is equivalent to the UC set $\{x, 1\}$ and we will choose the latter formulation for convenience of notation.

Example 3: The function $f(x) = x^3$ is a UC function. Now

$$f'(x) = 3x^2, \quad f''(x) = 6x, \quad f'''(x) = 6, \quad \text{and} \quad f^{(n)}(x) = 0, \text{ for all } n \geq 4,$$

so the linearly independent UC functions of which successive derivatives of f are either constant multiples or linear combinations are those given by x^2, x, and 1. Thus the UC set of x^3 is

$$\{x^3, x^2, x, 1\}.$$

Example 4: For the UC function $f(x) = \sin 2x$,

$$f'(x) = 2\cos 2x, \quad \text{and} \quad f''(x) = -4\sin 2x,$$

so the UC set of $\sin 2x$ is
$$\{\sin 2x, \cos 2x\}.$$

The table in Figure 4.10 gives specific examples of UC functions and their corresponding UC sets. The table in Figure 4.11 gives the general situation as it summarizes our results concerning UC functions.

UC function	UC set
1. x^n	$\{x^n, x^{n-1}, \ldots, x, 1\}$
2. e^{ax}	$\{e^{ax}\}$
3. $\sin(bx+c)$ or $\cos(bx+c)$	$\{\sin(bx+c), \cos(bx+c)\}$
4. $x^n e^{ax}$	$\{x^n e^{ax}, x^{n-1} e^{ax}, \ldots, x e^{ax}, e^{ax}\}$
5. $x^n \sin(bx+c)$ or $x^n \cos(bx+c)$	$\{x^n \sin(bx+c), x^n \cos(bx+c),$ $x^{n-1} \sin(bx+c), x^{n-1} \cos(bx+c), \ldots,$ $x \sin(bx+c), x \cos(bx+c),$ $\sin(bx+c), \cos(bx+c)\}$
6. $e^{ax} \sin(bx+c)$ or $e^{ax} \cos(bx+c)$	$\{e^{ax} \sin(bx+c), e^{ax} \cos(bx+c)\}$
7. $x^n e^{ax} \sin(bx+c)$ or $x^n e^{ax} \cos(bx+c)$	$\{x^n e^{ax} \sin(bx+c), x^n e^{ax} \cos(bx+c),$ $x^{n-1} e^{ax} \sin(bx+c), x^{n-1} e^{ax} \cos(bx+c),$ $\ldots, x e^{ax} \sin(bx+c), x e^{ax} \cos(bx+c),$ $e^{ax} \sin(bx+c), e^{ax} \cos(bx+c)\}$

FIGURE 4.11:   UC functions and sets.

4.5.1   The Method

Equipped with the table in Figure 4.11, we are able to give an outline of the method of undetermined coefficients for finding a particular solution y_p of (4.42) for which the nonhomogeneous term F is a linear combination of UC functions.

There are 5 steps to the method:

1. For each of the UC functions $u_1, u_2, \ldots, u_m$ of which the nonhomogeneous term F is a linear combination, form the corresponding UC set, that is

$$S_1, S_2, \ldots, S_m.$$

2. Suppose that one of the UC sets formed, say S_j, is identical (or contained in) another, say S_k. Then omit the set S_j from your list.

3. For the UC sets remaining after step **2**, examine the list to see if the complementary solution y_c is listed, say in S_ℓ. If this is the case, multiply <u>each</u> member of this UC set S_ℓ by the lowest positive integer power of

x so that the resulting revised set will contain no members that are solutions of the corresponding homogeneous equation.

4. In general, there remain (i) certain elements of the original UC sets and (ii) revised UC sets. From these UC sets, form a linear combination of <u>all</u> the elements of these sets in each of these categories. The linear combination is formed with unknown constant coefficients.

5. Determine the unknown coefficients by substituting the linear combination formed in step **4** into the differential equation and demanding it identically satisfy the differential equation.

The first 4 steps above are the easiest and create the correct general form of a particular solution. That is, after the first 4 steps of the UC method we end up with a linear combination of terms with unknown constant coefficients. Step 5 requires us to find these unknown (i.e., undetermined) coefficients so that the function will be a solution of the differential equation in question.

Example 5: Solve the differential equation

$$\frac{d^2y}{dx^2} - 2\frac{dy}{dx} - 3y = 2e^x - 10\sin x.$$

The corresponding homogeneous equation is

$$\frac{d^2y}{dx^2} - 2\frac{dy}{dx} - 3y = 0,$$

which has characteristic equation

$$r^2 - 2r - 3 = 0.$$

This equation has roots $r = 3$ and $r = -1$. The complementary solution is

$$y_c = c_1 e^{3x} + c_2 e^{-x}.$$

The nonhomogeneous term is the linear combination $2e^x - 10\sin x$ of the UC functions e^x and $\sin x$. We now apply the method just outlined to first obtain the correct general form of the particular solution (steps 1–4) and then to find the specific particular solution for this problem.

1. Form the UC set for each of these; we have

$$S_1 = \{e^x\}$$

and

$$S_2 = \{\sin x, \cos x\}.$$

2. Neither of these are identical or contained in each other, so both are retained.

3. The terms e^{3x} and e^{-x} from the complementary solution are not in S_1 and S_2, so there are no revisions required.

4. Form the linear combination

$$y_p = Ae^x + B\sin x + C\cos x$$

as a particular solution.

5. Find A, B, and C such that y_p is a solution of the differential equation. We differentiate to obtain

$$y_p' = Ae^x + B\cos x - C\sin x \text{ and } y_p'' = Ae^x - B\sin x - C\cos x.$$

Substitution gives

$$Ae^x - B\sin x - C\cos x - 2(Ae^x + B\cos x - C\sin x)$$
$$-3(Ae^x + B\sin x + C\cos x) = 2e^x - 10\sin x,$$

which simplifies as

$$-4Ae^x + (-4B + 2C)\sin x + (-4C - 2B)\cos x = 2e^x - 10\sin x.$$

This equation is true for all x, so that

$$-4A = 2, \quad -4B + 2C = -10, \text{ and } -4C - 2B = 0.$$

The system solves to give

$$A = -\frac{1}{2}, \ B = -2, \quad \text{and} \quad C = 1.$$

Thus, a particular solution is

$$y_p = -\frac{1}{2}e^x - 2\sin x + \cos x.$$

Hence, the general solution is

$$y = y_c + y_p = c_1 e^{3x} + c_2 e^{-x} - \frac{1}{2}e^x - 2\sin x + \cos x.$$

Example 6: Solve the differential equation

$$\frac{d^2y}{dx^2} - 3\frac{dy}{dx} + 2y = 2x^2 + e^x + 2xe^x + 4e^{3x}.$$

The corresponding homogeneous equation has the characteristic equation

$$r^2 - 3r + 2 = 0$$

which has roots $r = 2$ and $r = 1$. Thus, the corresponding solutions are e^{2x} and e^x, which gives the complementary solution as

$$y_c = c_1 e^x + c_2 e^{2x}.$$

The nonhomogeneous term is $2x^2 + e^x + 2xe^x + 4e^{3x}$, which corresponds to the UC functions x^2, e^x, xe^x, and e^{3x}.

1. For each of these functions, we form the corresponding UC set:

$$S_1 = \{x^2, x, 1\},$$

$$S_2 = \{e^x\},$$

$$S_3 = \{xe^x, e^x\}$$

and

$$S_4 = \{e^{3x}\}.$$

2. Note that S_2 is completely contained in S_3, so that S_2 is omitted from further consideration.

3. Further note that

$$S_3 = \{xe^x, e^x\}$$

includes e^x which is part of the complementary function, thus multiply each member of S_3 by x to obtain the revised set

$$S_3' = \{x^2 e^x, xe^x\}$$

which does not contain solutions of the corresponding homogeneous equation.

4. From the revised and remaining UC sets, we have the six UC functions $x^2, x, 1, e^{3x}, x^2 e^x$, and xe^x, from which we form the linear combination

$$Ax^2 + Bx + C + De^{3x} + Ex^2 e^x + Fxe^x.$$

5. A particular solution is thus of the form

$$y_p = Ax^2 + Bx + C + De^{3x} + Ex^2 e^x + Fxe^x.$$

In order to find these unknown coefficients, we differentiate y_p to obtain

$$y_p' = 2Ax + B + 3De^{3x} + 2Exe^x + Ex^2 e^x + Fe^x + Fxe^x$$

and

$$y_p'' = 2A + 9De^{3x} + Ex^2 e^x + 4Exe^x + 2Ee^x + Fxe^x + 2Fe^x.$$

Substituting for y'', y' and y gives

$$2A + 9De^{3x} + Ex^2 e^x + 4Exe^x + 2Ee^x + Fxe^x + 2Fe^x$$

$$- 3(2Ax + B + 3De^{3x} + 2Exe^x + Ex^2 e^x + Fe^x + Fxe^x)$$

$$+ 2(Ax^2 + Bx + C + De^{3x} + Ex^2 e^x + Fxe^x)$$

$$= 2x^2 + e^x + 2xe^x + 4e^{3x}.$$

This simplifies as

$$(2A - 3B + 2C) + (2B - 6A)x + 2Ax^2 + 2De^{3x} - 2Exe^x + (2E - F)e^x$$
$$= 2x^2 + e^x + 2xe^x + 4e^{3x}. \tag{4.43}$$

Comparing coefficients gives

$$2A - 3B + 2C = 0, \quad 2B - 6A = 0, \quad 2A = 2,$$
$$2D = 4, \quad -2E = 2, \quad 2E - F = 1. \tag{4.44}$$

Solving these six equations for the six unknown constants gives $A = 1$, $B = 3$, $C = 7/2$, $D = 2$, $E = -1$, and $F = -3$. A particular solution is

$$y_p = x^2 + 3x + \frac{7}{2} + 2e^{3x} - x^2 e^x - 3xe^x$$

and the general solution is

$$y = y_c + y_p = c_1 e^x + c_2 e^{2x} + x^2 + 3x + \frac{7}{2} + 2e^{3x} - x^2 e^x - 3xe^x.$$

Example 7: Solve the differential equation

$$\frac{d^4 y}{dx^4} + \frac{d^2 y}{dx^2} = 3x^2 + 4\sin x - 2\cos x.$$

The corresponding homogeneous equation is

$$\frac{d^4 y}{dx^4} + \frac{dy^2}{dx^2} = 0$$

which has characteristic equation

$$r^4 + r^2 = 0.$$

This polynomial factors as

$$r^2(r^2 + 1) = 0$$

so that $r = 0$ is a double root and $r = \pm i$. The complementary solution is thus

$$y_c = c_1 + c_2 x + c_3 \sin x + c_4 \cos x.$$

The nonhomogeneous term is the linear combination $3x^2 + 4\sin x - 2\cos x$ of the UC functions x^2, $\sin x$, and $\cos x$.
1. The UC sets are

$$S_1 = \{x^2, x, 1\},$$
$$S_2 = \{\sin x, \cos x\},$$
$$S_3 = \{\cos x, \sin x\}.$$

2. S_2 and S_3 are identical; we only retain one. Thus, the UC sets are

$$S_1 = \{x^2, x, 1\},$$

$$S_2 = \{\sin x, \cos x\}.$$

3. Notice

$$S_1 = \{x^2, x, 1\}$$

contains 1 and x, which <u>are</u> solutions of the corresponding homogeneous differential equation. Thus, we multiply <u>each</u> element by x^2 to obtain

$$S_1' = \{x^4, x^3, x^2\},$$

none of which are solutions. Similarly, $\sin x$ and $\cos x$ are solutions of the corresponding homogeneous differential equation. Thus, we multiply each by an x, so that

$$S_2' = \{x \sin x, x \cos x\}.$$

4. We form the linear combination of these five UC functions, i.e.,

$$Ax^4 + Bx^3 + Cx^2 + Ex \sin x + Fx \cos x,$$

where A, B, C, E, and F are undetermined.

5. Substitute the particular solution

$$y_p = Ax^4 + Bx^3 + Cx^2 + Ex \sin x + Fx \cos x$$

into the differential equation in order to determine the unknown coefficients.

Differentiating gives

$$y_p' = 4Ax^3 + 3Bx^2 + 2Cx + Ex \cos x + E \sin x - Fx \sin x + F \cos x,$$

$$y_p'' = 12Ax^2 + 6Bx + 2C - Ex \sin x + 2E \cos x - Fx \cos x - 2F \sin x,$$

$$y_p''' = 24Ax + 6B - Ex \cos x - 3E \sin x + Fx \sin x - 3F \cos x,$$

and

$$y_p^{(4)} = 24A + Dx \sin x - 4E \cos x + Fx \cos x + 4F \sin x.$$

Substitution gives

$$24A + Ex \sin x - 4E \cos x + Fx \cos x + 4F \sin x$$

$$+ 12Ax^2 + 6Bx + 2C - Ex \sin x + 2E \cos x - Fx \cos x - 2F \sin x$$

$$= 3x^2 + 4 \sin x - 2 \cos x.$$

Equating coefficients gives

$$24A + 2C = 0, \quad 6B = 0, \quad 12A = 3,$$
$$-2E = -2, \text{ and } 2F = 4. \tag{4.45}$$

Solving this system of equations gives

$$A = \frac{1}{4}, \ \ B = 0, \ \ C = -3, \ \ E = 1, \ \ F = 2,$$

so that a particular solution is

$$y_p = \frac{1}{4}x^4 - 3x^2 + x \sin x + 2x \cos x.$$

This gives the general solution

$$y = c_1 + c_2 x + c_3 \sin x + c_4 \cos x + \frac{1}{4}x^4 - 3x^2 + x \sin x + 2x \cos x.$$

It may often be easier to do steps 4 and 5 in the above method with the help of our computer programs. In this previous example, we begin with the assumption of the correct form for the particular solution:

Computer Code 4.4: **Obtaining coefficients for the particular solution of a nonhomogeneous equation**

Matlab, Maple, Mathematica

```
                            Matlab
>>  %This requires the Symbolic Math Toolbox
>>  syms x A B C E F
>>  eqyp=A*x^4+B*x^3+C*x^2+E*x*sin(x)+F*x*cos(x)
>>  eq1=diff(eqyp,x,4)+diff(eqyp,x,2)-3*x^2-4*sin(x)+2*cos(x)
>>  %The above eq1 is the original ode written as lhs-rhs=0,
>>  %where the unwritten '=0' at the end is understood by
>>  %matlab.  We now type the relationships of the
>>  %coeffs by inspection of the calculated ode eq1
>>  eq2a=-3+12*A %coefficient of x^2
>>  eq2b=6*B %coefficient of x
>>  eq2c=24*A+2*C %constant terms
>>  eq2d=-2*E+2 %coefficient of cos(x)
>>  eq2e=2*F-4 %coefficient of sin(x)
>>  [A,B,C,E,F]=solve(eq2a,eq2b,eq2c,eq2d,eq2e)
>>  yp=subs(eqyp)
```

Maple

```
> eqyp:=y[p](x)=A*x^4+B*x^3+C*x^2+E*x*sin(x)+F*x*cos(x);
> eqODE:=diff(y[p](x),x$4)+diff(y[p](x),x$2)=3*x^2+4*sin(x)
  -2*cos(x);
> eq1:=simplify(subs(eqyp,eqODE));
> eq2a:=coeff(lhs(eq1),cos(x))=coeff(rhs(eq1),cos(x));
> eq2b:=coeff(lhs(eq1),sin(x))=coeff(rhs(eq1),sin(x));
> eq2c:=coeff(rhs(eq1),x,2); #This should give you an error
  #Maple won't compute a coefficient of a power of x
  #with cos(x) and sin(x) still in the expression
> eq3:=subs(cos(x)=0,sin(x)=0,eq1); #Thus we ignore
  #since we already have their coefficients
> eq3a:=coeff(lhs(eq3),x,2)=coeff(rhs(eq3),x,2);
> eq3b:=coeff(lhs(eq3),x,1)=coeff(rhs(eq3),x,1);
> eq3c:=coeff(lhs(eq3),x,0)=coeff(rhs(eq3),x,0);
  #The above three lines have the coeffs of terms with
  #x^2, x^1, x^0=constant, respectively
> eq4:=solve({eq2a,eq2b,eq3a,eq3b,eq3c},{A,B,C,E,F});
> eq5:=y[p]=subs(eq4,eqyp);
```

Mathematica

```
yp[x_] = Ax^4+Bx^3+Cx^2+E1xSin[x]+FxCos[x]  (*E1 is a variable*)
dey[x_]=y''''[x]+y''[x]-3 x^2 - 4Sin[x] + 2Cos[x]
lhs=ReplaceAll[dey[x],y→ yp]
eq1= Coefficient[ReplaceAll[lhs,{Cos[x]→ 0,Sin[x]→ 0}],x,0]
  (*eq1 is the contant term*)
eq2=Coefficient[lhs,x]  (*coefficient of x term*)
eq3=Coefficient[lhs,x,2]  (*coefficient of x^2 term*)
eq4=Coefficient[lhs,Cos[x]]  (*coefficient of cos(x)*)
eq5=Coefficient[lhs,Sin[x]]  (*coefficient of sin(x)*)
Solve[{eq1==0,eq2==0,eq3==0,eq4==0, eq5==0},{A,B,C,E1,F}]
```

4.5.2   Resonance of a Forced Mass on a Spring

In the previous section, we showed the relevance of studying nonhomogeneous equations by using a forced mass on a spring as motivation. Equipped with our knowledge of the method of undetermined coefficients, we can take a closer look at the phenomenon of resonance. We will consider the equation

$$m\frac{d^2x}{dt^2} + b\frac{dx}{dt} + kx = F_0 \sin(\omega t), \tag{4.46}$$

where ω again denotes the forcing frequency. The method of undetermined coefficients applies in this situation and we thus assume the particular solution has the form

$$x_p(t) = A\sin(\omega t) + B\cos(\omega t).$$

We leave it as an exercise for the reader to show that substitution into the differential equation (4.46), with the substitutions $\alpha = b/(2m)$ and $\omega_0 = \sqrt{k/m}$, gives

$$A = \frac{F_0(\omega_0^2 - \omega^2)}{m[4\alpha^2\omega^2 + (\omega^2 - \omega_0^2)^2]} \tag{4.47}$$

$$B = \frac{-2\omega\alpha F_0}{m[4\alpha^2\omega^2 + (\omega^2 - \omega_0^2)^2]}. \tag{4.48}$$

With the details again left as an exercise, we use our now-favorite trigonometric identities to obtain[6]

$$x_p(t) = \frac{F_0/m}{\sqrt{4\alpha^2\omega^2 + (\omega^2 - \omega_0^2)^2}}\cos(\omega t - \phi), \tag{4.49}$$

where

$$\cos\phi = \frac{B}{\sqrt{A^2 + B^2}} \text{ and } \sin\phi = \frac{A}{\sqrt{A^2 + B^2}}.$$

We know that the homogeneous solution decays exponentially and it is only the steady-state solution that determines the long-term behavior of the system. The amplitude of these steady-state solutions are thus given by the factor

$$\frac{F_0/m}{\sqrt{4\alpha^2\omega^2 + (\omega^2 - \omega_0^2)^2}}. \tag{4.50}$$

For a given mass-spring system, we can only change the frequency of the forcing, that is, ω. The frequency that makes equation (4.50) as large as possible is the resonant frequency. Because we consider ω as the only variable, finding the resonant frequency reduces to the problem of finding the ω that makes this function have its maximum value. This can be done by taking the derivative, setting it equal to zero, and finding the critical points. We again leave it as an exercise for the reader to show that the resonant frequency is

$$\omega_{res} = \sqrt{\omega_0^2 - 2\alpha^2}, \tag{4.51}$$

where $\alpha = b/(2m)$, $\omega_0 = \sqrt{k/m}$. There is no longer guesswork in determining the resonant frequency for equation (4.46). For forcing functions that are exclusively in terms of cosine or in terms of both sine and cosine, the derivation

---

[6]As in Sections 4.2.1 and 4.2.2, we also could have used $\cos\phi = \frac{A}{\sqrt{A^2+B^2}}$ and $\sin\phi = \frac{B}{\sqrt{A^2+B^2}}$ to obtain $\sin(\omega t + \phi)$ instead of $\cos(\omega t - \phi)$ in (4.49).

is similar.

Example 7: Determine the resonant frequency for $3x'' + 2x' + 2x = F_0 \sin(\omega t)$.

For this problem, we have $m = 3$, $b = 2$, $k = 2$. Then $\omega_0^2 = 2/3$, $\alpha = 2/6$, and we see that

$$\omega_{res} = \frac{2}{3}.$$

Problems

In problems 1–8, write the form of the particular solution but do not solve the differential equation. Make sure to solve for y_c first!

1. $y'' + y = e^{-x} + x^2$
2. $y'' + 3y' = x^2 e^{2x} + \cos x$
3. $y'' + y = 4 \sin x + e^x \cos x$
4. $y'' + 4y' + y = \sin(2x) + \cos x$
5. $y''' + y = xe^{2x} \cos x + \sin x$
6. $y'' - y = e^x + xe^{-x}$
7. $y''' + 8y = 4 \sin x + \cos x$
8. $y^{(5)} + y'' = x^3 + xe^x \cos x$

   In problems 9–15, use the method of undetermined coefficients (step 5 only) to find the coefficients of the particular solution for the given equations.

9. $y'' + 4y' + 13y = e^{-2x}$, $\quad y_p = Ae^{-2x}$
10. $y'' + 4y' = x^2 - 3$, $\quad y_p = Ax^3 + Bx^2 + Cx$
11. $y'' - 4y' = \cos x + 3 \sin x$, $\quad y_p = A \cos x + B \sin x$
12. $9y'' + y = \cos x + \sin(2x)$, $\quad y_p = A \cos x + B \sin x + C \cos(2x) + E \sin(2x)$
13. $4y'' + 25y = x \cos x$, $\quad y_p = A \cos x + B \sin x + Cx \cos x + Ex \sin x$
14. $y'' + 2y' + 17y = e^x + 2$, $\quad y_p = Ae^x + B$
15. $y'' + 3y' - 10y = xe^x + 2x$, $\quad y_p = Ae^x + Bxe^x + Cx + E$

   For problems 16–32, find the general solution of the given differential equations. If instructed to do so, plot the analytical solution and compare with the solution obtained by numerically solving the equation for the given initial condition (or an initial condition of your choice if none is given).

16. $y'' - 3y' + 2y = 4x^2$
17. $y'' - 2y' - 8y = 4e^{2x} - 21e^{-3x}$
18. $y'' + 2y' + 5y = 6 \sin 2x + 7 \cos 2x$

19. $y''' + 10y'' + 34y' + 40y = xe^{-4x} + 2e^{-3x}\cos x$

20. $y'' - 4y = 32x$, $y(0) = 0, y'(0) = 6$

21. $y'' - 2y' + 2y = e^x + x\cos x$

22. $y'' + 6y' + 10y = 3xe^{-3x} - 2e^{3x}\cos x$, $y(0) = 1, y'(0) = -2$

23. $y'' - 8y' + 20y = 5xe^{4x}\sin 2x$

24. $y'' + 7y' + 10y = xe^{-2x}\cos 5x$, $y(1) = 3, y'(1) = 0$

25. $y'' - 2y' + 5y = 2xe^x + e^x\sin 2x$

26. $y'' - 2y' + y = 2xe^x + e^x\sin 2x$

27. $y'' - 3y' + 2y = e^x$, $y(0) = 1, y'(0) = 0$

28. $y'' - y = 4\sinh x$

29. $y'' + 4y' + 3y = \cosh x$

30. $y'' + 4y = (\sinh x)(\sin 2x)$

31. $y'' + 2y' + 2y = (\cosh x)(\sin x)$

32. $y^{(4)} - 18y'' + 81y = e^{3x}$

33. The method of undetermined coefficients can be used to solve first-order constant coefficient nonhomogeneous equations. Use this method to solve the following problems:
    a. $y' - 4y = x^2$
    b. $y' + y = \cos 2x$
    c. $y' - y = e^{4x}$

34. Derive equation (4.48), which give the constants of the particular solution.

35. Derive equation (4.49), thus obtaining the amplitude-phase form for the particular solution.

36. Derive the resonant frequency equation (4.51) for the forced mass-spring system.

37. Suppose that $y_1(x)$ and $y_2(x)$ are solutions of

$$a\frac{d^2y}{dx^2} + b\frac{dy}{dx} + cy = f(x),$$

where a, b, and c are positive constants.
    a. Show that

$$\lim_{x \to \infty} [y_2(x) - y_1(x)] = 0.$$

    b. Is the result of **a** true if $b = 0$?
    c. Suppose that $f(x) = k$, where k is a constant. Show that

$$\lim_{x \to \infty} y(x) = \frac{k}{c}.$$

for every solution $y(x)$ of

$$a\frac{d^2y}{dx^2} + b\frac{dy}{dx} + cy = k.$$

d. Determine the solution $y(x)$ of

$$a\frac{d^2y}{dx^2} + b\frac{dy}{dx} = k.$$

Find

$$\lim_{x\to\infty} y(x).$$

e. Determine the solution $y(x)$ of

$$a\frac{d^2y}{dx^2} = k.$$

Find

$$\lim_{x\to\infty} y(x).$$

38. a. Let $f(x)$ be a polynomial of degree n. Show that, if $b \neq 0$, there is always a solution that is a polynomial of degree n for the equation $y'' + ay' + by = f(x)$.
    b. Find a particular solution of $y'' + 3y' + 2y = 9 + 2x - 2x^2$.

39. In many physical applications, the nonhomogeneous term $F(x)$ is specified by different formulas in different intervals of x.
    a. Find a general solution of the equation

$$y'' + y = \begin{cases} x, & 0 \leq x \leq 1, \\ 1, & 1 \leq x. \end{cases}$$

Note that the solution is not differentiable at $x = 1$.
    b. Find a particular solution of

$$y'' + y = \begin{cases} x, & 0 \leq x \leq 1, \\ 1, & 1 \leq x \end{cases}$$

that satisfies the initial conditions $y(0) = 0$ and $y'(0) = 1$.

4.6   Method of Undetermined Coefficients via the Annihilator Method

We saw in the previous section how to obtain the particular solution by first assuming the correct general form (based on $F(x)$) and then substituting into the differential equation to obtain the specific particular solution that satisfies the equation. Here, we present an alternative method to obtaining the correct general form based on $F(x)$.[7] The operator notation gives us a useful idea, as can be seen in the following example.

Example 1: Consider the nonhomogeneous constant coefficient differential equation

$$\frac{d^2y}{dx^2} - y = x,$$

which can be written as

$$(D^2 - 1)y = x \qquad (4.52)$$

using operator notation. The general solution to this nonhomogeneous equation, $y = y_c + y_p$, must satisfy this equation upon substitution into it. If we differentiate the given equation twice, we obtain the fourth-order linear homogeneous equation with constant coefficients

$$\frac{d^4y}{dx^4} - \frac{d^2y}{dx^2} = 0, \qquad (4.53)$$

which can also be written as

$$D^2(D^2 - 1)y = D^2x = 0. \qquad (4.54)$$

The general solution to (4.52) must also satisfy the differential equation (4.54). We note that (4.54), or equivalently, (4.53), is *homogeneous* with constant coefficients. This equation has the characteristic polynomial

$$r^4 - r^2 = 0,$$

which can be further factored as

$$r^2(r + 1)(r - 1) = 0.$$

Thus the general solution of (4.54) must be of the form

$$y(x) = k_1 + k_2x + c_1e^{-x} + c_2e^x$$

for constants k_1, k_2, c_1, c_2. Substituting this form of the solution into the original equation gives

$$(c_1e^{-x} + c_2e^x) - (c_1e^{-x} + c_2e^x + k_1 + k_2x) = x$$

so that

$$k_1 = 0 \text{ and } k_2 = -1$$

---

[7]**Important Note:** If Section 4.5.1 was skipped, the reader is strongly encouraged to read Section 4.5.2 after finishing this current section. Section 4.5.2 deals with the phenomenon of resonance that can occur in the forced mass on a spring system.

by comparing coefficients. Thus,

$$y(x) = -x + c_1 e^{-x} + c_2 e^x.$$

This example motivates the idea of an *annihilator*. Notice we were able to obtain a homogeneous equation by "annihilating" the nonhomogeneous term x.

DEFINITION 4.2   *The linear differential operator $A(D)$ annihilates a function $y(x)$ if*
$$A(D)y(x) = 0$$
for all x. In this case, $A(D)$ is called an annihilator of $y(x)$.

Example 2: The linear differential operator $A_1(D) = D^3$ annihilates the function $y(x) = x^2$ because $D^3 x^2 = 0$. Similarly, the linear differential operator $A_2(D) = D - 2$ annihilates the function $y(x) = e^{2x}$ as

$$A_2(D)(e^{2x}) = De^{2x} - 2e^{2x} = 0.$$

Example 3: Using the functions of the previous example, we can also observe that
$$A_1(D)A_2(D)(x^2 + e^{2x}) = D^3(D - 2)(x^2 + e^{2x}) = 0.$$

This property holds in general. That is, if $A_1(D)$ and $A_2(D)$ are linear differential operators having constant coefficients with A_1 annihilating $y_1(x)$ and A_2 annihilating $y_2(x)$, then $A_1 A_2$ annihilates $c_1 y_1(x) + c_2 y_2(x)$, where c_1 and c_2 are constants. This is easy to see, as

$$A_1 A_2(c_1 y_1(x) + c_2 y_2(x)) = A_1 A_2(c_1 y_1(x)) + A_1 A_2(c_2 y_2(x))$$

$$= c_1 A_1 A_2(y_1(x)) + c_2 A_1 A_2(y_2(x))$$

$$= c_1 A_2 A_1(y_1(x)) + c_2 A_1 A_2(y_2(x))$$

$$= c_1 A_2(0) + c_2 A_1(0)$$

$$= 0.$$

It is important to note that $A_2 A_1$ also annihilates $c_1 y_1(x) + c_2 y_2(x)$ because the operator notation is commutative (in the constant coefficient case); see (3.24).

4.6.1   Annihilators of Familiar Functions

It will be useful for us to know annihilators for our familiar functions. The differential equation

$$\frac{d^n y}{dx^n} = 0$$

in operator form is

$$D^n y = 0,$$

and has solution

$$y = c_1 + c_2 x + c_3 x^2 + \ldots + c_{n-1} x^{n-1}.$$

Thus, D^n annihilates the functions

$$1, x, x^2, \ldots, x^{n-1}$$

as well as any linear combination of these functions. This result means that any nth degree polynomial in x will be annihilated by an $(n+1)$st differential operator.

A general solution of the differential equation

$$(D - r)^n y = 0$$

is

$$y = c_1 e^{rx} + c_2 x e^{rx} + c_3 x^2 e^{rx} + \ldots + c_{n-1} x^{n-1} e^{rx}$$

since we see that r is a root of the characteristic equation and the root occurs n times. Thus, the differential operator $(D - r)^n$ annihilates the functions

$$e^{rx}, x e^{rx}, x^2 e^{rx}, \ldots, x^{n-1} e^{rx}.$$

We show this for the case $n = 2$:

$$(D - r)^2 \left(c_1 e^{rx} + c_2 x e^{rx} \right)$$
$$= (D - r)^2 (c_1 e^{rx}) + (D - r)^2 (c_2 x e^{rx})$$
$$= c_1 (D - r)^2 (e^{rx}) + c_2 (D - r)^2 (x e^{rx})$$
$$= c_1 (D - r)(r e^{rx} - r e^{rx}) + c_2 (D - r)(r x e^{rx} + e^{rx} - r x e^{rx})$$
$$= c_2 (D - r)(e^{rx})$$
$$= 0 \tag{4.55}$$

If the characteristic equation of a differential equation has complex roots

$$r_{1,2} = a \pm bi$$

with $b \neq 0$, we can write the characteristic equation as

$$(r - (a + bi))(r - (a - bi)) = r^2 - 2ar + (a^2 + b^2) = 0$$

which corresponds to the differential equation

$$y'' - 2ay' + (a^2 + b^2)y = [D^2 - 2aD + (a^2 + b^2)]y = 0.$$

Thus, the differential operator

$$D^2 - 2aD + (a^2 + b^2)$$

annihilates the functions $e^{ax} \cos bx$ and $e^{ax} \sin bx$.

If the complex roots are repeated with multiplicity n, they correspond to the characteristic equation

$$((r - (a + bi))(r - (a - bi)))^n = (r^2 - 2ar + (a^2 + b^2))^n = 0,$$

but this corresponds to an order $2n$ differential equation, which in operator form is

$$(D^2 - 2aD + (a^2 + b^2))^n y = 0.$$

Thus, the differential operator

$$(D^2 - 2aD + (a^2 + b^2))^n$$

annihilates the functions

$$e^{ax} \cos bx, e^{ax} \sin bx, xe^{ax} \cos bx, xe^{ax} \sin bx, \ldots, x^{n-1}e^{ax} \cos bx, x^{n-1}e^{ax} \sin bx.$$

These results are summarized in the table in Figure 4.12.

Function	Annihilator
$1, x, x^2, \ldots, x^{n-1}$	D^n
$e^{rx}, xe^{rx}, \ldots, x^{n-1}e^{rx}$	$(D - r)^n$
$e^{ax} \cos bx, xe^{ax} \cos bx, \ldots x^{n-1}e^{ax} \cos bx$	$(D^2 - 2aD + (a^2 + b^2))^n$
$e^{ax} \sin bx, xe^{ax} \sin bx, \ldots x^{n-1}e^{ax} \sin bx$	$(D^2 - 2aD + (a^2 + b^2))^n$

FIGURE 4.12:   Annihiliators of familiar functions.

With the ideas of annihilators in place, we return to the problem of solving nonhomogeneous equations with constant coefficients. If the nth order linear nonhomogeneous equation with constant coefficients can be written as

$$P(D)y = F(x)$$

where $F(x)$ is one of the functions (or combinations of functions) listed in the table in Figure 4.12, then we can find another differential operator $A(D)$ that annihilates $F(x)$.

If the differential operator $A(D)$ annihilates $F(x)$, applying $A(D)$ to the nonhomogeneous equation gives

$$A(D)P(D)y = A(D)F(x) = 0,$$

which is a homogeneous equation.

Let's see how to apply these ideas.

Example 4: Solve the nonhomogeneous equation

$$y'' + y = x^2.$$

If we write this equation in operator notation we have

$$(D^2 + 1)y = x^2.$$

Now the annihilator of x^2 is D^3. We thus can rewrite our original equation as the homogeneous equation

$$D^3(D^2 + 1)y = 0,$$

which has characteristic equation

$$r^3(r^2 + 1) = 0.$$

This gives the eigenvalues $0, 0, 0, i, -i$ so that the solution to the homogeneous equation is

$$y(x) = c_1 + c_2 x + c_3 x^2 + c_4 \sin x + c_5 \cos x. \qquad (4.56)$$

Now if we solve the corresponding homogeneous equation

$$y'' + y = 0$$

we have

$$y_c(x) = b_1 \sin x + b_2 \cos x.$$

Eliminating these terms from solution (4.56) indicates the particular solution has the form

$$y_p(x) = c_1 + c_2 x + c_3 x^2.$$

Notice that we would have obtained the same form for y_p if we used steps 1–4 from the method introduced in the previous section. The name "method of undetermined coefficients" really applies to the methods once the correct general form of y_p has been assumed. At this point, we thus proceed with step 5 as before; see Section 4.5.1. In order to use the original differential equation, we will need to obtain the second derivative to y_p in order to substitute. Calculating the derivatives gives

$$y_p' = c_2 + 2c_3 x \quad \text{and} \quad y_p'' = 2c_3.$$

Substitution gives

$$y_p'' + y_p = 2c_3 + c_1 + c_2 x + c_3 x^2.$$

We know that $y_p'' + y_p = x^2$ and thus

$$x^2 = 2c_3 + c_1 + c_2 x + c_3 x^2.$$

Equating coefficients of like terms gives

$$c_1 = -2, c_2 = 0 \text{ and } c_3 = 1.$$

Thus a particular solution is

$$y_p(x) = x^2 - 2,$$

and the general solution to the nonhomogeneous equation is

$$y(x) = y_c(x) + y_p(x) = b_1 \sin x + b_2 \cos x + x^2 - 2.$$

Example 5: Solve the nonhomogeneous equation

$$y''' + y' = \cos x + x.$$

We first will find the annihilator of the nonhomogeneous term $\cos x + x$. Since $D^2 + 1$ annihilates $\cos x$ and D^2 annihilates x, the annihilator of $\cos x + x$ is

$$A(D) = D^2(D^2 + 1).$$

If we apply this annihilator to both sides of the nonhomogeneous equation, we have
$$D^2(D^2 + 1)(D^3 + D)y = D^2(D^2 + 1)(\cos x + x).$$

The right-hand side is thus zero and our equation simplifies to the homogeneous equation
$$D^2(D^2 + 1)(D^3 + D)y = 0.$$

The corresponding characteristic equation is

$$r^2(r^2 + 1)(r^3 + r) = 0,$$

which can be factored as
$$r^3(r^2 + 1)^2 = 0.$$

The eigenvalues are thus
$$0, 0, 0, -i, -i, i, i$$

so that the solution of this homogeneous equation is

$$y(x) = c_1 + c_2 x + c_3 x^2 + c_4 \cos x + c_5 \sin x + c_6 x \cos x + c_7 x \sin x.$$

As we want to assume the simplest form for the particular solution that will satisfy the nonhomogeneous equation, we ignore any factors that also occur in the homogeneous equation. Solving the corresponding homogeneous equation $y''' + y' = 0$, we find the characteristic polynomial is $r(r^2 + 1)$ and the complementary solution is thus

$$y_c(x) = b_1 + b_2 \cos x + b_3 \sin x.$$

Comparing $y(x)$ and $y_c(x)$ we determine that the particular solution must be of the form

$$y_p(x) = Ax + Bx^2 + Cx \cos x + Ex \sin x.$$

This is again the same form we could have obtained with steps 1–4 of Section 4.5.1. We again use the original nonhomogeneous equation to determine the constants A, B, C, and E. Differentiating y_p three times and substituting into the left-hand side of the nonhomogeneous equation gives

$$y_p''' + y_p' = -3C \cos x + Cx \sin x - 3E \sin x - Ex \cos x$$

$$+ A + 2Bx + C \cos x - Cx \sin x + E \sin x + Ex \cos x$$

$$= -2C \cos x - 2E \sin x + 2Bx + A.$$

This must equal the right-hand side of the nonhomogeneous equation and so we have

$$-2C \cos x - 2E \sin x + 2Bx + A = \cos x + x.$$

Comparing coefficients gives

$$A = 0, \ B = \frac{1}{2}, \ C = -\frac{1}{2}, \ \text{and } E = 0.$$

This gives the particular solution of the nonhomogeneous equation as

$$y_p(x) = \frac{1}{2}x^2 - \frac{1}{2}x \cos x.$$

Thus, the general solution of the nonhomogeneous equation is

$$y(x) = b_1 + b_2 \cos x + b_3 \sin x + \frac{1}{2}x^2 - \frac{1}{2}x \cos x.$$

Example 6: Solve $y'' + 2y' - 3y = 4e^x - \sin x$ with the conditions that $y(0) = 0$ and $y'(0) = 1$.

We first note that this problem differs from previous ones because of the presence of initial conditions, but these do not come into the problem until the last step. Proceeding as in previous problems, since $D - 1$ annihilates e^x and $D^2 + 1$ annihilates $\sin x$, the nonhomogeneous term $4e^x - \sin x$ is annihilated by $A(D) = (D - 1)(D^2 + 1)$. Thus if we apply this annihilator to both sides of the differential equation we have

$$(D - 1)(D^2 + 1)(D^2 + 2D - 3)y = (D - 1)(D^2 + 1)(4e^x - \sin x)$$

and our problem becomes that of solving the homogeneous equation

$$(D-1)^2(D^2+1)(D+3)y = 0.$$

The corresponding characteristic equation is

$$(r-1)^2(r^2+1)(r+3) = 0$$

which has roots $-3, 1, 1, -i, i$. The solution is then

$$y(x) = c_1 e^{-3x} + c_2 e^x + c_3 x e^x + c_4 \cos x + c_5 \sin x. \tag{4.57}$$

The solution to the corresponding homogeneous equation $y'' + 2y' - 3y = 0$ with characteristic equation $r^2 + 2r - 3 = (r+3)(r-1) = 0$ is

$$y_c(x) = b_1 e^{-3x} + b_2 e^x.$$

We eliminate these terms from (4.57) to obtain the form for the particular solution as

$$y_p(x) = Axe^x + B\cos x + C\sin x.$$

Differentiating $y_p(x)$ gives

$$y_p'(x) = Ae^x + Axe^x - B\sin x + C\cos x$$

and

$$y_p''(x) = 2Ae^x + Axe^x - B\cos x - C\sin x.$$

Substitution of these three equations into the original nonhomogeneous equation gives

$$4Ae^x + (2C - 4B)\cos x + (-4C - 2B)\sin x = 4e^x - \sin x,$$

so that

$$4A = 4, \ 2C - 4B = 0, \ \text{and} \ -4C - 2B = -1.$$

Thus, $A = 1, B = \frac{1}{10}$, and $C = \frac{1}{5}$. and the particular solution y_p is

$$y_p(x) = xe^x + \frac{1}{10}\cos x + \frac{1}{5}\sin x.$$

This gives the general solution as

$$y(x) = b_1 e^{-3x} + b_2 e^x + xe^x + \frac{1}{10}\cos x + \frac{1}{5}\sin x.$$

We now apply the initial conditions to determine the constants b_1 and b_2. Since

$$0 = y(0) = b_1 + b_2 + \frac{1}{10} \ \text{and} \ 1 = y'(0) = b_2 - 3b_1 + \frac{6}{5}$$

we can solve the two equations simultaneously to obtain

$$b_1 = \frac{1}{40} \ \text{and} \ b_2 = -\frac{1}{8}.$$

Thus the solution to the initial value problem is

$$y(x) = \frac{1}{40}e^{-3x} - \frac{1}{8}e^x + xe^x + \frac{1}{10}\cos x + \frac{1}{5}\sin x.$$

4.6.2   Summary of the Annihilator Method

We can summarize the steps of the annihilator method of solving the nth order linear nonhomogeneous equation with constant coefficients

$$P(D)y = F(x)$$

as follows:

1. Determine an annihilator of $F(x)$. That is, find $A(D)$ such that

$$A(D)F(x) = 0.$$

2. Apply the annihilator to both sides of the nonhomogeneous differential equation

$$A(D)P(D)y = A(D)F(x) = 0.$$

3. Solve the homogeneous equation

$$A(D)P(D)y = 0.$$

4. Find the corresponding solution $y_c(x)$ to the homogeneous equation

$$P(D)y = 0$$

which corresponds to the original nonhomogeneous equation.

5. Eliminate the terms of the homogeneous equation $y_c(x)$ from the general solution obtained in step **3.** The expression that remains is the correct form of a particular solution.

6. Solve for the unknown coefficients in the particular solution $y_p(x)$ by substituting it into the original nonhomogeneous equation

$$P(D)y = F(x).$$

7. A general solution of the nonhomogeneous equation is

$$y(x) = y_c(x) + y_p(x).$$

It may often be easier to do step 6 in the above summary/method with the help of our computer programs. In this Example 6, we begin with the assumption of the correct form for the particular solution:

Computer Code 4.5:   Obtaining coefficients for the particular solution of a nonhomogeneous equation

Matlab, Maple, Mathematica

Matlab

```
>>  %This requires the Symbolic Math Toolbox
>>  syms x A B C
>>  eqyp=A*x*exp(x)+B*cos(x)+C*sin(x)
>>  eq1=diff(eqyp,x,2)+2*diff(eqyp,x)-3*eqyp -4*exp(x)+sin(x)
>>  %The above eq1 is the original ode written as lhs-rhs=0,
>>  %where the unwritten '=0' at the end is understood by
>>  %matlab.  We now type the relationships of the
>>  %coeffs by inspection of the calculated ode eq1
>>  eq2a=4*A-4 %coefficient of exp(x)
>>  eq2b=2*C-4*B %coefficient of cos(x)
>>  eq2c=-4*C-2*B+1 %coefficient of sin(x)
>>  [A,B,C]=solve(eq2a,eq2b,eq2c)
>>  yp=subs(eqyp)
```

Maple

```
>  eqyp:=y[p](x)=A*x*exp(x)+B*cos(x)+C*sin(x);
>  eqODE:=diff(y[p](x),x$2)+2*diff(y[p](x),x)-3*y[p](x)=
   4*exp(x)-sin(x);
>  eq1:=simplify(subs(eqyp,eqODE));
   #We need to inspect the powers of like terms and
   #write down the expressions involving A,B,C.
>  eq2a:=subs(exp(x)=1,cos(x)=0,sin(x)=0,eq1); #coeff of exp
>  eq2b:=subs(exp(x)=0,cos(x)=1,sin(x)=0,eq1); #coeff of cos
>  eq2c:=subs(exp(x)=0,cos(x)=0,sin(x)=1,eq1); #coeff of sin
>  eq3:=solve({eq2a,eq2b,eq2c},{A,B,C});
>  eq4:=y[p]=subs(eq3,eqyp);
```

Mathematica

```
yp[x_] = A x e^x +B Cos[x] + C Sin[x]
dey[x_]=y''[x]+2 y'[x]-3 y[x]-4 e^x+ Sin[x]
lhs=Simplify[ReplaceAll[dey[x],y→ yp]]
eq1=Coefficient[lhs,e^x]
eq2=Coefficient[lhs,Cos[x]]
eq3=Coefficient[lhs,Sin[x]]
Solve[{eq1==0,eq2==0,eq3==0},{A,B,C}]
```

Important Note: If Section 4.5.1 was skipped, the reader is strongly encouraged to read Section 4.5.2, which deals with the phenomenon of resonance that can occur in the forced mass on a spring system.

Problems

In problems 1–8, use the annihilator method to write the form of the particular solution but do not solve the differential equation. You will need to do steps 1–5 of the summary; see Section 4.6.2.

1. $y'' + y = e^{-x} + x^2$
2. $y'' + 3y' = x^2 e^{2x} + \cos x$
3. $y'' + y = 4\sin x + e^x \cos x$
4. $y'' + 4y' + y = \sin(2x) + \cos x$
5. $y''' + y = xe^{2x}\cos x + \sin x$
6. $y'' - y = e^x + xe^{-x}$
7. $y''' + 8y = 4\sin x + \cos x$
8. $y^{(5)} + y'' = x^3 + xe^x \cos x$

In problems 9–15, use the method of undetermined coefficients (only step 6 of the summary) to find the coefficients of the particular solution for the given equations.

9. $y'' + 4y' + 13y = e^{-2x}$, $y_p = Ae^{-2x}$
10. $y'' + 4y' = x^2 - 3$, $y_p = Ax^3 + Bx^2 + Cx$
11. $y'' - 4y' = \cos x + 3\sin x$, $y_p = A\cos x + B\sin x$
12. $9y'' + y = \cos x + \sin(2x)$, $y_p = A\cos x + B\sin x + C\cos(2x) + E\sin(2x)$
13. $4y'' + 25y = x\cos x$, $y_p = A\cos x + B\sin x + Cx\cos x + Ex\sin x$
14. $y'' + 2y' + 17y = e^x + 2$, $y_p = Ae^x + B$
15. $y'' + 3y' - 10y = xe^x + 2x$, $y_p = Ae^x + Bxe^x + Cx + E$

For problems 16–32, find the general solution of the given differential equations using the annihilator method. If instructed to do so, plot the analytical solution and compare with the solution obtained by numerically solving the equation for the given initial condition (or an initial condition of your choice if none is given).

16. $y'' - 3y' + 2y = 4x^2$
17. $y'' - 2y' - 8y = 4e^{2x} - 21e^{-3x}$
18. $y'' + 2y' + 5y = 6\sin 2x + 7\cos 2x$
19. $y''' + 10y'' + 34y' + 40y = xe^{-4x} + 2e^{-3x}\cos x$
20. $y'' - 4y = 32x$, $y(0) = 0, y'(0) = 6$
21. $y'' - 2y' + 2y = e^x + x\cos x$

22. $y'' + 6y' + 10y = 3xe^{-3x} - 2e^{3x}\cos x, y(0) = 1, y'(0) = -2$

23. $y'' - 8y' + 20y = 5xe^{4x}\sin 2x$

24. $y'' + 7y' + 10y = xe^{-2x}\cos 5x, y(1) = 3, y'(1) = 0$

25. $y'' - 2y' + 5y = 2xe^x + e^x\sin 2x$

26. $y'' - 2y' + y = 2xe^x + e^x\sin 2x$

27. $y'' - 3y' + 2y = e^x, \ y(0) = 1, y'(0) = 0$

28. $y'' - y = 4\sinh x$

29. $y'' + 4y' + 3y = \cosh x$

30. $y'' + 4y = (\sinh x)(\sin 2x)$

31. $y'' + 2y' + 2y = (\cosh x)(\sin x)$

32. $y^{(4)} - 18y'' + 81y = e^{3x}$

33. The annihilator method can be used to solve first-order constant coefficient nonhomogeneous equations. Use this method to solve the following problems:

    a. $y' - 4y = x^2$
    b. $y' + y = \cos 2x$
    c. $y' - y = e^{4x}$

Section 4.5.2 shows how to use the problem of a forced mass on a spring and this is a specific example of a nonhomogeneous equation. Use the annihilator method to answer problems 34–36.

34. Derive equation (4.48), which gives the constants of the particular solution.

35. Derive equation (4.49), thus obtaining the amplitude-phase form for the particular solution.

36. Derive the resonant frequency, equation (4.51), for the forced mass-spring system.

37. Suppose that $y_1(x)$ and $y_2(x)$ are solutions of

$$a\frac{d^2y}{dx^2} + b\frac{dy}{dx} + cy = f(x),$$

where a, b, and c are positive constants.

a. Show that

$$\lim_{x\to\infty}[y_2(x) - y_1(x)] = 0.$$

b. Is the result of **a** true if $b = 0$?

c. Suppose that $f(x) = k$, where k is a constant. Show that

$$\lim_{x\to\infty} y(x) = \frac{k}{c}.$$

for every solution $y(x)$ of

$$a\frac{d^2y}{dx^2} + b\frac{dy}{dx} + cy = k.$$

d. Determine the solution $y(x)$ of

$$a\frac{d^2y}{dx^2} + b\frac{dy}{dx} = k.$$

Find

$$\lim_{x \to \infty} y(x).$$

e. Determine the solution $y(x)$ of

$$a\frac{d^2y}{dx^2} = k.$$

Find

$$\lim_{x \to \infty} y(x).$$

38. a. Let $f(x)$ be a polynomial of degree n. Show that, if $b \neq 0$, there is always a solution that is a polynomial of degree n for the equation $y'' + ay' + by = f(x)$.
   b. Find a particular solution of $y'' + 3y' + 2y = 9 + 2x - 2x^2$.

39. In many physical applications, the nonhomogeneous term $F(x)$ is specified by different formulas in different intervals of x.
   a. Find a general solution of the equation

$$y'' + y = \begin{cases} x, \ 0 \le x \le 1, \\ 1, \ 1 \le x. \end{cases}$$

Note that the solution is not differentiable at $x = 1$.
   b. Find a particular solution of

$$y'' + y = \begin{cases} x, \ 0 \le x \le 1, \\ 1, \ 1 \le x \end{cases}$$

that satisfies the initial conditions $y(0) = 0$ and $y'(0) = 1$.

4.7 Variation of Parameters

The method of undetermined coefficients (either via tables or annihilation) *does not* apply to functions $F(x)$ that are *not* already solutions to some linear constant-coefficient homogeneous differential equation, nor to equations with variable coefficients. We will now develop a general method that applies to all nonhomogeneous linear equations. The only catch is that it requires knowledge of the homogeneous solution yet does not give us a method for finding this solution. If the homogeneous part of the equation has constant coefficients, we can use the previous methods to find the complementary solution. If the homogeneous equation has *non-constant* coefficients, we must be given the solutions in order to apply this method.

Consider the nonhomogeneous equation

$$y'' + y = \tan x. \tag{4.58}$$

The nonhomogeneous term

$$F(x) = \tan x$$

is not of the form

$$x^j, x^j e^{rx}, x^j e^{ax} \cos bx \text{ or } x^j e^{ax} \sin bx$$

so our previous methods will not work here. The variation of parameters approach, which was discovered by Lagrange, is a method that finds a particular solution for the differential equation. The idea is to use the general solution of the corresponding homogeneous equation

$$y'' + y = 0$$

to find a particular solution. In this case, the characteristic polynomial is

$$r^2 + 1 = 0,$$

which gives the solution

$$y_c(x) = c_1 \sin x + c_2 \cos x.$$

We will seek a particular solution of the form

$$y_p(x) = u_1(x) \sin x + u_2(x) \cos x \tag{4.59}$$

where we have replaced the constants c_1 and c_2 by unknown functions $u_1(x)$ and $u_2(x)$.

Thus, we see how the name of this method applies. We *vary the parameters* c_1 and c_2 by allowing them to be *functions* of x instead of *constants*. Notice that there should be many choices for $u_1(x)$ and $u_2(x)$ because we have two unknown functions and only one equation, the nonhomogeneous differential equation, to use to find them. Because we know that y_p must be a solution to (4.58), we substitute it into this equation. Differentiating y_p gives

$$y_p'(x) = u_1(x)\cos x + u_1'(x)\sin x - u_2(x)\sin x + u_2'(x)\cos x.$$

Now to simplify the process of finding $u_1(x)$ and $u_2(x)$, we impose our *first restriction*

$$u_1'(x)\sin x + u_2'(x)\cos x = 0 \qquad (4.60)$$

on $u_1(x)$ and $u_2(x)$. We can think of this condition as arising from the assumed form of the particular solution in the case where the two functions $u_1(x)$ and $u_2(x)$ are actually constants (and thus we would have $u_1' = u_2' = 0$). Even though, in general, our two functions $u_1(x), u_2(x)$ are actually *not* constant, we are simply imposing a condition that will allow us to find suitable functions that will make y_p a solution.

Because of our restriction in (4.60), we have

$$y_p'(x) = u_1(x)\cos x - u_2(x)\sin x.$$

The second derivative of y_p is thus

$$y_p''(x) = -u_1(x)\sin x + u_1'(x)\cos x - u_2(x)\cos x - u_2'(x)\sin x.$$

Now we substitute y_p and y_p'' into the nonhomogeneous equation to obtain

$$y_p''(x) + y_p(x) = (-u_1(x)\sin x + u_1'(x)\cos x - u_2(x)\cos x - u_2'(x)\sin x)$$

$$+ (u_1(x)\sin x + u_2(x)\cos x)$$

$$= u_1'(x)\cos x - u_2'(x)\sin x.$$

Since y_p is a particular solution, we know that it must satisfy the nonhomogeneous equation. We thus have our *second restriction* on $u_1(x)$ and $u_2(x)$, namely,

$$u_1'(x)\cos x - u_2'(x)\sin x = \tan x. \qquad (4.61)$$

Our two restrictions, equations (4.60) and (4.61), give us a system of two equations in the two unknown functions $u_1'(x), u_2'(x)$,

$$\begin{cases} u_1'(x)\sin x + u_2'(x)\cos x = 0 \\ -u_1'(x)\cos x + u_2'(x)\sin x = \tan x, \end{cases} \qquad (4.62)$$

which we can solve for $u_1'(x)$ and $u_2'(x)$. We could use Cramer's rule (see Appendix C.2.4 for a self-contained review of it) to solve this or simply manipulate the equations to solve.

In the former case, we have that

$$u_1'(x) = \frac{\begin{vmatrix} 0 & \cos x \\ \tan x & -\sin x \end{vmatrix}}{\begin{vmatrix} \sin x & \cos x \\ \cos x & -\sin x \end{vmatrix}}$$

$$= \frac{(0)(-\sin x) - (\cos x)(\tan x)}{(\sin x)(-\sin x) - (\cos x)(\cos x)} = \frac{-\cos x \tan x}{-1}$$

and

$$u_2'(x) = \frac{\begin{vmatrix} \sin x & 0 \\ \cos x & \tan x \end{vmatrix}}{\begin{vmatrix} \sin x & \cos x \\ \cos x & -\sin x \end{vmatrix}}$$

$$= \frac{(\sin x)(\tan x) - (\cos x)(0)}{(\sin x)(-\sin x) - (\cos x)(\cos x)} = \frac{\sin x \tan x}{-1}.$$

To find the two functions, we simply need to integrate. This gives

$$u_1(x) = \int \frac{-\cos x \tan x}{-1}\, dx = \int \sin x\, dx = -\cos x$$

and

$$u_2(x) = \int \frac{\sin x \tan x}{-1}\, dx = \int \frac{-\sin^2 x}{\cos x}\, dx = \sin x - \ln|\sec x + \tan x|.$$

Note that *any* antiderivative of $u_1'(x)$ and $u_2'(x)$ is possible as a choice for $u_1(x)$ and $u_2(x)$, respectively. We have simply taken the one with constant zero. Substitution into (4.59) gives the particular solution

$$y_p(x) = u_1(x)y_1(x) + u_2(x)y_2(x)$$

$$= -\cos x \sin x + (\sin x - \ln|\sec x + \tan x|)\cos x.$$

Thus, the general solution is

$$y(x) = y_c(x) + y_p(x)$$

$$= c_1 \sin x + c_2 \cos x - \cos x \sin x + (\sin x - \ln|\sec x + \tan x|)\cos x.$$

In general, to solve the second-order linear nonhomogeneous differential equation

$$a_0(x)y''(x) + a_1(x)y'(x) + a_2(x)y(x) = F(x),$$

where

$$y_c(x) = c_1 y_1(x) + c_2 y_2(x)$$

is a general solution of the corresponding homogeneous equation

$$a_0(x)y''(x) + a_1(x)y'(x) + a_2(x)y(x) = 0,$$

we divide by the lead coefficient $a_0(x)$ to write

$$y''(x) + p(x)y'(x) + q(x)y(x) = G(x).$$

At this point, we need to be given or be able to calculate a fundamental solution set $\{y_1(x), y_2(x)\}$ for the homogeneous equation. If the homogeneous equation has constant coefficients this is done by our previous methods; if the coefficients are not constant, we must be given the fundamental set as the methods for calculating these are beyond the scope of this text. We next assume that a particular solution has a form similar to the general solution y_c by varying the parameters c_1 and c_2, that is, we let

$$y_p(x) = u_1(x)y_1(x) + u_2(x)y_2(x).$$

We need to determine $u_1(x)$ and $u_2(x)$, so we seek two equations by substituting y_p into the second-order linear nonhomogeneous differential equation

$$y''(x) + p(x)y'(x) + q(x)y(x) = G(x).$$

So, differentiating y_p gives

$$y_p'(x) = u_1(x)y_1'(x) + u_1'(x)y_1(x) + u_2(x)y_2'(x) + u_2'(x)y_2(x)$$

which simplifies to

$$y_p'(x) = u_1(x)y_1'(x) + u_2(x)y_2'(x)$$

by making the assumption

$$u_1'(x)y_1(x) + u_2'(x)y_2(x) = 0.$$

The second derivative is

$$y_p''(x) = u_1(x)y_1''(x) + u_1'(x)y_1'(x) + u_2(x)y_2''(x) + u_2'(x)y_2'(x).$$

Now we substitute these expressions into the second-order linear nonhomogeneous differential equation to obtain

$$y_p''(x) + p(x)y_p'(x) + q(x)y_p(x)$$

$$= u_1(x)y_1''(x) + u_1'(x)y_1'(x) + u_2(x)y_2''(x) + u_2'(x)y_2'(x)$$

$$+ p(x)\left[u_1(x)y_1'(x) + u_2(x)y_2'(x)\right]$$

$$+ q(x)\left[u_1(x)y_1(x) + u_2(x)y_2(x)\right]$$

$$= u_1(x)\left[y_1''(x) + p(x)y_1'(x) + q(x)y_1(x)\right]$$

$$+ u_2(x)\left[y_2''(x) + p(x)y_2'(x) + q(x)y_2(x)\right]$$

$$+ u_1'(x)y_1'(x) + u_2'(x)y_2'(x).$$

However, $y_1(x)$ and $y_2(x)$ are solutions of the corresponding homogeneous equation

$$y''(x) + p(x)y'(x) + q(x)y(x) = 0$$

so that

$$y_1''(x) + p(x)y_1'(x) + q(x)y_1(x) = 0$$

and

$$y_2''(x) + p(x)y_2'(x) + q(x)y_2(x) = 0.$$

Thus, since

$$y''(x) + p(x)y'(x) + q(x)y(x) = G(x),$$

it follows that

$$u_1'(x)y_1'(x) + u_2'(x)y_2'(x) = G(x).$$

Hence, we have the second equation of our system needed to determine $u_1(x)$ and $u_2(x)$. Specifically,

$$\begin{cases} u_1'(x)y_1(x) + u_2'(x)y_2(x) = 0 \\ u_1'(x)y_1'(x) + u_2'(x)y_2'(x) = G(x). \end{cases}$$

Using Cramer's rule to solve this system (see Appendix C.2.4 for a brief review), we have the unique solution

$$u_1'(x) = \frac{\begin{vmatrix} 0 & y_2(x) \\ G(x) & y_2'(x) \end{vmatrix}}{W(x)} = \frac{-y_2(x)G(x)}{W(x)}$$

and

$$u_2'(x) = \frac{\begin{vmatrix} y_1(x) & 0 \\ y_1'(x) & G(x) \end{vmatrix}}{W(x)} = \frac{y_1(x)G(x)}{W(x)},$$

where

$$W(x) = W(y_1(x), y_2(x)) = \begin{vmatrix} y_1(x) & y_2(x) \\ y_1'(x) & y_2'(x) \end{vmatrix}.$$

We note that $W(x) \neq 0$ because $\{y_1(x), y_2(x)\}$ is a fundamental solution set and thus we know that $y_1(x)$ and $y_2(x)$ are linearly independent.

4.7.1 Summary of Variation of Parameters Method

Rewrite the given second-order equation so that the leading coefficient is one:

$$y''(x) + p(x)y'(x) + q(x)y(x) = G(x).$$

Then follow the following 5 steps:

1. Find or be given a complementary (homogeneous) solution

$$y_c(x) = c_1 y_1(x) + c_2 y_2(x)$$

and fundamental solutions

$$S = \{y_1(x), y_2(x)\}$$

of the corresponding homogeneous equation

$$y''(x) + p(x)y'(x) + q(x)y(x) = 0.$$

2. Let

$$u_1'(x) = \frac{-y_2(x)G(x)}{W(x)} \quad \text{and} \quad u_2'(x) = \frac{y_1(x)G(x)}{W(x)}.$$

3. Integrate $u_1'(x)$ and $u_2'(x)$ to obtain $u_1(x)$ and $u_2(x)$.

4. A particular solution of $y''(x) + p(x)y'(x) + q(x)y(x) = G(x)$ is given by

$$y_p(x) = u_1(x)y_1(x) + u_2(x)y_2(x).$$

5. The general solution of $y''(x) + p(x)y'(x) + q(x)y(x) = G(x)$ is given by

$$y(x) = y_c(x) + y_p(x).$$

Example 1: Solve $y'' - 2y' + y = e^x \ln x$ for $x > 0$.

The corresponding homogeneous equation $y'' - 2y' + y = 0$ has the characteristic polynomial

$$m^2 - 2m + 1 = 0$$

so that the corresponding solution is

$$y_c(x) = c_1 e^x + c_2 x e^x.$$

Thus,

$$y_1(x) = e^x \text{ and } y_2(x) = xe^x.$$

So, $S = \{e^x, xe^x\}$ and

$$W(x) = \begin{vmatrix} e^x & xe^x \\ e^x & e^x + xe^x \end{vmatrix} = (e^x)(e^x + xe^x) - (e^x)(xe^x) = e^{2x}.$$

This gives

$$u_1(x) = \int \frac{-xe^x (e^x \ln x)}{e^{2x}} dx$$

and

$$u_2(x) = \int \frac{e^x (e^x \ln x)}{e^{2x}} dx.$$

Both of these integrals can be obtained by integrating by parts and are

$$u_1(x) = \frac{1}{4}x^2 - \frac{1}{2}x^2 \ln x$$

and

$$u_2(x) = x \ln x - x.$$

The particular solution is

$$y_p(x) = u_1(x)y_1(x) + u_2(x)y_2(x)$$

$$= \left(\frac{1}{4}x^2 - \frac{1}{2}x^2 \ln x \right) e^x + (x \ln x - x) xe^x$$

$$= \frac{1}{2}x^2 e^x \ln x - \frac{3}{4}x^2 e^x.$$

This gives the general solution as

$$y(x) = y_c(x) + y_p(x)$$

$$= c_1 e^x + c_2 xe^x + \frac{1}{2}x^2 e^x \ln x - \frac{3}{4}x^2 e^x.$$

We can use Matlab, Maple, or Mathematica to quickly calculate our determinants and evaluate the integrals.

Computer Code 4.6: **Obtaining a particular solution via variation of parameters**

Matlab, Maple, Mathematica

```
                        Matlab

≫  syms x
≫  y1=exp(x)
≫  y2=x*exp(x)
≫  Gx=exp(x)*log(x)
≫  A=[y1,y2; diff(y1,x),diff(y2,x)]
≫  Wronsk=simplify(det(A))
≫  numer1=[0,y2; Gx,diff(y2,x)]
≫  numer2=[y1,0; diff(y1,x),Gx]
≫  u1=int(det(numer1)/Wronsk,x)
≫  u2=int(det(numer2)/Wronsk,x)
```

```
                        Maple

>  y1:=exp(x);
>  y2:=x*exp(x);
>  Gx:=exp(x)*log(x);
>  with(linalg):
>  A:=matrix(2,2,[y1,y2,diff(y1,x),diff(y2,x)]);
>  Wronsk:=simplify(det(A));
>  numer1:=matrix(2,2,[0,y2,Gx,diff(y2,x)]);
>  numer2:=matrix(2,2,[y1,0,diff(y1,x),Gx]);
>  u1:=int(det(numer1)/Wronsk,x);
>  u2:=int(det(numer2)/Wronsk,x);
```

```
                     Mathematica

y1[x_]=eˣ
y2[x_]= x eˣ
G[x_]=eˣ Log[x]  (*right-hand-side of ODE*)
A={{y1[x],y2[x]},{y1'[x],y2'[x]}}
Wronsk[x_]=Det[A]
eq1[x_]=Det[{{0,y2[x]},{G[x],y2'[x]}}]
eq2[x_] = Det[{{y1[x],0},{y1'[x],G[x]}}]
```
$$u1[x\_]=\int \frac{eq1[x]}{Wronsk[x]}\, dx \quad (*\text{entered from palette}*)$$
$$u2[x\_]=\int \frac{eq2[x]}{Wronsk[x]}\, dx \quad (*\text{entered from palette}*)$$

Example 2: Solve the nonhomogeneous equation

$$y'' + \frac{1}{4}y = \sec \frac{x}{2} + \csc \frac{x}{2} \tag{4.63}$$

on the interval $0 < x < \pi$.

The characteristic equation is

$$r^2 + \frac{1}{4} = 0$$

with corresponding solution

$$y_c(x) = c_1 \cos \frac{x}{2} + c_2 \sin \frac{x}{2}.$$

We thus let

$$y_1(x) = \cos \frac{x}{2} \text{ and } y_2(x) = \sin \frac{x}{2}$$

and calculate

$$W(x) = \begin{vmatrix} \cos \frac{x}{2} & \sin \frac{x}{2} \\ -\frac{1}{2}\sin \frac{x}{2} & \frac{1}{2}\cos \frac{x}{2} \end{vmatrix}$$

$$= \left(\cos \frac{x}{2}\right)\left(\frac{1}{2}\cos \frac{x}{2}\right) - \left(\sin \frac{x}{2}\right)\left(-\frac{1}{2}\sin \frac{x}{2}\right) = \frac{1}{2}.$$

This gives

$$u_1(x) = \int -2 \sin \frac{x}{2}\left(\sec \frac{x}{2} + \csc \frac{x}{2}\right) dx$$

$$= -2x + 4\ln\left|\cos \frac{x}{2}\right|,$$

$$u_2(x) = \int 2 \cos \frac{x}{2}\left(\sec \frac{x}{2} + \csc \frac{x}{2}\right) dx$$

$$= 2x + 4\ln\left|\sin \frac{x}{2}\right|.$$

Thus, a particular solution is given by

$$y_p(x) = u_1(x) \cos \frac{x}{2} + u_2(x) \sin \frac{x}{2}$$

$$= \left(-2x + 4\ln\left|\cos \frac{x}{2}\right|\right)\cos \frac{x}{2} + \left(2x + 4\ln\left|\sin \frac{x}{2}\right|\right)\sin \frac{x}{2}$$

and the general solution is

$$y(x) = y_c(x) + y_p(x)$$

$$= c_1 \cos \frac{x}{2} + c_2 \sin \frac{x}{2}$$

$$+ \left(-2x + 4\ln\left|\cos \frac{x}{2}\right|\right)\cos \frac{x}{2} + \left(2x + 4\ln\left|\sin \frac{x}{2}\right|\right)\sin \frac{x}{2}.$$

Example 3: Solve the nonhomogeneous equation

$$x^2 \frac{d^2y}{dx^2} + x\frac{dy}{dx} - y = 3e^{2x} \tag{4.64}$$

for $x > 0$.

We first solve the homogeneous equation

$$x^2 \frac{d^2y}{dx^2} + x\frac{dy}{dx} - y = 0,$$

a Cauchy-Euler equation that we considered in Section 4.3. By making the change of variables $x = e^t$ we have the equation

$$\frac{d^2y}{dt^2} - y = 0.$$

The corresponding characteristic equation is

$$r^2 - 1 = 0$$

which has roots $r_1 = -1$ and $r_2 = 1$. This gives the solution to the homogeneous Cauchy-Euler equation as

$$y_c(x) = c_1 x^{-1} + c_2 x.$$

Letting

$$y_1(x) = x^{-1} \text{ and } y_2(x) = x$$

we have

$$W(x) = \begin{vmatrix} x^{-1} & x \\ -x^{-2} & 1 \end{vmatrix} = 2x^{-1},$$

which we note is not zero for any real x. This gives

$$u_1'(x) = \frac{-3}{2}x^2 e^{2x} \text{ and } u_2'(x) = \frac{-3}{2}e^{2x},$$

so

$$u_1(x) = \int \frac{-3}{2}x^2 e^{2x} \, dx$$

$$= -\frac{3}{8}(2x^2 - 2x + 1)e^{2x}$$

integrating by parts twice and

$$u_2(x) = \int \frac{-3}{2}e^{2x} \, dx = -\frac{3}{4}e^{2x}.$$

So a particular solution is given by

$$y_p(x) = u_1(x)x^{-1} + u_2(x)x$$

$$= -\frac{3}{8}(2x^2 - 2x + 1)e^{2x}x^{-1} + -\frac{3}{4}e^{2x}x$$

and the general solution is

$$y(x) = y_c(x) + y_p(x)$$

$$= c_1 x^{-1} + c_2 x + -\frac{3}{8}(2x - 2 + x^{-1})e^{2x} + -\frac{3}{4}xe^{2x}.$$

Higher-Order Equations

Variation of Parameters extends to higher dimensions. We consider an nth order equation with leading coefficient equal to one:

$$y^{(n)} + p_1(x)y^{(n-1)} + \ldots + p_{n-1}(x)y' + p_n(x)\,y = G(x). \tag{4.65}$$

If $\{y_1, y_2, \cdots, y_n\}$ is a fundamental set of solutions for the reduced equation, then we assume a particular solution of the form

$$y_p = u_1(x)y_1 + u_2(x)y_2 + \cdots + u_n(x)y_n$$

where the $u_i(x)$ are determined by the system of equations

$$
\begin{array}{ccccccc}
y_1 u_1' & + & y_2 u_2' & + \cdots + & y_n u_n' & = & 0 \\
y_1' u_1' & + & y_2' u_2' & + \cdots + & y_n' u_n' & = & 0 \\
\vdots & & \vdots & \ddots & \vdots & & \vdots \\
y_1^{(n-1)} u_1' & + & y_2^{(n-1)} u_2' & + \cdots + & y_n^{(n-1)} u_n', & = & G(x).
\end{array}
\tag{4.66}
$$

Using Cramer's rule, we can calculate the functions $u_i'(x)$ that we will integrate. In the case of the third-order equation with fundamental set of solutions $\{y_1, y_2, y_3\}$, we write a particular solution in the form

$$y_p = u_1(x)y_1 + u_2(x)y_2 + u_3(x)y_3, \tag{4.67}$$

where

$$
u_1'(x) = \frac{\begin{vmatrix} 0 & y_2 & y_3 \\ 0 & y_2' & y_3' \\ G(x) & y_2'' & y_3'' \end{vmatrix}}{\begin{vmatrix} y_1 & y_2 & y_3 \\ y_1' & y_2' & y_3' \\ y_1'' & y_3' & y_3'' \end{vmatrix}}, \quad
u_2'(x) = \frac{\begin{vmatrix} y_1 & 0 & y_3 \\ y_1' & 0 & y_3' \\ y_1'' & G(x) & y_3'' \end{vmatrix}}{\begin{vmatrix} y_1 & y_2 & y_3 \\ y_1' & y_2' & y_3' \\ y_1'' & y_3' & y_3'' \end{vmatrix}}, \quad
u_3'(x) = \frac{\begin{vmatrix} y_1 & y_2 & 0 \\ y_1' & y_2' & 0 \\ y_1'' & y_3' & G(x) \end{vmatrix}}{\begin{vmatrix} y_1 & y_2 & y_3 \\ y_1' & y_2' & y_3' \\ y_1'' & y_3' & y_3'' \end{vmatrix}}.
$$

We then integrate the u_i' and plug into (4.67) to obtain a particular solution. For higher-order equations, we again solve (4.66) and then integrate to obtain the $u_i(x)$. The particular solution is created and thus we are able to write the general solution.

Problems

Solve 1–17 using variation of parameters. If instructed to do so plot the analytical solution and compare with the solution obtained by numerically solving the equation for the given initial condition (or an initial condition of your choice if none is given).

1. $y'' = \dfrac{1}{1 + x^2}$

2. $y'' - 2y' + y = e^x \sqrt{x}$

3. $y'' + y = \sec x, y(0) = 1, y'(0) = 2$

4. $y'' + y = \tan x, y(0) = -1, y'(0) = 1$

5. $y'' + y = \sec^2 x, y(0) = 0, y'(0) = 1$

6. $y'' + 4y = \csc 2x$

7. $y'' - 2y' + y = \dfrac{e^x}{x^2}$

8. $y'' + 2y' + y = \dfrac{e^{-x}}{x^4}$

9. $y'' - 7y' + 10y = e^{3x}, y(0) = 1, y'(0) = 2$

10. $y'' + 5y' + 6y = e^{-x}, y(0) = 1, y'(0) = 2$

11. $y'' + y = 6x, \ y(0) = 1, y'(0) = 1$

12. $y'' + y = \sin^2 x, y(0) = 1, y'(0) = 0$

13. $(D - 1)^3(y) = \dfrac{e^x}{x}$

14. $D(D + 1)(D - 2)(y) = x^3$

15. $y'' + 27y = e^{3x}$

16. $y''' + y' = \tan x$

17. $y^{(4)} - 16y' = e^{4x}$

18. Solve $y'' + 4y' + 3y = 65 \cos 2x$ by
   a. the method of undetermined coefficients;
   b. the method of variation of parameters.
   Which method is more easily applied?

In problems 19–22, first check that the given fundamental set of solutions is actually a fundamental set of solutions. Then use variation of parameters to solve the equation.

19. $y'' + \dfrac{x-1}{x} y' = \dfrac{e^x}{1+x}$, $x > 0$ with fundamental set of solutions $\{1, (1 + x)e^{-x}\}$

20. $y'' - \dfrac{1}{x} y' + \dfrac{1}{x^2} y = \dfrac{1}{x}$, $x > 0$ with fundamental set of solutions $\{x, x \ln x\}$

21. $y'' + \dfrac{1}{x} y' + \dfrac{1}{x^2} y = \dfrac{1}{x}$, $x > 0$ with fundamental set of solutions $\{\sin(\ln x), \cos(\ln x)\}$

22. $2x^2 y''' + 6xy'' = 1$, $x > 0$ with fundamental set of solutions $\{1, 1/x, x\}$

23. Show that the solution of the initial-value problem

$$y''(x) + a_1(x)y'(x) + a_0(x)y(x) = f(x)$$

with $y(x_0) = y_0$, $y'(x_0) = y_0$ can be written as $y(x) = u(x) + v(x)$ where u is the solution of

$$u''(x) + a_1(x)u'(x) + a_0(x)u(x) = 0$$

with $u(x_0) = y_0$, $u'(x_0) = y_0$ and v is the solution of

$$v''(x) + a_1(x)v'(x) + a_0(x)v(x) = f(x)$$

with $v(x_0) = 0, v'(x_0) = 0$.

Solve the nonhomogenous Cauchy-Euler equations in problems 24–26.

24. $x^2 y'' + 4xy' + 2y = e^x$

25. $x^2 y'' + 3xy' + y = \ln x$

26. $x^2 y'' + xy' + 4y = 2$

4.8   Chapter 4: Additional Problems and Projects

ADDITIONAL PROBLEMS

In problems 1–5, determine whether the statement is true or false. If it is true, give reasons for your answer. If it is false, give a counterexample or other explanation of why it is false.

1. Every linear homogeneous equation has e^{rx} as a solution, where r is a constant that may be real or complex.

2. For a forced mass on a spring system, the transient solution dies off as $t \to \infty$ while the steady-state solution persists.

3. For a forced mass on a spring system, the steady-state solution corresponds to the complementary (or homogeneous) solution while the transient solution corresponds to the particular solution.

4. The method of undetermined coefficients (either via Tables or via the Annihilator Method) applies to any nonhomogeneous term as long as the associated homogeneous equation has constant coefficients.

5. Variation of parameters works for all differential equations.

Write the general solution of the equations given in 6–23.

6. $y'' - 5y' + 6y = 0$

7. $y'' - 3y' - 10y = 0$

8. $y'' + 6y' + 9y = 0$, $y(0) = 1, y'(0) = 0$

9. $y'' + 6y' + 8y = 0$, $y(0) = 1, y'(0) = -1$

10. $y'' + 2y' + 2y = 0$, $y(\pi) = 0, y'(\pi) = 1$

11. $y'' + 4y' + 13y = 0$ $y(\pi/2) = 2, y'(\pi/2) = 1$

12. $(D^2 + 1)(D - 2)(y) = 0$

13. $D^2(D + 1)(D - 4)(y) = 0$

14. $(D^2 + 1)(D - 2)(y) = 0$

15. $D(D + 1)^3(y) = 0$

16. $(D^3 + 4D)(y) = 0$, $y(0) = 1, y'(0) = 0, y''(0) = -1$

17. $(D^3 + 1)(y) = 0$, $y(0) = 2, y'(0) = -1, y''(0) = 0$

18. $(D - 1)^2(D + 2)(y) = 0$

19. $D(D + 7)(D^2 - 4)(y) = 0$

20. $(D^3 - 27)(y) = 0$

21. $2x^2y'' + 5xy' + y = 0$

22. $2x^2y'' + 3xy' - y = 0$

23. $3x^2y'' + 3xy' + y = 0$

Solve problems 24–32 using:
a. Method of Undetermined Coefficients
b. Variation of Parameters

24. $y'' + 4y' = xe^{-x}$

25. $y'' - 4y' = \cos x$

26. $y'' + 16y = 2\cos 4x$

27. $y'' + 4y' + 4y = e^{-2x}$

28. $9y'' + y = x$

29. $4y'' + 25y = 2e^{-x}$

30. $y'' + 2y' + 17y = e^{-x}\cos 4x$

31. $y'' + 3y' - 10y = x + e^{-3x}$

32. $y'' - y' - 2y = 16xe^{x/2}$

For problems 33–44, classify the motion of the mass on a spring as either underdamped, critically damped, or overdamped, or find the parameter value that will give the desired result. Assume $x = x(t)$, that is, x is a function of t.

33. $x'' + 4x' + x = 0$

34. $x'' + \frac{3}{2}x' + x = 0$

35. $x'' + 3x' + 2x = 0$

36. $3x'' + 4x' + 1x = 0$

37. $x'' + 10x' + 25x = 0$

38. $2x'' + 8x' + 8x = 0$

39. $x'' + bx' + 3x = 0$, undamped motion

40. $x'' + bx' + 4x = 0$, underdamped motion

41. $2x'' + x' + kx = 0$, overdamped motion

42. $x'' + 3x' + kx = 0$, underdamped motion

43. $mx'' + 2x' + 4x = 0$, critically damped motion

44. $mx'' + 7x' + 3x = 0$, overdamped motion

> *In Section 3.5, we learned how to numerically solve nonhomogeneous equations. Parts **b–d** in problems 45–52, ask you to compare your answers obtained with methods of this chapter with those obtained numerically using your computer program. In each case:*
>
> **a.** *Use the Method of Undetermined Coefficients (either via Tables or via the Annihilator Method) or Variation of Parameters to find the analytical solution for the given initial condition.*
>
> **b.** *Use your answer from a to obtain the value of the solution at x_f.*
>
> **c.** *For the step sizes $h = .1, .01$, calculate the approximate solution at this final value by using the fourth-order Runge-Kutta method.*
>
> **d.** *Plot the analytical solution and both approximate solutions on the same graph. Comment on the accuracy of the numerical method for each step size.*

45. $y'' + 2y' + y = e^x$, $y(0) = 1, y'(0) = 0$, $x_f = 4$

46. $y'' + 2y' + y = e^{-x}$, $y(0) = 1, y'(0) = 0$, $x_f = 4$

47. $y'' + 4y = \sin x$, $y(\pi) = 1, y'(\pi) = 0$, $x_f = 4$

48. $y'' - 4y = xe^{x^2}$, $y(0) = 1, y'(0) = 0$, $y_f = 3$

49. $y''' - 6y'' + 8y' = \ln x$, $y(1) = 1, y'(1) = 0$, $x_f = 5$

50. $2x^2 y'' + 5xy' + y = x$, $y(1) = 1, y'(1) = 0$, $x_f = 5$

51. $2x^2 y'' + 3xy' - y = \sqrt{x}$, $y(1) = 1, y'(1) = -1$, $x_f = 5$

52. $3x^2 y'' + 3xy' + y = 1$, $y(1) = 1, y'(1) = 1$, $x_f = 5$

PROJECTS FOR CHAPTER 4

Project 1: Forced Duffing Equation

One formulation of the forced Duffing equation is

$$x'' + bx' + kx + \delta x^3 = F_0 \sin(\omega t), \tag{4.68}$$

where $x = x(t)$. When $\delta = 0$, the equation reduces to that of the forced mass on a spring of Section 4.5.2. Thus we can think of the left-hand side of equation (4.68) as describing a *nonlinear* spring with attached mass $m = 1$. Let's take $k = \delta = 1$, for simplicity, and see what this means. For small x-values, i.e., the mass on the spring is not too far from rest position, the x^3 term is probably negligible and we can approximate the solution by dropping this term. However, for moderate or large x-values we cannot neglect this term.

1. The restoring force, $F = kx + \delta x^3$, is nonlinear. Give a physical interpretation of this term. For example, is the restoring force greater for small x or large x? What physical implications does this have?

2. Set $k = \delta = 1$, $F_0 = 0$, $b = 0.2$. Using Matlab, Maple, or Mathematica with $h = 0.01$, let $x'(0) = 1.0$ and begin with the five different initial conditions $x(0) = 0, 0.5, 1.0, 3.0, 4.0$ to plot five different solutions in the t-x plane. What are the transient and long-term behaviors of the solution? Numerically solve the equation to at least $t = 100$.

3. Now we will observe what happens as we force this nonlinear oscillator by changing F_0 and ω. Still keeping $k = \delta = 1$, $b = 0.2$, set $\omega = 1$ and plot solutions for $F_0 = .1, .5, 2, 10, 15, 20, 25, 30$. Use the initial condition $x(0) = 2$, $x'(0) = 1$. What are the transient and long-term behaviors of each solution? Comment on any difference observed in the plots.

4. Repeat the above step but now numerically solve Duffing's equation using the initial condition $x(0) = 1$, $x'(0) = 2$. What are the transient and long-term behaviors of each solution? Comment on any difference observed in the plots, both in this part and when compared with those of part 3.

5. Now fix $F_0 = 25$ (with $k = \delta = 1$, $b = .2$) and vary ω. Choose the values $\omega = .1, .2, .5, .8, 1, 1.2, 1.5, 2.0, 5.0$ and others, if you so desire. What are the transient and long-term behaviors of each solution? Comment on any difference observed in the pictures.

What effect does the nonlinearity and forcing have on the motion $x(t)$? Is it predictable?

For additional background material on this problem, the interested reader should consult [19] and [34] in the References.

Project 2: Stiff Differential Equations

We studied the equation for a damped, unforced mass on a spring:

$$mx'' + bx' + kx = 0,$$

where $m, b, k > 0$ are constant. There are many situations that arise where these constants are not close to each other in magnitude. One specific area of interest is when the spring constant k is large. Such systems are said to be *stiff*. Here we will consider the equation

$$x'' + x' + 1000x = 0. \tag{4.69}$$

1. Solve this constant coefficient ODE and write its analytical solution. Use the initial condition $x(0) = 1$, $x'(0) = 0$.

2. Using Matlab, Maple, or Mathematica solve this equation using the fourth-order Runge-Kutta method introduced in this chapter. Try seven different runs, each with a different step size, using step sizes of $h = 0.2, 0.1, 0.05$, $0.02, 0.01, 0.005, 0.002$ and estimating the solution at $t = 20$.

3. Compare the analytical solution from 1 with the numerical approximation from 2 Comment on the accuracy of the numerical answers.

Besides decreasing the step size, stiff differential equations are often handled with other methods, typically implicit ones. See Burden and Faires [9] for some potential methods and their implementation.

Chapter 5

Fundamentals of Systems of Differential Equations

Earlier we briefly considered systems of equations when converting an nth order equation to a system of n first-order equations. But systems of differential equations arise in their own right—whenever there is more than one dependent variable for an independent variable. For example, one might consider a system with two or more interacting species with the population sizes changing over time. This is known as a *predator–prey* system. Here, the rate at which the prey population grows depends upon the number of predators that kill the prey. Similarly, the rate at which the predator population grows depends on the size of their food supply, namely, the prey population. In general, these conditions produce nonlinear equations that are very difficult to solve analytically. This is just one scenario we can consider. In this chapter, we discuss methods of solution for these types of systems.

The objects of study of the first few sections of this chapter are linear systems of equations, which are differential equations of the form

$$\begin{aligned}
\frac{dx_1}{dt} &= a_{11}(t)x_1 + a_{12}(t)x_2 + \cdots + a_{1n}(t)x_n + f_1(t) \\
\frac{dx_2}{dt} &= a_{21}(t)x_1 + a_{22}(t)x_2 + \cdots + a_{2n}(t)x_n + f_2(t) \\
&\;\;\vdots \qquad\quad \vdots \qquad\quad \vdots \qquad\;\; \ddots \qquad\qquad \vdots \\
\frac{dx_n}{dt} &= a_{n1}(t)x_1 + a_{n2}(t)x_2 + \cdots + a_{nn}(t)x_n + f_n(t),
\end{aligned} \tag{5.1}$$

where the variables x_i, f_i and coefficients, a_{ij}, are all functions of t. The situation when an nth order system is derived from an nth order equation is simply a special case. In the event that the coefficients a_{ij} are constant and the f_i are zero, we refer to system (5.1) as an *autonomous* system of n first-order equations and there are special techniques that will apply. System (5.1) has aspects of matrix analysis, so that at this point, the reader is strongly encouraged to review the material in Appendix C.1 before proceeding.

5.1   Systems of Two Equations—Motivational Examples

This section considers an example where both analytical and graphical methods provide useful insight into the behavior of solutions. It serves as a motivation for the study of related topics in linear algebra, which can be found in Sections 5.3, 5.4, as well as Appendix C.

To begin our study, we will consider systems of two linear homogeneous first-order differential equations. These systems are easily solved by hand, and the behaviors of their solutions are readily categorized. They also lend themselves to a useful graphical interpretation. We begin by considering systems of the form

$$\frac{dx}{dt} = ax + by$$
$$\frac{dy}{dt} = cx + ey \tag{5.2}$$

where a, b, c, and e are constants.[1]

5.1.1   Vector Fields and the Phase Plane

To motivate our study of such a system, we consider an undamped mass-spring system, also known as a *simple harmonic oscillator*, which we studied previously in Section 4.2.1:

$$mx'' + kx = 0.$$

Using the methods of Section 3.5.1, we solve for the highest-order derivative to obtain

$$x'' = \frac{-k}{m}x.$$

Making the substitution $y = x'$ so that $y' = x''$ allows us to write the system in terms of the variables x, y:

$$\frac{dx}{dt} = y$$
$$\frac{dy}{dt} = \frac{-k}{m}x. \tag{5.3}$$

This is an example of a second-order system (5.2) with $a = e = 0$, $b = 1$, $c = -k/m$. At this point, we will *not* use numerical methods to analyze the behavior of solutions. Indeed, we could easily have solved this system

---

[1] We don't use the constant d in order to avoid any confusion as to whether dy is the product $d \times y$ or the differential dy.

because it is homogeneous with constant coefficients. Instead, we will use a graphical approach, analogous to phase line analysis of first-order autonomous equations. We view solutions in the *phase plane*, that is, the x-y plane where x and y are the two variables in this system. The right-hand sides of (5.3), and more generally (5.2), tell us how the system is changing at the given point and the change is given by (x', y'). Thus, to each point (x, y) in the phase plane, we will associate a vector (x', y') that describes the change. The collection of these vectors makes up what we call the *vector field*. Once we specify an initial condition, our motion is determined for all t.

For the equation of the simple harmonic oscillator, we can easily calculate slope lines according to the quadrant, keeping in mind that m and k are always positive. We refer the reader to Figure 5.1 as we go from quadrants 4 down to quadrant 1:

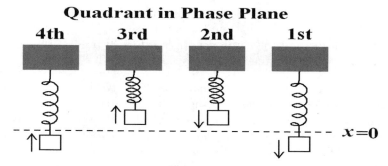

FIGURE 5.1: Simple harmonic oscillator and the phase plane.

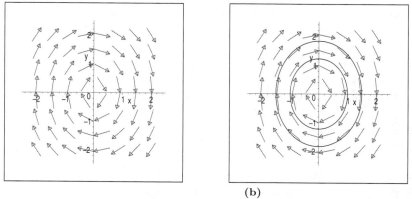

(a)   (b)

FIGURE 5.2: Simple harmonic oscillator with $m = 2$, $k = 3$: **(a)** Vector field; **(b)** vector field with orbits corresponding to the initial conditions $(1, 0)$ and $(1.5, 0)$.

4th quadrant: $x > 0$, $y < 0$. We have that $x' < 0$ and $y' < 0$ and the solution in the phase plane is moving to the left (because $x' < 0$) and down (because $y' < 0$). In terms of the physical picture, the spring length is extended past its rest length but is beginning to compress as the block moves up.

3rd quadrant: $x < 0$, $y < 0$. We have that $x' < 0$ and $y' > 0$ and the solution is moving to the left (because $x' < 0$) and up (because $y' > 0$). In terms of the physical picture, the spring length is shorter than its rest length and is continuing to compress as the block moves up.

2nd quadrant: $x < 0$, $y > 0$. We have that $x' > 0$ and $y' > 0$ and the solution is moving to the right (because $x' > 0$) and up (because $y' > 0$). In terms of the physical picture, the spring length is shorter than its rest length but is beginning to extend as the block moves down.

1st quadrant: $x > 0$, $y > 0$. We can substitute values to see that $x' > 0$ and $y' < 0$ and the solution is moving to the right (because $x' > 0$ and thus x is increasing) and down (because $y' < 0$). In terms of the physical picture, the spring length is extended past its rest length and is continuing to extend as the block moves down.

The vector field for the simple harmonic oscillator with $m = 2, k = 3$ is given in Figure 5.2a. Equally useful is the inclusion of solutions in this picture. We will often refer to the solutions as *trajectories*. We should be careful here, though, because we are technically *not* looking at solutions since solutions live in the three-dimensional $t - x - y$ space. What we are viewing are the *orbits* of this equation. As long as t does not explicitly appear in our equations, i.e., the system is autonomous (constant coefficient and homogeneous for linear equations), it is usually safe to think of the curves we draw as solutions. As we mentioned, given any initial condition, we can begin to trace out its path because we know that the vector field is tangent to the orbit at each point in the phase plane. Figure 5.2b shows two orbits of this system.

The phase plane with the orbits for various initial conditions drawn in is called the *phase portrait* of the system. Note that even if we do not super-impose the vector field, we still refer to it as the phase portrait. The reader should compare Figures 5.1 and 5.2 until convinced that they are describing the same situation. Remember that the horizontal axis represents the *position* and the vertical axis represents the *velocity* of the mass.

Every single point in the phase plane describes a part of a given orbit of our system. Although the origin may look boring (the vector there is the zero vector!), we actually give it a special name: *equilibrium point*. This word has the same context as it did when we discussed autonomous first-order equations. It simply means a point where there is no change in the x or y value of the system, that is, $x'(t) = 0$ and $y'(t) = 0$. We note that for any first-order system of the form given in equation (5.1), the origin is *always* an equilibrium point.

At this point, we give Matlab, Maple, and Mathematica code only for generating the vector fields of a system of first-order equations. We will give the code for the superimposing solutions after some additional discussion.

We consider our mass-spring example with $m = 2, k = 3$:

Computer Code 5.1: **Plotting the vector field for a system of two first-order differential equations**

Matlab, Maple, Mathematica

Matlab

```
>>  [X,Y]=meshgrid(-2:.5:2,-2:.5:2);
>>  DX=Y;
>>  DY=(-3/2)*X;
>>  DW=sqrt(DX.^2+DY.^2);
>>  quiver(X,Y,DX./DW,DY./DW,.5);
>>  xlabel('x'); ylabel('y');
>>  axis([-2 2 -2 2])
>>  title('vector field for mass-spring system');
```

We again normalized the length of the vectors. For the original unscaled vector field, we would replace the above quiver line with `quiver(X,Y,DX,DY)`.

Maple

```
> with(plots):
> with(DEtools):
> eq1:=diff(x(t),t)=y(t);
> eq2:=diff(y(t),t)=-(3/2)*x(t);
> DEplot([eq1,eq2],[x(t),y(t)],t=0..1,x=-2..2,y=-2..2,
    linecolor=black,dirgrid=[8,8],title="vector field
    for mass-spring system",arrows=medium);
```

Mathematica

```
<<Graphics`PlotField`
p1=PlotVectorField[{y,- 3x/2},{x,-2,2},{y,-2,2},
  ScaleFunction→ (2&),Axes→ Automatic,
  AxesLabel→ {"x","y"}]
```

5.1.2   Analytical Insight

With an understanding of the phase plane, we now proceed to an analytical method. Our approach, though not often used in practice for reasons that will become clear later, provides motivation for the usefulness and applicability of linear algebra to enrich the understanding of differential equations. The results that we state here can be restated in terms of eigenvalues, but because systems of two first-order equations often arise in practice, this section is useful in its own right.

We again consider system (5.2),

$$\frac{dx}{dt} = ax + by$$
$$\frac{dy}{dt} = cx + ey,$$

where a, b, c, and e are constants. We are going to rewrite this system as

$$\frac{d^2x}{dt^2} - \beta\frac{dx}{dt} + \gamma x = 0$$

where $\beta = a + e$ and $\gamma = ae - cd$. To do this, we differentiate

$$\frac{dx}{dt} = ax + by$$

with respect to t, to obtain

$$\frac{d^2x}{dt^2} = a\frac{dx}{dt} + b\frac{dy}{dt}.$$

Now we substitute

$$\frac{dy}{dt} = cx + ey$$

into this expression to obtain

$$\frac{d^2x}{dt^2} = a\frac{dx}{dt} + b(cx + ey)$$
$$= a\frac{dx}{dt} + bcx + eby.$$

But from the first equation of the system, $by = \frac{dx}{dt} - ax$ so

$$\frac{d^2x}{dt^2} = a\frac{dx}{dt} + bcx + e\left(\frac{dx}{dt} - ax\right)$$
$$= (a + e)\frac{dx}{dt} + (bc - ae)x.$$

Rearranging gives

$$\frac{d^2x}{dt^2} - (a + e)\frac{dx}{dt} + (ae - bc)x = 0,$$

which is a second-order linear homogeneous equation with constant coefficients.

Now if we notice that system (5.2) can be written in matrix form as

$$\begin{pmatrix} \dfrac{dx}{dt} \\ \dfrac{dy}{dt} \end{pmatrix} = \begin{pmatrix} a & b \\ c & e \end{pmatrix} \begin{pmatrix} x \\ y \end{pmatrix},$$

we have in this matrix notation $\beta = a + e$ as the *trace*, denoted Tr, of the coefficient matrix which is the sum of the diagonal entries, and $\gamma = ae - bc$ as the value of the determinant. The characteristic equation of the second-order differential equation is $\lambda^2 - \beta\lambda + \gamma = 0$. Using the quadratic formula we have

$$\lambda_{1,2} = \frac{\beta \pm \sqrt{\beta^2 - 4\gamma}}{2}.$$

Now with $x(t)$ found, we can find $y(t)$ by differentiating to find $x'(t)$ and substituting both x and x' into $\frac{dx}{dt} = ax + by$. It is important to note that y has the same roots of the characteristic equation as x did. The roots of the characteristic equation are also called *eigenvalues* and will be discussed in detail in Section 5.4. In this context, the eigenvalues, the roots of the characteristic equation, will also be referred to as the eigenvalues of the differential equation. Thus there are only two (possibly repeated) eigenvalues for this system of equations.

Example 1: Find the general solution to

$$\frac{dx}{dt} = -5x + 8y$$

$$\frac{dy}{dt} = -4x + 7y.$$

In matrix form this equation is

$$\mathbf{x'} = \begin{pmatrix} -5 & 8 \\ -4 & 7 \end{pmatrix} \mathbf{x} \quad \text{where} \quad \mathbf{x'} = \begin{pmatrix} \dfrac{dx}{dt} \\ \dfrac{dy}{dt} \end{pmatrix}, \mathbf{x} = \begin{pmatrix} x \\ y \end{pmatrix}$$

so that the coefficient matrix is

$$\mathbf{A} = \begin{pmatrix} -5 & 8 \\ -4 & 7 \end{pmatrix};$$

thus

$$\beta = \text{Tr}\begin{pmatrix} -5 & 8 \\ -4 & 7 \end{pmatrix} = -5 + 7 = 2$$

and

$$\gamma = \det \begin{pmatrix} -5 & 8 \\ -4 & 7 \end{pmatrix} = (-5)(7) - (8)(-4) = -3.$$

So

$$\frac{d^2 x}{dt^2} - \beta \frac{dx}{dt} + \gamma x = 0 \implies \frac{d^2 x}{dt^2} - 2\frac{dx}{dt} - 3x = 0,$$

which has roots of the characteristic equation

$$\lambda_1 = -1 \text{ and } \lambda_2 = 3.$$

(We also used r_1, r_2 in previous sections.) Thus,

$$x(t) = c_1 e^{-t} + c_2 e^{3t}.$$

Now,

$$x'(t) = -c_1 e^{-t} + 3c_2 e^{3t},$$

which we can compare with the first equation in the original system to see that

$$x'(t) = -5x(t) + 8y(t) = -5(c_1 e^{-t} + c_2 e^{3t}) + 8y(t).$$

Knowing the last two equations for $x'(t)$ must be the same, we see that

$$y(t) = \frac{1}{2} c_1 e^{-t} + c_2 e^{3t}.$$

In vector notation, we see that the solution is thus

$$\begin{pmatrix} x(t) \\ y(t) \end{pmatrix} = c_1 \begin{pmatrix} e^{-t} \\ \frac{1}{2} e^{-t} \end{pmatrix} + c_2 \begin{pmatrix} e^{3t} \\ e^{3t} \end{pmatrix} = c_1 e^{-t} \begin{pmatrix} 1 \\ \frac{1}{2} \end{pmatrix} + c_2 e^{3t} \begin{pmatrix} 1 \\ 1 \end{pmatrix}.$$

If we have an initial condition $x(0) = x_0, y(0) = y_0$, then

$$x(t) = (2x_0 - 2y_0) e^{-t} + (-x_0 + 2y_0) e^{3t}$$
$$y(t) = \frac{1}{2} (2x_0 - 2y_0) e^{-t} + (-x_0 + 2y_0) e^{3t}. \tag{5.4}$$

One term of the solution in x is growing exponentially while the other is decreasing exponentially; see Figure 5.3. The same is true for the solution in y. In terms of phase plane behavior, the origin is the only equilibrium solution and is called a *saddle point*. A little help from our computer programs gives the figure shown here and helps explain the word "saddle."

Classification of Equilibrium Solutions

System (5.2) has equilibrium points whenever there is no change in either variable, i.e., whenever both derivatives are zero. As long as the system is non-degenerate, the only equilibrium is the origin $(0, 0)$.

If all the trajectories move toward an equilibrium point, it is said to be *stable* and is sometimes called a *sink*; if they all move away, it is said to be

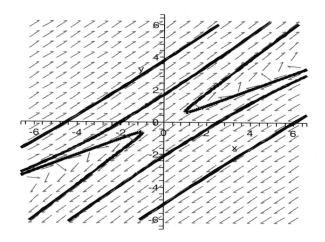

FIGURE 5.3: Phase portrait for Example 1. The origin is a saddle.

unstable and is sometimes called a *source*; if some trajectories move toward and others move away, it is called a *saddle*; and if all trajectories orbit around the equilibrium point, it is called a *center*. Further, the stable and unstable equilibria may exhibit spiraling behaviors in which case they are called *stable spirals* or *unstable spirals*.[2] Sometimes we will use the word *node* (e.g., stable node or unstable node) to indicate that no spiraling is occurring.[3] These results are summarized in the following theorem.

THEOREM 5.1.1 *Consider (5.2). Let* $\beta = a + e$ *and* $\gamma = ae - bc$; *then*
a. *If* $\gamma < 0$, *then the origin is a saddle.*
b. *If* $\gamma > 0$ *and* $\beta < 0$, *then the origin is stable.*
c. *If* $\gamma > 0$ *and* $\beta > 0$, *then the origin is unstable.*
d. *If* $\gamma > 0$ *and* $\beta = 0$, *then the origin is a center.*
e. *If* $\beta^2 - 4\gamma < 0$, *then a sink or source is a stable spiral or an unstable spiral, respectively.*

Figures 5.4 and 5.5 summarize these behaviors, based on the values of β and γ. We note that the straight lines in the case of the nodes and saddle are given by the *eigenvectors*, which will be discussed in detail in Section 5.4.

---

[2]Stable spirals are also called *spiral sinks* and unstable spirals are also called *spiral sources*.
[3]Stable nodes are also called *nodal sinks* and unstable nodes are also called *nodal sources*.

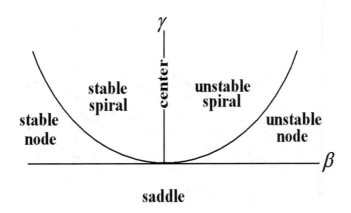

FIGURE 5.4:   Classification of equilibria in the β-γ plane.

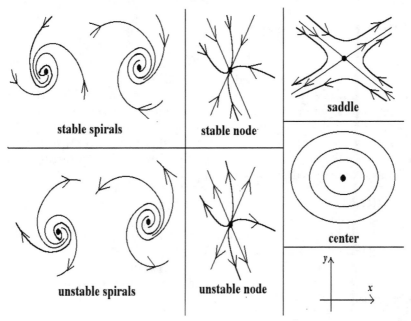

FIGURE 5.5:   Sample phase portraits from Figure 5.4. The coordinate axes are given in the lower right corner of this figure.

Often our primary interest is in the behaviors of solutions, rather than the solutions themselves. From the above discussion, notice that by looking at the system of equations and evaluating β and γ, we are able to predict the behavior of the system.

Example 2: Consider the system:

$$\frac{dx}{dt} = 2x + y$$

$$\frac{dy}{dt} = -3x + 4y.$$

Here $\beta = 6$ and $\gamma = 8 + 3 = 11$, so $\beta^2 - 4\gamma = 36 - 44 = -8$. The origin is an equilibrium point. From Figure 5.4 we note that since β and γ are both positive, we have either an unstable node or an unstable spiral. Since $\beta^2 - 4\gamma < 0$, the trajectories spiral away from the equilibrium and we see that we have an unstable spiral.

Example 3: Consider the simple harmonic oscillator from before:

$$\frac{dx}{dt} = y$$

$$\frac{dy}{dt} = \frac{-k}{m}x. \tag{5.5}$$

Here $\beta = 0$ and $\gamma = k/m > 0$. The origin is the equilibrium point and is a center with trajectories encircling it in closed curves.

5.1.3 A Slicker Approach

Although we were able to gain significant insight into two-dimensional autonomous systems, it will often be more convenient to study systems from a different perspective. We again consider the system of Example 1:

$$\frac{dx}{dt} = -5x + 8y$$

$$\frac{dy}{dt} = -4x + 7y. \tag{5.6}$$

Rather than convert the system to a second-order equation or examine it in the phase plane, we choose to find a *linear change of coordinates* that will make the system *uncoupled*, i.e., where each equation will only have functions and derivatives of one variable. Consider the following new variables u_1, u_2:

$$u_1 = x - y$$

$$u_2 = -x + 2y. \tag{5.7}$$

In terms of our original variables, we can solve to obtain [4]

$$x = 2u_1 + u_2$$
$$y = u_1 + u_2. \tag{5.8}$$

To find equations in terms of our new variables, we differentiate (5.7):

$$u_1' = x' - y'$$
$$u_2' = -x' + 2y'. \tag{5.9}$$

Substituting the original equations (5.6) for x', y', we have

$$u_1' = (-5x + 8y) - (-4x + 7y) = -x + y$$
$$u_2' = -(-5x + 8y) + 2(-4x + 7y) = -3x + 6y. \tag{5.10}$$

We substitute using (5.8) to put everything in terms of the new variables:

$$u_1' = -(2u_1 + u_2) + (u_1 + u_2)$$
$$u_2' = -3(2u_1 + u_2) + 6(u_1 + u_2), \tag{5.11}$$

which simplifies to

$$u_1' = -u_1$$
$$u_2' = 3u_2. \tag{5.12}$$

Although this change of variables was a lot of work, (5.12) can be easily solved since each is a very simple separable equation. We should also note that the roots obtained from the characteristic equation of Example 1 in Section 5.1.2 were $-1, 3$ and it is not a coincidence that these appear here as coefficients! The solution is

$$u_1 = c_1 e^{-t}$$
$$u_2 = c_2 e^{3t}. \tag{5.13}$$

The phase portrait of the original system can be found by plotting the new coordinate axes, $u_1 = 0, u_2 = 0$ and sketching the trajectories. The $u_1 = 0$ axis gives the line $y = x$ and the solution for u_2 shows that solutions are exponentially growing along this line. The $u_2 = 0$ axis gives the line $y = x/2$ and the solution for u_1 shows that solutions are exponentially decreasing along this line. (Later in Section 5.4, it is seen that these lines actually give the direction of the eigenvectors for the two eigenvalues of the system.) If our

---

[4]The reason for this choice is not arbitrary and is detailed in Appendix C.3.1. Using terminology that is covered in Section 5.4, we will see that the eigenvectors of the coefficient matrix of (5.6) are $(2 \quad 1)^T$ and $(1 \quad 1)^T$, which are exactly the columns of the coefficient matrix of (5.8). The inverse of this coefficient matrix gives the coefficients in (5.7).

initial condition in the original system is $x(0) = x_0, y(0) = y_0$, then (5.7) together with (5.13) gives

$$c_1 = x_0 - y_0$$
$$c_2 = -x_0 + 2y_0. \tag{5.14}$$

Thus

$$u_1 = (x_0 - y_0) e^{-t}$$
$$u_2 = (-x_0 + 2y_0) e^{3t}. \tag{5.15}$$

In terms of our original variables, (5.8) gives

$$x = 2(x_0 - y_0) e^{-t} + (-x_0 + 2y_0) e^{3t}$$
$$y = (x_0 - y_0) e^{-t} + (-x_0 + 2y_0) e^{3t}, \tag{5.16}$$

which is the same solution we obtained previously; see (5.4).

In studying differential equations from this approach, we typically will not convert back to our original variables. Why? In Section 5.1.2 we converted the system to a second-order equation and found the roots of the characteristic equation, i.e., the eigenvalues. In this section, our uncoupled system had the eigenvalues as the coefficients multiplying the variables. A proper choice of coordinates will always greatly simplify the system. Before we go into more details of solving systems of equations, we need to first establish terminology for a system of n linear differential equations. After a discussion about linear transformations, eigenvalues, and eigenvectors, we will then tie these concepts together as we solve systems of differential equations.

Problems

In problems 1–13, (i) use Theorem 5.1.1 to determine the stability of the origin (it's the only equilibrium solution) for the given system of equations; (ii) use a computer algebra system to draw the vector field; (iii) sketch some orbits by hand to verify your conclusion in (i).

1. a. $\begin{cases} x' = -3x + 6y \\ y' = -2x + 5y \end{cases}$ b. $\begin{cases} x' = x + 2y \\ y' = -x + 4y \end{cases}$

2. a. $\begin{cases} x' = x - 2y \\ y' = 3x - y \end{cases}$ b. $\begin{cases} x' = 3x - 2y \\ y' = 5x - 3y \end{cases}$

3. a. $\begin{cases} x' = -3x + y \\ y' = 2x + y \end{cases}$ b. $\begin{cases} x' = y \\ y' = 3x + 2y \end{cases}$

4. a. $\begin{cases} x' = 4x + y \\ y' = 3x + 2y \end{cases}$ b. $\begin{cases} x' = x - 2y \\ y' = 4x + 5y \end{cases}$

5. a. $\begin{cases} x' = x - 8y \\ y' = x + 3y \end{cases}$ b. $\begin{cases} x' = x - 5y \\ y' = x + 3y \end{cases}$

6. a. $\begin{cases} x' = -x - 5y \\ y' = x - 3y \end{cases}$ b. $\begin{cases} x' = -3x - 5y \\ y' = x - 7y \end{cases}$

7. a. $\begin{cases} x' = 5x - 6y \\ y' = 2x - y \end{cases}$ b. $\begin{cases} x' = 5x \\ y' = 2x - 3y \end{cases}$

8. a. $\begin{cases} x' = 7x + 6y \\ y' = 2x - 4y \end{cases}$ b. $\begin{cases} x' = 2x + y \\ y' = -x \end{cases}$

9. a. $\begin{cases} x' = -2x + y \\ y' = -x \end{cases}$ b. $\begin{cases} x' = -2x + 3y \\ y' = -5y \end{cases}$

10. a. $\begin{cases} x' = -4x + 7y \\ y' = -y \end{cases}$ b. $\begin{cases} x' = 3x - 2y \\ y' = -6y \end{cases}$

11. a. $\begin{cases} x' = x - 2y \\ y' = 4x - 5y \end{cases}$ b. $\begin{cases} x' = -5x + 2y \\ y' = -2x - 5y \end{cases}$

12. a. $\begin{cases} x' = -5x + 2y \\ y' = -2x - 3y \end{cases}$ b. $\begin{cases} x' = x - 3y \\ y' = 7y \end{cases}$

13. Consider the simple harmonic oscillator with damping:

$$x'' + bx' + kx = 0,$$

where $b, k > 0$. Convert this to a system of 2 first-order equations and then answer questions (i)–(iii) above.

In problems 14–18, use the given substitutions and the methods of Section 5.1.3 to rewrite the equations as an uncoupled system. Then write the solution in terms of u_1, u_2 and use this to find the solution in terms of x, y.

14. $u_1 = x - y, u_2 = -x + 2y$ for $\begin{cases} x' = x + 2y \\ y' = -x + 4y \end{cases}$

15. $u_1 = 2x - y, u_2 = -x + y$ for $\begin{cases} x' = x - 2y \\ y' = 4x - 5y \end{cases}$

16. $u_1 = \frac{1}{4}x, u_2 = -\frac{1}{4}x + y$ for $\begin{cases} x' = 5x \\ y' = 2x - 3y \end{cases}$

17. $u_1 = -\frac{1}{2}x + \frac{3}{2}y, u_2 = \frac{1}{2}x - \frac{1}{2}y$ for $\begin{cases} x' = 4x - 6y \\ y' = 2x - 4y \end{cases}$

18. $u_1 = \frac{3}{2}x - \frac{1}{2}y, u_2 = \frac{-1}{2}x + \frac{1}{2}y$ for $\begin{cases} x' = -2x + y \\ y' = -3x + 2y \end{cases}$

19. Show that

$$\frac{dx}{dt} = ax + by$$

$$\frac{dy}{dt} = cx + ey$$

can be rewritten as

$$\frac{d^2x}{dt^2} - \beta \frac{dx}{dt} + \gamma x = 0.$$

5.2 Useful Terminology

Many of the terms that we are about to encounter have arisen at earlier points in this book. The significance here is that they apply to a system of equations as well. We consider (5.1) written in matrix notation:

$$
\begin{pmatrix} \dfrac{dx_1}{dt} \\ \dfrac{dx_2}{dt} \\ \vdots \\ \dfrac{dx_n}{dt} \end{pmatrix} = \begin{pmatrix} a_{11}(t)x_1 + a_{12}(t)x_2 + \cdots + a_{1n}(t)x_n \\ a_{21}(t)x_1 + a_{22}(t)x_2 + \cdots + a_{2n}(t)x_n \\ \vdots \qquad \vdots \qquad \ddots \qquad \vdots \\ a_{n1}(t)x_1 + a_{n2}(t)x_2 + \cdots + a_{nn}(t)x_n \end{pmatrix} + \begin{pmatrix} f_1(t) \\ f_2(t) \\ \vdots \\ f_n(t) \end{pmatrix}, \quad (5.17)
$$

or more compactly, simply as

$$
\frac{d\mathbf{x}}{dt} = \mathbf{A}\mathbf{x} + \mathbf{F}, \tag{5.18}
$$

where $\mathbf{x}$, $\mathbf{A}$, $\mathbf{F}$ are possibly all functions of t. If $\mathbf{F}(t) = 0$, we say that the system is homogeneous and write it as

$$
\frac{d\mathbf{x}}{dt} = \mathbf{A}\mathbf{x}. \tag{5.19}
$$

Before continuing, we formally address the notion of the *derivative* of a vector.

DEFINITION 5.1 *The derivative of the $m \times n$ matrix $\mathbf{A}(t) = (a_{ij}(t))$, whose elements $a_{ij}(t)$ are differentiable on some interval I, is defined by taking the derivative of each component:*

$$
\frac{d\mathbf{A}}{dt} = \left(\frac{da_{ij}}{dt} \right).
$$

The integral of the $m \times n$ matrix $\mathbf{A}(t) = (a_{ij}(t))$, whose elements $a_{ij}(t)$ are continuous on some interval I, is defined by taking the integral of each component:

$$
\int_{t_0}^{t} \mathbf{A} = \left(\int_{t_0}^{t} a_{ij} \, ds \right).
$$

Example 1: Consider the matrix

$$
\mathbf{A} = \begin{pmatrix} 1 & 0 & \sin t \\ e^{2t} & -t & 3t^2 - 4 \end{pmatrix}.
$$

Find the derivative of $\mathbf{A}$. Then find the integral of $\mathbf{A}$ (from 0 to t).

To find the derivative, we differentiate componentwise. Thus

$$\frac{d\mathbf{A}}{dt} = \begin{pmatrix} \dfrac{d}{dt}(1) & \dfrac{d}{dt}(0) & \dfrac{d}{dt}(\sin t) \\ \dfrac{d}{dt}(e^{2t}) & \dfrac{d}{dt}(-t) & \dfrac{d}{dt}(3t^2 - 4) \end{pmatrix} = \begin{pmatrix} 0 & 0 & \cos t \\ 2e^{2t} & -1 & 6t \end{pmatrix}.$$

For the integral, we similarly integrate componentwise. We note that we use the dummy variable s to do this integration:

$$\int_0^t \mathbf{A} = \begin{pmatrix} \displaystyle\int_0^t 1\,ds & \displaystyle\int_0^t 0\,ds & \displaystyle\int_0^t \sin s\,ds \\ \displaystyle\int_0^t e^{2s}\,ds & \displaystyle\int_0^t -s\,ds & \displaystyle\int_0^t (3s^2 - 4)\,ds \end{pmatrix}$$

$$= \begin{pmatrix} t & 0 & -\cos t \\ \dfrac{1}{2}e^{2t} & \dfrac{-t}{2} & t^3 - 4t \end{pmatrix}.$$

Using this idea, some of our previous work can be extended. For example, if the components of a vector $\mathbf{x}$ are differentiable on an open interval (a, b) and satisfy the system of differential equations, then we say that $\mathbf{x}$ is a *solution* of (5.17).

Example 2: Verify that the vectors

$$\mathbf{x}_1 = \begin{pmatrix} 0 \\ 0 \\ 1 \end{pmatrix} e^{-2t}, \quad \mathbf{x}_2 = \begin{pmatrix} 1 \\ 0 \\ 1 \end{pmatrix} e^t, \quad \text{and} \quad \mathbf{x}_3 = \begin{pmatrix} 5 \\ 10 \\ 3 \end{pmatrix} e^{3t}$$

are each a solution to the system

$$\frac{d\mathbf{x}}{dt} = \mathbf{A}\mathbf{x}, \quad \text{where} \quad \mathbf{A} = \begin{pmatrix} 1 & 1 & 0 \\ 0 & 3 & 0 \\ 3 & 0 & -2 \end{pmatrix}.$$

This is done by substitution of each vector into the differential equation:

$$\frac{d\mathbf{x}_1}{dt} = \begin{pmatrix} 0 \\ 0 \\ 1 \end{pmatrix}(-2e^{-2t}) \quad \text{and} \quad \mathbf{A}\mathbf{x}_1 = \begin{pmatrix} 1 & 1 & 0 \\ 0 & 3 & 0 \\ 3 & 0 & -2 \end{pmatrix} \begin{pmatrix} 0 \\ 0 \\ 1 \end{pmatrix} e^{-2t} = \begin{pmatrix} 0 \\ 0 \\ -2 \end{pmatrix} e^{-2t};$$

$$\frac{d\mathbf{x}_2}{dt} = \begin{pmatrix} 1 \\ 0 \\ 1 \end{pmatrix}(e^t) \quad \text{and} \quad \mathbf{A}\mathbf{x}_2 = \begin{pmatrix} 1 & 1 & 0 \\ 0 & 3 & 0 \\ 3 & 0 & -2 \end{pmatrix} \begin{pmatrix} 1 \\ 0 \\ 1 \end{pmatrix} e^t = \begin{pmatrix} 1 \\ 0 \\ 1 \end{pmatrix} e^t;$$

$$\frac{d\mathbf{x}_3}{dt} = \begin{pmatrix} 5 \\ 10 \\ 3 \end{pmatrix} (3e^{3t}) \quad \text{and} \quad \mathbf{A}\mathbf{x}_3 = \begin{pmatrix} 1 & 1 & 0 \\ 0 & 3 & 0 \\ 3 & 0 & -2 \end{pmatrix} \begin{pmatrix} 5 \\ 10 \\ 3 \end{pmatrix} e^{3t} = \begin{pmatrix} 15 \\ 30 \\ 9 \end{pmatrix} e^{3t}.$$

In each case, we see that the vector $\mathbf{x}_i$ satisfies the differential equation for all t and thus is a solution for all t.

As we have seen, it is straightforward to check if a given vector is a solution. Just as we have done in previous chapters, it is useful to know *when* we can actually expect to have a solution. Because we are currently considering a linear system of equations, the following theorems are an asset.

THEOREM 5.2.1 Existence and Uniqueness *Consider the inital value problem for the system*

$$\frac{d\mathbf{x}}{dt} = \mathbf{A}\mathbf{x} + \mathbf{F}, \quad \mathbf{x}(t_0) = \mathbf{x}_0, \quad\quad\quad (5.20)$$

where $\mathbf{A}(t)$ is an $m \times n$ matrix and $\mathbf{F}(t)$ is an m-component vector of continuous real functions on the interval $a \leq t \leq b$. If t_0 is any point in the interval (a, b), then there exists a unique solution to (5.20) and this solution is defined over the entire interval $a \leq t \leq b$.

For the homogeneous case (5.19), we can consider linear combinations of these solutions.

THEOREM 5.2.2 *Let $\mathbf{x}_1, \mathbf{x}_2, \cdots, \mathbf{x}_k$ be* **any** *k solutions of the homogeneous system $\mathbf{x}' = \mathbf{A}(t)\mathbf{x}$. Then*

$$c_1 \mathbf{x}_1 + c_2 \mathbf{x}_2 + \cdots + c_k \mathbf{x}_k$$

is also a solution of the homogeneous system, where $c_1, c_2, \cdots, c_k$ are arbitrary constants.

Example 3: In Example 2, we showed the vectors

$$\mathbf{x}_1 = \begin{pmatrix} 0 \\ 0 \\ 1 \end{pmatrix} e^{-2t}, \quad \mathbf{x}_2 = \begin{pmatrix} 1 \\ 0 \\ 1 \end{pmatrix} e^t, \quad \text{and} \quad \mathbf{x}_3 = \begin{pmatrix} 5 \\ 10 \\ 3 \end{pmatrix} e^{3t}$$

are each a solution of the system

$$\frac{d\mathbf{x}}{dt} = \mathbf{A}\mathbf{x}, \quad \text{where} \quad \mathbf{A} = \begin{pmatrix} 1 & 1 & 0 \\ 0 & 3 & 0 \\ 3 & 0 & -2 \end{pmatrix}.$$

According to Theorem 5.2.2, we have that

$$\mathbf{x}(t) = c_1\mathbf{x}_1 + c_2\mathbf{x}_2 + c_3\mathbf{x}_3 = \begin{pmatrix} c_2e^t + c_35e^{3t} \\ c_310e^{3t} \\ c_1e^{-2t} + c_2e^t + c_33e^{3t} \end{pmatrix}$$

is also a solution of $\mathbf{x}' = \mathbf{Ax}$. To see this, we first take the derivative of this linear combination to get

$$\mathbf{x}'(t) = \begin{pmatrix} c_2e^t + c_315e^{3t} \\ c_330e^{3t} \\ c_1(-2)e^{-2t} + c_2e^t + c_39e^{3t} \end{pmatrix}.$$

We also evaluate the right-hand side, $\mathbf{Ax}(t)$, to get

$$\begin{pmatrix} 1 & 1 & 0 \\ 0 & 3 & 0 \\ 3 & 0 & -2 \end{pmatrix} \begin{pmatrix} c_2e^t + c_35e^{3t} \\ c_310e^{3t} \\ c_1e^{-2t} + c_2e^t + c_33e^{3t} \end{pmatrix} = \begin{pmatrix} c_2e^t + c_315e^{3t} \\ c_330e^{3t} \\ c_1(-2)e^{-2t} + c_2e^t + c_39e^{3t} \end{pmatrix}.$$

The equality of the results of these two calculations shows that $c_1\mathbf{x}_1 + c_2\mathbf{x}_2 + c_3\mathbf{x}_3$ is indeed a solution.

In Section 3.2, we considered the above theorems for the case of functions and not vectors. We wanted to write the general solution and we needed to know when functions were linearly independent. We can formulate a similar definition for vectors.

DEFINITION 5.2   *The k functions* $\mathbf{x}_1, \mathbf{x}_2, \cdots, \mathbf{x}_k$ *are* **linearly dependent** *on* $a \le t \le b$ *if there exist constants* $c_1, c_2, \cdots, c_k$, *not all zero, such that*

$$c_1\mathbf{x}_1 + c_2\mathbf{x}_2 + \cdots + c_k\mathbf{x}_k = \mathbf{0}$$

for all t in the interval (a, b). *We say that the k functions* $\mathbf{x}_1, \mathbf{x}_2, \cdots, \mathbf{x}_k$ *are* **linearly independent** *on* $a \le t \le b$ *if they are not linearly dependent there. That is,* $\mathbf{x}_1, \mathbf{x}_2, \cdots, \mathbf{x}_k$ *are linearly independent on* $a \le t \le b$ *if*

$$c_1\mathbf{x}_1 + c_2\mathbf{x}_2 + \cdots + c_k\mathbf{x}_k = \mathbf{0}$$

for all t in (a, b) *implies* $c_1 = c_2 = \ldots = c_n = 0$.

Example 4: The vectors

$$\mathbf{x}_1 = \begin{pmatrix} e^t \\ 0 \\ e^t \end{pmatrix}, \quad \mathbf{x}_2 = \begin{pmatrix} e^{-2t} \\ e^{-2t} \\ 0 \end{pmatrix}, \quad \mathbf{x}_3 = \begin{pmatrix} 3e^t - 2e^{-2t} \\ -2e^{-2t} \\ 3e^t \end{pmatrix}$$

are linearly dependent because

$$3\mathbf{x}_1 - 2\mathbf{x}_2 - \mathbf{x}_3 = 0.$$

Example 5: The vectors

$$\mathbf{x}_1 = \begin{pmatrix} 0 \\ 0 \\ 1 \end{pmatrix} e^t, \quad \mathbf{x}_2 = \begin{pmatrix} 1 \\ 0 \\ 1 \end{pmatrix} e^{-t}, \quad \text{and} \quad \mathbf{x}_3 = \begin{pmatrix} 1 \\ 2 \\ 3 \end{pmatrix} e^{3t}$$

are linearly independent because if

$$c_1\mathbf{x}_1 + c_2\mathbf{x}_2 + c_3\mathbf{x} = \begin{pmatrix} 0 & e^{-t} & e^{3t} \\ 0 & 0 & 2e^{3t} \\ e^t & e^{-t} & 3e^{3t} \end{pmatrix} \begin{pmatrix} c_1 \\ c_2 \\ c_3 \end{pmatrix} = \begin{pmatrix} 0 \\ 0 \\ 0 \end{pmatrix}$$

for all x, then we can solve this equation for the c_i since our matrix is invertible to obtain $c_1 = c_2 = c_3 = 0$. (We could have solved this by Cramer's rule (determinant is not zero), Gaussian elimination, or any other method to obtain this conclusion, too.)

With this concept of linear independence in place, we can state an extremely useful theorem.

THEOREM 5.2.3 *The homogeneous system (5.19) always possesses n solution vectors that are linearly independent. Further, if $\mathbf{x}_1, \mathbf{x}_2, \ldots, \mathbf{x}_n$ are n linearly independent solutions of (5.19), then every solution $\mathbf{x}$ of (5.19) can be expressed as a linear combination*

$$c_1\mathbf{x}_1(t) + c_2\mathbf{x}_2(t) + \ldots + c_n\mathbf{x}_n(t) \qquad (5.21)$$

of these n linearly independent solutions by proper choice of the constants $c_1, c_2, \ldots, c_n$. Expression (5.21) is called the **general solution** *of (5.19) and is defined on (a, b), the interval on which solutions exist and are unique.*

We thus have that the solutions $\mathbf{x}_1, \ldots, \mathbf{x}_n$ can be combined to give us any solution we desire. As before, the concept of a *fundamental set of solutions* gives us the set necessary to write the general solution.

Three Necessary and Sufficent Conditions for a Fundamental Set of Solutions of (5.19)

1. The number of vectors (elements) in this set must be the same as the number of first-order ODEs in system (5.19).
2. Each vector $\mathbf{x}_i$ in this set must be a solution to system (5.19).
3. The vectors must be linearly independent.

As before, we note that a fundamental set of solutions is *not* unique. Once we have a fundamental set of solutions, we can construct all possible solutions from it.

Testing for linear independence is usually the challenging part and we can use a familiar tool to help us.

DEFINITION 5.3   *Let*

$$\mathbf{x}_1 = \begin{pmatrix} x_{11}(t) \\ x_{21}(t) \\ \vdots \\ x_{n1}(t) \end{pmatrix}, \quad \mathbf{x}_2 = \begin{pmatrix} x_{12}(t) \\ x_{22}(t) \\ \vdots \\ x_{n2}(t) \end{pmatrix}, \quad \mathbf{x}_n = \begin{pmatrix} x_{1n}(t) \\ x_{2n}(t) \\ \vdots \\ x_{nn}(t) \end{pmatrix}$$

be n real vector functions of t. The determinant

$$W(t) = W(\mathbf{x}_1, \mathbf{x}_2, \ldots, \mathbf{x}_n)(t) = \begin{vmatrix} x_{11}(t) & x_{12}(t) & \ldots & x_{1n}(t) \\ x_{21}(t) & x_{22}(t) & \ldots & x_{2n}(t) \\ \vdots & \vdots & \ddots & \vdots \\ x_{n1}(t) & x_{n2}(t) & \ldots & x_{nn}(t) \end{vmatrix}$$

is called the Wronskian.

THEOREM 5.2.4 *Let $\mathbf{x}_1, \mathbf{x}_2, \ldots, \mathbf{x}_n$ be defined as in Definition 5.3.*
1.   If $W(t_0) \neq 0$ for some $t_0 \in (a, b)$, then it follows that $\mathbf{x}_1, \mathbf{x}_2, \ldots, \mathbf{x}_n$ are linearly independent on (a, b).
2.   If $\mathbf{x}_1, \mathbf{x}_2, \ldots, \mathbf{x}_n$ are linearly dependent on (a, b), then $W(t) = 0$, for all $t \in (a, b)$.

THEOREM 5.2.5 *Let $\mathbf{x}_1, \mathbf{x}_2, \ldots, \mathbf{x}_n$ be defined as in Definition 5.3. Suppose that the $\mathbf{x}_i$ are each a solution of (5.19). Then exactly one of the following statements is true:*
1. $W(t) \neq 0$, for all $t \in (a, b)$
2. $W(t) = 0$, for all $t \in (a, b)$
Moreover, $W(t) \neq 0$ for all $t \in (a, b)$ if and only if the $\{\mathbf{x}_i\}$ are linearly independent on (a, b). Similarly, $W(t) = 0$ for all $t \in (a, b)$ if and only if the $\{\mathbf{x}_i\}$ are linearly dependent on (a, b).

Example 6: The vectors

$$\mathbf{x}_1 = \begin{pmatrix} 0 \\ 0 \\ 1 \end{pmatrix} e^{-2t}, \quad \mathbf{x}_2 = \begin{pmatrix} 1 \\ 0 \\ 1 \end{pmatrix} e^t, \quad \text{and} \quad \mathbf{x}_3 = \begin{pmatrix} 5 \\ 10 \\ 3 \end{pmatrix} e^{3t}$$

are linearly independent because

$$\begin{vmatrix} 0 & e^t & 5e^{3t} \\ 0 & 0 & 10e^{3t} \\ e^{-2t} & e^t & 3e^{3t} \end{vmatrix} = 0 \cdot \begin{vmatrix} 0 & 10e^{3t} \\ e^t & 3e^{3t} \end{vmatrix} - e^t \cdot \begin{vmatrix} 0 & 10e^{3t} \\ e^{-2t} & 3e^{3t} \end{vmatrix} + 5e^{3t} \cdot \begin{vmatrix} 0 & 0 \\ e^{-2t} & e^t \end{vmatrix}$$

$$= -e^t(-10e^{3t}e^{-2t}) = 10e^{2t}, \qquad (5.22)$$

which is never zero.

All of the theory so far has given us the necessary information to state a theorem for the nonhomogeneous linear system (5.17).

THEOREM 5.2.6 *Let* $\mathbf{x}_p$ *be a solution of the nonhomogeneous system (5.17). Let*

$$\mathbf{x}_c = c_1\mathbf{x}_1 + \mathbf{x}_2\mathbf{x}_2 + \ldots + c_n\mathbf{x}_n$$

be the general solution of the corresponding homogeneous equation (5.19). Then every solution $\mathbf{\Phi}$ *of the nonhomogeneous system (5.17) can be expressed in the form*

$$\mathbf{\Phi} = \mathbf{x}_c + \mathbf{x}_p.$$

Problems

In problems 1–4, find the derivative and antiderivative of each matrix.

1. $\mathbf{A} = \begin{pmatrix} \sin t & e^t \\ t^2 & 3t \end{pmatrix}$, $\quad \mathbf{B} = \begin{pmatrix} \sin^2 t & te^{-t} \\ 0 & 3+t \end{pmatrix}$

2. $\mathbf{A} = \begin{pmatrix} \sin t & e^t \\ t^2 & 3t \end{pmatrix}$, $\quad \mathbf{B} = \begin{pmatrix} e^{2t}\cos t & t^2e^{-t} \\ 3 & \ln|t| \end{pmatrix}$

3. $\mathbf{A} = \begin{pmatrix} 1 & e^{-t} & e^{3t} \\ t & 0 & 2e^{3t} \\ e^t & e^{-t} & 3e^{3t} \end{pmatrix}$, $\quad \mathbf{B} = \begin{pmatrix} \sin t & te^{-t} & e^{-t} \\ 0 & 0 & 2e^{3t} \\ e^t & t^2+e^{-t} & \cos t \end{pmatrix}$

4. $\mathbf{A} = \begin{pmatrix} e^{-t}\sin t & -e^{3t} & e^{-3t} \\ \sqrt{t} & t^4 & \cos t \\ t^{4/3} & 0 & te^{3t} \end{pmatrix}$, $\quad \mathbf{B} = \begin{pmatrix} \tan t & te^{-t} & e^{-t} \\ 0 & t & 2e^{3t} \\ e^t & t^2+e^{-t} & \cos t \end{pmatrix}$

In problems 5–11, verify that given vectors are solutions to the equation $\frac{d\mathbf{x}}{dt} = \mathbf{A}\mathbf{x}.$

5. $\mathbf{x}_1 = \begin{pmatrix} -1 \\ -2 \end{pmatrix} e^{4t}$, $\quad \mathbf{x}_2 = \begin{pmatrix} 1 \\ 3 \end{pmatrix} e^t$, $\quad \mathbf{A} = \begin{pmatrix} 10 & -3 \\ 18 & -5 \end{pmatrix}$

6. $\mathbf{x}_1 = \begin{pmatrix} -4 \\ 2 \end{pmatrix} e^{-t}$, $\quad \mathbf{x}_2 = \begin{pmatrix} 3 \\ -1 \end{pmatrix} e^{2t}$, $\quad \mathbf{A} = \begin{pmatrix} 8 & 18 \\ -3 & -7 \end{pmatrix}$

7. $\mathbf{x}_1 = \begin{pmatrix} 2 \\ 3 \end{pmatrix} e^{-2t}$, $\quad \mathbf{x}_2 = \begin{pmatrix} 1 \\ 2 \end{pmatrix} e^{-3t}$, $\quad \mathbf{A} = \begin{pmatrix} 1 & -2 \\ 6 & -6 \end{pmatrix}$

8. $\mathbf{x}_1 = \begin{pmatrix} 2 \\ 5 \end{pmatrix} e^{2t}, \quad \mathbf{x}_2 = \begin{pmatrix} 1 \\ 3 \end{pmatrix} e^t, \quad \mathbf{A} = \begin{pmatrix} 7 & -2 \\ 15 & -4 \end{pmatrix}$

9. $\mathbf{x}_1 = \begin{pmatrix} 1 \\ 0 \\ 0 \end{pmatrix} e^t, \mathbf{x}_2 = \begin{pmatrix} 0 \\ 1 \\ 0 \end{pmatrix} e^{-t}, \mathbf{x}_3 = \begin{pmatrix} -1 \\ 0 \\ 1 \end{pmatrix} e^{2t}, \mathbf{A} = \begin{pmatrix} 1 & 0 & -1 \\ 0 & -1 & 0 \\ 0 & 0 & 2 \end{pmatrix}$

10. $\mathbf{x}_1 = \begin{pmatrix} 1 \\ 1 \\ 1 \end{pmatrix} e^{2t}, \mathbf{x}_2 = \begin{pmatrix} -1 \\ 2 \\ 1 \end{pmatrix} e^{-t}, \mathbf{x}_3 = \begin{pmatrix} 1 \\ 0 \\ 0 \end{pmatrix} e^{-2t}, \mathbf{A} = \begin{pmatrix} 2 & -5 & 9 \\ 0 & -4 & 6 \\ 0 & -3 & 5 \end{pmatrix}$

11. $\mathbf{x}_1 = \begin{pmatrix} 1 \\ -3 \\ 1 \end{pmatrix} e^{-2t}, \mathbf{x}_2 = \begin{pmatrix} -1 \\ 2 \\ 0 \end{pmatrix} e^{-t}, \mathbf{x}_3 = \begin{pmatrix} 1 \\ 0 \\ -1 \end{pmatrix} e^t,$

and with matrix $\mathbf{A} = \begin{pmatrix} 1 & 1 & 0 \\ 6 & 2 & 6 \\ -6 & -3 & -5 \end{pmatrix}$

12. If $\mathbf{x}_1, \mathbf{x}_2$ are both solutions to $\mathbf{x}' = \mathbf{A}\mathbf{x}$, show that $c_1\mathbf{x}_1 + c_2\mathbf{x}_2$ is also a solution.

13. Classify the following sets of vectors as linearly dependent or linearly independent:

a. $\left\{ \begin{pmatrix} 1 \\ 0 \end{pmatrix} e^t, \ \begin{pmatrix} 1 \\ 1 \end{pmatrix} e^t, \right\}$, b. $\left\{ \begin{pmatrix} 1 \\ 1 \end{pmatrix}, \ \begin{pmatrix} 2 \\ -1 \end{pmatrix} \right\}$, c. $\left\{ \begin{pmatrix} 3 \\ -1 \end{pmatrix}, \ \begin{pmatrix} -6 \\ 2 \end{pmatrix} \right\}$

14. Classify the following sets of vectors as linearly dependent or linearly independent:

a. $\left\{ \begin{pmatrix} 3 \\ -1 \end{pmatrix} e^t, \ \begin{pmatrix} -6 \\ 2 \end{pmatrix} e^{-t}, \right\}$, b. $\left\{ \begin{pmatrix} -1 \\ 1 \end{pmatrix}, \ \begin{pmatrix} 2 \\ -1 \end{pmatrix} \right\}$,

c. $\left\{ \begin{pmatrix} 3 \\ -1 \\ 0 \end{pmatrix}, \ \begin{pmatrix} 1 \\ 1 \\ 0 \end{pmatrix}, \ \begin{pmatrix} 0 \\ 0 \\ 1 \end{pmatrix} \right\}$

15. Classify the following sets of vectors as linearly dependent or linearly independent:

a. $\left\{ \begin{pmatrix} 3 \\ 1 \\ 0 \end{pmatrix} e^t, \ \begin{pmatrix} -3 \\ -1 \\ 0 \end{pmatrix} e^{2t}, \ \begin{pmatrix} 0 \\ 0 \\ 1 \end{pmatrix} e^{-t} \right\}$, b. $\left\{ \begin{pmatrix} 2 \\ -1 \\ 0 \end{pmatrix}, \ \begin{pmatrix} -2 \\ 1 \\ 1 \end{pmatrix}, \ \begin{pmatrix} 0 \\ 0 \\ 1 \end{pmatrix} \right\}$

16. Determine if the following set forms a fundamental set of solutions for the given differential equation.

$$\left\{ \begin{pmatrix} e^{3t}\sin(t) \\ e^{3t}\cos(t) \end{pmatrix}, \begin{pmatrix} e^{3t}\cos(t) \\ -e^{3t}\sin(t) \end{pmatrix}, \right\}, \quad \mathbf{x}' = \begin{pmatrix} 3 & 2 \\ -2 & 3 \end{pmatrix} \mathbf{x}$$

17. Determine if the following set forms a fundamental set of solutions for the given differential equation:

$$\left\{ \begin{pmatrix} e^{3t}\sin(2t) \\ e^{3t}\cos(2t) \end{pmatrix}, \begin{pmatrix} e^{3t}\cos(2t) \\ -e^{3t}\sin(2t) \end{pmatrix} \right\}, \quad \mathbf{x}' = \begin{pmatrix} 3 & 2 \\ -2 & 3 \end{pmatrix} \mathbf{x}$$

18. Determine if the following set forms a fundamental set of solutions for the given differential equation:

$$\left\{ \begin{pmatrix} -e^{5t} \\ e^{5t} \end{pmatrix}, \begin{pmatrix} e^t \\ e^t \end{pmatrix} \right\}, \quad \mathbf{x}' = \begin{pmatrix} 3 & -2 \\ -2 & 3 \end{pmatrix} \mathbf{x}$$

19. Determine if the following set forms a fundamental set of solutions for the given differential equation:

$$\left\{ \begin{pmatrix} e^{-t} \\ -e^{-t} \end{pmatrix}, \begin{pmatrix} 2e^{-t} \\ 2e^{-t} \end{pmatrix} \right\}, \quad \mathbf{x}' = \begin{pmatrix} 1 & 2 \\ 0 & -1 \end{pmatrix} \mathbf{x}$$

20. Determine if the following set forms a fundamental set of solutions for the given differential equation:

$$\left\{ \begin{pmatrix} 1 \\ 0 \\ 0 \end{pmatrix} e^t, \begin{pmatrix} -1 \\ 1 \\ 0 \end{pmatrix} e^{-t}, \begin{pmatrix} 0 \\ 0 \\ 1 \end{pmatrix} e^{-2t} \right\}, \quad \mathbf{x}' = \begin{pmatrix} 1 & 2 & 0 \\ 0 & -1 & 0 \\ 0 & 0 & -2 \end{pmatrix} \mathbf{x}$$

21. Determine if the following set forms a fundamental set of solutions for the given differential equation:

$$\left\{ \begin{pmatrix} 0 \\ 0 \\ 1 \end{pmatrix} e^{-t}, \begin{pmatrix} 1 \\ 0 \\ \frac{1}{3} \end{pmatrix} e^t, \begin{pmatrix} -1 \\ 1 \\ -1 \end{pmatrix} e^{-t} \right\}, \quad \mathbf{x}' = \begin{pmatrix} 1 & 2 & 0 \\ 0 & -1 & 0 \\ 1 & 0 & -2 \end{pmatrix} \mathbf{x}$$

5.3 Linear Transformations and the Fundamental Subspaces

In Section 5.1, we considered the equations

$$\frac{dx}{dt} = y$$
$$\frac{dy}{dt} = -\frac{k}{m}x \qquad (5.23)$$

and

$$\frac{dx}{dt} = -5x + 8y$$
$$\frac{dy}{dt} = -4x + 7y, \qquad (5.24)$$

and discussed how the vector field can be used to describe the motion of a trajectory (projection of the solution onto the *x-y* plane) through a given point (x, y). In the first system, equation (5.23), we can write

$$\begin{pmatrix} x' \\ y' \end{pmatrix} = \begin{pmatrix} 0 & 1 \\ -\frac{k}{m} & 0 \end{pmatrix} \begin{pmatrix} x \\ y \end{pmatrix}$$

and thus view the right-hand side as "the coefficient matrix $\mathbf{A}$ multiplied by the vector $\begin{pmatrix} x \\ y \end{pmatrix}$." The result gives the instantaneous rate of change of the trajectory through that point. For example, if we consider the trajectory that passes through $(1, 2)$ (for, say, $m = 2, k = 3$), we find its instantaneous rate of change is

$$\begin{pmatrix} 0 & 1 \\ -\frac{3}{2} & 0 \end{pmatrix} \begin{pmatrix} 1 \\ 2 \end{pmatrix} = \begin{pmatrix} 2 \\ -\frac{3}{2} \end{pmatrix}$$

and the motion of the trajectory is right and down as was shown in the picture of the vector field (see Figure 5.1).

For the second system, equation (5.24), we can again consider the trajectory through the point $(1, 2)$ and calculate the instantaneous rate of change as

$$\begin{pmatrix} -5 & 8 \\ -4 & 7 \end{pmatrix} \begin{pmatrix} 1 \\ 2 \end{pmatrix} = \begin{pmatrix} 11 \\ 10 \end{pmatrix}.$$

Thus the motion of the trajectory is right and up and we again saw this in the vector field (see Figure 5.3).

In both of these systems, multiplication of a vector by a coefficient matrix describes the slope of the solution through the given (x, y) pair. So we may think of *a matrix doing something to or acting on a vector*; this gives us a useful interpretation of matrix-vector multiplication.

With this as our motivation, we will now consider some useful ideas from linear algebra. For the purposes of this section we adopt the viewpoint of $\mathbf{A}$ *acting* on a vector $\mathbf{x}$. In the case of a square $n \times n$ matrix $\mathbf{A}$, the vector $\mathbf{x}$ gets moved from one location in $\mathbb{R}^n$ to another; geometrically, the vectors in $\mathbb{R}^n$ have been moved according to the "action" of $\mathbf{A}$, with perhaps some vectors (e.g., $\mathbf{0}$) being left unchanged. In the case of a rectangular $m \times n$ matrix $\mathbf{A}$, we have a situation where $\mathbf{A}$ takes a vector from one vector space ($\mathbb{R}^n$) to another vector space ($\mathbb{R}^m$). Note that the origin in $\mathbb{R}^n$ is mapped to the origin in $\mathbb{R}^m$.

Example 1: The matrix-vector multiplication

$$\begin{pmatrix} -1 & 3 & 2 \\ 1 & 1 & 0 \\ 2 & -3 & 4 \end{pmatrix} \begin{pmatrix} -1 \\ 2 \\ 3 \end{pmatrix} = \begin{pmatrix} 13 \\ 1 \\ 4 \end{pmatrix} \quad \text{and} \quad \begin{pmatrix} -1 & 3 & 2 \\ 1 & 1 & 0 \\ 2 & -3 & 4 \end{pmatrix} \begin{pmatrix} 0 \\ 0 \\ 0 \end{pmatrix} = \begin{pmatrix} 0 \\ 0 \\ 0 \end{pmatrix}$$

show that a 3×3 matrix takes a vector in $\mathbb{R}^3$ to a vector in $\mathbb{R}^3$, whereas

$$\begin{pmatrix} -1 & 3 & 2 \\ 1 & 1 & 0 \end{pmatrix} \begin{pmatrix} -1 \\ 2 \\ 3 \end{pmatrix} = \begin{pmatrix} 13 \\ 1 \end{pmatrix} \quad \text{and} \quad \begin{pmatrix} -1 & 3 & 2 \\ 1 & 1 & 0 \end{pmatrix} \begin{pmatrix} 0 \\ 0 \\ 0 \end{pmatrix} = \begin{pmatrix} 0 \\ 0 \end{pmatrix}$$

shows that a 2×3 matrix takes a vector in $\mathbb{R}^3$ to a vector in $\mathbb{R}^2$. Note that in both cases, the origin was mapped to the origin (in a different space in the case of a rectangular matrix).

Regardless of whether our matrix $\mathbf{A}$ is square or rectangular, our definition of matrix multiplication gives us the following:

PROPOSITION 5.3.1 *Let c_1, c_2 be any scalars, let $\mathbf{x}, \mathbf{y}$ be vectors in $\mathbb{R}^n$, and let $\mathbf{A}$ be an $m \times n$ matrix. Then*

$$\mathbf{A}(c_1 \mathbf{x} + c_2 \mathbf{y}) = c_1(\mathbf{A}\mathbf{x}) + c_2(\mathbf{A}\mathbf{y}) \tag{5.25}$$

and we say $\mathbf{A}$ is a linear transformation.

In this case, we say that $\mathbf{A}$ takes $\mathbb{R}^n$ into $\mathbb{R}^m$. A matrix is one example of a linear transformation but there are many other examples. When the domain and range are the same space, as is the case when $\mathbf{A}$ is a square matrix, we say that $\mathbf{A}$ is a *linear operator*.

Example 2: Let D denote the differentiation operator, d/dt, from Section 3.4. D is a linear transformation that takes polynomials of degree n to polynomials of degree $n - 1$:

$$D(a_n t^n + \cdots + a_1 t + a_0) = n a_n t^{n-1} + \cdots + a_1. \tag{5.26}$$

If $\mathbf{x}$ and $\mathbf{y}$ are polynomials then we can easily verify that

$$D(c_1 \mathbf{x} + c_2 \mathbf{y}) = c_1(D\mathbf{x}) + c_2(D\mathbf{y}).$$

It is a natural question to ask if all matrices are linear transformations, is it the case that all linear transformations can somehow be represented by a matrix? The answer is yes (for $n < \infty$) but we refer the reader to other texts for the details [33]. Here, instead, we focus on writing the above example in matrix notation and then examining linear transformations as matrix multiplications, as this gives us geometrical insight into our problem.

Example 3: In the last example it was shown that D is a linear transformation; can we represent this linear transformation by a matrix? For a fixed $n < \infty$, consider a polynomial $a_n t^n + \cdots + a_1 t + a_0$, writing the coefficients of the powers as the entries in a vector $\mathbf{x}$ of length $n + 1$

$$\mathbf{x} = \begin{pmatrix} a_n \\ a_{n-1} \\ \vdots \\ a_1 \\ a_0 \end{pmatrix}. \tag{5.27}$$

Differentiation can be represented with the matrix

$$
D = \begin{pmatrix}
0 & 0 & 0 & \cdots & 0 & 0 & 0 \\
n & 0 & 0 & \cdots & 0 & 0 & 0 \\
0 & n-1 & 0 & \cdots & 0 & 0 & 0 \\
0 & 0 & n-2 & \cdots & 0 & 0 & 0 \\
& \vdots & & \ddots & & \vdots & \\
0 & 0 & 0 & \cdots & 2 & 0 & 0 \\
0 & 0 & 0 & \cdots & 0 & 1 & 0
\end{pmatrix}.
\tag{5.28}
$$

For instance, consider the polynomial $\mathbf{x}(t) = 2t^4 - t^3 + \sqrt{2}t^2 + t - 1$. Then

$$
D\mathbf{x} = \begin{pmatrix}
0 & 0 & 0 & 0 & 0 \\
4 & 0 & 0 & 0 & 0 \\
0 & 3 & 0 & 0 & 0 \\
0 & 0 & 2 & 0 & 0 \\
0 & 0 & 0 & 1 & 0
\end{pmatrix}
\begin{pmatrix}
2 \\ -1 \\ \sqrt{2} \\ 1 \\ -1
\end{pmatrix}
=
\begin{pmatrix}
0 \\ 8 \\ -3 \\ 2\sqrt{2} \\ 1
\end{pmatrix},
\tag{5.29}
$$

and this last vector can then be rewritten as the polynomial $8t^3 - 3t^2 + 2\sqrt{2}t + 1$, which is the derivative of $2t^4 - t^3 + \sqrt{2}t^2 + t - 1$.

Some linear transformations with simple geometric interpretations

Interpreting a linear transformation as a matrix that takes vectors from one space to another is often useful. We present four examples that have a concrete geometrical interpretation. The vectors, as we will see, remain in the same space. For simplicity, we will consider vectors in the plane $\mathbb{R}^2$.

1. Stretch. In this situation, a vector is stretched (or shrunk) by a factor k. The matrix that will do this is one of the form

$$
\mathbf{A} = \begin{pmatrix} k & 0 \\ 0 & k \end{pmatrix}.
\tag{5.30}
$$

This matrix can be written more compactly as $\mathbf{A} = k\mathbf{I}$, where $\mathbf{I}$ is the 2×2 identity matrix. We know that multiplication by the identity leaves things unchanged. Also, multiplication of a scalar k times a vector (or matrix) requires us to multiply *every entry* by this factor of k. Thus $k\mathbf{I}$ will lengthen any vector by a factor of k if $k > 1$ (and will reverse the orientation and lengthen by a factor of k if $k < -1$) and will shorten any vector by a factor of k if $0 < k < 1$ (and will reverse the orientation and shorten by a factor of k if $-1 < k < 0$).

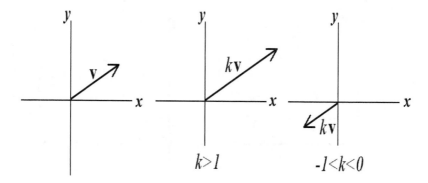

FIGURE 5.6:   Vector **v** stretched by a factor of k. Note that the direction may change depending on the sign of k.

2. Rotation. In this case, a vector is rotated counterclockwise about the origin by some angle ϕ. The matrix that will do this is of the form

$$\mathbf{A} = \begin{pmatrix} \cos\phi & -\sin\phi \\ \sin\phi & \cos\phi \end{pmatrix}.$$
 (5.31)

Consider the vector $(1 \ 0)^T$, which lies along the x-axis. If this vector through the origin is rotated through an angle of $\phi = \pi/4$ rad, the matrix simplifies to

$$\mathbf{A} = \begin{pmatrix} 1/\sqrt{2} & -1/\sqrt{2} \\ 1/\sqrt{2} & 1/\sqrt{2} \end{pmatrix}$$
 (5.32)

and thus

$$\begin{pmatrix} 1/\sqrt{2} & -1/\sqrt{2} \\ 1/\sqrt{2} & 1/\sqrt{2} \end{pmatrix} \begin{pmatrix} 1 \\ 0 \end{pmatrix} = \begin{pmatrix} 1/\sqrt{2} \\ 1/\sqrt{2} \end{pmatrix} = \frac{1}{\sqrt{2}} \begin{pmatrix} 1 \\ 1 \end{pmatrix} \approx \begin{pmatrix} .7071 \\ .7071 \end{pmatrix}.$$
 (5.33)

Recalling from multivariable calculus that the Euclidean length of a vector is the square root of the sum of the squares, we see that the resulting vector is again of length one. It should also be apparent that the vector is oriented at an angle of $\pi/4$ rad.

3. Projection. We can also write a matrix that will project any vector onto a given line. If we consider a vector and a line on which we wish to project, the projection is defined as the component of the vector in the direction of the line. In this situation, we can write a matrix in the form

$$\mathbf{A} = \begin{pmatrix} \cos^2\phi & \cos\phi\,\sin\phi \\ \cos\phi\,\sin\phi & \sin^2\phi \end{pmatrix},$$
 (5.34)

where ϕ (measured counterclockwise from the origin) is the angle of the line through the origin that we will project onto. Let us again look at a simple

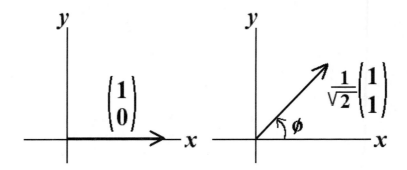

FIGURE 5.7:   Vector $(1 \ 0)^T$ rotated through an angle of $\phi = \pi/4$.

example by considering the vector $(1 \ 2)^T$ and projecting it onto the y-axis. The y-axis is $\pi/2$ rad counterclockwise from the positive x-axis and thus $\phi = \pi/2$ rad. Our matrix in this case becomes

$$\mathbf{A} = \begin{pmatrix} 0 & 0 \\ 0 & 1 \end{pmatrix}, \tag{5.35}$$

and applying this matrix to our vector gives

$$\begin{pmatrix} 0 & 0 \\ 0 & 1 \end{pmatrix} \begin{pmatrix} 1 \\ 2 \end{pmatrix} = \begin{pmatrix} 0 \\ 2 \end{pmatrix}. \tag{5.36}$$

This answer is what we expected given that the component in the direction of the y-axis is 2. The reader should again try a few additional examples so as to be convinced that the above matrix projects any given vector passing through the origin onto a line passing through the origin with angle ϕ.

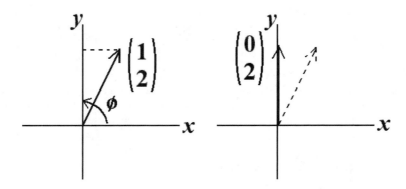

FIGURE 5.8:   Vector $(1 \ 2)^T$ projected onto the y-axis.

4. Reflection. Sometimes we wish to reflect a vector over a given line. We again consider a line through the origin that makes an angle ϕ with the x-axis, where ϕ is measured counterclockwise. The matrix that will do this reflection is

$$\mathbf{A} = \begin{pmatrix} 2\cos^2\phi - 1 & 2\cos\phi\,\sin\phi \\ 2\cos\phi\,\sin\phi & 2\sin^2\phi - 1 \end{pmatrix}. \tag{5.37}$$

As an example, let us reflect the vector $(1\ \ 1)^T$ across the $\pi/6$ rad line. Our linear transformation and vector multiply as

$$\mathbf{A} = \begin{pmatrix} 1/2 & \sqrt{3}/2 \\ \sqrt{3}/2 & -1/2 \end{pmatrix}\begin{pmatrix} 1 \\ 1 \end{pmatrix} = \begin{pmatrix} \sqrt{3}/2 + 1/2 \\ (\sqrt{3} - 1)/2 \end{pmatrix} \approx \begin{pmatrix} 1.366 \\ 0.366 \end{pmatrix}. \tag{5.38}$$

The angle that this vector makes with the positive x-axis is

$$\tan\theta = \frac{(\sqrt{3}-1)/2}{(\sqrt{3}+1)/2} \implies \theta = 15° \approx 0.2618 \text{ rad.}$$

The reader should again try a few additional examples so as to be convinced that the above matrix reflects any given vector passing through the origin across a line passing through the origin with angle ϕ.

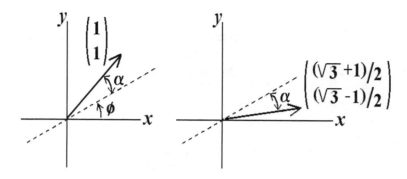

FIGURE 5.9:   Vector $(1\ \ 1)^T$ reflected across the line $\pi/6$.

Computer Code 5.2:   Calculating with rotation matrix, projection matrix, reflection matrix in $\mathbb{R}^2$

Matlab, Maple, Mathematica

```
                          Matlab
>>  syms phi
>>  %commands for rotation matrix, Ar in eq.(5.32)
>>  Ar=[cos(phi),-sin(phi); sin(phi),cos(phi)]
>>  Ar1=subs(Ar,phi,pi/4)
>>  xr=[1;0]
>>  Ar1*xr %resulting vector, see eq.(5.33)
>>  Ar2=sym(subs(Ar,phi,pi/4)) %symbolic calculation
>>  Ar2*xr %symbolic calculation
>>  eval(Ar2*xr)
```

```
                          Matlab
>>  %commands for projection matrix, Ap in eq.(5.34)
>>  Ap=[cos(phi)^2,cos(phi)*sin(phi); cos(phi)*sin(phi),
    sin(phi)^2]
>>  Ap1=subs(Ap,phi,pi/2)
>>  xp=[1;2]
>>  Ap1*xp %resulting vector, see eq.(5.36)
>>  Ap2=sym(subs(Ap,phi,pi/2)) %symbolic calculation
>>  Ap2*xp %symbolic calculation
>>  eval(Ap2*xp)
```

```
                          Matlab
>>  %commands for reflection matrix, Af in eq.(5.37)
>>  Af=[2*cos(phi)^2-1,2*cos(phi)*sin(phi);
    2*cos(phi)*sin(phi),2*sin(phi)^2-1]
>>  Af1=subs(Af,phi,pi/2)
>>  xf=[1;1]
>>  Af1*xf %resulting vector, see eq.(5.38)
>>  Af2=sym(subs(Af,phi,pi/6)) %symbolic calculation
>>  Af2*xf %symbolic calculation
>>  eval(Af2*xf)
```

```
                            ┌──────┐
                            │ Maple │
                            └──────┘
> with(linalg):
  #commands for rotation matrix, A[rot] in eq.(5.32)
> A[rot]:=matrix(2,2,[cos(phi),-sin(phi),sin(phi),cos(phi)]);
> A1[rot]:=subs(phi=Pi/4,evalm(A[rot]));
> evalf(evalm(A1[rot]));
> x[rot]:=matrix(2,1,[1,0]);
> eq1:=multiply(A1[rot],x[rot]);
  #resulting vector, see eq.(5.33)
> evalf(evalm(eq1));
```

```
                            ┌──────┐
                            │ Maple │
                            └──────┘
  #commands for projection matrix, A[proj] in eq.(5.34)
> A[proj]:=matrix(2,2,[cos(phi)^2,cos(phi)*sin(phi),
  cos(phi)*sin(phi),sin(phi)^2]);
> A1[proj]:=subs(phi=Pi/2,evalm(A[proj]));
> evalf(evalm(A1[proj]));
> x[proj]:=matrix(2,1,[1,2]);
> eq2:=multiply(A1[proj],x[proj]);
  #resulting vector, see eq.(5.36)
> evalf(evalm(eq2));
```

```
                            ┌──────┐
                            │ Maple │
                            └──────┘
  #commands for reflection matrix, A[ref] in eq.(5.37)
> A[ref]:=matrix(2,2,[2*cos(phi)^2-1,2*cos(phi)*sin(phi),
  2*cos(phi)*sin(phi),2*sin(phi)^2-1]);
> A1[ref]:=subs(phi=Pi/6,evalm(A[ref]));
> evalf(evalm(A1[ref]));
> x[ref]:=matrix(2,1,[1,1]);
> eq3:=multiply(A1[ref],x[ref]);
  #resulting vector, see eq.(5.38)
> evalf(evalm(eq3));
```

```
                         ┌────────────┐
                         │ Mathematica │
                         └────────────┘
(*commands for rotation matrix, Rot[φ_] in eq.(5.32)*)
Rot[φ_] = {{Cos[φ], -Sin[φ]}, {Sin[φ], Cos[φ]}}
Rot[π/4]//MatrixForm (*puts matrix in nicer visual form*)
eq1 = Rot[π/4].{{2}, {1}} (*resulting vector, see eq.(5.33)*)
N[eq1]
```

|Mathematica|

(*commands for projection matrix, Proj[ϕ_] in eq.(5.34)*)
Proj[ϕ_] = {{Cos[ϕ]^2, Cos[ϕ]Sin[ϕ]}, {Cos[ϕ]Sin[ϕ], Sin[ϕ]^2}}
Proj[π/2]//MatrixForm (*puts matrix in nicer visual form*)
eq2 = Proj[π/2].{{2}, {1}} (*resulting vector, see eq.(5.36)*)
N[eq2]

|Mathematica|

(*commands for reflection matrix, Ref[ϕ_] in eq.(5.37)*)
Ref[ϕ_] = {{2Cos[ϕ]^2 − 1, 2Cos[ϕ]Sin[ϕ]}, {2Cos[ϕ]Sin[ϕ], 2Sin[ϕ]^2 − 1}}
Ref[π/6]//MatrixForm (*puts matrix in nicer visual form*)
eq3 = Ref[π/6].{{2}, {1}} (*resulting vector, see eq.(5.38)*)
N[eq3]

5.3.1   The Four Fundamental Subspaces

We conclude this section with a consideration of linear transformations in relation to subspaces.[5] Consider the following matrix equations:

$$\begin{pmatrix} 2 & -2 & 2 \\ 2 & -2 & 2 \\ 1 & -1 & 1 \end{pmatrix} \begin{pmatrix} 1 \\ 1 \\ 0 \end{pmatrix} = \begin{pmatrix} 0 \\ 0 \\ 0 \end{pmatrix} \text{ and } \begin{pmatrix} 2 & -2 & 2 \\ 2 & -2 & 2 \\ 1 & -1 & 1 \end{pmatrix} \begin{pmatrix} -1 \\ 0 \\ 1 \end{pmatrix} = \begin{pmatrix} 0 \\ 0 \\ 0 \end{pmatrix}.$$
(5.39)

Thus, the matrix $\mathbf{A}$ takes both $(1 \ 1 \ 0)^T$ and $(-1 \ 0 \ 1)^T$ to the zero vector $(0 \ 0 \ 0)^T$. Is this a property of the vectors or the matrix? Actually, it's both. The structure of the given matrix is such that any vector that is a linear combination of the above two vectors is also sent to the zero vector. And it's not hard to construct similar examples for an $m \times n$ matrix as well. For example, we can also see that

$$\begin{pmatrix} 2 & -2 & 2 \\ -3 & 3 & -3 \end{pmatrix} \begin{pmatrix} 0 \\ 1 \\ 1 \end{pmatrix} = \begin{pmatrix} 0 \\ 0 \end{pmatrix} \text{ and } \begin{pmatrix} 2 & -2 & 2 \\ -3 & 3 & -3 \end{pmatrix} \begin{pmatrix} 1 \\ 2 \\ 1 \end{pmatrix} = \begin{pmatrix} 0 \\ 0 \end{pmatrix}.$$
(5.40)

The following definition will clarify this situation:

---

[5]See any of the numerous linear algebra books, for example, Strang [33], for a more in-depth discussion.

DEFINITION 5.4   *The* **nullspace** *of an* $m \times n$ *matrix* $\mathbf{A}$ *is defined as*

$$null(\mathbf{A}) = \{\mathbf{x} \in \mathbb{R}^n | \mathbf{Ax} = \mathbf{0}\}.$$

We note that null($\mathbf{A}$) is a subspace of $\mathbb{R}^n$. This is easily seen because it satisfies the three conditions of a subspace:

(i) We know the zero vector is in null($\mathbf{A}$) because $\mathbf{A0} = \mathbf{0}$.

(ii) If $\mathbf{x}, \mathbf{y}$ are two vectors in $null(\mathbf{A})$, then $\mathbf{A}(\mathbf{x} + \mathbf{y}) = \mathbf{Ax} + \mathbf{Ay}$ and thus $\mathbf{x} + \mathbf{y}$ is in null($\mathbf{A}$).

(iii) For any scalar c and any vector $\mathbf{x}$ in null($\mathbf{A}$), we have $\mathbf{A}(c\mathbf{x}) = c\mathbf{Ax} = c\mathbf{0} = \mathbf{0}$ and thus $c\mathbf{x}$ is in null($\mathbf{A}$).

Because the three conditions are satisfied, we conclude that null($\mathbf{A}$) is indeed a subspace of $\mathbb{R}^n$.

Sometimes the nullspace of $\mathbf{A}$ is referred to as the *kernel* of $\mathbf{A}$. We also observe that not all vectors are taken to zero. In the examples immediately preceding Definition 5.4, we can see that matrix A takes $(1 \ \ 0 \ \ 2)^T$ and $(1 \ \ 0 \ \ 1)^T$ to $(6 \ \ 6 \ \ 3)^T$ and $(4 \ \ -6)^T$, respectively. What about these vectors that are *not* taken to zero upon left multiplication by a matrix? We have another definition.

DEFINITION 5.5   *The* **column space** *or* **range** *of an* $m \times n$ *matrix* $\mathbf{A}$ *is defined as*

$$R(\mathbf{A}) = \{\mathbf{y} \in \mathbb{R}^m | \mathbf{Ax} = \mathbf{y}\}.$$

We note that R($\mathbf{A}$) is a subspace of $\mathbb{R}^m$. It is also called the column space because it is the subspace of $\mathbb{R}^m$ that is spanned by the columns of $\mathbf{A}$. We also show this:

(i) We know the zero vector is in R($\mathbf{A}$) because $\mathbf{A0} = \mathbf{0}$.

(ii) If $\mathbf{y}_1, \mathbf{y}_2$ are two vectors in R($\mathbf{A}$), then we know $\mathbf{Ax}_1 = \mathbf{y}_1$ and $\mathbf{Ax}_2 = \mathbf{y}_2$ for some vectors $\mathbf{x}_1, \mathbf{x}_2$ in $\mathbb{R}^n$. We then have $\mathbf{A}(\mathbf{x}_1 + \mathbf{x}_2) = \mathbf{Ax}_1 + \mathbf{Ax}_2 = \mathbf{y}_1 + \mathbf{y}_2$ and thus $\mathbf{y}_1 + \mathbf{y}_2$ is in R($\mathbf{A}$).

(iii) For any scalar c and any vector $\mathbf{y}$ in R($\mathbf{A}$), we have $\mathbf{A}(\mathbf{x}) = \mathbf{y}$ for some $\mathbf{x}$. Then $\mathbf{A}(c\mathbf{x}) = c\mathbf{Ax} = c\mathbf{y}$ and thus $c\mathbf{y}$ is in R($\mathbf{A}$).

Because the three conditions are satisfied, we conclude that R($\mathbf{A}$) is indeed a subspace of $\mathbb{R}^m$.

At this point, we pause to define another important concept of linear algebra.

DEFINITION 5.6   *The* **rank** *of a matrix* $\mathbf{A}$ *is the number of linearly independent columns of* $\mathbf{A}$.

This definition gives us the following useful result, where $\text{rank}(\mathbf{A}) = r$:

$$\dim(\text{R}(\mathbf{A})) = r. \tag{5.41}$$

We arrived at a subspace by considering the span of the columns of $\mathbf{A}$. Although the next two examples of subspaces are not used in the study of differential equations, we present them here for completeness.

Another subspace can be obtained by considering the rows of $\mathbf{A}$, which are the same as the columns of $\mathbf{A}^T$:

DEFINITION 5.7   *The* **row space** *of an* $m \times n$ *matrix* $\mathbf{A}$ *is defined as*

$$R(\mathbf{A}^T) = \{\mathbf{x} \in \mathbb{R}^n | \mathbf{A}^T\mathbf{y} = \mathbf{x}\}.$$

We note that $R(\mathbf{A}^T)$ is a subspace of $\mathbb{R}^n$. Likewise, we can examine the nullspace of $\mathbf{A}^T$:

DEFINITION 5.8   *The* **left nullspace** *of an* $m \times n$ *matrix* $\mathbf{A}$ *is defined as*

$$null(\mathbf{A}^T) = \{\mathbf{y} \in \mathbb{R}^m | \mathbf{y}^T\mathbf{A} = \mathbf{0}\}.$$

The left nullspace (i.e., the nullspace of $\mathbf{A}^T$) is a subspace of $\mathbb{R}^m$. We state, without proof, a very important result:

THEOREM 5.3.1

dimension of row space = dimension of column space = rank.

In other words, the number of linearly independent rows is the same as the number of linearly independent columns. We thus also have

$$\dim(\text{R}(\mathbf{A}^T)) = r.$$

There are straightforward ways to calculate bases for each of the four fundamental subspaces. We refer the reader to the references for these methods and instead turn to Matlab, Maple, or Mathematica to find the bases. Let's consider the matrix

$$\mathbf{A} = \begin{pmatrix} 2 & -2 & 2 & 2 \\ 2 & -2 & 2 & 2 \\ 1 & -1 & 1 & 0 \end{pmatrix}. \tag{5.42}$$

Bases from the four fundamental subspaces can be found with the following code:

Computer Code 5.3:   **Calculating bases for the four fundamental subspaces (nullspace, column space, row space, and left nullspace)**

Matlab, Maple, Mathematica

```
                          Matlab
>> A=[2, -2, 2, 2; 2, -2, 2, 2; 1, -1, 1, 0]
>> null(A) %orthonormal basis for nullspace, which is
   %useful for numerical computation
>> null(A,'r') %this is also a basis but is easier to
   %work with by hand
>> colspace(sym(A)) %finds a basis for the column space
>> B=A';
>> colspace(sym(B)) %finds a basis for the row space
>> null(B,'r') %this is the basis for the left nullspace
```

```
                          Maple
> with(linalg):
> A:=matrix(3,4,[2, -2, 2, 2, 2, -2, 2, 2, 1, -1, 1, 0]);
> nullspace(A); #basis for the nullspace
> kernel(A); #alternate syntax
> colspace(A); #basis for the column space
> rowspace(A); #basis for the row space
> colspace(transpose(A)); #alternate syntax
> nullspace(transpose(A)); #basis for the left nullspace
```

```
                       Mathematica
A={{2,-2,2,2},{2,-2,2,2},{1,-1,1,0}}
NullSpace[A] (*basis for the nullspace*)
MatrixFrom[RowReduce[Transpose[A]]]
   (*non-zero rows are the basis vectors for column space*)
MatrixForm[RowReduce[A]]
   (*non-zero rows are the basis vectors for the row space*)
NullSpace[Transpose[A]]
   (*non-zero rows are basis for the left nullspace*)
```

The above code gives us the following results:

$$\text{null}(\mathbf{A}) = \left\{ \begin{pmatrix} 1 \\ 1 \\ 0 \\ 0 \end{pmatrix}, \begin{pmatrix} -1 \\ 0 \\ 1 \\ 0 \end{pmatrix} \right\}, \quad \text{R}(\mathbf{A}) = \left\{ \begin{pmatrix} 0 \\ 0 \\ 1 \end{pmatrix}, \begin{pmatrix} 1 \\ 1 \\ 0 \end{pmatrix} \right\},$$

$$\text{R}(\mathbf{A}^T) = \left\{ \begin{pmatrix} 0 \\ 0 \\ 0 \\ 1 \end{pmatrix}, \begin{pmatrix} 1 \\ -1 \\ 1 \\ 0 \end{pmatrix} \right\}, \quad \text{null}(\mathbf{A}^T) = \left\{ \begin{pmatrix} -1 \\ 1 \\ 0 \end{pmatrix} \right\}. \tag{5.43}$$

For our matrix $\mathbf{A}$, we have $m = 3$, $n = 4$. We should observe that we do have $\text{null}(\mathbf{A}) \subset \mathbb{R}^4$, $\text{R}(\mathbf{A}) \subset \mathbb{R}^3$, $\text{R}(\mathbf{A}^T) \subset \mathbb{R}^4$, $\text{null}(\mathbf{A}^T) \subset \mathbb{R}^3$, as we showed earlier. There is another worthwhile observation for this example:

$$\dim(\text{R}(\mathbf{A})) + \dim(\text{null}(\mathbf{A})) = 4$$
$$\dim(\text{R}(\mathbf{A}^T)) + \dim(\text{null}(\mathbf{A}^T)) = 3$$

The following is true in general and is a very important result in linear algebra:

THEOREM 5.3.2 *For an $m \times n$ matrix $\mathbf{A}$, we have*

$$dim(R(\mathbf{A})) + dim(null(\mathbf{A})) = n \tag{5.44}$$
$$dim(R(\mathbf{A}^T)) + dim(null(\mathbf{A}^T)) = m. \tag{5.45}$$

Problems

1. Consider the vector $(1 \quad 3)^T$. Find the vectors that result from
   a. stretch by a factor of $\frac{3}{2}$ (sketch the original vector and the resulting vector)
   b. rotation by an angle of $\pi/2$ (sketch the original vector, the angle of rotation, and the resulting vector)
   c. projection onto the line that makes an angle $\pi/2$ with the x-axis (sketch the original vector, the line of projection, and the resulting vector)
   d. reflection through the line that makes an angle $\pi/2$ with the x-axis (sketch the original vector, the line of reflection, and the resulting vector)

2. Consider the vector $(-1 \quad -2)^T$. Find the vectors that result from
   a. stretch by a factor of -2 (sketch the original vector and the resulting vector)
   b. rotation by an angle of π (sketch the original vector, the angle of

rotation, and the resulting vector)

c. projection onto the line that makes an angle π with the x-axis (sketch the original vector, the line of projection, and the resulting vector)

d. reflection through the line that makes an angle π with the x-axis (sketch the original vector, the line of reflection, and the resulting vector)

3. Consider the vector $(1 \quad -3)^T$. Find the vectors that result from

a. stretch by a factor of $\frac{1}{2}$ (sketch the original vector and the resulting vector)

b. rotation by an angle of $\pi/3$ (sketch the original vector, the angle of rotation, and the resulting vector)

c. projection onto the line that makes an angle $\pi/3$ with the x-axis (sketch the original vector, the line of projection, and the resulting vector)

d. reflection through the line that makes an angle $\pi/3$ with the x-axis (sketch the original vector, the line of reflection, and the resulting vector)

4. Consider the vector $(-2 \quad 3)^T$. Find the vectors that result from

a. stretch by a factor of 3 (sketch the original vector and the resulting vector)

b. rotation by an angle of $\pi/3$ (sketch the original vector, the angle of rotation, and the resulting vector)

c. projection onto the line that makes an angle $\pi/3$ with the x-axis (sketch the original vector, the line of projection, and the resulting vector)

d. reflection through the line that makes an angle $\pi/3$ with the x-axis (sketch the original vector, the line of reflection, and the resulting vector)

5. Consider the vector $(2 \quad 1)^T$. Find the vectors that result from

a. stretch by a factor of -1 (sketch the original vector and the resulting vector)

b. rotation by an angle of $-\pi/4$ (sketch the original vector, the angle of rotation, and the resulting vector)

c. projection onto the line that makes an angle 0 with the x-axis (sketch the original vector, the line of projection, and the resulting vector)

d. reflection through the line that makes an angle 0 with the x-axis (sketch the original vector, the line of reflection, and the resulting vector)

6. Consider the vector $(-3 \quad 1)^T$. Find the vectors that result from

a. stretch by a factor of 2 (sketch the original vector and the resulting vector)

b. rotation by an angle of $-\pi/2$ (sketch the original vector, the angle of rotation, and the resulting vector)

c. projection onto the line that makes an angle $\pi/3$ with the x-axis (sketch the original vector, the line of projection, and the resulting vector)

d. reflection through the line that makes an angle $\pi/6$ with the x-axis (sketch the original vector, the line of reflection, and the resulting vector)

7. Let $\mathbf{A}$ be a 2×2 reflection matrix. Show that $\mathbf{A}^2 = I$. This shows that if $\mathbf{y}$ is a reflection of $\mathbf{x}$, then applying the reflection matrix again gives us the original vector.

Find bases for the column space and nullspace of the following matrices given in problems 8–11.

8. $\mathbf{A} = \begin{pmatrix} 1 & 3 \\ 2 & 1 \end{pmatrix}, \quad \mathbf{B} = \begin{pmatrix} -1 & 3 \\ 3 & -9 \end{pmatrix}$

9. $\mathbf{A} = \begin{pmatrix} 1 & 3 & 1 \\ 2 & 1 & -1 \end{pmatrix}, \quad \mathbf{B} = \begin{pmatrix} -1 & 3 & 2 \\ 2 & 0 & 2 \end{pmatrix}$

10. $\mathbf{A} = \begin{pmatrix} 3 & 1 & 1 & 1 \\ 1 & 3 & -1 & -3 \end{pmatrix}, \quad \mathbf{B} = \begin{pmatrix} 0 & 1 & -1 & 1 \\ 2 & 5 & -1 & -2 \end{pmatrix}$

11. $\mathbf{A} = \begin{pmatrix} -1 & 2 \\ 2 & 4 \\ -1 & 3 \end{pmatrix}, \quad \mathbf{B} = \begin{pmatrix} -1 & 3 \\ 2 & -6 \\ -1 & 3 \end{pmatrix}$

12. Use Matlab, Maple, or Mathematica to construct a 3×3 matrix that has a nullspace consisting only of the zero vector.

13. Use Matlab, Maple, or Mathematica to construct a 3×3 matrix that has a nullspace with a basis consisting of one non-zero vector.

14. Use Matlab, Maple, or Mathematica to construct a 3×3 matrix that has a nullspace with a basis consisting of two non-zero vectors.

15. Consider the matrix

$$\mathbf{A} = \begin{pmatrix} 2 & -2 & 2 \\ 2 & -2 & 2 \\ 1 & -1 & 1 \end{pmatrix} \tag{5.46}$$

from the beginning of this section. Use Matlab, Maple, or Mathematica to show that possible bases of the four fundamental subspaces are

$$\text{null}(\mathbf{A}) = \left\{ \begin{pmatrix} -1 \\ 0 \\ 1 \end{pmatrix}, \begin{pmatrix} 1 \\ 1 \\ 0 \end{pmatrix} \right\}, \quad \text{R}(\mathbf{A}) = \left\{ \begin{pmatrix} 1 \\ 1 \\ 1/2 \end{pmatrix} \right\},$$

$$\text{R}(\mathbf{A}^T) = \left\{ \begin{pmatrix} 1 \\ -1 \\ 1 \end{pmatrix} \right\}, \quad \text{null}(\mathbf{A}^T) = \left\{ \begin{pmatrix} -1 \\ 1 \\ 0 \end{pmatrix}, \begin{pmatrix} -1/2 \\ 0 \\ 1 \end{pmatrix} \right\}.$$

16. Consider the matrix

$$\mathbf{A} = \begin{pmatrix} 1 & 2 & 3 \\ 4 & 5 & 6 \\ 7 & 8 & 9 \end{pmatrix}.$$

Use Matlab, Maple, or Mathematica to find bases for the four fundamental subspaces of $\mathbf{A}$.

17. Consider the matrix

$$\mathbf{A} = \begin{pmatrix} -1 & 2 & 3 \\ 4 & 5 & 6 \\ 7 & 8 & 9 \end{pmatrix}.$$

Use Matlab, Maple, or Mathematica to find bases for the four fundamental subspaces of $\mathbf{A}$.

18. Show that the set of $m \times n$ matrices forms a vector space with the previously defined rules for addition and scalar multiplication.

19. Show that the dimension of the vector space of 2×3 matrices is six.

20. Consider the reflection matrix given in equation (5.37). Let m be the slope of the line of reflection. By setting $m = \tan \phi$ and using basic trig identities, show that the reflection matrix can also be written

$$\frac{1}{m^2 + 1} \begin{pmatrix} 1 - m^2 & 2m \\ 2m & m^2 - 1 \end{pmatrix}. \tag{5.47}$$

5.4 Eigenvalues and Eigenvectors

We have seen that we can write the linear system of equations

$$\frac{dx}{dt} = ax + by$$
$$\frac{dy}{dt} = cx + ey$$

in matrix-vector notation $\mathbf{x}' = \mathbf{A}\mathbf{x}$. If $x, A \in \mathbb{R}$, then this equation is easily solved by separation of variables, with solution $x = c_1 e^{At}$. When we considered nth order linear homogeneous equations with constant coefficients, we similarly guessed a solution of the form $y = e^{rx}$, substituted the result into the equation, and obtained the characteristic (or auxiliary) equation; solving for the roots gave r-values that made $y = e^{rx}$ a solution. Thus even though $\mathbf{x} \in \mathbb{R}^n, \mathbf{A} \in \mathbb{R}^{n \times n}$, we might still hope that we can find an exponential solution. Analogously to what we've done before, we assume a solution of the form

$$\mathbf{x} = \mathbf{v} e^{\lambda t},$$

where $\mathbf{v}$ is a vector (same size as $\mathbf{x}$) and λ is a scalar. (Either of these may be complex.) If $\mathbf{v}e^{\lambda t}$ is a solution, it must satisfy the original differential equation $\mathbf{x}' = \mathbf{A}\mathbf{x}$. Substitution gives

$$\lambda e^{\lambda t}\mathbf{v} = \mathbf{A}e^{\lambda t}\mathbf{v}. \tag{5.48}$$

Because $e^{\lambda t} \neq 0$, we can divide by it to obtain the equation

$$\mathbf{A}\mathbf{v} = \lambda \mathbf{v}. \tag{5.49}$$

Thus,

$\mathbf{v}e^{\lambda t}$ *is a solution to* $\mathbf{x}' = \mathbf{A}\mathbf{x}$ *if we can find* $\mathbf{v}, \lambda$ *such that* $\mathbf{A}\mathbf{v} = \lambda \mathbf{v}$.

In this section, we will consider the consequence of a square matrix acting on a vector $\mathbf{v}$ and yielding a constant multiple of the same vector $\mathbf{v}$. In symbols, we have

$$\mathbf{A}\mathbf{v} = \lambda \mathbf{v}, \tag{5.50}$$

where λ is a constant, called a *scale factor*. The scale factor λ modifies the length of the vector $\mathbf{v}$; for example,

$$\begin{pmatrix} -5 & 8 \\ -4 & 7 \end{pmatrix}\begin{pmatrix} 2 \\ 1 \end{pmatrix} = -1\begin{pmatrix} 2 \\ 1 \end{pmatrix} \quad \text{and} \quad \begin{pmatrix} -5 & 8 \\ -4 & 7 \end{pmatrix}\begin{pmatrix} 1 \\ 1 \end{pmatrix} = 3\begin{pmatrix} 1 \\ 1 \end{pmatrix} \tag{5.51}$$

are two situations where a matrix (the same matrix!) multiplies two different vectors and only changes them by scaling or stretching them. Note that multiplication of *any* vector by this matrix does *not* necessarily simply stretch it. For example,

$$\begin{pmatrix} -5 & 8 \\ -4 & 7 \end{pmatrix}\begin{pmatrix} 1 \\ -2 \end{pmatrix} = \begin{pmatrix} -21 \\ -18 \end{pmatrix},$$

which cannot be written as a product of a scalar and $(1 \quad -2)^T$.

A vector $\mathbf{v}$ that satisfies $\mathbf{A}\mathbf{v} = \lambda \mathbf{v}$ for a given matrix $\mathbf{A}$ is called an *eigenvector* of $\mathbf{A}$ and the factor by which it is multiplied, λ, is called the *eigenvalue* of the matrix corresponding to the particular eigenvector. (We note that we briefly mentioned these two words in Section 4.1 when we assumed solutions of the form e^{rx} and again in Section 5.1 as we gave an introduction into a system of two first-order equations.) In our previous example, the matrix

$$A = \begin{pmatrix} -5 & 8 \\ -4 & 7 \end{pmatrix}$$

(seen previously in Sections 5.1.2 and 5.1.3) has eigenvalue and eigenvector pairs of

$$\left\{-1, \begin{pmatrix} 2 \\ 1 \end{pmatrix}\right\} \quad \text{and} \quad \left\{3, \begin{pmatrix} 1 \\ 1 \end{pmatrix}\right\}.$$

The eigenvalues and eigenvectors of a given matrix give us tremendous insight into the behavior of the matrix in situations as seemingly different as raising a matrix to a power to solving a system of differential equations!

We will now consider how to find the eigenvalues and eigenvectors associated with a matrix $\mathbf{A}$. Let us consider this same matrix

$$\mathbf{A} = \begin{pmatrix} -5 & 8 \\ -4 & 7 \end{pmatrix}.$$

We want to find λ and $\mathbf{v}$ so that

$$\mathbf{A}\,\mathbf{v} = \lambda\mathbf{v}.$$

Equivalently,

$$(\mathbf{A} - \lambda\mathbf{I})\mathbf{v} = \mathbf{0}.$$

Non-trivial solutions occur only when the matrix $(\mathbf{A} - \lambda\mathbf{I})$ is singular. From Theorem C.1.2, this happens when

$$\det(\mathbf{A} - \lambda\mathbf{I}) = 0.$$

This determinant gives us a polynomial in λ, called the *characteristic polynomial for A*, and when we set it equal to zero, we have the *characteristic equation* for $\mathbf{A}$. In our specific example, calculating the characteristic equation gives

$$\lambda^2 - 2\lambda - 3 = 0.$$

This characteristic equation is solved to obtain the characteristic roots, or eigenvalues, keeping in mind that we allow for complex roots as well as real roots. Because we allow for complex roots, the Fundamental Theorem of Algebra guarantees that there are exactly n roots when the characteristic polynomial is of degree n. As you may be aware, there is no general formula that exists for finding the roots of a polynomial that is degree 5 or higher. Having the help of Matlab, Maple, or Mathematica to find the eigenvalues or the approximations of them will be very useful. Again referring to our specific example, the characteristic equation can be rewritten as

$$(\lambda - 3)(\lambda + 1) = 0,$$

which gives roots of $\lambda_1 = -1$, $\lambda_2 = 3$.

Once we have the eigenvalues, we take each one in turn and plug it into the equation $(\mathbf{A} - \lambda\mathbf{I})\mathbf{v} = \mathbf{0}$ and solve for $\mathbf{v}$. For this example,

$$(\mathbf{A} - \lambda_1\mathbf{I})\mathbf{v}_1 = \mathbf{0} \implies \begin{pmatrix} -5 - (-1) & 8 \\ -4 & 7 - (-1) \end{pmatrix} \begin{pmatrix} v_{11} \\ v_{21} \end{pmatrix} = \begin{pmatrix} 0 \\ 0 \end{pmatrix}$$

$$\implies -4v_{11} + 8v_{21} = 0.$$

You might think of any number of possible combinations that will make this last equality true. In fact, any v_{11}, v_{21} that satisfy $v_{11}/v_{21} = 2$ will work.

Thus the eigenvectors are not unique in magnitude. Another way to think of this is that *any non-zero multiple of an eigenvector is still an eigenvector.* For example, if **v** is an eigenvector, so are $-\mathbf{v}$ and $2\mathbf{v}$. The simplest choice is perhaps $v_{11} = 2$, $v_{21} = 1$ which then gives

$$\mathbf{v_1} = \begin{pmatrix} 2 \\ 1 \end{pmatrix}$$

as the eigenvector corresponding to the eigenvalue $\lambda_1 = -1$. We could do a similar calculation to obtain

$$\mathbf{v_2} = \begin{pmatrix} 1 \\ 1 \end{pmatrix}$$

as the eigenvector corresponding to the eigenvalue $\lambda_2 = 3$.

We note that the above method worked in a straightforward fashion because our eigenvalues were real and distinct. If they were complex, we can still find eigenvectors but it is a bit more work; see Example 5 in this section. If we have eigenvalues that are repeated, we can *try* to plug them as we did above, and this will give us at least one eigenvector but we may or may not be able to find more. Sometimes we will need to obtain a *generalized eigenvector*; this will be discussed in Section 6.2.1.

We revisit the equivalence theorem of the last section and add one more result. We will conclude our brief introduction to eigenvectors and eigenvalues with two important theorems. We will have occasion in later sections to make use of both of these results. The first theorem is an important and useful characterization of an invertible $n \times n$ matrix **A**.

THEOREM 5.4.1 *The following are equivalent characterizations of the* $n \times n$ *matrix* **A**:

a) **A** *is invertible.*
b) The system $\mathbf{Ax} = \mathbf{b}$ *has a unique solution* **x** *for each* **b** *in* $\mathbb{R}^n$.
c) The system $\mathbf{Ax} = \mathbf{0}$ *has* $\mathbf{x} = \mathbf{0}$ *as its unique solution.*
d) $\det(\mathbf{A}) \neq 0$.
e) The n *columns of* **A** *form a basis for* $\mathbb{R}^n$.
f) 0 is not an eigenvalue of **A**.

Regarding e) in this theorem, we note that because the columns of **A** form a basis of $\mathbb{R}^n$, they automatically span $\mathbb{R}^n$ *and* are linearly independent. We also remind the reader that there is never a unique basis for a given vector space. In fact, it will often be useful to convert from one basis to another and we will do so when we analyze the stability of solutions in future sections. We refer the reader to Appendix C for a discussion of change of bases and coordinates.

The next theorem connects the trace and determinant of an $n \times n$ matrix **A** with the eigenvalues of **A**. This theorem will be used in Chapter 6 when we wish to analyze the qualitative behavior of a linear system.

> **THEOREM 5.4.2** *Let λ_i denote the eigenvalues of an $n \times n$ matrix $\mathbf{A}$. Then*
>
> $$\mathrm{Tr}(\mathbf{A}) = \sum_i \lambda_i, \quad \det(\mathbf{A}) = \prod_i \lambda_i.$$

5.4.1 Eigenvalues and Eigenvectors with Matlab, Maple, and Mathematica

We can calculate the eigenvalues and eigenvectors easily with our computer programs. We will show two ways of doing it—first by calculating the characteristic equation and then finding its roots and second by using the built-in software of the given package to find the eigenvalues. If the exact value of an eigenvalue can be found, it will often be useful to write the characteristic equation and look for roots. However, there will be numerous times when we will only be able to find *accurate but still approximate* values for our eigenvalues. For example, we may be able to obtain eigenvalues with eight decimal places of accuracy but not more. In these situations, the subject of numerical analysis tells us that calculating the roots of a polynomial is more prone to numerical error than calculating the eigenvalues by certain other approaches (such as converting the matrix to Hessenberg form and then applying the QR algorithm [9]). The reader does not need to know the details of these algorithms but rather needs to understand that although in theory we can calculate the characteristic equation and then find its roots, in practice our computer software packages numerically find eigenvalues by a different method.

Computer Code 5.4: Calculating eigenvalues and eigenvectors

Matlab, Maple, Mathematica

```
                              Matlab
>>  A=[3, 1, 0; 2, 4, 0; 3, -1, 1]
>>  lambda=eig(A) %calculates eigenvalues of A
>>  [v,d]=eig(A) %calculates eigenvalues AND eigenvectors
    %of A; eigenvectors given as columns of v;
    %corresponding eigenvalues are on diagonal of d
>>  d(1) %first eigenvalue of A
>>  v(:,1) %corresponding first eigenvector of A
>>  A*v(:,1)
>>  d(1)*v(:,1) %check that A*v=lambda*v for first
    %eigenvalue-eigenvector pair; the last two
    %answers should be identical
```

The eigenvectors used by Matlab are normalized so the the length of each is equal to one. For the purposes of numerical computation, having unit length is extremely important. However, students often prefer to have "nicer looking" eigenvectors and eigenvalues. We can achieve these with

```
                              Matlab
>>  A=[3, 1, 0; 2, 4, 0; 3, -1, 1]
>>  lambda=eig(sym(A)) %calculates eigenvalues of A
>>  [v,d]=eig(sym(A)) %calculates eigenvalues AND
    %eigenvectors of A; eigenvectors given as columns of
    %v; corresponding eigenvalues are on diagonal of d
>>  d(1) %first eigenvalue of A
>>  v(:,1) %corresponding first eigenvector of A
>>  A*v(:,1)
>>  d(1)*v(:,1) %check that A*v=lambda*v for first
    % eigenvalue-eigenvector pair; the last two
    %answers should be identical
```

```
                              Maple
>  with(linalg):
>  A:=matrix(3,3,[1,0,3,0,1,2,-1,1,1]);
>  eq1:=charpoly(A,lambda); #characteristic polynomial
   # with variable lambda
>  solve(eq1,lambda); #finds the roots of this polynomial
>  eigenvals(A); #calculates eigenvalues more efficiently
>  eq2:=eigenvects(A); #calculates eigenvectors and
     #eigenvalues; answer is a set with the elements
     #(i) eigenvalue (ii) its multiplicity (iii) eigenvector
>  lambda1:=eq2[1][1]; #this is the first eigenvalue
>  eq2[1][3]; #this is the SET that contains the
     #corresponding eigenvector
>  v1:=eq2[1][3][1]; #this is the actual eigenvector
>  multiply(A,v1); #syntax for matrix-vector multiplication
>  evalm(lambda1*evalm(v1)); #checks A*v=lambda*v for first
     #eigenvalue-eigenvector pair; the last two
     #answers should be identical
```

```
                          Mathematica
A={{3,1,0},{2,4,0},{3,-1,1}}
MatrixForm[A] (*puts matrix in nicer visual form*)
eq1=CharacteristicPolynomial[A,λ]
   (*characteristic polynomial with variable λ*)
Solve[eq1==0,λ] (*finds the roots of this polynomial*)
Eigenvalues[A] (*calculates eigenvalues more efficiently*)
eq2=Eigensystem[A] (*calculates eigenvalues and *)
   (*eigenvectors; answer is a set with (i) each *)
   (*eigenvalue and (ii) set of eigenvectors*)
lambda1=eq2[[1]][[1]] (*this is the first eigenvalue*)
v1=eq2[[2]][[1]] (*this is the first eigenvector*)
A.v1 (*syntax for matrix-vector multiplication*)
lambda1 v1 (*checks Av=λv for first eigenvalue-eigenvector*)
   (*pair; the last two answers should be identical*)
```

Example 1: Use Matlab, Maple, or Mathematica to verify the conclusions of Theorem 5.4.2 for the following four matrices:

$$\mathbf{A}_1 = \begin{pmatrix} -5 & 8 \\ -4 & 7 \end{pmatrix}, \quad \mathbf{A}_2 = \begin{pmatrix} -1 & 0 \\ 5 & 3 \end{pmatrix},$$

$$\mathbf{A}_3 = \begin{pmatrix} 3 & 4 & 2 \\ 2 & 1 & -2 \\ -2 & -4 & -1 \end{pmatrix}, \quad \mathbf{A}_4 = \begin{pmatrix} 1 & 2 & 3 \\ 0 & -3 & 7 \\ 0 & 0 & 5 \end{pmatrix}.$$

We can now easily calculate that both $\mathbf{A}_1$ and $\mathbf{A}_2$ have eigenvalues of $\lambda_1 = -1$, $\lambda_2 = 3$. Thus, for both matrices we have $\lambda_1 + \lambda_2 = -1 + 3 = 2$ and $\lambda_1 \lambda_2 = (-1)(3) = -3$. For $\mathbf{A}_1$, we have

$$\text{Tr}(\mathbf{A}_1) = -5 + 7 = 2, \quad \det(\mathbf{A}_1) = -35 + 32 = -3,$$

which shows the theorem holds for $\mathbf{A}_2$. For $\mathbf{A}_2$, we have

$$\text{Tr}(\mathbf{A}_1) = -1 + 3 = 2, \quad \det(\mathbf{A}_1) = -3 + 0 = -3,$$

which shows the theorem holds for $\mathbf{A}_2$. For the other two matrices, we find that both have the same eigenvalues, $\lambda_1 = 1, \lambda_2 = -3, \lambda_3 = 5$. The computer gives us

$$\text{Tr}(\mathbf{A}_3) = 3 + 1 - 1 = 3, \quad \det(\mathbf{A}_3) = -15,$$

which shows the theorem holds for $\mathbf{A}_3$. The reader can easily check that this is also the case for $\mathbf{A}_4$. We conclude this example by noting that Theorem 5.4.2 does not say that the eigenvalues lie on the diagonal. Sometimes this will be the case (for example, with triangular or diagonal matrices) but usually the eigenvalues do not appear in the matrix. Nevertheless, we now have a way to calculate the trace and determinant in terms of eigenvalues.

5.4.2   Some Insights into Phase Portraits

Eigenvalues and eigenvectors give us tremendous insight into solving systems of differential equations, particularly in the case of a system of two equations (in part because it is simply easier to picture than in higher dimensions). We will consider systems where we have equilibrium solutions that are saddles and then spirals.

Example 2: Consider

$$\frac{dx}{dt} = -5x + 8y$$

$$\frac{dy}{dt} = -4x + 7y.$$

We calculated earlier that the eigenvalue and eigenvector pairs are

$$\left\{-1, \begin{pmatrix} 2 \\ 1 \end{pmatrix}\right\} \quad \text{and} \quad \left\{3, \begin{pmatrix} 1 \\ 1 \end{pmatrix}\right\}.$$

If we use the eigenvectors as our basis (see Appendix C.3.1 for the details of this *similarity transformation*), we can rewrite the matrix of our system as

$$\begin{pmatrix} -1 & 0 \\ 0 & 3 \end{pmatrix}.$$

The significance of the eigenvalues and eigenvectors in terms of linear transformations is then clear. The first pair says that along the vector $(2 \ \ 1)^T$, which lies along the line $y = x/2$, the solution is decreasing because $\lambda_1 = -1 < 0$. More specifically, if we let the u_1-axis be the line $y = x/2$, then solutions along this line are $u_1 = c_1 e^{-t}$. The second pair says that along the vector $(1 \ \ 1)^T$, which lies along the line $y = x$, the solution is increasing because $\lambda_2 = 3 > 0$. We again say more: if we let the u_2-axis be the line $y = x$, the solutions along this line are $u_2 = c_2 e^{3t}$. These lines were not drawn in the first picture but sketching the eigenvectors helps distinguish qualitatively different trajectories in the cases of saddles and nodes (both stable and unstable). We show the phase portrait again with the trajectories drawn along the eigenvectors. Solutions are attracted to the origin along the first eigenvector but are repelled from the origin along the second. Note that the Existence and Uniqueness theorem (Theorem 5.2.1) tells us that solutions do not cross each other. Because the system has constant coefficients, this means that trajectories cannot cross either. We can make an interesting mathematical observation: As $t \longrightarrow \infty$, all solutions approach the line $y = x$; as $t \longrightarrow -\infty$, all solutions approach the line $y = x/2$; see Figure 5.10.

Another way in which the eigenvalues and eigenvectors arise is in the solution of the system of equations. As we will see in Section 6.1, if a matrix

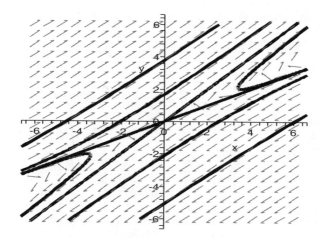

FIGURE 5.10: Phase portrait for Example 2. The origin is a saddle. The stable direction is along the $(2 \quad 1)^T$ eigenvector and the unstable direction is along the $(1 \quad 1)^T$ eigenvector.

has distinct real eigenvalues then we can write the solution in terms of the eigenvalues and eigenvectors.

Example 3: For the system

$$\mathbf{x}' = \begin{pmatrix} -5 & 8 \\ -4 & 7 \end{pmatrix} \mathbf{x}$$

of Example 2, show that $e^{\lambda_1}\mathbf{v}_1, e^{\lambda_2}\mathbf{v}_2$ are solutions.

We calculated the eigenvalue and eigenvector pairs before. Thus we need to show that

$$e^{-t}\begin{pmatrix} 2 \\ 1 \end{pmatrix} \quad \text{and} \quad e^{3t}\begin{pmatrix} 1 \\ 1 \end{pmatrix}$$

are solutions. The derivative of the first gives

$$-e^{-t}\begin{pmatrix} 2 \\ 1 \end{pmatrix}$$

and substitution into the differential equation gives

$$-e^{-t}\begin{pmatrix} 2 \\ 1 \end{pmatrix} = \begin{pmatrix} -5 & 8 \\ -4 & 7 \end{pmatrix} e^{-t}\begin{pmatrix} 2 \\ 1 \end{pmatrix} = e^{-t}\begin{pmatrix} -2 \\ -1 \end{pmatrix},$$

which shows that we indeed substituted a solution. For the second, we can take the derivative and substitute into the differential equation to see that

$$3e^{3t} \begin{pmatrix} 1 \\ 1 \end{pmatrix} = \begin{pmatrix} -5 & 8 \\ -4 & 7 \end{pmatrix} e^{3t} \begin{pmatrix} 1 \\ 1 \end{pmatrix} = e^{3t} \begin{pmatrix} 3 \\ 3 \end{pmatrix},$$

which again shows that we substituted a solution.

Now we consider the system

$$\frac{dx}{dt} = ax - by$$

$$\frac{dy}{dt} = bx + ay, \tag{5.52}$$

where a, b are real and $b \neq 0$. In this situation the eigenvalues of the coefficient matrix $\mathbf{A}$ are $\lambda_1 = a + ib$, $\lambda_2 = a - ib$. We will not solve the system now but simply want to examine the right-hand side $\mathbf{Ax}$, where $\mathbf{x} = (x \quad y)^T$. If we use polar coordinates and let

$$r = \sqrt{a^2 + b^2}, \quad \tan\theta = \frac{b}{a}, \tag{5.53}$$

then *multiplication of a vector* $\mathbf{v}$ *by* $\mathbf{A} = \begin{pmatrix} a & -b \\ b & a \end{pmatrix}$ *corresponds, in the case* $b > 0$, *to a counterclockwise rotation through θ rad, which is then followed by a stretching or shrinking of the length of the vector by a factor of r;* see Hirsch and Smale [16] for a more in-depth discussion. If $b < 0$, the rotation is clockwise.

The reader should recall that in Section 5.3, we defined a rotation matrix (now defined in terms of rotation through θ instead of ϕ)

$$\mathbf{R}_\theta = \begin{pmatrix} \cos\theta & -\sin\theta \\ \sin\theta & \cos\theta \end{pmatrix}. \tag{5.54}$$

Because we have $a = r\cos\theta, b = r\sin\theta$, we can write

$$\begin{pmatrix} a & -b \\ b & a \end{pmatrix} = \begin{pmatrix} r & 0 \\ 0 & r \end{pmatrix} \begin{pmatrix} \cos\theta & -\sin\theta \\ \sin\theta & \cos\theta \end{pmatrix}.$$

This last equality gives us the interpretation of a rotation followed by a stretch.

Example 4: Consider the system:

$$\frac{dx}{dt} = -x + \sqrt{3}y$$

$$\frac{dy}{dt} = -\sqrt{3}x - y.$$

We can calculate, either by hand or on the computer, that the eigenvalues are complex and are $\lambda_1 = -1 + \sqrt{3}i, \lambda_1 = -1 - \sqrt{3}i$. With

$$r = \sqrt{(-1)^2 + (-\sqrt{3})^2}, \quad \tan\theta = \frac{-\sqrt{3}}{-1},$$

we see that, for example, that the vector $(5 \quad 0)^T$ is rotated *clockwise* through the angle of $-2\pi/3$ and then stretched by a factor of 2. Keep in mind that this is not the solution but it is simply the result of a vector under the action of our given matrix (a linear transformation that rotates, then stretches). In differential equations, we know the right-hand side gives the vector field for our equation. Thus, we can expect our spiral to be a clockwise spiral; see Figure 5.11.

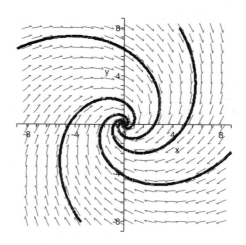

FIGURE 5.11: Spiral trajectories. Note how the slope at the initial condition $(5, 0)$ can be interpreted as a rotation followed by a stretch.

It is probably natural for the reader to question the usefulness of the interpretation of Example 4 in the case when our matrix is not initially of the form

$$\begin{pmatrix} a & -b \\ b & a \end{pmatrix},$$

where the eigenvalues of the original matrix are $a \pm ib$. For a general matrix with complex (nonreal) eigenvalues, we can again use a similarity transformation to convert the original matrix to the above form. The details of this can again be found in Appendix C.3.1.

We finish with an example of finding the eigenvectors of a matrix with complex (nonreal) eigenvalues.

Example 5: Consider the matrix

$$\begin{pmatrix} 2 & -3 \\ 3 & 2 \end{pmatrix}.$$

From our above discussion, we know that the eigenvalues are $2 \pm 3i$. But if we didn't know/see this, we would proceed just as before. That is, we would calculate

$$\det(\mathbf{A} - \lambda \mathbf{I}) = 0 \Longrightarrow (2 - \lambda)^2 + 9 = 0 \Longrightarrow \lambda^2 - 4\lambda + 13 = 0.$$

This equation is quadratic and so we obtain

$$\lambda = \frac{4 \pm \sqrt{4^2 - 4(1)(13)}}{2(1)} = \frac{4 \pm \sqrt{-36}}{2} = \frac{4 \pm 6i}{2} = 2 \pm 3i.$$

To calculate the eigenvectors, we again use the equation $(\mathbf{A} - \lambda \mathbf{I})\mathbf{v} = \mathbf{0}$ to find the eigenvector by solving for $\mathbf{v}$. Unlike the real case, however, we only need to find *one* of the eigenvectors—it turns out that the second eigenvector is the complex conjugate of the first! For this example, let's use $\lambda_1 = 2 + 3i$ to find $\mathbf{v}_1$:

$$(\mathbf{A} - \lambda_1 \mathbf{I})\mathbf{v}_1 = \mathbf{0} \Longrightarrow \begin{pmatrix} 2 - (2 + 3i) & -3 \\ 3 & 2 - (2 + 3i) \end{pmatrix} \begin{pmatrix} v_{11} \\ v_{21} \end{pmatrix} = \begin{pmatrix} 0 \\ 0 \end{pmatrix}$$

$$\begin{pmatrix} -3i & -3 \\ 3 & -3i \end{pmatrix} \begin{pmatrix} v_{11} \\ v_{21} \end{pmatrix} = \begin{pmatrix} 0 \\ 0 \end{pmatrix}. \tag{5.55}$$

As before, the two rows of $(\mathbf{A} - \lambda_1 \mathbf{I})$ are constant multiples of each other—the second row is just the first row multiplied by $-i$. Thus we have

$$-3iv_{11} - 3v_{21} = 0 \Longrightarrow -iv_{11} = v_{21}.$$

We are free to choose any values for v_{11}, v_{21} that will make this equation true as long as both are not zero. It is probably easiest to choose $v_{11} = 1$, which immediately gives $v_{21} = -i$. Thus

$$\lambda_1 = 2 + 3i \text{ has eigenvector } \mathbf{v}_1 = \begin{pmatrix} 1 \\ -i \end{pmatrix}.$$

The second eigenvalue-eigenvector pair is just the complex conjugate of the first:

$$\lambda_2 = 2 - 3i \text{ has eigenvector } \mathbf{v}_2 = \begin{pmatrix} 1 \\ i \end{pmatrix}.$$

The reader should verify that this is the case. In Section 6.2.2 we will be using complex eigenvalues and eigenvectors to obtain the general solution to systems such as equation (5.52).

Problems

For problems 1–6, calculate the characteristic equation of the given matrix by hand. Then find the eigenvalues and corresponding eigenvectors. The eigenvalues are real.

1. a. $\begin{pmatrix} 4 & -5 \\ 2 & -3 \end{pmatrix}$, b. $\begin{pmatrix} 3 & -2 \\ 0 & -6 \end{pmatrix}$, c. $\begin{pmatrix} 1 & -2 \\ 4 & -5 \end{pmatrix}$

2. a. $\begin{pmatrix} -2 & 3 \\ 0 & -5 \end{pmatrix}$, b. $\begin{pmatrix} -4 & 7 \\ 0 & -1 \end{pmatrix}$, c. $\begin{pmatrix} 1 & -3 \\ 0 & 7 \end{pmatrix}$

3. a. $\begin{pmatrix} 2 & 1 \\ 1 & 2 \end{pmatrix}$, b. $\begin{pmatrix} 0 & 2 \\ 2 & 0 \end{pmatrix}$, c. $\begin{pmatrix} 1 & 3 \\ 3 & 1 \end{pmatrix}$

4. a. $\begin{pmatrix} -3 & 6 \\ -2 & 5 \end{pmatrix}$, b. $\begin{pmatrix} 1 & 2 \\ -1 & 4 \end{pmatrix}$, c. $\begin{pmatrix} -2 & 1 \\ -1 & 0 \end{pmatrix}$

5. a. $\begin{pmatrix} 0 & 1 \\ 3 & 2 \end{pmatrix}$, b. $\begin{pmatrix} 4 & 1 \\ 3 & 2 \end{pmatrix}$, c. $\begin{pmatrix} -3 & 1 \\ 2 & 1 \end{pmatrix}$

6. a. $\begin{pmatrix} 5 & 0 \\ 2 & -3 \end{pmatrix}$, b. $\begin{pmatrix} 7 & 6 \\ 2 & -4 \end{pmatrix}$, c. $\begin{pmatrix} 2 & 1 \\ -1 & 0 \end{pmatrix}$

For problems 7–11, calculate the characteristic equation of the given matrix by hand. Then find the eigenvalues and corresponding eigenvectors. The eigenvalues are complex (nonreal).

7. a. $\begin{pmatrix} 0 & -1 \\ 1 & 0 \end{pmatrix}$, b. $\begin{pmatrix} -1 & -2 \\ 2 & -1 \end{pmatrix}$

8. a. $\begin{pmatrix} 3 & -2 \\ 5 & -3 \end{pmatrix}$, b. $\begin{pmatrix} 1 & -2 \\ 4 & 5 \end{pmatrix}$

9. a. $\begin{pmatrix} -1 & -5 \\ 1 & -3 \end{pmatrix}$, b. $\begin{pmatrix} 1 & -5 \\ 1 & 3 \end{pmatrix}$

10. a. $\begin{pmatrix} -3 & -5 \\ 1 & -7 \end{pmatrix}$, b. $\begin{pmatrix} 1 & -2 \\ 3 & -1 \end{pmatrix}$

11. a. $\begin{pmatrix} -5 & 2 \\ -2 & -5 \end{pmatrix}$, b. $\begin{pmatrix} 1 & -8 \\ 1 & 3 \end{pmatrix}$

In problems 12–14, use Matlab, Maple, or Mathematica to find the eigenvalues and eigenvectors of the given matrix.

12. a. $\begin{pmatrix} 3 & 1 & 0 \\ 2 & 4 & 0 \\ 2/3 & -1/3 & 1 \end{pmatrix}$, b. $\begin{pmatrix} -2 & 2 & 0 \\ 4 & 0 & 0 \\ -10/3 & 5/3 & 1 \end{pmatrix}$

13. a. $\begin{pmatrix} -1 & -2 & 3 \\ 2 & -1 & 1 \\ 0 & 0 & -1 \end{pmatrix}$, b. $\begin{pmatrix} 1 & -5 & -1 \\ 1 & 3 & 1 \\ 0 & 0 & 1 \end{pmatrix}$

14. a. $\begin{pmatrix} 3 & 0 & 0 \\ 0 & 3 & 0 \\ 8/3 & -4/3 & -1 \end{pmatrix}$, b. $\begin{pmatrix} -2/3 & -2/3 & 0 \\ -4/3 & -4/3 & 0 \\ 2/3 & -1/3 & -1 \end{pmatrix}$

5.5   Matrix Exponentials

As with most of the sections in this chapter, we have been considering the first-order linear system (5.1), which can be written in matrix notation as

$$\mathbf{x}' = \mathbf{A}\mathbf{x}. \tag{5.56}$$

In Section 5.4, we saw that $\mathbf{x} = e^{\lambda t}\mathbf{v}$ is a solution to (5.56) provided that λ is an eigenvalue of $\mathbf{A}$ with $\mathbf{v}$ as the corresponding eigenvector. We know that solutions are unique, but is this the only form in which we can write the solution? The answer is "no" for reasons we will see shortly. For the moment let us *suppose that it makes sense to write*

$$e^{\mathbf{A}t}, \tag{5.57}$$

where $\mathbf{A}$ is the *matrix* in (5.56). And further suppose that it makes sense to take the derivative of this function in the same way we take a derivative of the function e^{at} when a is a constant; that is, $\left(e^{at}\right)' = ae^{at}$. We thus *suppose that it also makes sense to take the derivative as follows:*

$$\frac{d}{dt}\left(e^{\mathbf{A}t}\right) = \mathbf{A}e^{\mathbf{A}t}. \tag{5.58}$$

If we assume a solution of the form $\mathbf{x} = e^{\mathbf{A}t}$, then

$$\mathbf{x}' = \mathbf{A}e^{\mathbf{A}t}$$

and substitution into (5.56) shows that both sides of the equation are indeed the same for all t. By our previous definitions, this means that $e^{\mathbf{A}t}$ is a solution. We thus have another form of a solution provided it makes sense to exponentiate a matrix and then take its derivative.

5.5.1   Formal Definition and Properties

We make the following definition to help us.

DEFINITION 5.9   *For any square matrix $\mathbf{A}$, define*

$$e^{\mathbf{A}} = \mathbf{I} + \frac{\mathbf{A}}{1!} + \frac{\mathbf{A}^2}{2!} + \cdots = \sum_{k=0}^{\infty} \frac{\mathbf{A}^k}{k!}. \tag{5.59}$$

This definition actually holds for any operator (a linear transformation from $\mathbb{R}^n$ to $\mathbb{R}^n$). For our purposes, we often consider $e^{\mathbf{A}t}$ and because a scalar times a matrix is just another matrix, the above definition gives us

$$e^{\mathbf{A}t} = \mathbf{I} + \frac{\mathbf{A}t}{1!} + \frac{\mathbf{A}^2 t^2}{2!} + \cdots = \sum_{k=0}^{\infty} \frac{\mathbf{A}^k t^k}{k!}. \tag{5.60}$$

The reader should note that this definition of the *matrix exponential* is in the form of an infinite series. It turns out that this series converges for any matrix $\mathbf{A}$ and thus it makes sense to talk about $e^{\mathbf{A}}$ or $e^{\mathbf{A}t}$ for any square matrix $\mathbf{A}$. We state a few useful results that will help give us insight into the solution of a differential equation.

THEOREM 5.5.1 *Let* $\mathbf{A}, \mathbf{B}, \mathbf{P}$ *be* $n \times n$ *matrices. Then*
a) $e^{-\mathbf{A}} = (e^{\mathbf{A}})^{-1}$.
b) If $\mathbf{A} = \mathbf{P}\mathbf{B}\mathbf{P}^{-1}$, *then* $e^{\mathbf{A}} = \mathbf{P}e^{\mathbf{B}}\mathbf{P}^{-1}$.
c) If $\mathbf{A}$ *and* $\mathbf{B}$ *commute (that is,* $\mathbf{A}\mathbf{B} = \mathbf{B}\mathbf{A}$*), then* $e^{\mathbf{A}+\mathbf{B}} = e^{\mathbf{A}}e^{\mathbf{B}}$.

Example 1: Show that $e^{\mathbf{A}(t-t_0)} = e^{\mathbf{A}t}e^{-\mathbf{A}t_0}$.

Part (c) of the above theorem applies if we can show that $\mathbf{A}t$ and $-\mathbf{A}t_0$ commute. Using some of the properties of matrices, we have

$$(\mathbf{A}t)(-\mathbf{A}t_0) = (\mathbf{A}\mathbf{A})(-tt_0) = (\mathbf{A}\mathbf{A})(-t_0t) = (-\mathbf{A}t_0)(\mathbf{A}t),$$

which shows that the matrices do commute. Part (c) of the above theorem then gives us our desired conclusion.

Although the theorem gives some useful properties, it does not tell us how to compute the series for the matrix exponential. The following theorem is useful for these kinds of computations.

THEOREM 5.5.2 *Let* $\mathbf{A}$ *be an* $n \times n$ *matrix.*
a) There exist functions $\alpha_1(t), \alpha_2(t), \ldots, \alpha_n(t)$, *such that*

$$e^{\mathbf{A}t} = \alpha_1(t)\mathbf{A}^{n-1}t^{n-1} + \alpha_2(t)\mathbf{A}^{n-2}t^{n-2} + \cdots + \alpha_{n-1}(t)\mathbf{A}t + \alpha_n(t)\mathbf{I}. \quad (5.61)$$

b) For the polynomial (in r*)*

$$p(r) = \alpha_1(t)r^{n-1} + \alpha_2(t)r^{n-2} + \cdots + \alpha_{n-1}(t)r + \alpha_n(t), \quad (5.62)$$

if λ *is an eigenvalue of* $\mathbf{A}$, *then*

$$e^{\lambda} = p(\lambda),$$

so that $e^{\lambda t} = p(\lambda t)$.
c) If λ *is an eigenvalue of multiplicity* k, *then*

$$e^{\lambda} = \left.\frac{dp(r)}{dr}\right|_{r=\lambda}, \quad e^{\lambda} = \left.\frac{d^2 p(r)}{dr^2}\right|_{r=\lambda}, \quad \cdots, \quad e^{\lambda} = \left.\frac{d^{k-1}p(r)}{dr^{k-1}}\right|_{r=\lambda}. \quad (5.63)$$

We calculate $e^{\mathbf{A}t}$ by first applying parts (b) and (c) of the above theorem to generate a set of linear equations in α_i. We then solve for these α_i and substitute into the formula in (a).

COROLLARY 5.5.1 *For a* 2×2 *matrix* $\mathbf{A} = \begin{pmatrix} a & -b \\ b & a \end{pmatrix}$, *we have*

$$e^{\mathbf{A}} = e^a \begin{pmatrix} \cos b & -\sin b \\ \sin b & \cos b \end{pmatrix} \quad \text{and thus} \quad e^{\mathbf{A}t} = e^a t \begin{pmatrix} \cos bt & -\sin bt \\ \sin bt & \cos bt \end{pmatrix}.$$

Note that the form of the matrix that results is a rotation matrix, discussed in Section 5.3.

Example 2: Find $e^{\mathbf{A}t}$ if $\mathbf{A} = \begin{pmatrix} -1 & -2 & 3 \\ 0 & 2 & -1 \\ 0 & 0 & 2 \end{pmatrix}$.

This is the case of $n = 3$ in part (a) of Theorem 5.5.2:

$$e^{\mathbf{A}t} = \alpha_1(t)\mathbf{A}^2 t^2 + \alpha_2(t)\mathbf{A}t + \alpha_3(t)\mathbf{I}$$

$$= \alpha_1 t^2 \begin{pmatrix} -1 & -2 & 3 \\ 0 & 2 & -1 \\ 0 & 0 & 2 \end{pmatrix}^2 + \alpha_2 t \begin{pmatrix} -1 & -2 & 3 \\ 0 & 2 & -1 \\ 0 & 0 & 2 \end{pmatrix} + \alpha_3 \begin{pmatrix} 1 & 0 & 0 \\ 0 & 1 & 0 \\ 0 & 0 & 1 \end{pmatrix}$$

$$= \begin{pmatrix} \alpha_1 t^2 - \alpha_2 t + \alpha_3 & -2\alpha_1 t^2 - 2\alpha_2 t & 5\alpha_1 t^2 + 3\alpha_2 t \\ 0 & 4\alpha_1 t^2 + 2\alpha_2 t + \alpha_3 & -4\alpha_1 t^2 - \alpha_2 t \\ 0 & 0 & 4\alpha_1 t^2 + 2\alpha_2 t + \alpha_3 \end{pmatrix}, \quad (5.64)$$

where we have stopped writing the dependence of α_i on t for notational convenience. The eigenvalues of $\mathbf{A}$ are easily seen to be $\lambda_1 = -1, \lambda_2 = 2, \lambda_3 = 2$. We thus set up a system of three equations to calculate $\alpha_1, \alpha_2, \alpha_3$: part (b) of Theorem 5.5.2 for λ_1 and part (c) for the repeated eigenvalues λ_2, λ_3. For the latter situation, we have

$$p(r) = \alpha_1 r^2 + \alpha_2 r + \alpha_3, \quad p'(r) = 2\alpha_1 r + \alpha_2.$$

This gives us

$$e^{2t} = \alpha_1(2t)^2 + \alpha_2(2t) + \alpha_3$$
$$e^{2t} = 2\alpha_1(2t) + \alpha_2.$$

For the non-repeated eigenvalue, we have

$$e^{-t} = \alpha_1(-t)^2 + \alpha_2(-t) + \alpha_3.$$

We solve this system of equations, for example using Matlab, Maple, or Mathematica, to obtain

$$\alpha_1 = \frac{e^{-t} + 3te^{2t} - e^{2t}}{3t^2}$$

$$\alpha_2 = -\frac{3te^{2t} + 2e^{-t} - 2e^{2t}}{3t}$$

$$\alpha_3 = e^{2t} - 2te^{2t}.$$

Substituting these values into (5.64), we see that

$$e^{\mathbf{A}t} = \frac{1}{3} \begin{pmatrix} 3e^{-t} & 2e^{-t} - 2e^{2t} & -e^{-t} + 6te^{2t} + e^{2t} \\ 0 & 3e^{2t} & -2e^{-t} - 9te^{2t} + 2e^{2t} \\ 0 & 0 & 3e^{2t} \end{pmatrix}. \qquad (5.65)$$

We will consider one more example before showing how to use the matrix exponential to solve systems of differential equations.

Example 3: Find $e^{\mathbf{A}t}$ if $\mathbf{A} = \begin{pmatrix} -1 & -2 \\ 2 & -1 \end{pmatrix}$.

The above corollary gives us

$$e^{\mathbf{A}t} = e^{-t} \begin{pmatrix} \cos 2t & -\sin 2t \\ \sin 2t & \cos 2t \end{pmatrix}.$$

Computer Code 5.5: **Solving a system of equations with complex numbers**

Matlab, Maple, Mathematica

Matlab

```
>>  soln1=solve('exp((-1-2*i)*t)=alpha1*(-1-2*i)*t+alpha2',
    'exp((-1+2*i)*t)=alpha1*(-1+2*i)*t+alpha2','alpha1,alpha2')
>>  a1=soln1.alpha1
>>  a2=soln1.alpha2
>>  maple('evalc',a1)
>>  maple('evalc',a2)
```

These last two answers are the expressions for α_1, α_2 (now called a1, a2, respectively). We note that the simplification involved in this required a call to the *Maple kernel*. The command `evalc` is a Maple command and we will again see its use below.

Maple

```
> eq1:=exp((-1-2*I)*t)=alpha[1]*(-1-2*I)*t+alpha[2];
> eq2:=exp((-1+2*I)*t)=alpha[1]*(-1+2*I)*t+alpha[2];
> eq3:=solve({eq1,eq2},{alpha[1],alpha[2]});
> evalc(eq3[1]);
> evalc(eq3[2]);
```

Mathematica

(*Note that both *e* and *i* (below) are entered from palette*)
$\text{eq1} = e^{(-1-2i)\,t} - \alpha 1\,(-1-2i)\,t - \alpha 2$
$\text{eq2} = e^{(-1+2i)\,t} - \alpha 1\,(-1+2i)\,t - \alpha 2$
$\text{soln} = \text{Solve}[\{\text{eq1} == 0, \text{eq2} == 0\}, \{\alpha 1, \alpha 2\}]$
$\text{FullSimplify}[\text{soln}]$

The computer gives us the same answers as the corollary:

$$e^{\mathbf{A}t} = e^{-t}\begin{pmatrix} \cos 2t & -\sin 2t \\ \sin 2t & \cos 2t \end{pmatrix}.$$

Built-in Computer Commands for the Matrix Exponential

In determining the exponential of a matrix in the above discussion, we used our computer programs to solve the system of equations that resulted by application of Theorem 5.5.2. For pedagogical reasons, it is useful and important to understand how the exponential of a matrix can actually be computed. However, if our ultimate goal is to solve and/or gain some insight into the solution of a given differential equation, the matrix exponential is simply a tool that is at our disposal. It is thus worthwhile to use the built-in computer commands to actually calculate the exponential of a given matrix. We use the matrix from Example 2 to illustrate this.

Example 4: Use the built-in Matlab, Maple, and Mathematica functions to compute the matrix exponential for

$$\mathbf{A} = \begin{pmatrix} -1 & -2 & 3 \\ 0 & 2 & -1 \\ 0 & 0 & 2 \end{pmatrix}.$$

Computer Code 5.6: Calculating the exponential of a matrix, $e^{\mathbf{A}t}$, with the built-in command

Matlab, Maple, Mathematica

Matlab

```
>>  A=[-1,-2,3;0,2,-1;0,0,2]    %defines the matrix A
>>  syms t;  %defines t as a symbolic variable
>>  expm(A*t)  %calculates the matrix exponential
```

```
                          ┌─────────┐
                          │  Maple  │
                          └─────────┘
>  with(linalg):
>  A:=matrix(3,3,[-1,-2,3,0,2,-1,0,0,2]);#defines matrix A
>  exponential(A*t);  #calculates the matrix exponential
```

```
                       ┌───────────────┐
                       │  Mathematica  │
                       └───────────────┘
A={{-1,-2,3},{0,2,-1},{0,0,2}}
MatrixForm[A] (*puts matrix in nicer visual form*)
MatrixExp[A t]//MatrixForm (*calculates the matrix*)
   (*exponential and then puts it in nicer visual form*)
```

We can easily check that the output is identical to that obtained by applying Theorem 5.5.2. The reader and instructor should discuss how to balance the use of these built-in commands with the need to understand how the matrix exponential is calculated in this section. We will take full advantage of these commands in examining solutions in the next section because it will allow us to go significantly further in our study of systems of differential equations.

5.5.2   The Derivative of the Matrix Exponential

We motivated the matrix exponential by saying that if we could define $e^{\mathbf{A}t}$ and the derivative of it (in the "normal" manner), we would have found another expression of the solution of $\mathbf{x}' = \mathbf{A}\mathbf{x}$. We formalized $e^{\mathbf{A}t}$ and we now show that we can take its derivative in the "normal" manner.

THEOREM 5.5.3

$$\frac{d}{dt}\left(e^{\mathbf{A}t}\right) = \mathbf{A}e^{\mathbf{A}t} = e^{\mathbf{A}t}\mathbf{A}.$$

Remark: Note that this theorem shows that $\mathbf{A}$ commutes with $e^{\mathbf{A}t}$.

Proof: The proof of this is straightforward and uses the definition of limits and series formulation of the matrix exponential discussed in the previous section.

$$\frac{d}{dt}\left(e^{\mathbf{A}t}\right) = \lim_{h\to 0}\left[\frac{e^{\mathbf{A}(t+h)} - e^{\mathbf{A}t}}{h}\right]$$

$$= \lim_{h\to 0}\left[\frac{e^{\mathbf{A}t}e^{\mathbf{A}h} - e^{\mathbf{A}t}}{h}\right] \quad \text{by Theorem 5.5.1}$$

$$= e^{\mathbf{A}t}\lim_{h\to 0}\left[\frac{e^{\mathbf{A}h} - \mathbf{I}}{h}\right]$$

$$= e^{\mathbf{A}t}\lim_{h\to 0}\left[\frac{\left(\mathbf{I} + \dfrac{\mathbf{A}h}{1!} + \dfrac{\mathbf{A}^2h^2}{2!} + \dfrac{\mathbf{A}^3h^3}{3!} + \cdots\right) - \mathbf{I}}{h}\right]$$

$$= e^{\mathbf{A}t}\lim_{h\to 0}\left[\frac{\mathbf{A}}{1!} + \frac{\mathbf{A}^2h}{2!} + \frac{\mathbf{A}^3h^2}{3!} + \cdots\right]$$

$$= e^{\mathbf{A}t}\mathbf{A}. \tag{5.66}$$

This proves the first equality. For the second, we only need to observe that $\mathbf{A}$ commutes with each term of the series of $e^{\mathbf{A}t}$. That is,

$$\mathbf{A}e^{\mathbf{A}t} = \mathbf{A}\left(\mathbf{I} + \frac{\mathbf{A}h}{1!} + \frac{\mathbf{A}^2h^2}{2!} + \frac{\mathbf{A}^3h^3}{3!} + \cdots\right)$$

$$= \left(\mathbf{A}\mathbf{I} + \frac{\mathbf{A}^2h}{1!} + \frac{\mathbf{A}^3h^2}{2!} + \frac{\mathbf{A}^4h^3}{3!} + \cdots\right)$$

$$= \left(\mathbf{I}\mathbf{A} + \frac{\mathbf{A}h\mathbf{A}}{1!} + \frac{\mathbf{A}^2h^2\mathbf{A}}{2!} + \frac{\mathbf{A}^3h^3\mathbf{A}}{3!} + \cdots\right)$$

$$= \left(\mathbf{I} + \frac{\mathbf{A}h}{1!} + \frac{\mathbf{A}^2h^2}{2!} + \frac{\mathbf{A}^3h^3}{3!} + \cdots\right)\mathbf{A}$$

$$= e^{\mathbf{A}t}\mathbf{A}. \tag{5.67}$$

The theorem is thus proved. □

Example 5: Show that $e^{\mathbf{A}t}$ is a solution of $\mathbf{x}' = \mathbf{A}\mathbf{x}$ for $\mathbf{A} = \begin{pmatrix} -1 & -2 \\ 2 & -1 \end{pmatrix}$.

This matrix was considered in Example 3 of this section and we found that

$$e^{\mathbf{A}t} = e^{-t}\begin{pmatrix} \cos 2t & -\sin 2t \\ \sin 2t & \cos 2t \end{pmatrix}.$$

For the derivative, we have

$$\left(e^{\mathbf{A}t}\right)' = -e^{-t}\begin{pmatrix} \cos 2t & -\sin 2t \\ \sin 2t & \cos 2t \end{pmatrix} + e^{-t}\begin{pmatrix} -2\sin 2t & -2\cos 2t \\ 2\cos 2t & -2\sin 2t \end{pmatrix}$$

$$= e^{-t}\begin{pmatrix} -\cos 2t - 2\sin 2t & \sin 2t - 2\cos 2t \\ -\sin 2t + 2\cos 2t & -\cos 2t - 2\sin 2t \end{pmatrix}. \tag{5.68}$$

Calculating $\mathbf{A}e^{\mathbf{A}t}$, we have

$$\mathbf{A}e^{\mathbf{A}t} = \begin{pmatrix} -1 & -2 \\ 2 & -1 \end{pmatrix} e^{-t} \begin{pmatrix} \cos 2t & -\sin 2t \\ \sin 2t & \cos 2t \end{pmatrix}$$

$$= e^{-t} \begin{pmatrix} -\cos 2t - 2\sin 2t & \sin 2t - 2\cos 2t \\ -\sin 2t + 2\cos 2t & -\cos 2t - 2\sin 2t \end{pmatrix}, \qquad (5.69)$$

which shows that $e^{\mathbf{A}t}$ is a solution.

Thus for a homogeneous constant coefficient system of equations $\mathbf{x}' = \mathbf{A}\mathbf{x}$, we have another method of finding a solution. In the next section we will see how this relates to our previously obtained solutions (in Sections 6.1 and 6.2), and then will extend this knowledge to handle situations where the coefficient matrix is not constant or is nonhomogeneous.

Problems

In problems 1–4, for the given matrix A, calculate $e^{\mathbf{A}t}$ without using the built-in computer commands.

1. a. $\begin{pmatrix} 3 & -2 \\ 0 & -6 \end{pmatrix}$,   b. $\begin{pmatrix} 1 & -2 \\ 4 & -5 \end{pmatrix}$

2. a. $\begin{pmatrix} 2 & 1 \\ 1 & 2 \end{pmatrix}$,   b. $\begin{pmatrix} 0 & 1 \\ 1 & 0 \end{pmatrix}$

3. a. $\begin{pmatrix} 1 & -1 \\ 1 & 1 \end{pmatrix}$,   b. $\begin{pmatrix} 3 & 2 \\ -2 & 3 \end{pmatrix}$

4. a. $\begin{pmatrix} 5 & 0 \\ 2 & -3 \end{pmatrix}$,   b. $\begin{pmatrix} 1 & -2 \\ 1 & 3 \end{pmatrix}$

5. Use the built-in computer commands to calculate $e^{\mathbf{A}t}$ for the following matrices:
$$\begin{pmatrix} -1 & 1 & 2 \\ 0 & 1 & 0 \\ 1 & 0 & 0 \end{pmatrix}, \quad \begin{pmatrix} 1 & 0 & 0 \\ 4 & 3 & -3 \\ 2 & -1 & 1 \end{pmatrix}$$

6. Use the built-in computer commands to calculate $e^{\mathbf{A}t}$ for
$$\begin{pmatrix} 2 & 3 & 2 & -3 \\ -1 & 2 & 1 & 2 \\ 0 & 0 & -3 & 0 \\ -1 & 3 & 0 & 1 \end{pmatrix}$$

In problems 7–10, for the given matrix A, obtain $e^{\mathbf{A}t}$ (either from problems 1–4 or with the built-in computer commands). Then verify that this answer is a solution to the equation $\mathbf{x}' = \mathbf{A}\mathbf{x}$.

7. a. $\begin{pmatrix} 3 & -2 \\ 0 & -6 \end{pmatrix}$,   b. $\begin{pmatrix} 1 & -2 \\ 4 & -5 \end{pmatrix}$

8. a. $\begin{pmatrix} 2 & 1 \\ 1 & 2 \end{pmatrix}$,   b. $\begin{pmatrix} 0 & 1 \\ 1 & 0 \end{pmatrix}$

9. a. $\begin{pmatrix} 1 & -1 \\ 1 & 1 \end{pmatrix}$,   b. $\begin{pmatrix} 3 & 2 \\ -2 & 3 \end{pmatrix}$

10. a. $\begin{pmatrix} 5 & 0 \\ 2 & -3 \end{pmatrix}$,   b. $\begin{pmatrix} 1 & -2 \\ 1 & 3 \end{pmatrix}$

11. The **Cayley-Hamilton theorem** states that any square matrix satisfies its own characteristic equation. That is, if

$$\det(\mathbf{A} - \lambda\mathbf{I}) = a_0\lambda^n + a_1\lambda^{n-1} + \cdots + a_{n-2}\lambda^2 + a_{n-1}\lambda + a_n \quad (5.70)$$

is the characteristic polynomial of $\mathbf{A}$, then

$$a_0\mathbf{A}^n + a_1\mathbf{A}^{n-1} + \cdots + a_{n-2}\mathbf{A}^2 + a_{n-1}\mathbf{A} + a_n\mathbf{I} = \mathbf{0}. \quad (5.71)$$

(Theorem 5.5.2 was obtained from this theorem.) Show the Cayley-Hamilton theorem holds for the matrices of problems 1–4:

(i) a. $\begin{pmatrix} 3 & -2 \\ 0 & -6 \end{pmatrix}$, b. $\begin{pmatrix} 1 & -2 \\ 4 & -5 \end{pmatrix}$ (ii) a. $\begin{pmatrix} 2 & 1 \\ 1 & 2 \end{pmatrix}$, b. $\begin{pmatrix} 0 & 1 \\ 1 & 0 \end{pmatrix}$

(iii) a. $\begin{pmatrix} 1 & -1 \\ 1 & 1 \end{pmatrix}$, b. $\begin{pmatrix} 3 & 2 \\ -2 & 3 \end{pmatrix}$ (iv) a. $\begin{pmatrix} 5 & 0 \\ 2 & -3 \end{pmatrix}$, b. $\begin{pmatrix} 1 & -2 \\ 1 & 3 \end{pmatrix}$

5.6   Chapter 5: Additional Problems and Projects

ADDITIONAL PROBLEMS

In problems 1–6, determine whether the statement is true or false. If is true, give reasons for your answer. If it is false, give a counterexample or other explanation of why it is false.

1. Every multiplication of a 2×2 matrix $\mathbf{A}$ with a vector $\mathbf{x}$ can be thought of as a stretch, a rotation, a reflection, or a projection.

2. The phase plane can be drawn for any autonomous system that can be written in the form $\mathbf{x}' = \mathbf{A}\mathbf{x}$.

3. One calculates the derivative of a matrix by taking the derivative of each element of the matrix.

4. If the vectors $\mathbf{x}_1, \mathbf{x}_2, \ldots, \mathbf{x}_n$ are constant vectors (i.e., with no dependence on t), then the Wronskian equal to zero implies that the set of vectors is linearly dependent.

5. Every 3×2 matrix with linearly independent columns has only the zero vector in its nullspace.

6. Every 2×3 matrix has only the zero vector as the only element in its column space.

For each of problems 7–13, do the following three parts: (i) consider the given matrix as the coefficient matrix $\mathbf{A}$ of the system $\mathbf{x}' = \mathbf{A}\mathbf{x}$ and use Theorem 5.1.1 to determine the classification of the origin; (ii) use one of the computer programs to draw the vector field for each system $\mathbf{x}' = \mathbf{A}\mathbf{x}$; and then (iii) find the eigenvalues and eigenvectors of the matrix.

7. a. $\begin{pmatrix} 3 & 1 \\ 1 & 3 \end{pmatrix}$,   b. $\begin{pmatrix} 1 & 0 \\ 4 & 3 \end{pmatrix}$

8. a. $\begin{pmatrix} 5 & 1 \\ -1 & 5 \end{pmatrix}$,   b. $\begin{pmatrix} 2 & 2 \\ -1 & 0 \end{pmatrix}$

9. a. $\begin{pmatrix} 4 & 1 \\ -10 & -2 \end{pmatrix}$,   b. $\begin{pmatrix} -1 & 2 \\ 1 & 0 \end{pmatrix}$

10. a. $\begin{pmatrix} 5 & -6 \\ 3 & -4 \end{pmatrix}$,   b. $\begin{pmatrix} -10 & 24 \\ -4 & 10 \end{pmatrix}$

11. a. $\begin{pmatrix} 5 & 1 \\ -17 & -3 \end{pmatrix}$,   b. $\begin{pmatrix} 4 & 1 \\ -20 & -4 \end{pmatrix}$

12. a. $\begin{pmatrix} 2 & 3 \\ -1 & 2 \end{pmatrix}$,   b. $\begin{pmatrix} 1 & -1 \\ 2 & 4 \end{pmatrix}$

13. a. $\begin{pmatrix} 5 & -6 \\ 2 & -1 \end{pmatrix}$,   b. $\begin{pmatrix} -5 & 2 \\ -2 & -3 \end{pmatrix}$

In problems 14–16, use the given substitutions and the methods of Section 5.1.3 to rewrite the equations as an uncoupled system. Then write the solution in terms of u_1, u_2 and use this to find the solution in terms of x, y.

14. $u_1 = 6x - 2y, u_2 = -5x + 2y$ for $\begin{cases} x' = -7x + 2y \\ y' = -15x + 4y \end{cases}$

15. $u_1 = x - y, u_2 = -x + 2y$ for $\begin{cases} x' = -6x + 8y \\ y' = -4x + 6y \end{cases}$

16. $u_1 = 2x - y, u_2 = -x + y$ for $\begin{cases} x' = 4x - y \\ y' = 2x + y \end{cases}$

In problems 17–22, find the eigenvalues and eigenvectors of the following matrices. Get clarification from your instructor as to whether you should find these by hand or by using Matlab, Maple, or Mathematica.

17. $\begin{pmatrix} 4 & -8 & -10 \\ -1 & 6 & 5 \\ 1 & -8 & -7 \end{pmatrix}$

18. $\begin{pmatrix} 0 & -2 & 0 \\ 1 & 3 & 0 \\ -1 & -2 & 1 \end{pmatrix}$

19. $\begin{pmatrix} 2 & -2 & 1 \\ -3 & 3 & 1 \\ 3 & -2 & 0 \end{pmatrix}$

20. $\begin{pmatrix} 1 & -2 & 2 \\ -4 & 3 & 2 \\ 4 & -2 & -1 \end{pmatrix}$

21. $\begin{pmatrix} 4 & -2 & -1 \\ -1 & 3 & -1 \\ 1 & -2 & 2 \end{pmatrix}$

22. $\begin{pmatrix} -17 & 4 & 2 \\ 5 & -25 & 1 \\ 3 & 12 & -12 \end{pmatrix}$

In problems 23–30, determine whether the given set forms a fundamental set of solutions for the given differential equation.

23. $\left\{ \begin{pmatrix} e^t \\ 2e^t \end{pmatrix}, \begin{pmatrix} e^{-t} \\ 3e^{-t} \end{pmatrix} \right\}$, $\mathbf{x}' = \begin{pmatrix} 5 & -2 \\ 12 & -5 \end{pmatrix} \mathbf{x}$

24. $\left\{ \begin{pmatrix} 3e^{2t} \\ 2e^{2t} \end{pmatrix}, \begin{pmatrix} e^{-3t} \\ e^{-3t} \end{pmatrix} \right\}$, $\mathbf{x}' = \begin{pmatrix} 1 & -2 \\ 2 & -3 \end{pmatrix} \mathbf{x}$

25. $\left\{ \begin{pmatrix} e^{3t} \\ 2e^{3t} \end{pmatrix}, \begin{pmatrix} -e^{6t} \\ e^{6t} \end{pmatrix} \right\}$, $\mathbf{x}' = \begin{pmatrix} 5 & -1 \\ -2 & 4 \end{pmatrix} \mathbf{x}$

26. $\left\{ \begin{pmatrix} 1 \\ 1 \end{pmatrix} e^{-4t}, \begin{pmatrix} -1 \\ 1 \end{pmatrix} e^{2t} \right\}$, $\mathbf{x}' = \begin{pmatrix} -1 & -3 \\ -3 & -1 \end{pmatrix} \mathbf{x}$

27. $\left\{ \begin{pmatrix} 2 \\ 1 \end{pmatrix} e^t, \begin{pmatrix} 1 \\ 3 \end{pmatrix} e^{-2t} \right\}$, $\mathbf{x}' = \begin{pmatrix} 3 & 1 \\ 2 & -1 \end{pmatrix} \mathbf{x}$

28. $\left\{ \begin{pmatrix} 1 \\ 2 \end{pmatrix} e^{-4t}, \begin{pmatrix} 1 \\ -2 \end{pmatrix} e^{-8t} \right\}$, $\mathbf{x}' = \begin{pmatrix} -6 & 1 \\ 4 & -6 \end{pmatrix} \mathbf{x}$

29. $\left\{ \begin{pmatrix} 1 \\ -1 \\ 1 \end{pmatrix} e^{5t}, \begin{pmatrix} 1 \\ 1 \\ 1 \end{pmatrix} e^t, \begin{pmatrix} 1 \\ 1 \\ -1 \end{pmatrix} e^{-t} \right\}$, $\mathbf{x}' = \begin{pmatrix} 2 & -2 & 1 \\ -3 & 3 & 1 \\ 3 & -2 & 0 \end{pmatrix} \mathbf{x}$

30. $\left\{ \begin{pmatrix} 1 \\ -1 \\ 1 \end{pmatrix} e^{5t}, \begin{pmatrix} 1 \\ 1 \\ 1 \end{pmatrix} e^t, \begin{pmatrix} -1 \\ 1 \\ 1 \end{pmatrix} e^{2t} \right\}$, $\mathbf{x}' = \begin{pmatrix} 4 & -2 & -1 \\ -1 & 3 & -1 \\ 1 & -2 & 2 \end{pmatrix} \mathbf{x}$

31. Consider the vector $(-1 \quad -1)^T$. Find the vectors that results from
    a. stretch by a factor of $\frac{5}{2}$ (sketch the original vector and the resulting vector)
    b. rotation by an angle of $\pi/3$ (sketch the original vector, the angle of rotation, and the resulting vector)
    c. projection onto the line that makes an angle π with the x-axis (sketch

the original vector, the line of projection, and the resulting vector)

d. reflection through the line that makes an angle π with the x-axis (sketch the original vector, the line of reflection, and the resulting vector)

32. Consider the vector $(2 \ 0)^T$. Find the vectors that results from

a. stretch by a factor of 3 (sketch the original vector and the resulting vector)

b. rotation by an angle of $-\pi/4$ (sketch the original vector, the angle of rotation, and the resulting vector)

c. projection onto the line that makes an angle $\pi/3$ with the x-axis (sketch the original vector, the line of projection, and the resulting vector)

d. reflection through the line that makes an angle $\pi/6$ with the x-axis (sketch the original vector, the line of reflection, and the resulting vector)

33. Consider the vector $(3 \ 2)^T$. Find the vectors that results from

a. stretch by a factor of 2 (sketch the original vector and the resulting vector)

b. rotation by an angle of π (sketch the original vector, the angle of rotation, and the resulting vector)

c. projection onto the line that makes an angle $\pi/2$ with the x-axis (sketch the original vector, the line of projection, and the resulting vector)

d. reflection through the line that makes an angle $\pi/2$ with the x-axis (sketch the original vector, the line of reflection, and the resulting vector)

34. Consider the matrix

$$\mathbf{A} = \begin{pmatrix} -1 & 3 \\ 2 & -6 \\ -1 & 1 \end{pmatrix}$$

Use Matlab, Maple, or Mathematica to find bases for the four fundamental subspaces of $\mathbf{A}$.

35. Consider the matrix

$$\mathbf{A} = \begin{pmatrix} -2 & 4 \\ 2 & -4 \\ -1 & 2 \end{pmatrix}$$

Use Matlab, Maple, or Mathematica to find bases for the four fundamental subspaces of $\mathbf{A}$.

36. Consider the matrix

$$\mathbf{A} = \begin{pmatrix} -2 & 0 & 2 \\ -4 & -2 & 2 \\ 1 & 4 & 3 \end{pmatrix}$$

Use Matlab, Maple, or Mathematica to find bases for the four fundamental subspaces of $\mathbf{A}$.

37. Consider the matrix
$$\mathbf{A} = \begin{pmatrix} -1 & 2 & 2 \\ -1 & 1 & 2 \end{pmatrix}$$

Use Matlab, Maple, or Mathematica to find bases for the four fundamental subspaces of **A**.

$$\mathbf{A} = \begin{pmatrix} 5 & -2 & 3 \\ -2 & 0 & 2 \end{pmatrix}$$

Use Matlab, Maple, or Mathematica to find bases for the four fundamental subspaces of **A**.

38. Consider the matrix
$$\mathbf{A} = \begin{pmatrix} -21 & 2 & 2 \\ -1 & 7 & 8 \\ 2 & 1 & 2 \end{pmatrix}$$

Use Matlab, Maple, or Mathematica to find bases for the four fundamental subspaces of **A**.

In problems 39–41, obtain $e^{\mathbf{A}t}$ (either by hand or with the built-in computer commands). Then verify that this answer is a solution to the equation $\mathbf{x}' = \mathbf{A}\mathbf{x}$.

39. a. $\begin{pmatrix} 1 & -2 \\ 1 & -1 \end{pmatrix}$,   b. $\begin{pmatrix} 2 & 1 \\ 0 & 1 \end{pmatrix}$

40. a. $\begin{pmatrix} -2 & 0 \\ 1 & 2 \end{pmatrix}$,   b. $\begin{pmatrix} 2 & 1 \\ 1 & 2 \end{pmatrix}$

41. a. $\begin{pmatrix} 1 & -2 \\ 1 & -1 \end{pmatrix}$,   b. $\begin{pmatrix} 1 & 1 \\ -1 & 1 \end{pmatrix}$

PROJECT FOR CHAPTER 5

Project 1: Another Application of Eigenvectors

In Appendix C.3.1, we learn how to diagonalize a matrix (if it is possible) as follows:

- Find the eigenvalues and eigenvectors of the $n \times n$ matrix $\mathbf{A}$. If it does not have a full set of eigenvectors (that is, the number of eigenvectors $< n$), then the matrix cannot be diagonalized.

- Put the eigenvectors as columns of a new matrix $\mathbf{V}$. Then

$$\mathbf{\Lambda} = \mathbf{V}^{-1}\mathbf{A}\mathbf{V}$$

  is a diagonal matrix with the eigenvalues of $\mathbf{A}$ as its diagonal elements. Moreover, the eigenvalue in $\mathbf{\Lambda}_{i,i}$ has corresponding eigenvector in the ith column of $\mathbf{V}$.

For a matrix that can be diagonalized, think of a computationally efficient way to calculate A^n. This would be useful, e.g., in Section 5.5.[6]
We now consider a *transition matrix* in stochastic processes, which describes the (transition) probability of one state going to the next. We make the assumption that the system going from one state to the next only depends on the current state of the system and not on the myriad of possibilities for how one could have arrived at the given configuration. This is a reasonable assumption and is a key characteristic of *Markov chains*, studied in the theory of probability. If we consider a system with 3 states, we write

$$\mathbf{T} = \begin{pmatrix} p_{11} & p_{21} & p_{31} \\ p_{12} & p_{22} & p_{32} \\ p_{13} & p_{23} & p_{33} \end{pmatrix}.$$

where

p_{ij} is the probability of moving from state i to state j.

For example, p_{23} is the probability of the system moving from state 2 to state 3. Note that the probability p_{ij} is located in the $(\mathbf{T})_{ji}$ position of the matrix. If we begin with an initial state vector $\mathbf{x}_0$, we can calculate the next state as $\mathbf{x}_1 = \mathbf{T}\mathbf{x}_0$. Similarly, the next state of the system is $\mathbf{x}_2 = \mathbf{T}\mathbf{x}_1 = \mathbf{T}^2\mathbf{x}_0$, and so on. The long-term behavior of the system can be described by examining $\mathbf{x}_n = \mathbf{T}^n\mathbf{x}_0$ for large n.

---

[6]Our three computer programs do not raise a matrix to a power in the naive way of `A*A*...*A` but instead do something similar to this efficient way that you are seeking.

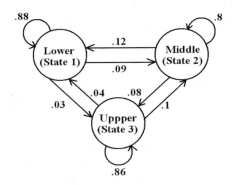

FIGURE 5.12:   General state diagram for income class model.

Application:
Consider a town of 30,000 families (economic units) who have been grouped
into three economic brackets: lower, middle, and upper. Each year there is
a 9% chance that the lower move into the middle, and 12% chance that the
middle move back to the lower; there is an 8% chance that the middle move to
the upper, and 10% chance that the upper move down to the middle; finally
there is a 3% chance that the lower move directly to the upper, and 4% chance
that the upper move down to the lower. These transitions occur continuously
throughout the year as people are hired, laid off, fired, promoted, retire, and
change careers.

Assume that initially there are 10,000 lower, 12,000 middle, and 8,000 upper
income families and consider the state diagram for this situation shown in
Figure 5.12.

Write the transition matrix of this system. Assuming that n is measured
in years, find the number of families in each state after 10 years; after 20
years; after 50 years. (Remember to use the hint above for efficiently raising
a matrix to a high power.)

Now find the eigenvalues and eigenvectors of the transition matrix. *Cal-
culate* $\mathbf{T}^\infty \mathbf{x}_0$. Give a mathematical explanation for what is happening. See
Example 3 in Section 6.1 for a differential equation approach to this problem.

Chapter 6

Techniques of Systems of Differential Equations

6.1 A General Method, Part I: Solving Systems with Real, Distinct Eigenvalues

In this section, we look at a general method for solving systems of first-order homogeneous linear differential equations with constant coefficients. It also provides a theoretical framework and makes one aware of what types of solutions are expected from such a linear system. Practically speaking, however, solving systems of three or more equations involves finding roots of polynomial equations of degree 3 or higher. Except in special cases, we need to use some technological tool to find these solutions. If we are using a computer algebra system, then we might as well just use it to solve the system and skip the method of this section. This is especially true if the graph of the solution is the only item of interest. On the other hand, if one is interested in the functional form of the solution, frequently the form of the solution given by a computer algebra system is large, cumbersome, and hard to study. Even simplification routines may not help much. In this case, the method of this section is valuable, even if we are using a machine to perform the individual steps. From a philosophical point of view, it is important that we know how to solve a problem, even though we may let a machine do the work.

We stated that the methods of the previous chapter generalize for n equations and now we consider the situation

$$
\begin{aligned}
\frac{dx_1}{dt} &= a_{11}x_1 + a_{12}x_2 + \cdots + a_{1n}x_n \\
\frac{dx_2}{dt} &= a_{21}x_1 + a_{22}x_2 + \cdots + a_{2n}x_n \\
\vdots \qquad & \qquad \vdots \qquad \vdots \qquad \ddots \qquad \vdots \\
\frac{dx_n}{dt} &= a_{n1}x_1 + a_{n2}x_2 + \cdots + a_{nn}x_n.
\end{aligned}
\tag{6.1}
$$

We have seen that this can be written in matrix notation as

$$
\mathbf{x}' = \mathbf{Ax}
\tag{6.2}
$$

where

$$\mathbf{x} = \begin{pmatrix} x_1 \\ x_2 \\ \vdots \\ x_n \end{pmatrix}, \qquad \mathbf{x}' = \begin{pmatrix} \frac{dx_1}{dt} \\ \frac{dx_2}{dt} \\ \vdots \\ \frac{dx_n}{dt} \end{pmatrix}, \quad \text{and} \quad A = \begin{pmatrix} a_{11} & a_{12} & \cdots & a_{1n} \\ a_{21} & a_{22} & \cdots & a_{2n} \\ \vdots & \vdots & \vdots & \vdots \\ a_{n1} & a_{n2} & \cdots & a_{nn} \end{pmatrix}.$$

We saw that exponential solutions arose in the solution of the nth order linear equations. In the examples that we considered in Section 5.1 we also saw that exponential solutions arose in the solution of a system of two first-order equations. Let us assume that solutions of (6.1) are of the form

$$\mathbf{x} = c \begin{pmatrix} v_1 \\ v_2 \\ \vdots \\ v_n \end{pmatrix} e^{\lambda t} = c\mathbf{v}e^{\lambda t} \tag{6.3}$$

for some constant $c \neq 0$, and consider (6.1) in its vector form, Then the solution must satisfy the differential equation so that

$$c\lambda\mathbf{v}e^{\lambda t} = c\mathbf{A}\mathbf{v}e^{\lambda t}.$$

We can divide by $ce^{\lambda t}$ (because it is never zero) and then rewrite the equation as

$$\mathbf{A}\mathbf{v} = \lambda\mathbf{v}. \tag{6.4}$$

Now we are in the realm of an eigenvalue problem. In particular, the only way to have solutions other than $\mathbf{v} = \mathbf{0}$ is to have the matrix $A - \lambda I$ be singular, which means that $\det(A - \lambda I) = 0$. Evaluating this determinant yields an nth degree polynomial in λ which has n, possibly repeating and possibly complex (non-real) eigenvalues $\lambda_1, \ldots, \lambda_n$. Each of these eigenvalues is substituted for λ in equation (6.4), which in turn is solved for its corresponding eigenvector $\mathbf{v}_i$. Based on our discussion in Section 5.4, we know that this last equation says that $\mathbf{v}e^{\lambda t}$ will be a solution of $\mathbf{x}' = \mathbf{A}\mathbf{x}$ when λ is an eigenvalue of $\mathbf{A}$ and $\mathbf{v}$ is the corresponding eigenvector. Thus one solution is

$$\mathbf{x}(t) = \mathbf{v}_i e^{\lambda_i t} c_i$$

where c_i is an arbitrary constant. In the case when all the eigenvalues are real and distinct, we can state a general theorem.

THEOREM 6.1.1 *Consider the system* $\mathbf{x}' = \mathbf{A}\mathbf{x}$, *where the coefficient matrix* $\mathbf{A}$ *has n distinct real eigenvalues. Let* $\lambda_1, \lambda_2, \cdots, \lambda_n$ *be these eigenvalues with* $\mathbf{v}_1, \mathbf{v}_2, \cdots, \mathbf{v}_n$ *as the corresponding eigenvectors. Then the* $c_i\mathbf{v}_ie^{\lambda_i t}$, $i = 1, 2, \cdots, n$ *are linearly independent and the general solution is given by*

$$\mathbf{x} = c_1\mathbf{v}_1 e^{\lambda_1 t} + c_2\mathbf{v}_2 e^{\lambda_2 t} + \cdots + c_n\mathbf{v}_n e^{\lambda_n t}$$

and is defined for $t \in (-\infty, \infty)$.

In the situation where we have

$$\mathbf{x}(0) = \mathbf{x_0},$$

then we set up a system of n equations in the unknowns $c_1, \ldots, c_n$ and solve for the c_i. Equivalently, the general solution is sometimes written as

$$\mathbf{x}(t) = \Phi(t)\mathbf{c}$$

where

$$\Phi(t) = \left(\mathbf{v}_1 e^{\lambda_1 t} \middle| \mathbf{v}_2 e^{\lambda_2 t} \middle| \cdots \middle| \mathbf{v}_n e^{\lambda_n t} \right) \quad \text{and} \quad \mathbf{c} = \{c_1, c_2, \ldots, c_n\}.$$

(Here the vertical lines separate the columns.) Thus given the initial condition $\mathbf{x}(0) = \mathbf{X_0}$, we write the matrix equation $\Phi(0)\mathbf{c} = \mathbf{X_0}$, which, if $\Phi^{-1}(0)$ exists, has a solution

$$\mathbf{c} = \Phi^{-1}(0)\mathbf{X_0}.$$

Example 1: Consider the system

$$x' = 5x - y$$

$$y' = 3y.$$

Find the general solution. Based on Theorem 5.1.1, determine the type and stability of the equilibrium solution. Sketch the phase portrait.

In matrix form this equation is

$$\mathbf{x}' = \begin{pmatrix} 5 & -1 \\ 0 & 3 \end{pmatrix} \mathbf{x}$$

so that the eigenvalues of

$$A = \begin{pmatrix} 5 & -1 \\ 0 & 3 \end{pmatrix}$$

are found from

$$\det(A - \lambda I) = \begin{vmatrix} 5 - \lambda & -1 \\ 0 & 3 - \lambda \end{vmatrix} = (5 - \lambda)(3 - \lambda) = 0.$$

The eigenvalues are

$$\lambda_1 = 3 \text{ and } \lambda_2 = 5.$$

Now an eigenvector

$$\mathbf{v}_1 = \begin{pmatrix} x_1 \\ y_1 \end{pmatrix}$$

corresponding to $\lambda_1 = 3$ is found from

$$\begin{pmatrix} 2 & -1 \\ 0 & 0 \end{pmatrix} \begin{pmatrix} x_1 \\ y_1 \end{pmatrix} = \begin{pmatrix} 0 \\ 0 \end{pmatrix}$$

which gives

$$y_1 = 2x_1.$$

Letting $x_1 = 1$, we obtain the eigenvector

$$\mathbf{v}_1 = \begin{pmatrix} 1 \\ 2 \end{pmatrix}.$$

Similarly, an eigenvector

$$\mathbf{v}_2 = \begin{pmatrix} x_2 \\ y_2 \end{pmatrix}$$

corresponding to $\lambda = 5$ satisfies

$$\begin{pmatrix} 0 & -1 \\ 0 & -2 \end{pmatrix} \begin{pmatrix} x_2 \\ y_2 \end{pmatrix} = \begin{pmatrix} 0 \\ 0 \end{pmatrix}$$

so that

$$y_2 = 0.$$

This gives

$$\mathbf{v}_2 = \begin{pmatrix} x_2 \\ 0 \end{pmatrix}$$

so that if we let $x_2 = 1$, then

$$\mathbf{v}_2 = \begin{pmatrix} 1 \\ 0 \end{pmatrix}.$$

So the general solution is

$$x(t) = c_1 \begin{pmatrix} 1 \\ 2 \end{pmatrix} e^{3t} + c_2 \begin{pmatrix} 1 \\ 0 \end{pmatrix} e^{5t}.$$

In the case of a 2×2 system, it is easy to use Theorem 5.1.1 to characterize the stability of the equilibria (the origin). For a general system of n first-order equations, we can again speak of equilibria of the system and we again have the types mentioned in Theorem 5.1.1. Except in very special cases, the origin will again be the only equilibrium solution that we need to consider. In the two-dimensional case, we talked about viewing the trajectories in the phase plane. For a system of n equations, the trajectories now live in n dimensional phase-space and we can only see (at most) three dimensions at a time. We can, however, characterize the stability according to the eigenvalues of the matrix.

THEOREM 6.1.2 *Let* $\{\lambda_1, \lambda_2, \cdots, \lambda_n\}$ *be the n (real or complex [non-real], possibly repeated) eigenvalues of the coefficient matrix* **A** *of a given linear constant-coefficient homogeneous system.*
a. *If the real part of the eigenvalue* $\Re(\lambda_i) < 0$ *for all i, then the equilibrium point is stable.*
b. *If the real part of the eigenvalue* $\Re(\lambda_i) < 0$ *for at least one i and the real part of the eigenvalue* $\Re(\lambda_j) > 0$ *for at least one j, then the equilibrium point is a saddle.*
c. *If the real part of the eigenvalue* $\Re(\lambda_i) > 0$ *for all i, then the equilibrium point is unstable.*
d. *If any of the eigenvalues are complex, then the stable or unstable equilibria is a spiral; if all of the eigenvalues are real, it is a node;*
e. *If a pair of complex conjugate eigenvalues* λ_i, $\overline{\lambda_i}$ *satisfy* $\Re(\lambda_i) = 0$, *then the equilibrium is a center in the plane containing the corresponding eigenvectors.*

As mentioned in Section 5.1, stable spirals and stable nodes are sometimes called spiral sinks and nodal sinks, respectively. Similarly, unstable spirals and unstable nodes are sometimes called spiral sources and nodal sources, respectively.

This theorem also applies to a system of two equations and it was straightforward to view the phase portraits in the phase plane. In higher dimensions, we can have combinations of the different behaviors. For example, we may have an unstable spiral in one plane even though solutions may be attracted to this plane! Sketching phase portraits becomes much more difficult although the pictures from Figure 5.5 are still the generic ones that may result.

Example 2: Consider the following system:

$$\frac{dx}{dt} = -2x - y - 2z$$

$$\frac{dy}{dt} = -4x - 5y + 2z$$

$$\frac{dz}{dt} = -5x - y + z.$$

Calculate the eigenvalues and use Theorem 6.1.2 to determine the stability of the equilibrium point (the origin). Then find the eigenvectors and use Theorem 6.1.1 to write the general solution.

We can calculate the eigenvalues either by hand or with the computer. Doing so gives

$$\lambda_1 = 3, \ \lambda_2 = -6, \ \lambda_3 = -3.$$

Part (b) of Theorem 6.1.2 shows that the origin is a saddle. We can similarly find the eigenvectors either by hand or with the computer and we obtain

$$\mathbf{v_1} = \begin{pmatrix} -1 \\ 1 \\ 2 \end{pmatrix}, \quad \mathbf{v_2} = \begin{pmatrix} 1 \\ 2 \\ 1 \end{pmatrix}, \quad \mathbf{v_3} = \begin{pmatrix} 1 \\ -1 \\ 1 \end{pmatrix}$$

as the respective eigenvectors. We can then write the general solution as

$$\mathbf{x} = c_1 e^{3t} \begin{pmatrix} -1 \\ 1 \\ 2 \end{pmatrix} + c_2 e^{-6t} \begin{pmatrix} 1 \\ 2 \\ 1 \end{pmatrix} + c_3 e^{-3t} \begin{pmatrix} 1 \\ -1 \\ 1 \end{pmatrix}$$

or, alternatively, we could write the general solution as

$$x(t) = -c_1 e^{3t} + c_2 e^{6t} + c_3 e^{-3t}$$
$$y(t) = c_1 e^{3t} + 2c_2 e^{6t} - c_3 e^{-3t}$$
$$z(t) = 2c_1 e^{3t} + c_2 e^{6t} + c_3 e^{-3t}.$$

Project 1 in Chapter 5 examined a probabilistic approach to a model using transition matrices and Markov Chains. We now reconsider the problem from a differential equation viewpoint. The reader should compare answers and approaches and note any similarities and differences.

Example 3: Social Mobility [27] Consider a town of 30,000 families (economic units) who have been grouped into three economic brackets: lower, middle, and upper. Each year 9% of the lower move into the middle, and 12% of the middle move back to the lower; 8% of the middle move to the upper, and 10% of the upper move down to the middle; finally 3% of the lower move directly to the upper, and 4% of the upper move down to the lower. These transitions occur continuously throughout the year as people are hired, laid off, fired, promoted, retire, and change careers.

We assume that initially there are 10,000 lower, 12,000 middle, and 8,000 upper income families and consider the compartmental diagram for this situation is shown in Figure 6.1.

The system of equations is

$$\frac{dx}{dt} = -0.12x + 0.12y + 0.04z$$

$$\frac{dy}{dt} = 0.09x - 0.2y + 0.1z$$

$$\frac{dz}{dt} = 0.03x + 0.08y - 0.14z.$$

Thus

$$\mathbf{A} = \begin{pmatrix} -0.12 & 0.12 & 0.04 \\ 0.09 & -0.2 & 0.1 \\ 0.03 & 0.08 & -0.14 \end{pmatrix},$$

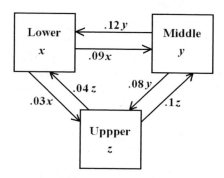

FIGURE 6.1: Compartmental diagram for income class model.

which has eigenvalues $\lambda_1 = -0.16597$, $\lambda_2 = -0.29403$, $\lambda_3 = 0$ with corresponding eigenvectors

$$\mathbf{v}_1 = \begin{pmatrix} .78441 \\ -0.058524 \\ -.72589 \end{pmatrix}, \mathbf{v}_2 = \begin{pmatrix} -.43404 \\ .72718 \\ -.29314 \end{pmatrix}, \mathbf{v}_3 = \begin{pmatrix} .75199 \\ .58655 \\ .49631 \end{pmatrix}.$$

(We have only listed five significant figures but more are kept in the intermediate calculations.) With two negative eigenvalues and one 0 eigenvalue, we know that the system eventually stabilizes (in proportions of the eigenvector associated with the 0 eigenvalue).

The general solution is thus

$$\mathbf{x} = c_1 e^{-0.16597t} \begin{pmatrix} .78441 \\ -0.058524 \\ -.72589 \end{pmatrix} + c_2 e^{-0.29403t} \begin{pmatrix} -.43404 \\ .72718 \\ -.29314 \end{pmatrix} + c_3 \begin{pmatrix} .75199 \\ .58655 \\ .49631 \end{pmatrix},$$

which can also be written as

$$\mathbf{x}(t) = \Phi(t)\mathbf{c} = \mathbf{x} = \begin{pmatrix} .78441e^{-0.16597t} & -.43404e^{-0.29403t} & .75199 \\ -0.058524e^{-0.16597t} & .72718e^{-0.29403t} & .58655 \\ -.72589e^{-0.16597t} & -.29314e^{-0.29403t} & .49631 \end{pmatrix} \begin{pmatrix} c_1 \\ c_2 \\ c_3 \end{pmatrix}.$$

In its expanded form, this is

$$x(t) = c_1(.78441)e^{-0.16597t} + c_2(-.43404)e^{-0.29403t} + c_3(.75199),$$
$$y(t) = c_1(-0.058524)e^{-0.16597t} + c_2(.72718)e^{-0.29403t} + c_3(.58655),$$
$$z(t) = c_1(-.72589)e^{-0.16597t} + c_2(-.29314)e^{-0.29403t} + c_3(.49631).$$

The initial conditions are given as $\mathbf{x}(0) = (10{,}000 \quad 12{,}000 \quad 8{,}000)^T$, so we find the c_i's by solving

$$10{,}000 = c_1(.78441) + c_2(-.43404) + c_3(.75199),$$
$$12{,}000 = c_1(-0.058524) + c_2(.72718) + c_3(.58655),$$
$$8{,}000 = c_1(-.72589) + c_2(-.29314) + c_3(.49631).$$

This system of three equations in three unknowns can be written using matrix-vector form as

$$\mathbf{x}(t) = \Phi(0)\mathbf{c} \begin{pmatrix} 10,000 \\ 12,000 \\ 8,000 \end{pmatrix} = \begin{pmatrix} .78441 & -.43404) & .75199 \\ -0.058524 & .72718 & .58655 \\ -.72589 & -.29314 & .49631 \end{pmatrix} \begin{pmatrix} c_1 \\ c_2 \\ c_3 \end{pmatrix}$$

which can be solved to give $\mathbf{c} = (-1{,}433.07 \quad 3{,}221.95 \quad 16{,}350.1)^T$. Our solution to the initial value problem is thus

$$x(t) = -896.64e^{-.16597t} - 1398.4e^{-.29403t} + 12295.1$$
$$y(t) = 66.898e^{-.16597t} + 2342.9e^{-.29403t} + 9590.2$$
$$z(t) = 829.74e^{-.16597t} - 944.50e^{-.29403t} + 8114.8. \tag{6.5}$$

Because of the decaying exponentials in these solutions, it is evident that the system stabilizes with 12,295 lower income families, 9,590 middle income families, and 8,115 upper income families. This type of information may be useful for builders trying to determine the number of houses to build in certain price ranges or for politicians trying to argue why the other party's economic policies don't work. A graph of the solution curves is shown in Figure 6.2.

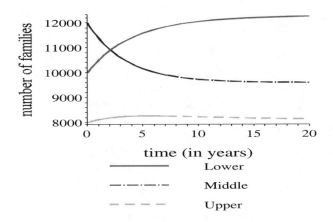

FIGURE 6.2: Solutions of income model of Example 3.

Computer Code 6.1: **Solving a system of differential equations and plotting this solution with axis labels, title, legend**

Matlab, Maple, Mathematica

```
                              Matlab
>>  %This requires the Symbolic Math Toolbox
>>  format long
>>  A=[-.12,.12,.04; .09,-.2,.1; .03,.08,-.14]
>>  [v,d]=eig(A)
>>  syms t
>>  genmatrix=[exp(0*t)*v(:,1),exp(d(2,2)*t)*v(:,2),
    exp(d(3,3)*t)*v(:,3)]
>>  %general soln is genmatrix*[c1; c2; c3]
>>  %Note the zero eigenvalue appeared first in d
>>  genmatrix0=subs(genmatrix,t,0)
>>  x0=[10000; 12000; 8000]
>>  cvals=genmatrix0\x0
>>  v1n=v(:,1)*cvals(1)
>>  v2n=v(:,2)*cvals(2)
>>  v3n=v(:,3)*cvals(3)
>>  Vmat=[v1n,v2n,v3n]
>>  xvec=[exp(0*t); exp(d(2,2)*t); exp(d(3,3)*t)]
>>  %the solution is Vmat*xvec
>>  soln=Vmat*xvec
>>  ezplot(soln(1),[0,20])
>>  hold on
>>  ezplot(soln(2),[0,20])
>>  ezplot(soln(3),[0,20])
>>  axis([0 20 8000 12500])
>>  title('Income Model')
>>  xlabel('years')
>>  ylabel('Number of families')
>>  legend('lower','middle','upper')
>>  hold off

% the user can go to the figure, click on the
%show plot tools icon and edit the curves
%in order to distinguish them.
```

```
                            ┌─────────┐
                            │  Maple  │
                            └─────────┘
> with(linalg):
> Digits:=15; #gives more precision in calculations
> eqA:=matrix(3,3,[-.12,.12,.04,.09,-.2,.1,.03,.08,-.14]);
> eq1:=eigenvects(eqA);
> #Observe that the first eigenvalue is 10^{-15}, which
> #would be zero if not for rounding errors
> soln1a:=exp(0*t)*(eq1[1][3][1]);
> soln1b:=exp(eq1[2][1]*t)*(eq1[2][3][1]);
> soln1c:=exp(eq1[3][1]*t)*(eq1[3][3][1]);
> gensoln:=simplify(evalm(c[1]*soln1a+c[2]*soln1b+
   c[3]*soln1c));
> ICeqn:=evalf(subs(t=0,evalm({gensoln[1]=10000,
   gensoln[2]=12000,gensoln[3]=8000})));
> Cvals:=solve(ICeqn,{c[1],c[2],c[3]});
> xsoln:=subs(Cvals,gensoln[1]);
> ysoln:=subs(Cvals,gensoln[2]);
> zsoln:=subs(Cvals,gensoln[3]);
> plot([xsoln,ysoln,zsoln],t=0..20,linestyle=[1,2,3],legend=
   ["lower","middle","upper"],labels=["years","Number
   of families"],title="Income Model");

#If your legend doesn't show up:  right click on
#the picture, then left-click show legend under Legend
```

```
                        ┌───────────────┐
                        │  Mathematica  │
                        └───────────────┘
A={{-.12,.12,.04},{.09,-.2,.1},{.03,.08,-.14}}
MatrixForm[A] (*puts matrix in nicer visual form*)
eq1=Eigensystem[A]
   (*Observe that the third eigenvalue is 10^{-17}, which*)
   (*would be zero if not for rounding errors*)
soln1a = e^{0 t}eq1[[2]][[3]]
soln1b = e^{eq1[[1]][[2]] t} eq1[[2]][[2]]
soln1c = e^{eq1[[1]][[1]] t} eq1[[2]][[1]]
gensoln[t_]=c1 soln1a+c2 soln1b +c3 soln1c
Cvals=Solve[{gensoln[0]=={10000,12000,8000}},{c1,c2,c3}]
xsoln[t_]=ReplaceAll[gensoln[t][[1]],Cvals]
ysoln[t_]=ReplaceAll[gensoln[t][[2]],Cvals]
zsoln[t_]=ReplaceAll[gensoln[t][[3]],Cvals]
Plot[{xsoln[t],ysoln[t],zsoln[t]},{t,0,20}, PlotStyle→
   {Dashing[{.01}],Dashing[{.03}], Dashing[{.06}]},
   AxesLabel→ {"years","number of families"}]
```

Problems

For problems 1–8, find the eigenvalues of the coefficient matrix and use Theorem 6.1.2 to determine the stability of the origin. Then find the corresponding eigenvectors (they are all real) and write the general solution.

1. a. $\begin{cases} x' = -x - 2y \\ y' = -x \end{cases}$ b. $\begin{cases} x' = -3x + 2y \\ y' = -4x + 3y \end{cases}$

2. a. $\begin{cases} x' = -4x - y \\ y' = 6x + y \end{cases}$ b. $\begin{cases} x' = 7x - 2y \\ y' = 12x - 3y \end{cases}$

3. a. $\begin{cases} x' = 5x - y \\ y' = 3x + y \end{cases}$ b. $\begin{cases} x' = 3x + 4y \\ y' = 2x + y \end{cases}$

4. a. $\begin{cases} x' = -7x + 5y \\ y' = -10x + 8y \end{cases}$ b. $\begin{cases} x' = -9x - 10y \\ y' = 5x + 6y \end{cases}$

5. a. $\begin{cases} x' = 10x - 6y \\ y' = 12x - 8y \end{cases}$ b. $\begin{cases} x' = -4x + y \\ y' = -2x - y \end{cases}$

6. a. $\begin{cases} x' = -5x + 6y \\ y' = -3x + 4y \end{cases}$ b. $\begin{cases} x' = 2x + y \\ y' = x + y \end{cases}$

7. a. $\begin{cases} x' = 4x + y \\ y' = 3x + 2y \end{cases}$ b. $\begin{cases} x' = 2x + 3y \\ y' = x + 2y \end{cases}$

8. a. $\begin{cases} x' = -17x + 36y \\ y' = -6x + 13y \end{cases}$ b. $\begin{cases} x' = -2x + 2y \\ y' = 3x - 2y \end{cases}$

For problems 9–13, find the eigenvalues of the coefficient matrix and use Theorem 6.1.2 to determine the stability of the origin. Do not attempt to find the general solution (the eigenvalues are all complex [non-real]).

9. a. $\begin{cases} x' = 3x - y \\ y' = x + 3y \end{cases}$ b. $\begin{cases} x' = -2y \\ y' = 2x \end{cases}$

10. a. $\begin{cases} x' = -2x + y \\ y' = -5x + 2y \end{cases}$ b. $\begin{cases} x' = -4x + 2y \\ y' = -10x + 4y \end{cases}$

11. a. $\begin{cases} x' = -5x - 2y \\ y' = 20x + 7y \end{cases}$ b. $\begin{cases} x' = -7x - 4y \\ y' = 10x + 5y \end{cases}$

12. a. $\begin{cases} x' = -6x - 2y \\ y' = 5x \end{cases}$ b. $\begin{cases} x' = -3x - 4y \\ y' = x - 3y \end{cases}$

13. a. $\begin{cases} x' = -8x - y \\ y' = -4x - 8y \end{cases}$ b. $\begin{cases} x' = y \\ y' = -10x + 6y \end{cases}$

In problems 14–17, use Matlab, Maple, or Mathematica to find the eigenvalues of the coefficient matrix and use Theorem 6.1.2 to determine the stability of the origin. If the eigenvalues are real, write the general solution.

14. a. $\begin{cases} x' = 3x + y \\ y' = 2x + 4y \\ z' = 3x - y + z \end{cases}$ b. $\begin{cases} x' = -2x + 2y \\ y' = 4x \\ z' = 3y + z \end{cases}$

15. a. $\begin{cases} x' = -x - 2y + z \\ y' = 2x - y + z \\ z' = z \end{cases}$ b. $\begin{cases} x' = 2x - y - 2z \\ y' = x + y + z \\ z' = z \end{cases}$

16. a. $\begin{cases} x' = -x + z \\ y' = -y + z \\ z' = -y - z \end{cases}$ b. $\begin{cases} x' = x + 2z \\ y' = -y \\ z' = x - y + 11z \end{cases}$

17. a. $\begin{cases} x' = -x + y - z \\ y' = -y \\ z' = x - 2z \end{cases}$ b. $\begin{cases} x' = -x + y - z \\ y' = -y \\ z' = x - z \end{cases}$

18. Use Theorem 6.1.2 to find values for a, b, c, and d so that the equilibrium solution (the origin) of

$$\frac{dx}{dt} = ax + by$$

$$\frac{dy}{dt} = cx + dy$$

is an

a. unstable node
b. stable node
c. unstable spiral
d. stable spiral
e. saddle point

19. Consider the equation $x''(t) + 4x'(t) + 3x(t) = 0$. Solve it using the methods of Section 4.1. Then use the methods of Section 3.5.1 to convert it to a system of two first-order equations in the variables $x(t)$ and $y(t) = x'(t)$. Solve the resulting system using the methods of this current Section 6.1. Compare your two answers.

20. Consider the equation $x''(t) + 2x'(t) - 8x(t) = 0$. Solve it using the methods of Section 4.1. Then use the methods of Section 3.5.1 to convert it to a system of two first-order equations in the variables $x(t)$ and $y(t) = x'(t)$. Solve the resulting system using the methods of this current Section 6.1. Compare your two answers.

6.2   A General Method, Part II: Solving Systems with Repeated Real or Complex Eigenvalues

We now continue our discussion of $\mathbf{x}' = \mathbf{A}\mathbf{x}$, begun in Section 6.1. We had assumed a solution of the form $c\mathbf{v}e^{\lambda t}$ and had concluded that λ, $\mathbf{v}$ must satisfy the eigenvalue equation

$$\mathbf{A}\mathbf{v} = \lambda \mathbf{v}.$$

We discussed the situation where we have distinct real eigenvalues. What if this is not the case? In this section, we consider the situations where we have repeated eigenvalues and then complex (non-real) eigenvalues.

6.2.1 Repeated Real Eigenvalues

We first consider the case of repeated real eigenvalues and use two specific matrices to show different scenarios.

Example 1: Find the eigenvalues and eigenvectors of

$$\mathbf{A} = \begin{pmatrix} 3 & 0 \\ 0 & 3 \end{pmatrix}, \quad \mathbf{B} = \begin{pmatrix} 3 & 2 \\ 0 & 3 \end{pmatrix}.$$

For the matrix $\mathbf{A}$, we can easily calculate the eigenvalues as $\lambda_{1,2} = 3, 3$. Substituting into $(\mathbf{A} - \lambda \mathbf{I})\mathbf{v} = \mathbf{0}$ gives

$$(\mathbf{A} - \lambda \mathbf{I})\mathbf{v} = \begin{pmatrix} 3 - 3 & 0 \\ 0 & 3 - 3 \end{pmatrix} \begin{pmatrix} v_{11} \\ v_{21} \end{pmatrix} = \mathbf{0}.$$

Because this is true for any choice of v_{11}, v_{21}, we can choose, for example, $v_{11} = 1, v_{21} = 0$ to give one eigenvector and then $v_{12} = 0, v_{22} = 1$ to give the other. Thus

$$\mathbf{v}_1 = \begin{pmatrix} 1 \\ 0 \end{pmatrix}, \quad \mathbf{v}_2 = \begin{pmatrix} 0 \\ 1 \end{pmatrix} \quad \text{for} \quad \mathbf{A} = \begin{pmatrix} 3 & 0 \\ 0 & 3 \end{pmatrix}.$$

In this situation, we see that we have *two* eigenvectors even though our eigenvalue was repeated. In general, if we can find n eigenvectors corresponding to n real eigenvalues (either distinct or repeated) for the coefficient matrix of a system of differential equations, we get lucky because we can write the general solution with the help of the following theorem.

THEOREM 6.2.1 *Consider the system* $\mathbf{x}' = \mathbf{A}\mathbf{x}$, *where the coefficient matrix* $\mathbf{A}$ *has real eigenvalue* λ *repeated* n *times. Suppose we can find* n *eigenvectors* $\mathbf{v}_1, \mathbf{v}_2, \cdots, \mathbf{v}_n$ *corresponding to this eigenvalue. Then the* $c_i \mathbf{v}_i e^{\lambda t}$, $i = 1, 2, \cdots, n$ *are linearly independent and the general solution is given by*

$$\mathbf{x} = c_1 \mathbf{v}_1 e^{\lambda t} + c_2 \mathbf{v}_2 e^{\lambda t} + \cdots + c_n \mathbf{v}_n e^{\lambda t}$$

and is defined for $t \in (-\infty, \infty)$.

Combined with Theorem 6.1.1, we see that as long as we can find n eigenvectors for a set of n real eigenvalues (regardless or whether they are distinct or repeated), we can obtain n linearly independent solutions with $c_i \mathbf{v}_i e^{\lambda t}$, $i = 1, 2, \cdots, n$. If we *cannot* find the same number of eigenvectors as the multiplicity of a repeated eigenvalue, then things get complicated. For our matrix

B, we again have $\lambda_{1,2} = 3, 3$ and substitution into $(\mathbf{B} - \lambda\mathbf{I})\mathbf{v} = \mathbf{0}$ gives

$$\begin{pmatrix} 3-3 & 2 \\ 0 & 3-3 \end{pmatrix} \begin{pmatrix} v_{11} \\ v_{21} \end{pmatrix} = \begin{pmatrix} 0 & 2 \\ 0 & 0 \end{pmatrix} \begin{pmatrix} v_{11} \\ v_{21} \end{pmatrix} = \mathbf{0}.$$

The second row of the system gives an equation that is always true, whereas the first row gives us $0 \cdot v_{11} + 2 \cdot v_{21} = 0$. This can only happen when $v_{21} = 0$. We are free to choose v_{11} to be anything (non-zero) and so we let $v_{11} = 1$. Thus we have *one* eigenvector

$$\mathbf{v}_1 = \begin{pmatrix} 1 \\ 0 \end{pmatrix} \quad \text{for} \quad \mathbf{B} = \begin{pmatrix} 3 & 2 \\ 0 & 3 \end{pmatrix}.$$

Can we find another eigenvector? The short answer is "no." We can, however, still find another vector that will help us.

If we put the above two matrices in the context of differential equations, Theorem 6.2.1 allows us to write the general solution of $\mathbf{x}' = \mathbf{A}\mathbf{x}$ as $\mathbf{x} = c_1 e^{3t}\mathbf{v}_1 + c_2 e^{3t}\mathbf{v}_2$ because we have two linearly independent eigenvectors $\mathbf{v}_1, \mathbf{v}_2$ but we cannot do the same for **B**. The reader should recall that when we had a repeated root of the characteristic equation in solving an nth order linear homogeneous constant-coefficient equation, we found a second solution by multiplying by t. Thus we had e^{rt}, te^{rt} as linearly independent solutions. Is it still the analogous situation here? To see if this is the case, let's assume that $\mathbf{x} = c_2 te^{3t}\mathbf{v}_1$ is a solution to $\mathbf{x}' = \mathbf{B}\mathbf{x}$, then substitution gives

$$c_2(e^{3t} + 3te^{3t})\mathbf{v}_1 = \mathbf{B}(c_2 te^{3t}\mathbf{v}_1)$$

$$\implies \frac{1}{te^{3t}}(e^{3t} + 3te^{3t})\mathbf{v}_1 = \mathbf{B}\mathbf{v}_1 \implies \mathbf{B}\mathbf{v}_1 = \left(\frac{1}{t} + 3\right)\mathbf{v}_1.$$

We know that $\mathbf{v}_1$ is an eigenvector of **B** and this last equation is not true because the corresponding eigenvalue is 3. Thus $c_2 te^{3t}\mathbf{v}_1$ is *not* a solution to $\mathbf{x}' = \mathbf{B}\mathbf{x}$. It turns out, however, that $te^{3t}\mathbf{v}_1$ is part of an additional solution but it is only a part of it.

If we consider the general situation where λ is a repeated eigenvalue of $\mathbf{x}' = \mathbf{B}\mathbf{x}$ and assume a solution of the form $\mathbf{x} = te^{\lambda t}\mathbf{v}_1 + f(t)\mathbf{u}_1$, where $f(t), \mathbf{u}_1$ are to be determined, we would need it to satisfy the differential equation. Substitution gives

$$(e^{\lambda t} + \lambda te^{\lambda t})\mathbf{v}_1 + f'(t)\mathbf{u}_1 = \mathbf{B}(te^{\lambda t}\mathbf{v}_1 + f(t)\mathbf{u}_1).$$

Rearranging this gives

$$(e^{\lambda t})\mathbf{v}_1 + f'(t)\mathbf{u}_1 + \lambda te^{\lambda t}\mathbf{v}_1 = \mathbf{B}(te^{\lambda t}\mathbf{v}_1) + \mathbf{B}(f(t)\mathbf{u}_1)$$

$$\implies (e^{\lambda t})\mathbf{v}_1 + f'(t)\mathbf{u}_1 - \mathbf{B}(f(t)\mathbf{u}_1) = \mathbf{B}(te^{\lambda t} - \lambda te^{\lambda t})\mathbf{v}_1$$

$$\implies (e^{\lambda t})\mathbf{v}_1 - (\mathbf{B}f(t) - f'(t)\mathbf{I})\mathbf{u}_1 = te^{\lambda t}(\mathbf{B} - \lambda\mathbf{I})\mathbf{v}_1$$

(note the right-hand side above is zero)

$$\implies (\mathbf{B}f(t) - f'(t)\mathbf{I})\mathbf{u}_1 = (e^{\lambda t})\mathbf{v}_1.$$

If we set $f(t) = e^{\lambda t}$ then this equation becomes

$$(\mathbf{B} - \lambda\mathbf{I})\mathbf{u}_1 = \mathbf{v}_1,$$

and finding a $\mathbf{u}_1$ that satisfies this last equation would then give us an additional solution, which we could verify is linearly independent. This result can be summarized in the following procedure. For simplicity, we consider only the case where we have *one* repeated eigenvalue that cannot generate the same number of eigenvectors as its multiplicity.

Algorithm for Calculating Generalized Eigenvectors

Consider an $n \times n$ matrix $\mathbf{A}$ that has an eigenvalue λ of multiplicity n with k corresponding eigenvectors $\mathbf{v}_1, \mathbf{v}_2, \cdots, \mathbf{v}_k$. We construct $n - k$ generalized eigenvectors, $\mathbf{u}_m^j$, as follows:

Set $j = 1$.

(a) Solve $(\mathbf{A} - \lambda\mathbf{I})\mathbf{u}_1^{(j)} = \mathbf{v}_j$ for $\mathbf{u}_1^{(j)}$, if possible.
If successful in finding $\mathbf{u}_1^{(j)}$, then go to part **(b)**; if not, increment j and repeat part **(a)** for this new j-value.

(b) Solve $(\mathbf{A} - \lambda\mathbf{I})\mathbf{u}_2^{(j)} = \mathbf{u}_1^{(j)}$ for $\mathbf{u}_2^{(j)}$, if possible. If successful in finding $\mathbf{u}_2^{(j)}$, then go to part **(c)**; if not, increment j and repeat part **(a)** for this new j-value.

(c) Repeat, solving $(\mathbf{A} - \lambda\mathbf{I})\mathbf{u}_{p+1}^{(j)} = \mathbf{u}_p^{(j)}$ for $\mathbf{u}_{p+1}^{(j)}$ (note that $p \geq 2$) until the equation can no longer be solved. When the equation can no longer be solved, increment j and repeat part **(a)** for this new j-value.

These steps are carried out until $n-k$ generalized eigenvectors are found.

Remark 1: It is possible that $(\mathbf{A} - \lambda\mathbf{I})\mathbf{u}_1^{(j)} = \mathbf{v}_j$ cannot be solved for one or more of the $\mathbf{v}_j$. However, there will always be at least one $\mathbf{v}_j$ that will allow the equation to be solved and all the $\mathbf{v}_j$ that do work generate a total of $n - k$ generalized eigenvectors.

Remark 2: The generalized eigenvectors are sometimes called *chain vectors*.

Remark 3: If we had additional (repeated) eigenvalues that also did not give rise to enough eigenvectors, we would simply apply the procedure to each eigenvalue in turn.

We can now use this method to show how to solve differential equations when we have repeated real eigenvalues.

THEOREM 6.2.2 *Consider the system* $\mathbf{x}' = \mathbf{A}\mathbf{x}$ *where (i)* λ *is an eigenvalue of multiplicity* n, *(ii)* $\mathbf{v}_1, \mathbf{v}_2, \cdots, \mathbf{v}_k$ *are* k *corresponding eigenvectors, and (iii)* $\mathbf{u}_1^{(k)}, \cdots, \mathbf{u}_{n-k}^{(k)}$ *are* $n - k$ *generalized eigenvectors (arising from the eigenvector* $\mathbf{v}_k$*). Set*

$$\mathbf{x}_{k+1} = te^{\lambda t}\mathbf{v}_k + e^{\lambda t}\mathbf{u}_1^{(k)}$$

$$\mathbf{x}_{k+2} = \frac{t^2}{2!}e^{\lambda t}\mathbf{v}_k + te^{\lambda t}\mathbf{u}_1^{(k)} + e^{\lambda t}\mathbf{u}_2^{(k)}$$

$$\vdots$$

$$\mathbf{x}_n = \frac{t^{n-k}}{(n-k)!}e^{\lambda t}\mathbf{v}_k + \frac{t^{n-k-1}}{(n-k-1)!}e^{\lambda t}\mathbf{u}_1^{(k)} \cdots + te^{\lambda t}\mathbf{u}_{n-k-1}^{(k)} + e^{\lambda t}\mathbf{u}_{n-k}^{(k)}.$$

Then $e^{\lambda t}\mathbf{v}_1, \cdots, e^{\lambda t}\mathbf{v}_k, \mathbf{x}_{k+1}, \cdots, \mathbf{x}_n$ *are linearly independent and the general solution can be written as*

$$\mathbf{x} = c_1 e^{\lambda t}\mathbf{v}_1 + c_2 e^{\lambda t}\mathbf{v}_2 + \cdots + c_k e^{\lambda t}\mathbf{v}_k + c_{k+1}\mathbf{x}_{k+1} + c_{k+2}\mathbf{x}_{k+2} + \cdots + c_n\mathbf{x}_n.$$

Remark 1: If we had multiple eigenvectors that gave rise to the full set of generalized eigenvectors, then the above theorem would apply to each set of eigenvectors and corresponding generalized eigenvectors.

Remark 2: If we also have distinct real eigenvalues, Theorem 6.1.1 applies to those eigenvalues and corresponding eigenvectors.

Example 2: Find the general solution of

$$\mathbf{x}' = \begin{pmatrix} -1 & 2 & -4 \\ 0 & -1 & 0 \\ 0 & 0 & -1 \end{pmatrix} \mathbf{x}.$$

Because the matrix is *upper triangular*, we can read the eigenvalues immediately from the diagonal and see that $\lambda = -1$ occurs three times. To find eigenvectors, we again need to solve $(\mathbf{A} - \lambda\mathbf{I})\mathbf{v} = \mathbf{0}$.

$$\begin{pmatrix} -1 - (-1) & 2 & -4 \\ 0 & -1 - (-1) & 0 \\ 0 & 0 & -1 - (-1) \end{pmatrix} \begin{pmatrix} v_{11} \\ v_{21} \\ v_{31} \end{pmatrix}$$

$$= \begin{pmatrix} 0 & 2 & -4 \\ 0 & 0 & 0 \\ 0 & 0 & 0 \end{pmatrix} \begin{pmatrix} v_{11} \\ v_{21} \\ v_{31} \end{pmatrix} = \begin{pmatrix} 0 \\ 0 \\ 0 \end{pmatrix}. \qquad (6.6)$$

The last two rows do not restrict our choices of the entries of $\mathbf{v}_1$. The first row gives

$$2v_{21} - 4v_{31} = 0 \implies v_{21}/v_{31} = 2.$$

We can, for example, choose $v_{21} = 2, v_{31} = 1$. We are free to do as we wish with the first entry v_{11}, say, $v_{11} = 0$. We could have also chosen $v_{21} = v_{31} =$

0 and then $v_{11} = 1$ (or any other non-zero number). Thus, we have two eigenvectors

$$\mathbf{v}_1 = \begin{pmatrix} 0 \\ 2 \\ 1 \end{pmatrix}, \quad \mathbf{v}_2 = \begin{pmatrix} 1 \\ 0 \\ 0 \end{pmatrix}.$$

We obtain the generalized eigenvector (or chain vector) by solving $(\mathbf{A} - (-1)\mathbf{I})\mathbf{u}_1 = \mathbf{v}_2$ for $\mathbf{u}_1$. (Note that we cannot solve this system if we put $\mathbf{v}_1$ on the right-hand side. Thus, part (a) of the generalized eigenvalue procedure does not yield anything when $j = 1$.) This gives us

$$2u_{21} - 4u_{31} = 1 \quad \text{where} \quad \Longrightarrow u_{21} = 1/2, u_{31} = 0$$

is one possibility. We are again free to choose u_{11} as anything, so we let $u_{11} = 0$. Thus

$$\mathbf{u}_1 = \begin{pmatrix} 0 \\ 1/2 \\ 0 \end{pmatrix}.$$

We can thus write the general solution as

$$\mathbf{x} = c_1 e^{-t}\mathbf{v}_1 + c_2 e^{-t}\mathbf{v}_2 + c_3(te^{-t}\mathbf{v}_2 + e^{-t}\mathbf{u}_1)$$

$$= c_1 e^{-t} \begin{pmatrix} 0 \\ 2 \\ 1 \end{pmatrix} + c_2 e^{-t} \begin{pmatrix} 1 \\ 0 \\ 0 \end{pmatrix} + c_3 \left(te^{-t} \begin{pmatrix} 1 \\ 0 \\ 0 \end{pmatrix} + e^{-t} \begin{pmatrix} 0 \\ 1/2 \\ 0 \end{pmatrix} \right). \quad (6.7)$$

Computer Code 6.2:   Solving a system with repeated real eigenvalues

Matlab, Maple, Mathematica

```
                          Matlab
>>  syms c1 c2 c3 t
>>  A=sym([-1,2,-4; 0,-1,0; 0,0,-1])
>>  [V,E]=eig(A)
    %output for Matlab 7.0.1 was V(:,1)=[1,0; 0,2; 0,1]
>>  eqA1=A-E(1,1)*eye(3,3)
>>  equ_v1=eqA1\V(:,1) %soln to eqA1*equ_v1=V(:,1)
>>  equ_v2=eqA1\V(:,2) %soln to eqA1*equ_v2=V(:,2)
    %only V(:,1) yields an answer and so we work with equ_v1
>>  soln=c1*exp(E(1,1)*t)*V(:,2)+c2*exp(E(1,1)*t)*V(:,1)
    +c3*(t*exp(E(1,1)*t)*V(:,1)+exp(E(1,1)*t)*(equ_v1))
```

Maple

```
> with(linalg):
> eqA:=matrix(3,3,[-1,2,-4,0,-1,0,0,0,-1]);
> eq1:=eigenvects(eqA); #output for Maple 9 was
  #[-1, 3, {[1, 0, 0], [0, 2, 1]}]
> eqA1:=evalm(eqA-eq1[1]*diag(1,1,1));
> eqv1:=matrix(3,1,eq1[3][1]); #aesthetic line only to
  #make column vector look like a column vector
  #system to be solved is eqA1*equ_v1=eqv1
> equ_v1:=linsolve(eqA1,eqv1);
> eqv2:=matrix(3,1,eq1[3][2]);
  #system to be solved is eqA1*equ_v2=eqv2
> equ_v2:=linsolve(eqA1,eqv2);
  #only equ_v1 yields an answer and we work with it
  #we can choose the constants but need to be sure
  #that a vector of all zeros doesn't result
  #The constants in Maple 9 appear as _t[1][1] and _t[1][2]
> equ1:=subs(_t[1][1]=0,_t[1][2]=0,evalm(equ_v1));
> soln:=evalm(c[1]*exp(eq1[1]*t)*eqv2+c[2]*exp(eq1[1]*t)*eqv1
  +c[3]*(t*exp(eq1[1]*t)*eqv1+exp(eq1[1]*t)*(equ1)));
```

Mathematica

```
A={{-1,2,-4},{0,-1,0},{0,0,-1}}
MatrixForm[A] (*puts matrix in nicer visual form*)
eq1=Eigensystem[A] (*output for Mathematica 5.2 was*)
   (*{{-1,-1,-1},{{0,2,1},{1,0,0},{0,0,0}}}*)
eqA1=A-eq1[[1]][[1]] IdentityMatrix[3]
eqv1=eq1[[2]][[1]]
   (*system to be solved is eqA1*equv1=eqv1*)
equv1=Solve[{eqA1.{u1,u2,u3}==eqv1},{u1,u2,u3}]
eqv2=eq1[[2]][[2]]
   (*system to be solved is eqA1*equv2=eqv2*)
equv2=Solve[{eqA1.{u1,u2,u3}==eqv2},{u1,u2,u3}]
   (*only equv2 yields an answer and we work with it*)
   (*we can choose the constants but need to be sure*)
   (*that a vector of all zeros doesn't result*)
equ1=ReplaceAll[{ReplaceAll[{u1,u2,u3},{u1-> 0,
   equv2[[1]][[1]]}]},{u3-> 0}]
(*As usual, e is entered from the palette*)
soln=c1 e^{eq1[[1]][[1]] t} eq1[[2]][[1]] + c2 e^{eq1[[1]][[2]] t} eq1[[2]][[2]] +
   c3 (t e^{eq1[[1]][[2]] t} eq1[[2]][[2]] + e^{eq1[[1]][[2]] t} equ1[[1]])
```

6.2.2 Complex Eigenvalues

In continuing our discussion of the solution of $\mathbf{x}' = \mathbf{Ax}$, we look at the final case when the eigenvalues of $\mathbf{A}$ are complex. We start off by considering

$$\mathbf{x}' = \begin{pmatrix} -1 & -2 \\ 2 & -1 \end{pmatrix} \mathbf{x}$$

and we let the coefficient matrix be denoted $\mathbf{A}$. We can easily calculate the eigenvalues as $\lambda_1 = -1 + 2i, \lambda_2 = 1 - 2i$. How do we find the corresponding eigenvectors? Just as we did when we had real eigenvalues! The first eigenvalue thus gives

$$\begin{aligned} \mathbf{0} &= (\mathbf{A} - \lambda_1 \mathbf{I})\mathbf{v}_1 \\ &= \begin{pmatrix} -1 - (-1 + 2i) & -2 \\ 2 & -1 - (-1 + 2i) \end{pmatrix} \begin{pmatrix} v_{11} \\ v_{21} \end{pmatrix} \\ &= \begin{pmatrix} -2i & -2 \\ 2 & -2i \end{pmatrix} \begin{pmatrix} v_{11} \\ v_{21} \end{pmatrix}. \end{aligned}$$

At this point, in the case of real distinct eigenvalues, we always had the situation where both rows gave us the same information. That is also the case now and we can see this by multiplying the second row by $-i$, for example. Thus it doesn't matter which row we choose, so let's consider the first row:

$$-2iv_{11} - 2v_{21} = 0.$$

Our only restriction is that we can't have $v_{11} = v_{21} = 0$. One possibility is thus $v_{11} = 1, v_{21} = -i$ and this gives us the eigenvector ($\mathbf{v}_1$) corresponding to $\lambda_1 = -1 + 2i$. If we repeat the above steps for $\lambda_2 = -1 - 2i$, we could obtain $v_{12} = 1, v_{22} = i$. Thus we have

$$\mathbf{v}_1 = \begin{pmatrix} 1 \\ -i \end{pmatrix}, \quad \mathbf{v}_2 = \begin{pmatrix} 1 \\ i \end{pmatrix}.$$

As you might hope/expect, it is *not* a coincidence that the eigenvectors are complex conjugates of each other. Because we had assumed a solution of the form $c\mathbf{v}e^{\lambda t}$, we thus have that $c_1 \mathbf{v}_1 e^{\lambda_1 t}, c_2 \mathbf{v}_2 e^{\lambda_2 t}$ are solutions. That is,

$$c_1 \begin{pmatrix} 1 \\ -i \end{pmatrix} e^{(-1+2i)t}, \quad c_2 \begin{pmatrix} 1 \\ i \end{pmatrix} e^{(-1-2i)t}$$

are two linearly independent solutions of

$$\mathbf{x}' = \begin{pmatrix} -1 & -2 \\ 2 & -1 \end{pmatrix} \mathbf{x}.$$

Although this is true, the non-appealing aspect is that both solutions are complex-valued even though our original system had only real entries. We had a similar situation occurring in Section 4.1 when we obtained complex-valued solutions to a second-order equation. In that section, we used Euler's formula to write a real-valued solution in terms of sines and cosines and that is exactly what we will do now.

To keep things in general terms, let us suppose that a given system has eigenvalues $\lambda_1 = \alpha + i\beta, \lambda_2 = \alpha - i\beta$ and the corresponding (complex) eigenvectors are $\mathbf{v}_1, \mathbf{v}_2$, where α, β are both real. We know that $e^{\lambda_1 t}\mathbf{v}_1$ and $e^{\lambda_2 t}\mathbf{v}_2$ are both solutions to $\mathbf{x}' = \mathbf{A}\mathbf{x}$ and we will use Euler's formula to rewrite them:

$$e^{\lambda_1 t}\mathbf{v}_1 = e^{(\alpha+i\beta)t}\mathbf{v}_1 = e^{\alpha t}(\cos \beta t + i \sin \beta t)\mathbf{v}_1$$
$$e^{\lambda_2 t}\mathbf{v}_2 = e^{(\alpha-i\beta)t}\mathbf{v}_2 = e^{\alpha t}(\cos \beta t - i \sin \beta t)\mathbf{v}_2.$$

Theorem 5.2.2 tells us that *any* linear combination of solutions to $\mathbf{x}' = \mathbf{A}(t)\mathbf{x}$ is still a solution. We also introduce real-valued vectors $\mathbf{a}, \mathbf{b}$ that represent the real and complex parts of the eigenvectors, i.e., $\mathbf{v}_1 = \mathbf{a} + i\mathbf{b}$ (and thus $\mathbf{v}_2 = \mathbf{a} - i\mathbf{b}$). Using these facts we can rewrite a complex-valued solution:

$$k_1 e^{(\alpha+i\beta)t}\mathbf{v}_1 + k_2 e^{(\alpha-i\beta)t}\mathbf{v}_2$$

$$= k_1 e^{\alpha t}(\cos \beta t + i \sin \beta t)\mathbf{v}_1 + k_2 e^{\alpha t}(\cos \beta t - i \sin \beta t)\mathbf{v}_2$$

$$= k_1 e^{\alpha t}(\cos \beta t + i \sin \beta t)(\mathbf{a} + i\mathbf{b}) + k_2 e^{\alpha t}(\cos \beta t - i \sin \beta t)(\mathbf{a} - i\mathbf{b})$$

$$= k_1 e^{\alpha t}\left[\mathbf{a}(\cos \beta t) + i\mathbf{a}(\sin \beta t) + i\mathbf{b}(\cos \beta t) - \mathbf{b}(\sin \beta t)\right]$$
$$\quad + k_2 e^{\alpha t}\left[\mathbf{a}(\cos \beta t) - i\mathbf{a}(\sin \beta t) - i\mathbf{b}(\cos \beta t) - \mathbf{b}(\sin \beta t)\right]$$

$$= e^{\alpha t}\{[k_1\mathbf{a}(\cos \beta t) - k_1\mathbf{b}(\sin \beta t) + k_2\mathbf{a}(\cos \beta t) - k_2\mathbf{b}(\sin \beta t)]$$
$$\quad + i\left[k_1\mathbf{a}(\sin \beta t) + k_1\mathbf{b}(\cos \beta t) - k_2\mathbf{a}(\sin \beta t) - k_2\mathbf{b}(\cos \beta t)\right]\}$$

$$= e^{\alpha t}\{[(k_1 + k_2)\mathbf{a}(\cos \beta t) + (-k_2 - k_1)\mathbf{b}(\sin \beta t)]$$
$$\quad + i\left[(k_1 - k_2)\mathbf{a}(\sin \beta t) + (k_1 - k_2)\mathbf{b}(\cos \beta t)\right]\}$$

$$= e^{\alpha t}\{[c_1\mathbf{a}(\cos \beta t) - c_1\mathbf{b}(\sin \beta t)] + [c_2\mathbf{a}(\sin \beta t) + c_2\mathbf{b}(\cos \beta t)]\},$$

where the constants c_1, c_2 are related to the k_i by $c_1 = k_1 + k_2, c_2 = i(k_1 - k_2)$. The constants are arbitrary and we thus have a real-valued formulation of the original solution. We state a theorem for the 2×2 case summarizing these results.

THEOREM 6.2.3 *Consider the* 2×2 *system* $\mathbf{x}' = \mathbf{A}\mathbf{x}$ *whose coefficient matrix has eigenvalues* $\lambda_1 = \alpha + i\beta, \lambda_2 = \alpha - i\beta,$ *with* α, β *real numbers. Choose one of the eigenvectors and write it as* $\mathbf{v}_j = \mathbf{a} + i\mathbf{b},$ *where* $\mathbf{a}, \mathbf{b}$ *are real-valued vectors. Then*

$$\mathbf{x}_1 = e^{\alpha t} \left(\mathbf{a}(\cos \beta t) - \mathbf{b}(\sin \beta t) \right)$$
$$\mathbf{x}_2 = e^{\alpha t} \left(\mathbf{a}(\sin \beta t) + \mathbf{b}(\cos \beta t) \right) \tag{6.8}$$

are two linearly independent solutions, defined for $-\infty < t < \infty$. *The general solution is a linear combination of these two:*

$$\mathbf{x} = e^{\alpha t} \left\{ c_1 \left(\mathbf{a}(\cos \beta t) - \mathbf{b}(\sin \beta t) \right) + c_2 \left(\mathbf{a}(\sin \beta t) + \mathbf{b}(\cos \beta t) \right) \right\} \tag{6.9}$$

Example 3: Write the real-valued general solution to

$$\mathbf{x}' = \begin{pmatrix} -1 & -2 \\ 2 & -1 \end{pmatrix} \mathbf{x}.$$

This was the example that we considered at the beginning of this discussion. We found that $\lambda_1 = -1 + 2i, \lambda_2 = -1 - 2i,$ with corresponding eigenvectors

$$\mathbf{v}_1 = \begin{pmatrix} 1 \\ -i \end{pmatrix}, \quad \mathbf{v}_2 = \begin{pmatrix} 1 \\ i \end{pmatrix}.$$

We rewrite one of the eigenvectors in terms of the real vectors $\mathbf{a}, \mathbf{b}$:

$$\mathbf{v}_1 = \mathbf{a} + i\mathbf{b} \implies \mathbf{a} = \begin{pmatrix} 1 \\ 0 \end{pmatrix}, \quad \mathbf{b} = \begin{pmatrix} 0 \\ -1 \end{pmatrix}.$$

With $\alpha = -1, \beta = 2$, Theorem 6.2.3 states that the general solution is

$$\mathbf{x} = e^{-t} \left\{ c_1 \left(\mathbf{a}(\cos 2t) - \mathbf{b}(\sin 2t) \right) + c_2 \left(\mathbf{a}(\sin 2t) + \mathbf{b}(\cos 2t) \right) \right\}$$
$$= e^{-t} \left\{ c_1 \left(\begin{pmatrix} 1 \\ 0 \end{pmatrix} (\cos 2t) - \begin{pmatrix} 0 \\ -1 \end{pmatrix} \right) (\sin 2t) \right.$$
$$\left. + c_2 \left(\begin{pmatrix} 1 \\ 0 \end{pmatrix} (\sin 2t) + \begin{pmatrix} 0 \\ -1 \end{pmatrix} (\cos 2t) \right) \right\}$$

Computer Code 6.3:   **Solving a system with complex (non-real) eigenvalues**

Matlab, Maple, Mathematica

Matlab

```
>> syms t c1 c2
>> A=[-1,-2; 2,-1]
>> [v,d]=eig(sym(A))
>> real(v(:,1))
>> a_vec=real(v(:,1))
>> b_vec=imag(v(:,1))
>> alpha=real(d(1,1))%real part of 1st eigenvalue
>> beta=imag(d(1,1))%imaginary part of 1st eigenvalue
>> solnx1=exp(alpha*t)*(a_vec*cos(beta*t)-b_vec*sin(beta*t))
>> solnx2=exp(alpha*t)*(a_vec*sin(beta*t)+b_vec*cos(beta*t))
>> %solnx1 and solnx2 are from eq.(6.8)
>> soln=c1*solnx1+c2*solnx2
```

Maple

```
> with(linalg):
> eqA:=matrix(2,2,[-1,-2,2,-1]);
> eq1:=eigenvects(eqA);
> eq1[1][3][1][1];#first entry of first eigenvector
> eq1[1][3][1][2];#second entry of first eigenvector
> avec:=matrix(2,1,[Re(eq1[1][3][1][1]),Re(eq1[1][3][1][2])]);
> bvec:=matrix(2,1,[Im(eq1[1][3][1][1]),Im(eq1[1][3][1][2])]);
> alpha:=Re(eq1[1][1]);#real part of 1st eigenvalue
> beta:=Im(eq1[1][1]);#imaginary part of 1st eigenvalue
> solnx1:=evalm(exp(alpha*t)*(avec*cos(beta*t)
   -bvec*sin(beta*t)));
> solnx2:=evalm(exp(alpha*t)*(avec*sin(beta*t)
   +bvec*cos(beta*t)));
> #solnx1 and solnx2 are from eq.(6.8)
> soln:=evalm(c[1]*solnx1+c[2]*solnx2);
```

Mathematica

```
A={{-1,-2},{2,-1}}
MatrixForm[A] (*puts matrix in nicer visual form*)
eq1=Eigensystem[A]
eq1[[2]][[1]][[1]] (*first entry of eigenvector*)
eq1[[2]][[1]][[2]] (*second entry of eigenvector*)
avec=Re[eq1[[2]][[1]]]
bvec=Im[eq1[[2]][[1]]]
```
α =Re[eq1[[1]][[1]]] (*real part of 1st eigenvalue*)
β =Im[eq1[[1]][[1]]] (*imaginary part of 1st eigenvalue*)
$\text{solnx1} = e^{\alpha t} \, \text{avec} \, \text{Cos}[\beta t] - \text{bvec} \, \text{Sin}[\beta t]$
$\text{solnx2} = e^{\alpha t} \, \text{avec} \, \text{Sin}[\beta t] + \text{bvec} \, \text{Cos}[\beta t]$
```
(*solnx1 and solnx2 are from eq.(6.8)*)
```
$\text{soln} = \text{c1} \, \text{solnx1} + \text{c2} \, \text{solnx2}$

We finish this section by tying together the techniques we have employed.

THEOREM 6.2.4 *Suppose that* $\mathbf{x}_1, \mathbf{x}_2, \cdots, \mathbf{x}_n$ *form a fundamental set of solutions for the linear homogeneous system* $\mathbf{x}' = \mathbf{A}\mathbf{x}$. *The general solution to this system is a linear combination of n linearly independent solutions*

$$\mathbf{x} = c_1\mathbf{x}_1 + c_2\mathbf{x}_2 + \cdots + c_n\mathbf{x}_n.$$

We saw an analogous statement in Theorem 5.2.3 but at that time we did not know how to compute any solutions. The previous theorems show us how to obtain a fundamental set of solutions and then we apply Theorem 6.2.4. As we have seen, obtaining a fundamental set of solutions may be very easy or quite complicated.

Problems

For problems 1–5, find the eigenvalues and eigenvectors of the coefficient matrix by hand (the eigenvalues are all complex). Then use Theorem 6.2.3 and Theorem 6.2.4 to find the general solution.

1. a. $\begin{cases} x' = 3x - y \\ y' = x + 3y \end{cases}$, b. $\begin{cases} x' = -2y \\ y' = 2x \end{cases}$

2. a. $\begin{cases} x' = -2x + y \\ y' = -5x + 2y \end{cases}$, b. $\begin{cases} x' = -4x + 2y \\ y' = -10x + 4y \end{cases}$

3. a. $\begin{cases} x' = -5x - 2y \\ y' = 20x + 7y \end{cases}$, b. $\begin{cases} x' = -7x - 4y \\ y' = 10x + 5y \end{cases}$

4. a. $\begin{cases} x' = -6x - 2y \\ y' = 5x \end{cases}$, b. $\begin{cases} x' = -3x - 4y \\ y' = x - 3y \end{cases}$

5. a. $\begin{cases} x' = -8x - y \\ y' = -4x - 8y \end{cases}$,   b. $\begin{cases} x' = y \\ y' = -10x + 6y \end{cases}$

For problems 6–10, find the eigenvalues and eigenvector of the coefficient matrix by hand (the eigenvalues are all repeated with only one eigenvector). Use the algorithm in Section 6.2.1 to obtain a generalized eigenvector. Then use Theorem 6.2.2 and Theorem 6.2.4 to write the general solution.

6. a. $\begin{cases} x' = x + 2y \\ y' = y \end{cases}$,   b. $\begin{cases} x' = -2x + 3y \\ y' = -2y \end{cases}$

7. a. $\begin{cases} x' = 7x - 2y \\ y' = 8x - y \end{cases}$,   b. $\begin{cases} x' = x - y \\ y' = 4x - 3y \end{cases}$

8. a. $\begin{cases} x' = -2x + y \\ y' = -x \end{cases}$,   b. $\begin{cases} x' = 2x + y \\ y' = -x \end{cases}$

9. a. $\begin{cases} x' = -x + 4y \\ y' = -y \end{cases}$,   b. $\begin{cases} x' = x - 2y \\ y' = 2x - 3y \end{cases}$

10. a. $\begin{cases} x' = -x - 2y \\ y' = 2x + 3y \end{cases}$,   b. $\begin{cases} x' = 8x + 9y \\ y' = -4x - 4y \end{cases}$

In problems 11–15, use Matlab, Maple, or Mathematica to find the eigenvalues and eigenvectors of the coefficient matrix (there is a complex pair). Then use Theorem 6.2.3 and Theorem 6.2.4 to find the general solution.

11. $\begin{cases} x' = -x - 2y + z \\ y' = 2x - y + z \\ z' = z \end{cases}$

12. $\begin{cases} x' = 2x - y - 2z \\ y' = x + y + z \\ z' = z \end{cases}$

13. $\begin{cases} x' = -x + y - z \\ y' = -y \\ z' = x - z \end{cases}$

14. $\begin{cases} x' = -x + z \\ y' = -y + z \\ z' = -y - z \end{cases}$

15. $\begin{cases} x' = -x + y - z \\ y' = -y \\ z' = x - 2z \end{cases}$

For problems 16–18, use Matlab, Maple, or Mathematica to find the eigenvalues and eigenvector of the coefficient matrix (there is a repeated eigenvalue with only one corresponding eigenvector). Use the algorithm in Section 6.2.1 to obtain a generalized eigenvector. Then use Theorem 6.2.2 and Theorem 6.2.4 to write the general solution.

16. $\begin{cases} x' = x - y - z \\ y' = x + 3y + z \\ z' = z \end{cases}$

17. $\begin{cases} x' = 3x \\ y' = 3y \\ z' = \frac{8}{3}x - \frac{4}{3}y - z \end{cases}$

18. $\begin{cases} x' = -x + y + z \\ y' = 3y + z \\ z' = -4y - z \end{cases}$

19. Consider the system

$$\frac{d\mathbf{x}}{dt} = \begin{pmatrix} 1 & 2 & 0 & 0 & 0 & 0 \\ 0 & 2 & 1 & 0 & 0 & 0 \\ 0 & 0 & 2 & 1 & 0 & 0 \\ 0 & 0 & 0 & 2 & 0 & 0 \\ 0 & 0 & 0 & 0 & -1 & 2 \\ 0 & 0 & 0 & 0 & -2 & -1 \end{pmatrix} \mathbf{x}.$$

Write the general solution for the given system of differential equations.

6.3 Solving Linear Homogeneous and Nonhomogeneous Systems of Equations

The reader should recall that Section 3.5.1 showed that **any** nth order equation can be converted to a system of n first-order equations. Thus the methods that we consider in this section apply to any linear differential equation, regardless of whether it has constant or variable coefficients or whether it is homogeneous or nonhomogeneous. The matrix exponential will play a key role in this but we will also see two other familiar techniques arise.

6.3.1 Fundamental Matrices and the Matrix Exponential

We consider the homogeneous system in the form

$$\mathbf{x}' = \mathbf{A}\mathbf{x}.$$

We learned in Theorem 6.2.4 that if the vectors $\mathbf{x}_1(t), \mathbf{x}_2(t), \cdots, \mathbf{x}_n(t)$ form a fundamental set of solutions on (a, b), we can write the general solution as a linear combination of these vectors. If we put these vectors as the columns of a matrix, with $\mathbf{x}_i = (x_{1i} \quad x_{2i} \quad \cdots \quad x_{ni})^T$, we have

$$\boldsymbol{\Phi}(t) = \begin{pmatrix} x_{11}(t) & x_{12}(t) & \dots & x_{1n}(t) \\ x_{21}(t) & x_{22}(t) & \dots & x_{2n}(t) \\ \vdots & & \ddots & \vdots \\ x_{n1}(t) & x_{n2}(t) & \dots & x_{nn}(t) \end{pmatrix}, \tag{6.10}$$

which we call a **fundamental matrix**. If we let $\mathbf{c} = (c_1 \quad c_2 \quad \dots \quad c_n)^T$, we can write the general solution of our homogeneous linear system as

$$\mathbf{x} = \boldsymbol{\Phi}(t)\mathbf{c}. \tag{6.11}$$

If we are given an initial condition for the system $\mathbf{x}(t_0) = \mathbf{x}_0$, where $a < t_0 < b$, we must have that

$$\mathbf{x}_0 = \boldsymbol{\Phi}(t_0)\mathbf{c}.$$

Because the matrix has linearly independent columns regardless of t (because the columns are the vectors in the fundamental set of solutions), we know $\boldsymbol{\Phi}(t)$ is always invertible (nonsingular). Thus

$$\mathbf{c} = \boldsymbol{\Phi}^{-1}(t_0)\mathbf{x}_0.$$

Substituting into (6.11) gives

$$\mathbf{x} = \boldsymbol{\Phi}(t)\boldsymbol{\Phi}^{-1}(t_0)\mathbf{x}_0. \tag{6.12}$$

What is the significance of this? In solving systems of equations using the matrix exponential, we had not yet considered a problem with an initial condition. Noting that $\mathbf{c}e^{\mathbf{A}t}$ is also a solution to $\mathbf{x}' = \mathbf{A}\mathbf{x}$ for any constant vector $\mathbf{c}$ and also noting that $e^{\mathbf{A}0} = \mathbf{I}$, we can see that

$$\mathbf{x}' = \mathbf{A}\mathbf{x}, \ \mathbf{x}(t_0) = \mathbf{x}_0 \text{ has solution } e^{\mathbf{A}t}\mathbf{x}_0.$$

Because solutions must be unique for the constant coefficient system, we have

$$e^{\mathbf{A}t} = \boldsymbol{\Phi}(t)\boldsymbol{\Phi}^{-1}(t_0). \tag{6.13}$$

This is another way to calculate the matrix exponential, even though we would want to avoid using this method in practice due to the computation involved in calculating a matrix inverse. More importantly, this equality shows us how the vectors in a fundamental set of solutions relate to the matrix exponential.

6.3.2   Constant Coefficient Systems

In this subsection, we consider first-order systems of the form

$$\mathbf{x}' = \mathbf{A}\mathbf{x} + \mathbf{f}(t), \tag{6.14}$$

where $\mathbf{A}$ has constant entries and $\mathbf{f}$ is a vector with continuous functions as its entries.

Diagonalization

Appendix C.3.1 discusses constant matrices that can be *diagonalized*. In the situations where this can be done, we write the eigenvectors as columns of a matrix $\mathbf{V}$ and observe that $\mathbf{\Lambda} = \mathbf{V}^{-1}\mathbf{A}\mathbf{V}$ is a diagonal matrix. If we define a new variable $\mathbf{y}$ by

$$\mathbf{x} = \mathbf{V}\mathbf{y},$$

then the constant coefficient nonhomogeneous system (6.14) can be written as

$$\mathbf{V}\mathbf{y}' = \mathbf{A}\mathbf{V}\mathbf{y} + \mathbf{f}(t) \quad \Longrightarrow \quad \mathbf{y}' = \mathbf{V}^{-1}\mathbf{A}\mathbf{V}\mathbf{y} + \mathbf{V}^{-1}\mathbf{f}(t) = \mathbf{\Lambda}\mathbf{y} + \mathbf{V}^{-1}\mathbf{f}(t).$$

This gives a system of *uncoupled equations*, each of which is linear and first-order and thus can be solved by methods of Section 1.5.

Example 1: Solve

$$\mathbf{x}' = \begin{pmatrix} 1 & -4 \\ -2 & -1 \end{pmatrix} \mathbf{x} + \begin{pmatrix} -\sin t \\ e^t \end{pmatrix}.$$

We can use our computer programs to find the eigenvalues and eigenvectors as

$$\lambda_1 = -3, \quad \lambda_2 = 3, \quad \mathbf{v}_1 = \begin{pmatrix} 1 \\ 1 \end{pmatrix}, \quad \mathbf{v}_2 = \begin{pmatrix} -2 \\ 1 \end{pmatrix}.$$

Creating our matrix whose columns are these eigenvectors and calculating its inverse gives

$$\mathbf{V} = \begin{pmatrix} 1 & -2 \\ 1 & 1 \end{pmatrix} \quad \Longrightarrow \quad \mathbf{V}^{-1} = \frac{1}{3}\begin{pmatrix} 1 & 2 \\ -1 & 1 \end{pmatrix}.$$

By defining a new variable $\mathbf{y} = \mathbf{V}^{-1}\mathbf{x}$, our system becomes

$$\begin{aligned} \mathbf{y}' &= \mathbf{V}^{-1}\mathbf{A}\mathbf{V}\mathbf{y} + \mathbf{V}^{-1}\mathbf{f}(t) \\ &= \begin{pmatrix} -3 & 0 \\ 0 & 3 \end{pmatrix} \mathbf{y} + \frac{1}{3}\begin{pmatrix} 1 & 2 \\ -1 & 1 \end{pmatrix}\begin{pmatrix} -\sin t \\ e^t \end{pmatrix} \\ &= \begin{pmatrix} -3 & 0 \\ 0 & 3 \end{pmatrix} \mathbf{y} + \frac{1}{3}\begin{pmatrix} -\sin t + 2e^t \\ \sin t + e^t \end{pmatrix}. \end{aligned} \qquad (6.15)$$

This gives the uncoupled system of linear first-order equations

$$y_1' + 3y = \frac{1}{3}(-\sin t + 2e^t)$$

$$y_2' - 3y = \frac{1}{3}(\sin t + e^t).$$

This gives

$$y_1 = \frac{1}{30}\left(\cos t - 3\sin t\right) + \frac{1}{2}e^t + c_1 e^{-3t}$$

$$y_2 = \frac{1}{30}\left(-\cos t - 3\sin t\right) - \frac{1}{2}e^t + c_2 e^{3t}.$$

We still need to write this answer in terms of the original variables. Doing so gives us

$$
\begin{aligned}
\mathbf{x} = \mathbf{V}\mathbf{y} &= \begin{pmatrix} 1 & -2 \\ 1 & 1 \end{pmatrix}\begin{pmatrix} y_1 \\ y_2 \end{pmatrix} \\
&= \begin{pmatrix} y_1 - 2y_2 \\ y_1 + y_2 \end{pmatrix} \\
&= \begin{pmatrix} \frac{1}{10}\left(\cos t + \sin t\right) + \frac{1}{2}e^t + e^{-3t}c_1 - 2e^{3t}c_2 \\ \frac{-1}{5}\sin t + e^{-3t}c_1 + e^{3t}c_2 \end{pmatrix} \\
&= c_1 e^{-3t}\begin{pmatrix} 1 \\ 1 \end{pmatrix} + c_2 e^{3t}\begin{pmatrix} -2 \\ 1 \end{pmatrix} \\
&\quad + e^t \begin{pmatrix} 1/2 \\ 0 \end{pmatrix} + \cos t\begin{pmatrix} 1/10 \\ 0 \end{pmatrix} + \sin t\begin{pmatrix} 1/10 \\ -1/5 \end{pmatrix}.
\end{aligned}
\tag{6.16}
$$

The form of this solution should look familiar in that it is composed of the complementary solution (i.e., the solution to the system when $\mathbf{f}(t) = 0$) and a particular solution that is determined by the function $\mathbf{f}(t)$.

Matrix Exponential

In Section 1.5 we saw that a linear system of the form $x' + ax = f(t)$ has the solution

$$ce^{at} + e^{at}\int e^{-at}f(t)dt.$$

It can be shown that an analogous formulation of this solution exists in the case of (6.14). Thus $\mathbf{x}' = \mathbf{A}\mathbf{x} + \mathbf{f}(t)$ has solution

$$\mathbf{x} = e^{\mathbf{A}t}\mathbf{c} + e^{\mathbf{A}t}\int e^{-\mathbf{A}t}\mathbf{f}(t)dt. \tag{6.17}$$

(It is easiest to understand the derivation of this solution in terms of a general fundamental matrix and the method of variation of parameters, which we examine next.) If we also have an initial condition, the solution can be written to take this into account. Thus

$$\mathbf{x}' = \mathbf{A}\mathbf{x} + \mathbf{f}(t), \quad \mathbf{x}(t_0) = \mathbf{x}_0 \tag{6.18}$$

has solution

$$\mathbf{x} = e^{\mathbf{A}(t-t_0)}\mathbf{x}_0 + e^{\mathbf{A}t}\int_{t_0}^{t} e^{-\mathbf{A}s}\mathbf{f}(s)ds. \tag{6.19}$$

We again note that the form of this solution is composed of a complementary solution and a particular solution.

Example 2: Use the matrix exponential to solve

$$\mathbf{x}' = \begin{pmatrix} 1 & -4 \\ -2 & -1 \end{pmatrix}\mathbf{x} + \begin{pmatrix} -\sin t \\ e^t \end{pmatrix}.$$

We can calculate the matrix exponential by the methods of the previous section or by using the computer. If we choose the latter, we see that

$$e^{\mathbf{A}t} = \frac{1}{3}\begin{pmatrix} e^{-3t} + 2e^{3t} & 2e^{-3t} - 2e^{3t} \\ e^{-3t} - e^{3t} & 2e^{-3t} + e^{3t} \end{pmatrix}.$$

We also need to calculate $\int e^{-\mathbf{A}t}\mathbf{f}(t)dt$. We note that $e^{-\mathbf{A}t}$ is easily calculated from $e^{\mathbf{A}t}$ by simply replacing t with $-t$. Then we have

$$\int e^{-\mathbf{A}t}\mathbf{f}(t)dt$$

$$= \int \frac{1}{3}\begin{pmatrix} e^{3t} + 2e^{-3t} & 2e^{3t} - 2e^{-3t} \\ e^{3t} - e^{-3t} & 2e^{3t} + e^{-3t} \end{pmatrix}\begin{pmatrix} -\sin t \\ e^t \end{pmatrix}$$

$$= \int \frac{1}{3}\begin{pmatrix} -2e^{-3t}\sin t - e^{3t}\sin t + 2e^{4t} - 2e^{-2t} \\ -e^{3t}\sin t + e^{-3t}\sin t + e^{-2t} + 2e^{4t} \end{pmatrix}$$

$$= \frac{1}{30}\begin{pmatrix} 2e^{-3t}\cos t + 6e^{-3t}\sin t + e^{3t}\cos t - 3e^{3t}\sin t + 5e^{4t} + 10e^{-2t} \\ e^{3t}\cos t - 3e^{3t}\sin t - e^{-3t}\cos t - 3e^{-3t}\sin t - 5e^{-2t} + 5e^{4t} \end{pmatrix}.$$

We then left multiply by $e^{\mathbf{A}t}$ and obtain, after much simplification,

$$\frac{1}{10}\begin{pmatrix} \cos t + \sin t + 5e^t \\ -2\sin t \end{pmatrix}.$$

We add this last vector to the product of $e^{\mathbf{A}t}$ with an arbitrary vector. In the formula, we gave the arbitrary constant vector as $\mathbf{c}$ but for comparison purposes (with the previous example), we let $\mathbf{k}$ be our arbitrary constant vector.

$$\mathbf{x} = \frac{1}{3}\begin{pmatrix} e^{-3t} + 2e^{3t} & 2e^{-3t} - 2e^{3t} \\ e^{-3t} - e^{3t} & 2e^{-3t} + e^{3t} \end{pmatrix}\begin{pmatrix} k_1 \\ k_2 \end{pmatrix} + \frac{1}{10}\begin{pmatrix} \cos t + \sin t + 5e^t \\ -2\sin t \end{pmatrix}$$

$$= \frac{1}{3}\begin{pmatrix} (k_1 + 2k_2)e^{-3t} + (2k_1 - 2k_2)e^{3t} \\ (k_1 + 2k_2)e^{-3t} + (-k_1 + k_2)e^{3t} \end{pmatrix} + \frac{1}{10}\begin{pmatrix} \cos t + \sin t + 5e^t \\ -2\sin t \end{pmatrix}.$$

This solution is completely correct. However, so as to have the solution in the form given in Example 1, we let $c_1 = (k_1 + 2k_2)/3$ and $c_2 = (-k_1 + k_2)/3$.

Then our answer can be written as

$$\mathbf{x} = c_1 e^{-3t} \begin{pmatrix} 1 \\ 1 \end{pmatrix} + c_2 e^{3t} \begin{pmatrix} -2 \\ 1 \end{pmatrix}$$

$$+ e^t \begin{pmatrix} 1/2 \\ 0 \end{pmatrix} + \cos t \begin{pmatrix} 1/10 \\ 0 \end{pmatrix} + \sin t \begin{pmatrix} 1/10 \\ -1/5 \end{pmatrix}. \qquad (6.20)$$

6.3.3   Non-Constant Coefficient Systems

When we considered an nth order equation with non-constant coefficients, we stated that there was no general formula for calculating a fundamental set of solutions. However, *if we somehow obtained* a fundamental set of solutions, we could solve the nonhomogeneous equation by using variation of parameters. The *method of variation of parameters* extends to first-order systems of equations. Unfortunately, so does the fact that there is no general formula for obtaining a fundamental set of solutions. Thus we will be able to solve a nonhomogeneous problem only if we somehow can obtain a fundamental set of solutions. In the case when the coefficients of the matrix are periodic, there is a technique called *Floquet theory* that can help us in this quest for a fundamental set. But in general, we will not be so lucky. Let us begin with a system of the form

$$\mathbf{x}' = \mathbf{A}(t)\mathbf{x} + \mathbf{f}(t). \qquad (6.21)$$

We will assume that we somehow are able to obtain a fundamental matrix $\mathbf{\Phi}(t)$ for the homogeneous system (i.e., when $\mathbf{f}(t) = \mathbf{0}$). Analogous to Section 4.7, we use variation of parameters to try to obtain a particular solution. That is, we try to find a vector $\mathbf{u}(t)$ such that

$$\mathbf{x}_p = \mathbf{\Phi}(t)\mathbf{u}(t)$$

is a (particular) solution. We note that the dimensions of $\mathbf{\Phi}$ and $\mathbf{u}$ require that the order of the multiplication is as given here. We will need the derivative of this to substitute into (6.21). Calculating this gives

$$\mathbf{x}'_p = \mathbf{\Phi}'(t)\mathbf{u}(t) + \mathbf{\Phi}(t)\mathbf{u}'(t),$$

and we again note that the order of the multiplication matters. Substitution into (6.21) gives us

$$\mathbf{\Phi}'(t)\mathbf{u}(t) + \mathbf{\Phi}(t)\mathbf{u}'(t) = \mathbf{A}(t)\mathbf{\Phi}(t)\mathbf{u}(t) + \mathbf{f}(t). \qquad (6.22)$$

We can arrange this as

$$\left(\mathbf{\Phi}'(t) - \mathbf{A}(t)\mathbf{\Phi}(t)\right)\mathbf{u}(t) + \mathbf{\Phi}(t)\mathbf{u}'(t) = \mathbf{f}(t).$$

Because $\mathbf{\Phi}(t)$ is a solution, this simplifies to

$$\mathbf{\Phi}(t)\mathbf{u}'(t) = \mathbf{f}(t), \qquad (6.23)$$

which we can solve for $\mathbf{u}'(t)$ since $\mathbf{\Phi}(t)$ is a fundamental matrix (and thus is invertible):

$$\mathbf{u}'(t) = \mathbf{\Phi}^{-1}(t)\mathbf{f}(t). \tag{6.24}$$

We can, in theory, integrate both sides of this last equation and thus obtain $\mathbf{u}(t)$. In practice, this integral may be difficult to evaluate but we will still be able to write our solution with an integral in it. Integrating gives

$$\mathbf{u}(t) = \int \mathbf{\Phi}^{-1}(t)\mathbf{f}(t) \ dt, \tag{6.25}$$

and thus the particular solution is of the form

$$\mathbf{x}_p(t) = \mathbf{\Phi}(t) \int \mathbf{\Phi}^{-1}(t)\mathbf{f}(t) \ dt. \tag{6.26}$$

The general solution to (6.21) can be written as

$$\mathbf{x}(t) = \mathbf{\Phi}(t)\mathbf{c} + \mathbf{\Phi}(t) \int \mathbf{\Phi}^{-1}(t)\mathbf{f}(t) \ dt. \tag{6.27}$$

We now use variation of parameters to solve the problems of Examples 1 and 2.

Example 3: Use variation of parameters to solve

$$\mathbf{x}' = \begin{pmatrix} 1 & -4 \\ -2 & -1 \end{pmatrix} \mathbf{x} + \begin{pmatrix} -\sin t \\ e^t \end{pmatrix}.$$

In Example 1, we found the eigenvalues and eigenvectors to be

$$\lambda_1 = -3, \quad \lambda_2 = 3, \quad \mathbf{v}_1 = \begin{pmatrix} 1 \\ 1 \end{pmatrix}, \quad \mathbf{v}_2 = \begin{pmatrix} -2 \\ 1 \end{pmatrix}.$$

We can use these to create a fundamental matrix

$$\mathbf{\Phi} = \begin{pmatrix} e^{-3t} & -2e^{3t} \\ e^{-3t} & e^{3t} \end{pmatrix}.$$

The inverse is easily calculated because we are in the 2×2 case:

$$\mathbf{\Phi}(t) = \frac{1}{3} \begin{pmatrix} e^{3t} & 2e^{3t} \\ -e^{-3t} & e^{-3t} \end{pmatrix}.$$

Applying the formula for the solution and getting some help from our computer programs, we have

$$\mathbf{\Phi}^{-1}(t)\mathbf{f}(t) = \frac{1}{3} \begin{pmatrix} -e^{3t} \sin t + 2e^{4t} \\ e^{-3t} \sin t + e^{-2t} \end{pmatrix}$$

$$\implies \int \mathbf{\Phi}^{-1}(t)\mathbf{f}(t) \ dt = \frac{1}{30} \begin{pmatrix} e^{3t} \cos t - 3e^{3t} \sin t + 5e^{4t} \\ -e^{-3t} \cos t - 3e^{-3t} \sin t - 5e^{-2t} \end{pmatrix}.$$

Left multiplication of this result by $\Phi(t)$ and lots of simplification give us

$$\Phi(t) \int \Phi^{-1}(t)\mathbf{f}(t) \; dt = \frac{1}{10} \begin{pmatrix} \cos t + \sin t + 5e^t \\ -2\sin t \end{pmatrix}.$$

Thus, from (6.27), we can write our general solution as

$$\mathbf{x} = \frac{1}{3} \begin{pmatrix} e^{3t} & 2e^{3t} \\ -e^{-3t} & e^{-3t} \end{pmatrix} \begin{pmatrix} c_1 \\ c_2 \end{pmatrix} + \frac{1}{10} \begin{pmatrix} \cos t + \sin t + 5e^t \\ -2\sin t \end{pmatrix}$$

$$= c_1 e^{-3t} \begin{pmatrix} 1 \\ 1 \end{pmatrix} + c_2 e^{3t} \begin{pmatrix} -2 \\ 1 \end{pmatrix}$$

$$+ \cos t \begin{pmatrix} 1/10 \\ 0 \end{pmatrix} + \sin t \begin{pmatrix} 1/10 \\ -1/5 \end{pmatrix} + e^t \begin{pmatrix} 1/2 \\ 0 \end{pmatrix}. \tag{6.28}$$

As expected, this is the same answer as we obtained in Examples 1 and 2.

Problems

In problems 1–7, we consider the system $\mathbf{x}' = \mathbf{A}\mathbf{x}$.
   (i) Solve using the fundamental matrix.
   (ii) Solve using diagonalization.
   (iii) Solve using the matrix exponential.
   (iv) Solve using variation of parameters.

1. $\mathbf{A} = \begin{pmatrix} -3 & -2 \\ 3 & 4 \end{pmatrix}$

2. $\mathbf{A} = \begin{pmatrix} -5 & -2 \\ 1 & -2 \end{pmatrix}$

3. $\mathbf{A} = \begin{pmatrix} 4 & 0 \\ -1 & 2 \end{pmatrix}$

4. $\mathbf{A} = \begin{pmatrix} 3 & 2 \\ -1 & 0 \end{pmatrix}$

5. $\mathbf{A} = \begin{pmatrix} 1 & 0 & -2 \\ 1 & 2 & 5 \\ 0 & 0 & -1 \end{pmatrix}$

6. $\mathbf{A} = \begin{pmatrix} -1 & 2 & 0 \\ 0 & 1 & 0 \\ 4 & 0 & 3 \end{pmatrix}$

7. $\mathbf{A} = \begin{pmatrix} 0 & 2 & 1/2 \\ 1 & 1 & 1/2 \\ 2 & 0 & 2 \end{pmatrix}$

6.4 Nonlinear Equations and Phase Plane Analysis

Up to this point, we have studied a system of first-order linear differential equations with constant coefficients. We will now consider the more general situation where the equations are first order but are *nonlinear*. There are numerous books devoted to the study of such equations and we are merely "scratching the surface" of this topic. For an extremely accessible yet thorough excursion into *nonlinear dynamics*, the interested reader should examine the book by Strogatz given in the references [34].

Let's consider the following two differential equations:

$$\frac{dx}{dt} = f(t, x, y)$$
$$\frac{dy}{dt} = g(t, x, y). \tag{6.29}$$

When f and g are nonlinear functions, it will be rare when we can actually find an exact solution to these equations. In such cases we must resort to graphical or numerical analysis and interpretation of the behavior of the solutions. To better understand this new approach, we will consider an *autonomous* nonlinear system of the form

$$\frac{dx}{dt} = f(x, y)$$
$$\frac{dy}{dt} = g(x, y), \tag{6.30}$$

where time, t, is not explicit. We begin with an existence and uniqueness theorem for a general system.

THEOREM 6.4.1 *Consider the system given in (6.29) with initial condition $x(t_0) = x_0, y(t_0) = y_0$. If $\partial f/\partial x$, $\partial f/\partial y$, $\partial g/\partial x$, $\partial g/\partial y$ are all continuous on some rectangular region $R = \{(x,y) | a < x < b, \ c < y < d\}$ containing the point (t_0, x_0, y_0), then there exists a unique solution to (6.29) defined on $(t_0 - \tau, t_0 + \tau)$ for some $\tau > 0$.*

We shall assume that all partial derivatives are continuous for the remainder of this section.

Equilibria

We begin our analysis by finding the equilibria[1] of the system (6.30) of first-order differential equations. The idea is intuitive. When do we say that a

---
[1]Recall that equilibria are also referred to as constant solutions, critical points, fixed points, and steady-state solutions.

solution (x, y) is at equilibrium? Clearly, when it is not changing over time. In other words, when its rates of change with respect to time are zero. More precisely, the solution (x, y) is at equilibrium when $\frac{dx}{dt} = 0$ and $\frac{dy}{dt} = 0$. In order to find equilibria, we need to consider the curves $f(x, y) = 0$ and $g(x, y) = 0$ in the phase plane. Any curve of the form $h(x, y) = k$, where k is a constant, is called an *isocline* or *level curve* of the function h. In other words, a curve is an isocline or level curve of a function if the function takes the same value at every point on the curve. In the special case when $k = 0$, the curve $h(x, y) = k = 0$ is called a *nullcline*. In the system (6.30), $f(x, y) = 0$ and $g(x, y) = 0$ are the nullclines since they take the value zero. From the discussion above we see that the intersection point of the two nullclines is an equilibrium. More formally (x^*, y^*) is an equilibrium of (6.30) if $f(x^*, y^*) = 0$ and $g(x^*, y^*) = 0$. The equilibria (or constant solutions) are found by setting the right-hand side of the equations equal to zero and solving them simultaneously.

Example 1: Consider

$$\frac{dx}{dt} = y - x^2$$

$$\frac{dy}{dt} = y - x. \tag{6.31}$$

Here, $f(x, y) = y - x^2$ and $g(x, y) = y - x$. Therefore the nullclines are $y = x^2$ and $y = x$. These curves intersect at two points $(0, 0)$ and $(1, 1)$. From the definition we can conclude that these are the equilibrium solutions (x^*, y^*).

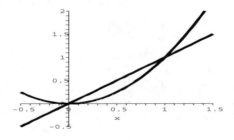

FIGURE 6.3:   Nullclines of system (6.31).

Directions of flow

From calculus of a single variable, we know that a function changes sign only at zero crossings and discontinuities. This is why we mark zeros and disconti-nuities when beginning sign charts. Similar ideas hold in two dimensions, and we assume that our functions are continuous so we only need to worry about zeros. We can view the nullcline as a surface, $z = f(x, y)$, that changes sign only at places where z is zero. By marking the f and g nullclines we have divided the plane up into regions where $\frac{dx}{dt}$ and $\frac{dy}{dt}$ are positive and negative. From equations (6.30), we see that $f(x, y) = 0$ corresponds to the curve where there is no change in the x-direction (i.e., the horizontal direction) and only change in the y (or vertical) direction. On this curve $\frac{dy}{dt}$ is positive or negative and $\frac{dx}{dt} = 0$. Similarly $g(x, y) = 0$ corresponds to the curve where $\frac{dy}{dt} = 0$ and $\frac{dx}{dt}$ is positive or negative. We know from calculus that when $\frac{dx}{dt} > 0$, x is increasing and when $\frac{dx}{dt} < 0$, x is decreasing.

Using these ideas we mark arrows to indicate the direction of derivatives. In each region where $\frac{dx}{dt}$ is positive we mark an arrow to the right (direction of growth for x), and where $\frac{dx}{dt}$ is negative we mark an arrow to the left. Similarly for y, we mark the regions where $\frac{dy}{dt}$ is positive or negative with upward or downward arrows, respectively. A region where $\frac{dx}{dt} > 0$ and $\frac{dy}{dt} > 0$ is marked with an arrow pointing in the northeast direction; see Figure 6.3. These arrows indicate the direction a trajectory moves. One may think of the vector field as a flowing body of water. If we drop a stick in the water, it moves in a path determined by the flow vectors. A computer-generated vector field gives a more accurate sense of the flow of trajectories; it's easier to observe, e.g., that as a trajectory crosses a nullcline, the direction it moves changes; see Figure 6.4(a). Figure 6.4(b) shows the nullclines and vector field superimposed to illustrate horizontal and vertical flow along each nullcline. In practice, we will not superimpose nullclines and vector fields because only trajectories are plotted on vector fields and having nullclines over these would be very confusing.

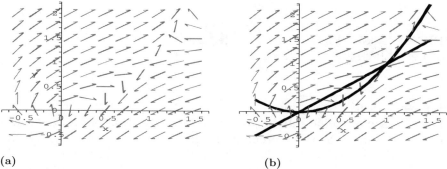

(a)   (b)

FIGURE 6.4: **(a)** Vector field of system (6.31). **(b)** Nullclines superimposed on vector field to show the horizontal or vertical flow along each nullcline. In practice we will not superimpose the nullclines because we will only plot trajectories (solutions) on the vector field; see Figure 6.5.

From a phase plane drawing, saddle points are evident; see Figure 6.5. However, it is difficult to tell if the equilbrium is a spiral sink, a center, or a spiral source. We discuss two methods used to classify our equilibrium points. The first is to use a differential-equations solver to find either exact or approximate solutions for initial points in various locations (i.e., for different initial conditions) and observe the behavior these solutions exhibit. For a review of the behavior near equilibrium points, see Figure 5.5. The second method is to do a local analysis of the solutions around the equilibrium point to classify the stability of the equilibrium. Such an approach requires us to linearize our equation or system and consider the behavior of the linearized system. The topology of the linearized system will be the same as the original system as long as the eigenvalues of the linearized system do not have zero real part.

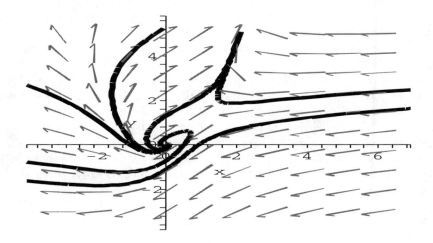

FIGURE 6.5:   Phase portrait: trajectories superimposed onto vector field of system (6.31).

Linearization

Linearization of a differential equation is the replacement of the equation of a curve or surface by the appropriate tangent line or tangent plane centered at a point. From calculus, we know that a differentiable function $F(x)$ can be approximated near a point by its tangent line, $F(x) \approx F(a) + F'(a)(x - a)$. In calculus of several variables, this is extended to a plane tangent to a surface at a point (x, y). The surface $h(x, y)$ is approximated by the linear Taylor expansion:

$$h(x, y) \approx h(x^*, y^*) + \frac{\partial h}{\partial x}(x^*, y^*)(x - x^*) + \frac{\partial h}{\partial y}(x^*, y^*)(y - y^*).$$

This is usually done to study the behavior of a system near that point, in our case the equilibrium point. Linearization of a set of equations yields a system of linear differential equations. Therefore, we use the methods learned in Section 5.1 to analyze and determine the behavior of solutions close to the point of interest.

Even though linearization can be done at any point, we linearize only around equilibria. Let (x^*, y^*) be equilibria of the system (6.30),

$$\frac{dx}{dt} = f(x, y)$$
$$\frac{dy}{dt} = g(x, y),$$

and let $(x^*, y^*) = 0$ be an equilibrium, i.e., $f(x^*, y^*) = 0$ and $g(x^*, y^*) = 0$. We use the notation $f_x(a, b)$ to denote the partial derivative of f with respect to x evaluated at (a, b), and $f_y(a, b)$ is the partial of f with respect to y. Similar notation is used for the function g.

The functions f and g at (x^*, y^*) are approximated by Taylor expanding $f(x, y)$ about (x^*, y^*) and omitting any higher-order terms.

$$\frac{dx}{dt} = f(x, y)$$
$$\approx f(x^*, y^*) + f_x(x^*, y^*)(x - x^*) + f_y(x^*, y^*)(y - y^*)$$
$$\frac{dy}{dt} = g(x, y)$$
$$\approx g(x^*, y^*) + g_x(x^*, y^*)(x - x^*) + g_y(x^*, y^*)(y - y^*).$$

Notice $f(x^*, y^*) = 0$ and $g(x^*, y^*) = 0$. Also

$$\frac{d(x - x^*)}{dt} = \frac{dx}{dt}$$

and

$$\frac{d(y - y^*)}{dt} = \frac{dy}{dt}$$

so we have the following pair of equations:

$$\frac{d(x - x^*)}{dt} \approx f_x(x^*, y^*)(x - x^*) + f_y(x^*, y^*)(y - y^*)$$
$$\frac{d(y - y^*)}{dt} \approx g_x(x^*, y^*)(x - x^*) + g_y(x^*, y^*)(y - y^*).$$

It is simpler to change coordinate systems and translate the equilibrium to the origin. Let $u = x - x^*$ and $v = y - y^*$. Then we have

$$\frac{du}{dt} \approx f_x(x^*, y^*)u + f_y(x^*, y^*)v$$

$$\frac{dv}{dt} \approx g_x(x^*, y^*)u + g_y(x^*, y^*)v.$$

This is a linear system which we rewrite as

$$\begin{pmatrix} \frac{du}{dt} \\ \frac{dv}{dt} \end{pmatrix} = \begin{pmatrix} f_x(x^*, y^*) & f_y(x^*, y^*) \\ g_x(x^*, y^*) & g_y(x^*, y^*) \end{pmatrix} \begin{pmatrix} u \\ v \end{pmatrix}.$$

The matrix of partial derivatives is sometimes called the *Jacobian matrix*, and we use the notation

$$J(a, b) = \begin{pmatrix} f_x(a, b) & f_y(a, b) \\ g_x(a, b) & g_y(a, b) \end{pmatrix}.$$

This matrix is used with Theorem 5.1.1 to determine the nature of the solution.

This linearization process may seem involved, but much of the complication is in the development of the equations. In practice it is only necessary to compute the four partial derivatives at the equilibrium and then use Theorem 6.1.2 to determine the nature of the solution.

Example 2: Consider the system (6.31) from Example 1:

$$\frac{dx}{dt} = y - x^2$$

$$\frac{dy}{dt} = y - x.$$

Find the equilibria and determine their stability via linearization.

We need to solve the right-hand sides of the equations simultaneously. Solving the second equation gives $y = x$. We then substitute into the the first equation to get

$$x - x^2 = 0 \Longrightarrow x = 0, 1.$$

Thus we have two equilibria, $(0, 0)$ and $(1, 1)$. Calculating their stability via linearization requires use of the Jacobian. We can easily calculate it as

$$\mathbf{J} = \begin{pmatrix} -2x & 1 \\ -1 & 1 \end{pmatrix}.$$

Evaluating the Jacobian at the respective equilibria gives

$$J(0, 0) = \begin{pmatrix} 0 & 1 \\ -1 & 1 \end{pmatrix},$$

so $\beta = \text{Tr}(J(0,0)) = 1$ and $\gamma = \det(J(0,0)) = 1$; thus $(0,0)$ is a spiral source. We also have

$$J(1,1) = \begin{pmatrix} -2 & 1 \\ -1 & 1 \end{pmatrix},$$

so $\beta = \text{Tr}(J(1,1)) = -1$ and $\gamma = \det(J(1,1)) = -1$; thus $(1,1)$ is a saddle point. See Figure 6.5.

6.4.1   Phase Plane with Matlab, Maple, and Mathematica

It may seem obvious how we could plot nullclines using Matlab, Maple, or Mathematica to help us in the phase plane analysis that we have been discussing. But the Runge-Kutta methods that we considered in the previous chapter apply to the system of two first-order equations we have been considering. The interested reader could also use the root finding and linear algebra capabilities of both to help in the analysis of equilibria but we refer the reader to Appendix A for this. We give code for numerically solving the example

$$\frac{dx}{dt} = y - x^2$$

$$\frac{dy}{dt} = y - x. \tag{6.32}$$

Computer Code 6.4:   **Generating phase portraits by numerically solving a system of two first-order equations, superimposing solutions on the vector field**

Matlab, Maple, Mathematica

Matlab

We first create the m-file `ExamplePP.m` which contains the equations:

```
                          Matlab
function f=ExamplePP(xn,yn)
%
%The original system is
%x'(t)=y-x^2 and y'(t)=y-x
%We let yn(1)=x, yn(2)=y
%
f= [yn(2)-yn(1).^2; yn(2)-yn(1)];
```

Then in the command window, we can use `ode45` or `RK4.m` to numerically solve the system for a given set of initial conditions.

```
                            Matlab
>>  t0=0; tf=10;
>>  IC1=[3,1];
>>  [t,y]=RK4('ExamplePP',[t0,tf],IC1,.05);
>>  IC2=[-1,0];
>>  [t2,y2]=RK4('ExamplePP',[t0,tf],IC2,.05);
>>  IC3=[-0.5,0.5];
>>  [t3,y3]=RK4('ExamplePP',[t0,tf],IC3,.05);
>>  IC4=[3,2];
>>  [t4,y4]=RK4('ExamplePP',[t0,tf],IC4,.05);
>>  IC5=[-0.15,0];
>>  [t5,y5]=RK4('ExamplePP',[t0,tf],IC5,.05);
>>  IC6=[-0.8,0];
>>  [t6,y6]=RK4('ExamplePP',[t0,tf],IC6,.05);
```

We now have all the solutions generated and we plot them with the following code:

```
                            Matlab
>>  plot(y(:,1),y(:,2));
>>  axis([-4  3   -4   6]);
>>  hold on
>>  plot(y2(:,1),y2(:,2));
>>  plot(y3(:,1),y3(:,2));
>>  plot(y4(:,1),y4(:,2));
>>  plot(y5(:,1),y5(:,2));
>>  plot(y6(:,1),y6(:,2));
>>  [X,Y]=meshgrid(-4:.5:3,-4:.5:6);
>>  DX=Y-X.^2;
>>  DY=Y-X;
>>  DW=sqrt(DX.^2+DY.^2);
>>  quiver(X,Y,DX./DW,DY./DW,.5);
>>  xlabel('x'); ylabel('y');
>>  title('phase plane example')
>>  hold off
```

Maple

As with Matlab, we need to calculate solutions through many different initial conditions (through trial and error as to which give "nice" pictures). An alternative form to produce a phase portrait is given in the last line.

Maple

```
> with(plots): with(DEtools):
> eq1:=diff(x(t),t)=y(t)-x(t)^2;
> eq2:=diff(y(t),t)=y(t)-x(t);
> initcond:=[[x(0)=3,y(0)=1],[x(0)=-1,y(0)=0],
    [x(0)=-0.5,y(0)=0.5],[x(0)=3,y(0)=2],[x(0)=-0.15,y(0)=0],
    [x(0)=-1.5,y(0)=2]];
> DEplot([eq1,eq2],[x(t),y(t)],t=0..10,x=-4..3,y=-4..6,
    initcond,stepsize=.05, title="phase plane example",
    linecolor=black,method=classical[rk4]);
> phaseportrait([eq1,eq2],[x(t),y(t)],t=0..10,initcond,
    stepsize=.05,scene=[x(t),y(t)],title="phase plane
    example",linecolour=black,method=classical[rk4]);
```

Mathematica

```
xde[t_]=y[t]-x[t]^2
yde[t_]=y[t]-x[t]
ICx={x[0]==3,x[0]==-1,x[0]==-.5,x[0]==3,x[0]==-.15,x[0]==-.8}
ICy={y[0]==1,y[0]==0,y[0]==.5,y[0]==2,y[0]==0,y[0]==0}
soln1=NDSolve[{x'[t]==xde[t],ICx[[1]],y'[t]==yde[t],
  ICy[[1]]},{x,y},{t,0,3},StartingStepSize→ .05,
  Method→ {FixedStep,Method→ ExplicitRungeKutta}]
soln2=NDSolve[{x'[t]==xde[t],ICx[[2]],y'[t]==yde[t],
  ICy[[2]]},{x,y},{t,-1,1},StartingStepSize→ .05,
  Method→ {FixedStep,Method→ ExplicitRungeKutta}]
soln3=NDSolve[{x'[t]==xde[t],ICx[[3]],y'[t]==yde[t],
  ICy[[3]]},{x,y},{t,-1,10},StartingStepSize→ .05,
  Method→ {FixedStep,Method→ ExplicitRungeKutta}]
soln4=NDSolve[{x'[t]==xde[t],ICx[[4]],y'[t]==yde[t],
  ICy[[4]]},{x,y},{t,0,10},StartingStepSize→ .05,
  Method→ {FixedStep,Method→ ExplicitRungeKutta}]
soln5=NDSolve[{x'[t]==xde[t],ICx[[5]],y'[t]==yde[t],
  ICy[[5]]},{x,y},{t,-1,8},StartingStepSize→ .05,
  Method→ {FixedStep,Method→ ExplicitRungeKutta}]
<<Graphics'PlotField'
p1=ParametricPlot[{Evaluate[{x[t],y[t]}/.soln1],
  Evaluate[{x[t],y[t]}/.soln2],Evaluate[{x[t],y[t]}/.soln3],
  Evaluate[{x[t],y[t]}/.soln4],Evaluate[{x[t],y[t]}/.soln5]},
  {t,0,7.9},PlotStyle→ {Thickness[0.015]},
  PlotRange→ {{-3,4},{-3,5}}]
p2=PlotVectorField[{y - x^2,y - x}, {x, -3, 4}, {y, -3, 5},
  ScaleFunction→ (1&),Axes→ Automatic,HeadLength→ 0,
  AxesLabel→ {"x","y"}]
Show[p1,p2]
```

Pplane in Matlab

For a system of only two equations, a very user-friendly software supplement called **pplane** exists for Matlab and it is freely available for educational use. See <http://math.rice.edu/~dfield/>. There are two to three programs that you will need to download (depending on your version of Matlab) and install in your working directory. Pplane is much easier to implement than either of the above methods. It can also numerically find equilibria, determine stability of them, plot nullclines, and many other things. The drawback is that it only works for a system of two autonomous equations, whereas the above methods work for any number of equations. We give a brief introduction here.

Once you have placed the relevant programs in your working directory, type `pplane6`. A new window should pop up; see Table 6.5. (Note that this is for Matlab 6.x. Download the appropriate files for other versions of Matlab, e.g., `pplane7` is for Matlab 7.)

We again consider the equations of the previous example and enter them in the `pplane` window as

```
x' = y-x^2
y' = y-x
```

Computer Code 6.5: **Pop-up window for `pplane` program used to obtain numerical solutions of two first-order equations. Note this was done with Matlab 6.5 and hence the program is `pplane6`.**

Matlab only

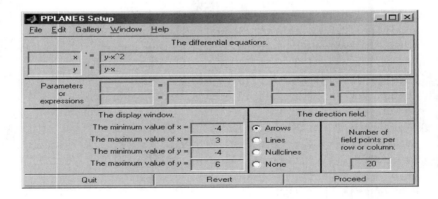

In the display window, set minimum $x = -4$, maximum $x = 3$, minimum $y = -4$, and maximum $y = 6$.

In the direction field box, make sure *arrows* is marked. Click *proceed* and observe the direction field. Click once to trace a trajectory forward and backward in time. Repeat a few times.

Now go to *solutions* $\longrightarrow$ *find an equilibrium point*. Then click with the mouse where you expect to see an equilibrium. Observe the Jacobian is given as are the corresponding eigenvalues and eigenvectors. (You can click *display the linearization* to see what happens near the fixed point.)

In the display window, graph t vs. x and t vs. y by going to *graph* $\longrightarrow$ *both*. Click on a trajectory that you want to see plotted.

Now re-plot the picture but with *lines, nullclines*, and *none* checked (three separate plots). Experiment with *number of field points per row or column* to observe how this changes the display window.

6.4.2   Bifurcations

The system of equations (6.30),

$$x' = f(x, y)$$
$$y' = g(x, y),$$

can have bifurcations, just as we did with first-order autonomous equations in Section 2.3. We can have saddlenode bifurcations in which a saddle point and a node (either stable or unstable) coalesce and disappear as a parameter is varied; we can have transcritical bifurcations in which a saddle and a node exchange stability as one passes through the other as a parameter is varied; we can have subcritical pitchfork (involving the birth/death of two unstable equilibria) and supercritical (involving the birth/death of two stable equilibria) pitchfork bifurcations in which two equilibria are born out of one but with different stability than the one from which they were born. These types of bifurcations occur when one eigenvalue of the point (i.e., equilibrium solution) undergoing the bifurcation is zero.

We consider the basic form for a supercritical pitchfork bifurcation:

$$x' = rx - x^3$$
$$y' = -y \tag{6.33}$$

where $r \in \mathbb{R}$. The equilibria are $(x^*, y^*) = (0, 0), (\sqrt{r}, -1), (\sqrt{r}, 1)$, where the last two equilibria only exist when $r > 0$. As in the first-order equation, we have three cases to consider: $r < 0, r = 0$, and $r > 0$. We could check the stability analytically using the linearization previously discussed. The Jacobian is easily calculated as

$$\mathbf{J} = \begin{pmatrix} r - 3x^2 & 0 \\ 0 & -1 \end{pmatrix}.$$

Evaluating at the equilibrium points gives for $r < 0$

$$J(0, 0) = \begin{pmatrix} r & 0 \\ 0 & -1 \end{pmatrix}.$$

Thus $\beta = \text{Tr}(J(0,0)) = r - 1$ and $\gamma = \det(J(0,0)) = -r$ so $(0,0)$ is a spiral source. Because we are in the case when $r < 0$, we see that $\beta < 0$ and $\gamma > 0$ and thus the origin is stable.

For $r = 0$, we have

$$J(0,0) = \begin{pmatrix} 0 & 0 \\ 0 & -1 \end{pmatrix}.$$

Thus $\beta = \text{Tr}(J(0,0)) = -1$ and $\gamma = \det(J(0,0)) = 0$ and the prediction is for a linear center. This is a borderline case and thus the conclusion may be affected by the nonlinear terms that were ignored.

For $r > 0$, we have three equilibria:

$$J(0,0) = \begin{pmatrix} r & 0 \\ 0 & -1 \end{pmatrix} \quad \text{and} \quad J(\pm\sqrt{r}, 0) = \begin{pmatrix} r - 3r & 0 \\ 0 & -1 \end{pmatrix}.$$

Thus $\beta = \text{Tr}(J(0,0)) = r - 1$ and $\gamma = \det(J(0,0)) = -r$ and because $r > 0$, we know this is a saddle point. The Jacobian is the same for the other two equilibria and we have that $\beta = \text{Tr}(J(\pm\sqrt{r}, 0)) = -2r - 1$ and $\gamma = \det(J(0,0)) = 2r$, which shows that both points are stable equilibrium points.

Thus, for $r < 0$ the origin is the only equilibrium solution and it is stable. For $r > 0$, two additional equilibria exist that were born out of the origin (when $r = 0$, they are located at $(0,0)$) and both are stable, while the origin has now changed its stability.

The above four bifurcations occur when $\lambda = 0$, as mentioned above. To see this, we calculate the characteristic equation of the Jacobian evaluated at one of the equilibria (in this example, all three give the same answer):

$$\det(\mathbf{J}(0,0) - \lambda\mathbf{I}) = (r - \lambda)(-1 - \lambda) = \lambda^2 + \lambda(1 - r) - r = 0.$$

We substitute $\lambda = 0$ and see that $r = 0$ is the parameter value for one of the above four bifurcations. We could then use linearization near the equilibria before and after the bifurcation value to see if there was a qualitative change in the number or stability of the equilibria.

There is also another type of bifurcation that can now occur: a *Hopf bifurcation*. These also come in supercritical and subcritical flavors. This bifurcation occurs when the real part of a complex pair of eigenvalues becomes zero as a parameter is varied. Either before or after the bifurcation value, we have the presence of a *limit cycle*, which is an isolated closed (and hence periodic) trajectory. (By isolated, we simply mean that we can find an $\epsilon > 0$ such that if we begin within ϵ of the limit cycle, we will not encounter another closed orbit.) See problems 20 and 21 for a closer look at this type of bifurcation.

6.4.3   Systems of More Than Two Equations

Nonlinear systems of equations are an active area of research as these equations occur in many areas of the sciences.

$$x_1' = f_1(t, x_1, x_2, \ldots, x_n)$$
$$x_2' = f_2(t, x_{1,2}, \ldots, x_n)$$
$$\vdots$$
$$x_n' = f_n(t, x_{1,2}, \ldots, x_n). \tag{6.34}$$

It will be extremely rare when we can actually find an exact solution to these equations. As with two equations, we must then rely heavily on graphical, approximation, or numerical techniques. The existence and uniqueness theorem for this general system is analogous to the one for two equations—we need to check that each of the functions f_i and partial derivatives $\partial f_i / \partial x_j$ are continuous in order to guarantee existence and uniqueness. We will often be concerned with these equations when they are *autonomous*, that is, there is not explicit t in the problem. The following theorem gives the *linear* stability in a neighborhood of an equilibrium solution. If a system has multiple equilibrium solutions, the theorem may be applied to each equilibrium solution. We often write (6.34) in its vector notation for convenience:

$$\mathbf{x}' = \mathbf{f}(\mathbf{x}).$$

THEOREM 6.4.2 *Let $\mathbf{x}' = \mathbf{f}(\mathbf{x})$ be a nonlinear system of n first-order equations with $\mathbf{x}^*$ as an equilibrium solution and $\mathbf{f}$ a sufficiently smooth vector function. Let $\mathbf{J}$ be the Jacobian (the matrix of partial derivatives) evaluated at this equilibrium solution:*

$$\mathbf{J}(\mathbf{x}^*) = \begin{pmatrix} \partial f_1/\partial x_1 & \partial f_1/\partial x_2 & \cdots & \partial f_1/\partial x_n \\ \partial f_2/\partial x_1 & \partial f_2/\partial x_2 & \cdots & \partial f_2/\partial x_n \\ \vdots & & \ddots & \\ \partial f_n/\partial x_1 & \partial f_n/\partial x_2 & \cdots & \partial f_n/\partial x_n \end{pmatrix}_{\mathbf{x}=\mathbf{x}^*} \tag{6.35}$$

Let $\{\lambda_1, \lambda_2, \cdots, \lambda_n\}$ be the n (real or complex, possibly repeated) eigenvalues of the Jacobian matrix.

a. *If the real part of the eigenvalue $\Re(\lambda_i) < 0$ for all i, then the equilibrium is stable.*

b. *If the real part of the eigenvalue $\Re(\lambda_i) < 0$ for at least one i and $\Re(\lambda_j) > 0$ for at least one j, then the equilibrium is a saddle.*

c. *If the real part of the eigenvalue $\Re(\lambda_i) > 0$ for all i, then the equilibrium is unstable.*

d. *If any of the eigenvalues are complex, then the stable or unstable equilibria is a spiral; if all of the eigenvalues are real, it is a node.*

e. *If a pair of complex conjugate eigenvalues λ_i, $\overline{\lambda_i}$ satisfy $\Re(\lambda_i) = 0$, then the equilibrium is a linear center in the plane containing the corresponding eigenvectors.*

THEOREM 6.4.3 *Let* $\mathbf{x}' = \mathbf{f}(\mathbf{x})$ *be a nonlinear system of n first-order equations with* $\mathbf{x}^*$ *as an equilibrium solution and* $\mathbf{f}$ *a sufficiently smooth vector function. If* $\Re(\lambda_i) \neq 0$ *for all i, then the predictions given by the linear stability results of Theorem 6.4.2 hold for the equilibrium solution in the nonlinear system.*

The significance of Theorem 6.4.3 cannot be understated. We found an equilibrium solution $\mathbf{x}^*$ and linearized about it. That is, we considered only the linear terms near this equilibrium solution. The results of the theorem allow us to conclude that only looking at linear terms near the equilibrium solution is sufficient to give us accurate stability predictions, as long as the real part of all eigenvalues is nonzero. This should be believable because adding nonlinear terms could possibly change the stability in these borderline cases. Alternative techniques beyond the scope of this book are needed to address these situations.

Thus the techniques used for autonomous systems are very familiar: finding equilibrium solutions, linearizing the system about the equilibria, determining the linear stability of the equilibria, and constructing phase portraits with the help of a computer program. Indeed, our plan of attack in order to understand the behavior of the solutions was identical for a system of two equations. The main difference here has to do with the structure of the space in which trajectories live. Things were very nice in two dimensions in that we could characterize many things about equilibria. Once we introduce a third (or more) dimension(s), very strange things can happen. The mathematical subject of *chaos* arose because of the kind of this strange behavior that can occur. We refer the interested reader to other books for an introduction.

Example 2: Find equilibrium solutions for the system

$$x' = -6x + 6y$$
$$y' = 36x - y - xz$$
$$z' = -3z + xy. \tag{6.36}$$

Then use Theorems 6.4.2 and 6.4.3 to classify the stability of the equilibrium solutions.

We use our three computer software packages to help us with these and give the code at the end of the example. We find three equilibria in the system:

$$(x^*, y^*, z^*) = (0,0,0), (\sqrt{105}, \sqrt{105}, 35), (-\sqrt{105}, -\sqrt{105}, 35).$$

In order to determine the stability of the equilibria, we first need to calculate the Jacobian matrix of the system:

$$\mathbf{J} = \begin{pmatrix} -6 & 6 & 0 \\ 36 - z & -1 & -x \\ y & x & -3 \end{pmatrix}, \tag{6.37}$$

and then substitute in the respective equilibrium points. We have

$$(0,0,0): \quad \lambda_1 = -3, \ \lambda_{2,3} = \frac{-7 \pm \sqrt{889}}{2},$$

which shows that the origin is a saddle point. According to Theorem 6.4.3, we can conclude that the origin is also a saddle in the original system (since $\Re(\lambda_i) \neq 0$). For the second equilibrium, we have

$$(\sqrt{105}, \sqrt{105}, 35): \quad \lambda_1 = -10, \ \lambda_{2,3} = \pm i3\sqrt{14}.$$

According to Theorem 6.4.2, this second equilibrium solution is a linear center. According to Theorem 6.4.3, we can only conclude that we have a linear center—it is possible that we have a nonlinear center in the full nonlinear system but it is also possible that the inclusion of the nonlinear terms makes this equilibrium solution either a stable spiral or an unstable spiral. For the third equilibrium, we have

$$(-\sqrt{105}, -\sqrt{105}, 35): \quad \lambda_1 = -10, \ \lambda_{2,3} = \pm i3\sqrt{14}.$$

This again gives the prediction of a linear center and doesn't allow us to conclude anything about the full system.

The system of the previous example is actually a very well-known system that has been studied extensively due to the seemingly unpredictable behavior of solutions with close initial conditions. Depending on the coefficients of the original equations, we can have between one and three equilibria and we have the possibility of trajectories wandering endlessly without approaching an equilibrium solution. See problem 24 for another look at this system.

Computer Code 6.6:   **Linear stability analysis for nonlinear system of equations**

Matlab, Maple, Mathematica

```
                            Matlab
>> syms x y z %defines the variables as symbolic
>> [x1,y1,z1]=solve(-6*x+6*y,36*x-y-x*z,-3*z+x*y)
>> %Note that x1 has the 3 x-coordinates, y1 has
>> %the 3 y-coords, and z1 has the 3 z-coords
>> equil1=[x1(1,1), y1(1,1), z1(1,1)] %1st equil
>> equil2=[x1(2,1), y1(2,1), z1(2,1)] %2nd equil
>> equil3=[x1(3,1), y1(3,1), z1(3,1)] %3rd equil
>> jac1=jacobian([-6*x+6*y; 36*x-y-x*z; -3*z+x*y],[x,y,z])
    %calculates jacobian
>> J1=subs(jac1,{x,y,z},equil1) %substitutes
    %1st equil into jacobian
>> J2=subs(jac1,{x,y,z},equil2) %substitutes
    %2nd equil into jacobian
>> J3=subs(jac1,{x,y,z},equil3) %substitutes
    %3rd equil into jacobian
>> eigenvals1=eig(J1) #computes eigenvalues for 1st equil
>> eigenvals2=eig(J2) #computes eigenvalues for 2nd equil
>> eigenvals3=eig(J3) #computes eigenvalues for 3rd equil
```

```
                            Maple
> with(linalg):
> eq1a:=-6*x+6*y; # the right-hand-side of 1st eqn
> eq1b:=36*x-y-x*z; # the rhs of 2nd eqn
> eq1c:=-3*z+x*y; # the rhs of 3rd eqn
> eq2:=solve({eq1a,eq1b,eq1c},{x,y,z}); # equil solns
> eq2a:=allvalues(eq2[2]); # expanded solns
> eqJ:=jacobian([eq1a,eq1b,eq1c],[x,y,z]); #calculates
    #jacobian of the system
> eq3a:=subs(eq2a[1],evalm(eqJ)); #substitutes 1st equil
    #into jacobian
> eq3b:=subs(eq2a[2],evalm(eqJ));#substitutes 2nd equil
    #into jacobian
> eq3c:=subs(eq2a[3],evalm(eqJ));#substitutes 3rd equil
    #into jacobian
> eq4a:=eigenvals(eq3a); #computes eigenvalue for 1st equil
> eq4b:=eigenvals(eq3b); #computes eigenvalue for 2nd equil
> eq4c:=eigenvals(eq3c); #computes eigenvalue for 3rd equil
```

```
                        Mathematica
eq1a=-6x +6y (*the right-hand side of the 1st eqn*)
eq1b=36x-y-xz (*the rhs of the 2nd eqn*)
eq1c=-3z+xy (*the rhs of the 3rd eqn*)
eq2=Solve[{eq1a==0,eq1b==0,eq1c==0},{x,y,z}] (*eqil solns*)
eqJ[x_,y_,z_]=D[{eq1a,eq1b,eq1c},{{x,y,z}}]
   (*calculates the jacobian of the system*)
eq3a=eqJ[x,y,z]/.eq2[[1]] (*substitutes 1st equil*)
eq3b=eqJ[x,y,z]/.eq2[[2]] (*substitutes 2nd equil*)
eq3c=eqJ[x,y,z]/.eq2[[3]] (*substitutes 3rd equil*)
eq4a=Eigenvalues[eq3a] (*computes eigenvalue for 1st equil*)
eq4b=Eigenvalues[eq3b] (*computes eigenvalue for 2nd equil*)
eq4c=Eigenvalues[eq3c] (*computes eigenvalue for 3rd equil*)
```

Problems

For problems 1–13, (i) find the equilibria of the given system; (ii) use linearization and Theorem 6.4.2 to classify the stability of the equilibria; (iii) use Matlab, Maple, or Mathematica to draw the vector field of the system; (iv) sketch trajectories on the vector field for various initial conditions (either by hand or with the computer). You should verify that your answers from parts (iii) and (iv) agree with your predictions in parts (i) and (ii).

1. $\begin{cases} x' = y \\ y' = 4 - x^2 \end{cases}$

2. $\begin{cases} x' = y - 1 \\ y' = x^2 - y \end{cases}$

3. $\begin{cases} x' = y - x \\ y' = x^2 + 2y \end{cases}$

4. $\begin{cases} x' = y^2 - x \\ y' = x - 3y \end{cases}$

5. $\begin{cases} x' = y \\ y' = x^3 - x \end{cases}$

6. $\begin{cases} x' = \sin(x) - y \\ y' = y^2 - \frac{1}{4} \end{cases}$

7. $\begin{cases} x' = y^2 - 1 \\ y' = x^2 - y \end{cases}$

8. $\begin{cases} x' = y + x \\ y' = x^3 - 8y \end{cases}$

9. $\begin{cases} x' = x(3 - x) - xy \\ y' = y(2 - y) - xy \end{cases}$

10. $\begin{cases} x' = x(3 - x) - 2xy \\ y' = y(2 - y) - xy \end{cases}$

11. $\begin{cases} x' = x(3 - x) - xy \\ y' = -2y + xy \end{cases}$

12. A well-known equation is the van der Pol oscillator, which models a triode valve where the resistance depended on the applied current [12]:

$$x'' + \epsilon x'(x^2 - 1) + x = 0, \tag{6.38}$$

where $x = x(t)$ and $\epsilon > 0$ is a constant. Using the methods of Section 3.5.1, convert this equation to a system of two first-order equations. Do part (i) and (ii) for a general ϵ. Then do parts (iii) and (iv) with $\epsilon = 0.1$. Repeat steps (iii) and (iv) for $\epsilon = 10$ and compare your phase portraits.

13. Another famous nonlinear differential equation is the double-well oscillator

$$x'' + \delta x' - x + x^3 = 0.$$

Using the methods of Section 3.5.1, convert this equation to a system of two first order equations. Do part (i) and (ii) for a general δ. Then do parts (iii) and (iv) with $\delta = 0.25$. Repeat steps (iii) and (iv) for $\delta = 1$ and compare your phase portraits.

14. *We now reconsider problem 2 from Section 2.7.* In a first physics course, students derive the equation of motion for a frictionless simple pendulum as

$$m\theta'' + g\sin\theta = 0, \tag{6.39}$$

where θ is the angle that the pendulum makes with the vertical; however, the next step is to assume the angle is small and use the small angle approximation ($\sin\theta \approx \theta + \cdots$) to rewrite this equation as

$$m\theta'' + g\theta = 0$$

which is simply the equation for simple harmonic motion. This approximation obviously fails if θ becomes too large. Previously, we derived an equation for the total energy. Now we analyze (6.39) by more recent methods. For convenience, set $m = g$.

a. Convert (6.39) to a system of two first-order equations.
b. Find the equilibria for $\theta \in [-2\pi, 2\pi]$.
c. Graph the nullclines on the phase plane.
d. Use Matlab, Maple, or Mathematica to sketch the vector field for the system of two first-order equations.
e. Plot trajectories for various initial conditions and obtain a phase portrait similar to Figure 2.22 in Section 2.7. Again interpret the three qualitatively different motions of the pendulum, keeping in mind that the pendulum is allowed to whirl over the top.

15. Now consider the simple pendulum with damping:

$$\theta'' + 0.3\theta' + \theta = 0.$$

Repeat parts a–d in the previous problem. Then plot trajectories for various initial conditions and compare with the phase portrait of the undamped motion. Interpret your picture and the differences between the two phase portraits.

Problems 16–21 involve bifurcations that can occur in higher-order systems. See Strogatz [34] for a more detailed coverage.

16. Consider the system

$$x' = r - x^2$$
$$y' = -y.$$

Find the equilibria and determine their stability. You will have to consider the cases $r < 0$, $r = 0$, and $r > 0$. For each of these cases, draw a phase portrait. Based on your knowledge from Section 2.3.3, classify the bifurcation that occurs.

17. Consider the system

$$x' = rx - x^2$$
$$y' = -y.$$

Find the equilibria and determine their stability. You will have to consider the cases $r < 0$, $r = 0$, and $r > 0$. For each of these cases, draw a phase portrait. Based on your knowledge from Section 2.3.3, classify the bifurcation that occurs.

18. Consider the system

$$x' = rx - x^3$$
$$y' = -y.$$

Find the equilibria and determine their stability. You will have to consider the cases $r < 0$, $r = 0$, and $r > 0$. For each of these cases, draw a phase portrait. Based on your knowledge from Section 2.3.3, classify the bifurcation that occurs.

19. Consider the system

$$x' = rx + x^3$$
$$y' = -y.$$

Find the equilibria and determine their stability. You will have to consider the cases $r < 0$, $r = 0$, and $r > 0$. For each of these cases, draw a phase portrait. Based on your knowledge from Section 2.3.3, classify the bifurcation that occurs.

20. Consider the system

$$x' = rx + 2y$$
$$y' = -2x + ry - x^2y.$$

Verify that the origin is an equilibrium point and determine its stability. Draw a phase portrait for the cases $r = -.25$ and $r = .25$, using a viewing window $x \in [-2, 2]$, $y \in [-2, 2]$. Qualitatively describe the difference between the two phase portraits. In particular describe the trajectory for the initial conditions $(0.5, 0)$ and $(1.5, 0)$ for $r = -.25$ and then again for $r = .25$. This type of bifurcation is called a *supercritical Hopf* bifurcation because a *stable limit cycle* is involved in the origin's stability switch.

21. Consider the system

$$x' = rx + 2y$$
$$y' = -2x + ry + x^2 y.$$

Verify that the origin is an equilibrium point and determine its stability. Draw a phase portrait for the cases $r = -.25$ and $r = .25$, using a viewing window $x \in [-2, 2]$, $y \in [-2, 2]$. Qualitatively describe the difference between the two phase portraits. In particular describe the trajectory for the initial conditions $(0.5, 0)$ and $(1.5, 0)$ for $r = -.25$ and then again for $r = .25$. This type of bifurcation is called a *subcritical Hopf* bifurcation because an *unstable limit cycle* is involved in the origin's stability switch.

22. *We now reconsider Project 1 of Chapter 4.* One formulation of the forced Duffing equation is

$$x'' + bx' + kx + \delta x^3 = F_0 \sin(\omega t), \tag{6.40}$$

where $x = x(t)$. When $\delta = 0$, the equation reduces to that of the forced mass on a spring of Section 4.5.2. Repeat steps 2-5 of this project but now also plot the trajectories in the phase plane. Because this system is nonautonomous, you will not be able to use `pplane`. Besides your explanations, be sure to address why the apparent crossing of solutions in the phase plane is not a violation of the Existence and Uniqueness theorem.

23. By Taylor expanding about the equilibrium point (x^*, y^*, z^*) and keeping only linear terms show that the Jacobian of the three-dimensional system (cf. Theorem 6.4.2)

$$x' = f(x, y, z)$$
$$y' = g(x, y, z) \quad \text{is} \quad \mathbf{J} = \begin{pmatrix} \partial f/\partial x & \partial f/\partial y & \partial f/\partial z \\ \partial g/\partial x & \partial g/\partial y & \partial g/\partial z \\ \partial h/\partial x & \partial h/\partial y & \partial h/\partial z \end{pmatrix}_{(x^*, y^*, z^*)}.$$
$$z' = h(x, y, z)$$

24. The Lorenz system can be written in the form

$$x' = -\sigma x + \sigma y$$
$$y' = rx - y - xz$$
$$z' = -bz + xy, \tag{6.41}$$

where σ, r, b are positive parameters. The system arose as a model of convective rolls in the atmosphere. Lorenz studied the parameter values $\sigma = 10, b = -8/3$ and examined how the behavior of solutions changed as r increased.

a. Determine the equilibria and their stability for $r = 1$.

b. Determine the equilibria and their stability for $r = 20$.

c. Determine the equilibria and their stability for $r = 25$.

d. Plot trajectories in x-y-z *phase space* for parts a, b, c. Go from $t = 0$ to $t = 100$ for three different initial conditions.

25. Consider the Lorenz system (6.41) given in the previous problem with $\sigma = 10, b = -8/3$.

a. Show that the origin is always an equilibrium solution with eigenvalues

$$\lambda_1 = -b, \lambda_{2,3} = -\frac{(\sigma + 1) \pm \sqrt{\sigma^2 - 2\sigma + 1 + 4r\sigma}}{2}.$$

b. Show that the origin undergoes a supercritical pitchfork bifurcation at $r = 1$.

c. Show that the two additional equilibria are given by $x^* = \pm\sqrt{b(r-1)}$, $z^* = r - 1$.

d. Show that each of these additional equilibria undergoes a subcritical Hopf bifurcation when $r_c = \dfrac{\sigma(\sigma + b + 3)}{\sigma - b - 1}$. It may be hard to see the unstable limit cycle in this three-dimensional system; however, if you start close to the equilibrium solution for r slightly smaller than r_c, you will see that initial conditions close enough to the point will approach it while those just slightly bigger will go away from it.

6.5 Epidemiological Models

This section deals with the interaction of groups of people in an effort to better understand the spread of a certain disease. There are a myriad of mathematical models that can be used to describe the spread and transmission of a wide range of diseases. The interested reader should examine Hethcote [15], Daley and Gani [11], and Brauer and Castillo-Chavez [8] for additional details of the subject of epidemiology. Most of the framework in epidemiology consists of dividing the population under consideration into various classes. Each class defines the state of the individual in reference to the disease being modeled. These classes typically consist of three groups: those that are *susceptible* to the disease, those that are *infected* with the disease, and those that are *recovered* from the disease. A model with only these three classes is called a Susceptible-Infected-Recovered (SIR) model. The type of mathematics that is used or implemented in the model is determined by the rules governing or

describing the movement of individuals from one class to another. In the SIR
model, individuals move from the susceptible class to the infected class to the
recovered class depending on their interactions with infected individuals and
on their bodies' ability to fight the disease. There are many variations of this
model. For example, the recovered individuals may be permanently immune
from the disease (e.g., measles) or they may have temporary immunity or no
immunity and be susceptible again to the disease (e.g., syphilis). In the latter
case, there are no recovered individuals as everyone who has been cured becomes immediately susceptible. However, in the case of temporary immunity
the recovered individuals eventually become susceptible again.

As with all mathematical models, there are limitations and not all details
can be incorporated into the model. We need to make many assumptions
in order to create a manageable mathematical model that captures the key
components of the disease under consideration and allows for new insight of
the disease. Some of these assumptions will be discussed a bit later but we
take a moment to consider alternative branches of *mathematics* that can be
used to study the spread of epidemics in a population.

- *Ordinary differential equations* assume a continuous population variable
  and thus work well when there is a large population of people. Much
  information can be obtained from these models but answers do *not* have
  to be whole numbers. This is the specific branch that we will study in
  this section.

- *Partial differential equations* are commonly used for large populations
  when we want to keep track of multiple characteristics of a person that
  may change. For example, we may consider an SIR model with *age
  structure* in which we also keep track of the age of the individuals within
  each class.

- *Difference equation* models are often used to describe epidemics in which
  either the population or time is considered discrete (e.g., may be sea-
  sonal). For example, one may look at generations of mosquitoes that
  carry a certain disease and examine how the disease passes through the
  population over many generations. These models can often be used if
  we do not want to consider a continuous population variable but want
  whole numbers instead. These models can also arise as a *discretization*
  of one of the continuous models above. For example, if we wrote down
  the formulas for solving the system using the fourth-order Runge-Kutta
  method for an SIR model, we would have a difference equation model.

- *Stochastic processes* may be used for models in which we want to consider a discrete population and time. A key difference between this type
  of mathematical framework and the previous is the assumption of *ran-
  domness* in one or more of the parameters. That is, at each step there
  is a probability that something will happen but it is not guaranteed. In
  this type of model, beginning with the same initial state of the system
  may yield different results every time. This is in direct contrast to the

previous three *deterministic* approaches in which a given initial state will always result in the same outcome.

- *Network* models use graph theory to understand how individuals are connected with each other and thereby gain a better understanding of how the disease spreads through a given community. By focusing on key individuals that are well-connected to others in the population, the spread of the disease can be controlled.

- *Agent-based* models are becoming increasingly common because of the large-scale computer simulations that arise from them and their ability to differentiate between individuals in the same class. In this type of model, a set of characteristics is used to describe each individual and thus each movement and interaction can be tracked. Based on these descriptions, a large-scale computer simulation can be done by considering a city of 500,000 people and watching how the interactions of the people can affect the spread of disease. These simulations have become popular with the threatened use of biological weapons in heavily populated areas.

Each type of modeling has its place and one needs to carefully choose the appropriate mathematical framework from which to model and analyze a given epidemic. In this book, we consider only models of ordinary differential equations. These are perhaps the simplest to understand and the ones in which we can carry the mathematics the furthest. Even though a number of assumptions will be required to formulate these models, the results can often shed tremendous insight on the behavior of individuals, their interactions with each other, and the overall effect of the disease on the population. These models can also give researchers insight on how to control the spread of a disease; mathematical models can sometimes indicate whether it is possible to eradicate a disease from the population (e.g., as happened with smallpox) or whether there is little or no chance of doing so.

For the remainder of this section, we will present in detail two variations of the well-known SIR model. We will formulate the model, discuss the rates of transmission and recovery, and mathematically analyze the resulting model. We will find the equilibria of the system, determine their stability, and examine whether any bifurcations are possible. These topics from Section 6.4 will help us understand the effect of a disease on the population and give us insight into the key parameters driving the spread of this disease.

6.5.1 A Model without Vital Dynamics

We consider a simple model in which individuals do not enter or leave the system through natural birth or death (or immigration or emmigration). A disease that spreads rapidly through a population is a good candidate for this framework. Pioneering work in epidemiology was done in 1927 by Kermack and McKendrick [18] in their study of the Bombay plague of 1906. They divided the population into $S(t)$ = number of susceptible individuals, $I(t)$ = number of recovered individuals, and $R(t)$ = number of removed (or deceased) individuals. They made three basic assumptions about the disease:

1. It traveled quickly through the population and thus no people were able to leave or enter the system. There are no births, deaths, and no immigration or emmigration. In epidemiological terms, this is a model *without vital dynamics.*

2. When an infected individual encounters a susceptible individual, there is a probability, β, that the susceptible individual will get the disease. This occurs in proportion to the numbers of individuals in the infected and susceptible classes.

3. Infected individuals recover at a constant rate, α.[2]

With these assumptions, we can write the following set of equations:

$$\frac{dS}{dt} = -\beta \frac{I}{N} S$$

$$\frac{dI}{dt} = \beta \frac{I}{N} S - \alpha I$$

$$\frac{dR}{dt} = \alpha I, \tag{6.42}$$

where $N = S + I + R$ is the total population and $S(t), I(t), R(t) \geq 0$; see Figure 6.6.

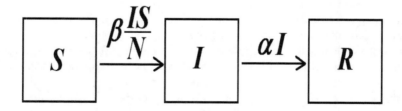

FIGURE 6.6: Flowchart for SIR model without vital dynamics. This diagram illustrates the movement of individuals from one class to another.

Another way of interpreting β is that it is the average number of contacts per unit time sufficient for transmission of the disease, sometimes referred

---

[2]If we let $u(t)$ represent the proportion of infected individuals remaining at t time units, then we can write $u' = -\alpha u$. The solution is $u(t) = u(0)e^{-\alpha t}$ and thus the fraction that is still infective after t time units is $e^{-\alpha t}$. If we think of this in terms of probability, we see this is an exponential distribution and thus the average length of the infective period is $1/\alpha$.

to as the average number of *adequate contacts*. Thus the average number of contacts that a susceptible individual has with infectives per unit time is given by $\beta I / N$. The number of new infections per unit time in a susceptible community with S individuals is $\left(\frac{\beta I}{N}\right) S$. Since α is the recovery rate, the average length of time an individual will remain infected is given by $1/\alpha$. If we multiply the average number of adequate contacts β by the average infectious period $1/\alpha$, we get the "average number of adequate contacts of a typical infective during the infectious period" (from Hethcote [15]). It is sometimes useful to consider only fractions of a population and we can do so with this model by introducing new variables[3]: $s = S/N, \iota = I/N, r = R/N$. In the current model (6.42), we can divide both sides of the three equations by the total population N and thus obtain

$$\frac{ds}{dt} = -\beta s\iota$$

$$\frac{d\iota}{dt} = \beta\iota s - \alpha\iota$$

$$\frac{dr}{dt} = \alpha\iota. \qquad (6.43)$$

The variables under consideration will satisfy $0 \le s, \iota, r \le 1$ and $s + \iota + r = 1$. Model (6.43) is often easier to deal with both in terms of interpreting the results and from a numerical point of view. It allows us to consider relative changes in the population size. Mathematically, the two formulations are equivalent. Homework problem 6 requires the reader to go through this derivation.

Since the total population is constant we have

$$s' + \iota' + r' = 0,$$

and we can reduce the dimension of our system to two by substituting $r = 1 - s - \iota$. We thus arrive at the equivalent yet simpler formulation

$$\frac{ds}{dt} = -\beta s\iota$$

$$\frac{d\iota}{dt} = \beta\iota s - \alpha\iota. \qquad (6.44)$$

Our solution can now be examined in $\mathbb{R}^2$ rather than $\mathbb{R}^3$. We could solve these equations analytically by considering $d\iota/ds = \iota'/s'$ but we will instead use the linearization methods covered in Section 6.4. We can observe that there is one equilibrium point at

$$s^* = 1, \iota^* = 0,$$

---

[3]We use the Greek letter "iota" (written as the letter "i" without the dot) so as not to confuse it with the imaginary $i = \sqrt{-1}$, which will sometimes arise as an eigenvalue in these problems.

which corresponds to the entire population being susceptible and none are infected. Note that any s^* would work but we are interested in understanding how the disease spreads across a susceptible population, hence the assumption that $s^* = 1$ is the appropriate choice for the equilibrium value. This equilibrium point in which there are no infected individuals is known as the *disease-free equilibrium (DFE)*. We calculate the Jacobian matrix of (6.44) as

$$\begin{pmatrix} -\beta\iota & -\beta s \\ \beta & \beta s - \alpha \end{pmatrix} \tag{6.45}$$

and evaluate it at the *DFE* to obtain

$$\begin{pmatrix} 0 & -\beta \\ \beta & \beta - \alpha \end{pmatrix}. \tag{6.46}$$

We can use the trace and determinant to determine its stability. Here we have

$$\beta = \mathrm{Tr}(J(1,0)) = \beta - \alpha$$

and

$$\gamma = \det(J(1,0)) = \beta^2 > 0$$

so $(0,0)$. The latter condition is always true and thus the

$$DFE \text{ is stable when } \beta < \alpha$$
$$DFE \text{ is unstable when } \beta > \alpha. \tag{6.47}$$

Recalling that β is average number of *adequate contacts* for disease transmission and α is the recovery rate of infected individuals, we see that this stability condition for the *DFE* is believable. *The disease will spread if the average number of adequate contacts β is larger than the recovery rate α.* Mathematical epidemiologists often consider the *basic reproductive number*, R_0, which gives the average number of infections caused by one infected individual over his/her period of infection as an equivalent method of determining if a disease will spread. Thus $R_0 < 1$ says that, on average, an infected individual infects less than 1 individual (for example, 10 infected individuals might only infect 7 others before recovering). Thus the disease will eventually die out. Similarly, $R_0 > 1$ says that, on average, an infected individual infects more than 1 individual (for example, 5 infected individuals might infect 8 others before recovering). In this latter case, the disease will spread through the population. If $R_0 > 1$, we say that there is an *epidemic*.

The mathematical epidemiologists examine the stability conditions for the *DFE* and determine which condition or conditions will first cause it to become unstable. When considering R_0, the stability conditions are rewritten in terms of R_0 such that

$$R_0 < 1 \iff DFE \text{ equilibrium is stable}$$
$$R_0 > 1 \iff DFE \text{ equilibrium is unstable}. \tag{6.48}$$

In our case, there is only one condition that will cause a change in stability. Our goal is to manipulate that expression so that $R_0 = 1$ corresponds to the switch in stability. In our current example, we can manipulate (6.47) to obtain

$$R_0 = \frac{\beta}{\alpha}.$$

Following the conditions in (6.48), the DFE is stable if $\frac{\beta}{\alpha} < 1$ and unstable if $\frac{\beta}{\alpha} > 1$, which is just what we previously found.

In the exercises, you will show that when a small number of infective individuals are introduced into a susceptible population,

$$\frac{\beta}{\alpha} \approx \frac{\ln\left(\frac{s(0)}{s^*}\right)}{1 - s^*}. \tag{6.49}$$

While it is usually not hard to determine the average length of the infection for a given disease, without this approximation it can often be difficult to determine the number of adequate contacts.

Example 1: Brauer and Castillo-Chavez [8] Consider a geographically isolated college campus and suppose 95% of the students are susceptible to the influenza virus at the beginning of the school year. By the end of the year, many had become sick with the flu and only 42% were still susceptible after the flu had run its course on campus. Estimate the basic reproductive number for this flu and determine if there was an epidemic. How might this epidemic have been prevented?

Since 42% of the individuals were still susceptible after the flu was gone, we see that $s^* = .42$. Similarly, we see that $s(0) = .95$. Thus we see that

$$R_0 = \frac{\beta}{\alpha} \approx \frac{\ln\frac{s(0)}{s^*}}{1 - s^*} = \frac{\ln\frac{.95}{.42}}{1 - .42} = 1.41,$$

and there was indeed an epidemic on the campus. Preventing the epidemic would require us to alter some of the values in the formula for R_0. For example, if we were to restrict the interactions of the students (which may or may not be practical), we could change the contact rate, thereby lowering R_0. We also could have *vaccinated* individuals at the start of the year. This would lower the initial number of susceptible individuals and would also reduce the number of students that had the flu. In this latter situation, we should note that we can't selectively vaccinate those students who will end up getting sick because we don't know in advance who will get sick.

6.5.2 A Model with Vital Dynamics

We now consider a somewhat more realistic model by allowing natural births and deaths in the population to occur. We again let S represent the number of susceptible individuals, I represent the number of infected individuals, and R represent the number of recovered individuals. We assume that the death rate and birth rate are the same so that there is no change in the overall population. Denote this rate as $\mu > 0$ and let $N = S + I + R$ again represent the total population. We assume that all individuals are born susceptible to the disease even when they are born from an infected individual. The number of births from the susceptible class, the infected class, and the recovered class are defined as μS, μI, and μR, respectively. Thus μN is the total number of births or newcomers into the susceptible class. Similarly the number of deaths in each class per unit time are given by μS, μI, and μR, so that the population level remains constant. If we again make assumptions 2 and 3 from the previous model, our diagram is shown in Figure 6.7 and the corresponding equations are

$$\frac{dS}{dt} = \mu N - \mu S - \beta \frac{I}{N} S$$
$$\frac{dI}{dt} = \beta \frac{I}{N} S - \alpha I - \mu I$$
$$\frac{dR}{dt} = \alpha I - \mu R. \tag{6.50}$$

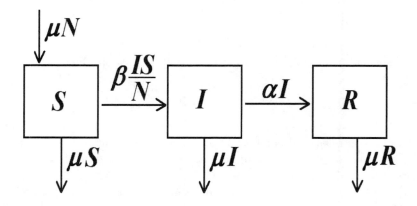

FIGURE 6.7: Flowchart for SIR model with vital dynamics.

This is called an *SIR model with vital dynamics*. Because N is constant, we can reduce the system to only two equations. If we consider the normalized populations of $s = S/N, \iota = I/N, r = R/N$, we can rewrite the governing equations as

$$\frac{ds}{dt} = \mu - \mu s - \beta \iota s$$

$$\frac{d\iota}{dt} = \beta \iota s - (\alpha + \mu)\iota. \qquad (6.51)$$

(The reader should convince him/herself that this can be done; see homework problem 7.) We have two equilibria (s^*, ι^*) for this system,

$$(1,0) \text{ and } \left(\frac{\alpha + \mu}{\beta}, \frac{\mu}{\alpha + \mu} - \frac{\mu}{\beta}\right). \qquad (6.52)$$

We have seen the first equilibrium point before—it's the *DFE* in which there is no disease and everyone is susceptible. The latter is new to us—it is called an *endemic equilibria (EE)* and represents the population levels if the disease persists.

It is very important to observe that the *EE* only makes sense biologically if $\iota^* > 0$, i.e., if $\frac{\mu}{\alpha + \mu} > \frac{\mu}{\beta}$. The Jacobian of the system can be calculated as

$$J(s^*, \iota^*) = \begin{pmatrix} -\beta \iota^* - \mu & -\beta s^* \\ \beta \iota^* & \beta s^* - (\alpha + \mu) \end{pmatrix}. \qquad (6.53)$$

Evaluating the Jacobian at the *DFE* gives

$$J_{DFE} \begin{pmatrix} -\mu & -\beta \\ 0 & \beta - (\alpha + \mu) \end{pmatrix}. \qquad (6.54)$$

We calculate the eigenvalues here instead of using the trace and determinant because the former is easier in this case since we have a triangular matrix. For systems with dimension higher than two, the latter method is not applicable. The eigenvalues for the *DFE* are

$$\lambda_1 = -\mu, \quad \lambda_2 = \beta - (\alpha + \mu).$$

Since $\mu > 0$ the *DFE* is a stable node if $\beta < (\alpha + \mu)$ and is a saddle point if $\beta > (\alpha + \mu)$. Based on these stability conditions the *DFE* switches stability when $\beta = \alpha + \mu$. From this we can conclude that the basic reproductive number is

$$R_0 = \frac{\beta}{\alpha + \mu}. \qquad (6.55)$$

This says that the disease will die out as long as the contact rate is less than the recovery rate plus the death rate (i.e., $R_0 < 1$). This should be believable because in order for the disease to persist, there should be at least as many people coming into the infective class as there are leaving it (due to death or recovery).

We noted above that the EE will only exist if $\iota^* > 0$, i.e., if $\frac{\mu}{\alpha+\mu} > \frac{\mu}{\beta}$. This condition can be rearranged to say that

$$\iota^* > 0 \iff \beta > \alpha + \mu.$$

This is exactly the condition for the instability of the DFE. It is no coincidence that the EE becomes biologically relevant at the instant when the DFE becomes unstable! We leave it as an exercise for the reader to show that

EE is a saddle point (and not biologically relevant) when $R_0 < 1$, and

EE is stable (and biologically relevant) when $R_0 > 1$.   (6.56)

We can take sample parameter values to show some of the plots. In Figure 6.8, the parameter values are chosen as $\beta = .3, \alpha = .4, \mu = .02$. This would correspond to an average infectious period of $\frac{1}{\alpha} = 2.5$ days and an average lifetime of $\frac{1}{\mu} = 50$ days.[4] These values give a basic reproductive number that is less than 1 and thus the EE is not biologically relevant. We note in Figure 6.8 that while mathematically we can choose any values for s and ι, their biological meaning requires us to only choose values where $s + \iota \leq 1$.

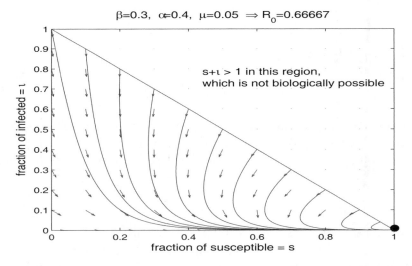

FIGURE 6.8: Only the DFE is biologically relevant. $R_0 < 1$ and the DFE is stable.

If we increase the contact rate to $\beta = .6$, we obtain an EE that is biologically relevant and stable. The DFE is a saddle for these parameter values

---

[4]The value $\mu = .02$ is an unrealistically low value that is used only to be consistent with Figure 6.9. The reason will be justified then.

and $R_0 > 1$. The previously used value of $\mu = .02$ is necessary in order to graphically see the equilibria above the horizontal axis. (The reader should refer to equation (6.52) to calculate the location of the EE for these parameter values. However, even for realistic values of μ, the EE will exist and be stable whenever $R_0 > 1$. See Figure 6.9, which contains the solutions in the phase plane and the corresponding time series plots for one of the solution curves. We must again impose the mathematical restriction that $s + \iota \leq 1$

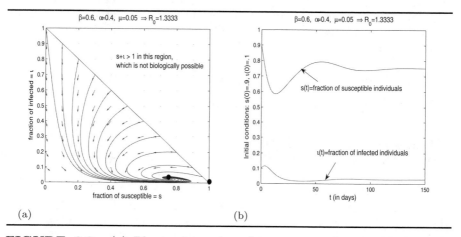

(a) (b)

FIGURE 6.9: (a) Phase plane plot of numerous solutions; the DFE is located at $(s^*, \iota^*) = (1, 0)$; the EE is at $(s^*, \iota^*) = (= .75, .027778)$. (b) Time series plot for the initial condition $s(0) = .9, \iota(0) = .1$.

since our total normalized population cannot exceed 1 (and we know that $r = 1 - s - \iota$). In Figure 6.9(a), we see that solutions are spiraling into the EE. Mathematically, we can show that the EE is a spiral sink. Biologically, the damped oscillations we see in Figure 6.9(b) can be interpreted as the effect of the mass action law together with the vital dynamics: the more susceptibles there are, the greater the likelihood of adequate contacts thus resulting in more infections. This results in a decrease of susceptibles in the next time period. The reduction in the fraction of susceptibles that results from more infections makes it less likely for an infective individual to have an adequate contact with a susceptible and infect. This will then allow for more susceptibles in the next time period and so on until a steady state is reached.

We go back again to the basic reproductive number to give one additional interpretation of it. Let's rewrite this as

$$R_0 = \beta \cdot \left(\frac{1}{\alpha + \mu} \right).$$

We said that β can be interpreted as the average number of susceptibles infected by infectious individuals per unit time. By a similar argument given in the footnote in the derivation of equation (6.42), we can interpret

$$\frac{1}{\alpha + \mu}$$

as the average length of the infectious period. Thus R_0 is the number of infections caused by an infected individual during her/his period of infectiousness. The interested reader is again encouraged to examine Hethcote [15], Daley and Gani [11] and Brauer and Castillo-Chavez [8] for a more in depth look at mathematical epidemiology.

6.5.3 Bifurcations in Epidemiological Systems

Recall that a bifurcation is a qualitative change in the system, often due to a change in the number or the stability of equilibria. Using the information in Sections 2.3.3 and 6.4.2, we see that the *DFE* and *EE* underwent a transcritical bifurcation where the *EE* passed through the DFE as it became biologically relevant and the two switched stability. This is typically what happens in an epidemiological system. For larger epidemiological systems involving more classes, the *EE* is often very difficult to obtain as a closed-form expression. In such cases, the existence of an *EE* is often deduced from the switch in stability of the *DFE* through a transcritical bifurcation! The analytical results can then be confirmed numerically.

In plotting the transcritical bifurcation curves, we plot each one of the equilibrium points' coordinates, s^* or ι^*, as a function of one specific parameter. When it is feasible, we consider R_0 as a function of the parameter we are varying and plot R_0 as the horizontal component (instead of a specific parameter of the model). This allows us to see that the transcritical bifurcation occurs when R_0 goes through 1, with the *EE* becoming biologically relevant and stable. For example, consider the coordinates of the *EE* in the model given in equation (6.51),

$$(s^*, \iota^*) = \left(\frac{\alpha + \mu}{\beta}, \frac{\mu}{\alpha + \mu} - \frac{\mu}{\beta} \right).$$

We can rearrange terms and manipulate the expression of i^* to make it an explicit function of R_0:

$$\iota^* = \frac{\mu}{\beta} (R_0 - 1). \tag{6.57}$$

This gives a line in the R_0-ι plane. If we also plot the curve produced by the ι^* coordinate of the DFE, $\iota^* = 0$ (a horizontal line in the R_0-ι plane), we obtain the plot in Figure 6.10.

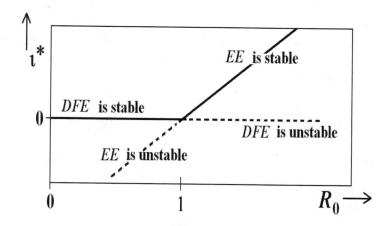

FIGURE 6.10: Bifurcation diagram plotted in the R_0-ι plane. A transcritical bifurcation occurs because the DFE and EE switch stability as the EE becomes biologically relevant.

While the transcritical bifurcation is the typical way for the EE to gain stability, it is sometimes part of another bifurcation curve (a saddlenode bifurcation curve). In these situations, we can have a *backward bifurcation* and the EE can exist even when $R_0 < 1$. This means that continuously varying the parameters back to their pre-epidemic values may not be enough to get rid of the epidemic. This phenomenon is known as *hysteresis* and is also observed in many physical systems [34].

Problems

1. In this exercise, we use (6.44) to obtain (6.49).
   a. Find an implicit solution to (6.44) in which time is not explicit by solving $d\iota/ds = \iota'/s'$. Write this solution in the form $f(s, \iota) = c$.
   b. Observe that your implicit solution $f(s, \iota) = c$ describes the evolution of the disease for a given initial condition. Show that $f(s(0), \iota(0)) = f(s^*, \iota^*)$.
   c. Use the initial approximations $s(0) \approx 1$ and $\iota(0) \approx 0$, and the limiting values (after the disease has passed) $\iota^* = 0$, $s^* = 0$, along with your answers in parts a. and b. to obtain

   $$\beta/\alpha \approx \frac{\ln\left(\frac{s(0)}{s^*}\right)}{1 - s^*}.$$

2. Consider Example 1. Assuming that the infective period of the flu is

about 3.5 days, determine the adequate number of contacts required for the disease to spread.

3. Again consider Example 1 and suppose that a vaccination strategy is implemented at the beginning of the school year. Determine what percentage of students must receive the vaccination in order to make $R_0 < 1$. You may assume, for example, that if 10% of the students are vaccinated, then $.95 \times .1$ were initially susceptible and $.42 \times .1$ would not have developed the flu even though they received the vaccination.

It is important to check that a given model is a well-posed one. That is, there should exist an invariant set in which solution can enter or leave a given region of phase space. In the case of a fixed population we need to have the total number of individuals remain constant and none of the variables should ever become negative-valued. Use this information to answer the following two questions.

4. a. By substituting $s = 0$ into (6.44), show that no individuals will be able to become susceptible.
   b. By substituting $\iota = 0$ into (6.44), show that no individuals will be able to become infected.
   c. Use $s + \iota \leq 1$ and your results from a and b to conclude that

   $$\{(s, \iota)|0 \leq s, 0 \leq \iota, s + \iota \leq 1\}$$

   is a positively invariant set.

5. a. By substituting $s = 0$ into (6.51), show that no individuals will be able to become susceptible.
   b. By substituting $\iota = 0$ into (6.51), show that no individuals will be able to become infected.
   c. Use $s + \iota \leq 1$ and your results from a and b to conclude that

   $$\{(s, \iota)|0 \leq s, 0 \leq \iota, s + \iota \leq 1\}$$

   is a positively invariant set.

6. Derive the normalized model without vital dynamics, (6.43).

7. Derive the normalized (reduced) model with vital dynamics, (6.51).

8. Consider an illness that is passing through an isolated college campus that is on the quarter system. Assume that students that get the illness can function okay for a while but then often have to drop out for the remainder of the quarter because they are falling too behind in their work. Thus we have susceptible individuals, infected individuals, and drop-outs (that are removed from the population and are assumed to no longer have contact with any enrolled student). Our model can be

written as the SIR model without vital dynamics

$$\frac{dS}{dt} = -\beta\frac{I}{N}S$$

$$\frac{dI}{dt} = \beta\frac{I}{N}S - \alpha I$$

$$\frac{dR}{dt} = \alpha I, \qquad\qquad (6.58)$$

where S and I have their normal meaning and R stands for those students that are removed from the population, i.e., that have dropped out. Our goal is to normalize (6.58).

a. Define $N = S + I$ as the currently enrolled population and $\tilde{N} = S + I + R$ as the total population. Decide if the variables N and $\tilde{N}$ are constant or dependent on time. Explain why $s = S/N$, $\iota = I/N$, and $r = R/\tilde{N}$ are the appropriate choices for the new variables. (Note that the denominator in the r-variable is different.)

b. Explain why $N/\tilde{N} = (1 - r)$.

c. Use the definitions from part a to rewrite model (6.58) as

$$\frac{ds}{dt} = -\beta s\iota$$

$$\frac{d\iota}{dt} = \beta\iota s - \alpha\iota$$

$$\frac{dr}{dt} = \alpha\iota(1 - r). \qquad\qquad (6.59)$$

Note that this is *not* the same as (6.43).

d. Analyze this new system, (6.59).

This normalization technique has also been used in a model of college drinking by Almada et al. [2].

9. Consider the model given by (6.50) with a constant influx, Λ, of people into the population

$$\frac{dS}{dt} = \Lambda - \mu S - \beta\frac{I}{N}S$$

$$\frac{dI}{dt} = \beta\frac{I}{N}S - \alpha I - \mu I$$

$$\frac{dR}{dt} = \alpha I - \mu R. \qquad\qquad (6.60)$$

By adding the equations together, obtain an expression for $N'(t)$, where $N = S + I + R$. Solve this equation and describe the population level as $t \longrightarrow \infty$.

10. Consider the model given by (6.50) in which susceptible individuals can

be vaccinated at a rate ν

$$\frac{dS}{dt} = \mu N - (\mu + \nu)S - \beta\frac{I}{N}S$$

$$\frac{dI}{dt} = \beta\frac{I}{N}S - \alpha I - \mu I$$

$$\frac{dR}{dt} = \alpha I - \mu R + \nu S. \qquad (6.61)$$

a. Show that the population level remains constant.
b. By considering only the first two equations (justified by your result in part a), compute the DFE.
c. Determine the stability of the DFE.
d. From your results of part c, find an expression for R_0 and give an interpretation of it.

11. Many diseases, such as tuberculosis and HIV, have an *exposed period* in which the susceptibles have contracted the disease but have not yet developed symptoms and cannot transmit the disease. After a period of time, they become infectious and are then able to transmit the diseases to susceptible people.

$$\frac{dS}{dt} = -\beta S\frac{I}{N} + \mu N - \mu S$$

$$\frac{dE}{dt} = \beta S\frac{I}{N} - \mu E - \delta E$$

$$\frac{dI}{dt} = \delta E - \mu I - \gamma I$$

$$\frac{dR}{dt} = \gamma I - \mu R. \qquad (6.62)$$

a. Show that the total population $N = S + E + I + R$ remains constant.
b. Calculate a simpler system involving only three equations.
c. Find R_0 for this simpler system and give an interpretation of it.

6.6　Chapter 6: Additional Problems and Projects

ADDITIONAL PROBLEMS

In problems 1–5, determine whether the statement is true or false. If is true, give reasons for your answer. If it is false, give a counterexample or other explanation of why it is false.

1. All linear systems of differential equations can be solved by first finding the eigenvalues and eigenvectors of the coefficient matrix and then taking a linear combination of $e^{\lambda_i t} \mathbf{v}_i$.

2. Linearization of a nonlinear system about an equilibrium solution reduces it to a linear system that can be solved by methods of Sections 6.1 and 6.2.

3. A linear system of differential equations with repeated complex eigenvalues cannot be solved.

4. Consider an nth order linear differential equation (studied in Chapter 4) that is written as a system of first-order equations (via the methods of Section 3.5.1). The solution obtained to the nth order equation by the methods of Chapter 4 is mathematically equivalent to the solution obtained by solving the system via methods of Chapter 6.

5. A basis can be found so that every matrix can be diagonalized or written in Jordan normal form with respect to this basis.

In problems 6–13, (i) find the eigenvalues and eigenvector of the coefficient matrix by hand; (ii) use Theorem 6.1.2 to classify the stability of the origin (the only equilibrium solution); (iii) find two linearly independent solutions; (iv) then use Theorem 6.2.4 to find the general solution. Part a has real, distinct roots; part b has complex (non-real) roots.

6. a. $\begin{cases} x' = 2x \\ y' = x + y \end{cases}$ b. $\begin{cases} x' = 8y \\ y' = -2x \end{cases}$

7. a. $\begin{cases} x' = x + y \\ y' = 3x - y \end{cases}$ b. $\begin{cases} x' = x + 2y \\ y' = 2x - 3y \end{cases}$

8. a. $\begin{cases} x' = 6x - y \\ y' = 5x \end{cases}$ b. $\begin{cases} x' = 7x - 4y \\ y' = 4x + 7y \end{cases}$

9. a. $\begin{cases} x' = x + 2y \\ y' = -6x - 6y \end{cases}$ b. $\begin{cases} x' = -y \\ y' = 16x \end{cases}$

10. a. $\begin{cases} x' = -11x + 6y \\ y' = -18x + 10y \end{cases}$ b. $\begin{cases} x' = -2x - y \\ y' = 13x + 4y \end{cases}$

11. a. $\begin{cases} x' = y \\ y' = -2x - 3y \end{cases}$ b. $\begin{cases} x' = 5x + 10y \\ y' = -x - y \end{cases}$

12. a. $\begin{cases} x' = -10x - y \\ y' = -4x - 10y \end{cases}$ b. $\begin{cases} x' = -x + 4y \\ y' = -2x + 3y \end{cases}$

13. a. $\begin{cases} x' = -6x + y \\ y' = -5x \end{cases}$ b. $\begin{cases} x' = 3x + 2y \\ y' = -2x + 5y \end{cases}$

For problems 14–16, (i) find the eigenvalues and eigenvector of the coefficient matrix by hand (the eigenvalues are all repeated with only one eigenvector); (ii) use Theorem 6.1.2 to classify the stability of the origin (the only equilibrium solution); (iii) use the algorithm in Section 6.2.1 to obtain a generalized eigenvector; (iv) then use Theorem 6.2.2 and Theorem 6.2.4 to write the general solution.

14. a. $\begin{cases} x' = -x + y \\ y' = -y \end{cases}$ b. $\begin{cases} x' = x + y \\ y' = -9x - 5y \end{cases}$

15. a. $\begin{cases} x' = 7x - 4y \\ y' = 9x - 5y \end{cases}$ b. $\begin{cases} x' = -4x + 12y \\ y' = -3x + 8y \end{cases}$

16. a. $\begin{cases} x' = 3x + 2y \\ y' = 3y \end{cases}$ b. $\begin{cases} x' = x + 4y \\ y' = -x + 5y \end{cases}$

In problems 17–21, we consider the system $\mathbf{x}' = \mathbf{Ax}$.
*   (i) Solve using the fundamental matrix.*
*   (ii) Solve using diagonalization.*
*   (iii) Solve using the matrix exponential.*
*   (iv) Solve using variation of parameters.*

17. $\begin{cases} x' = 2x \\ y' = x + y \end{cases}$

18. $\begin{cases} x' = x + y \\ y' = 3x - y \end{cases}$

19. $\begin{cases} x' = 6x - y \\ y' = 5x \end{cases}$

20. $\begin{cases} x' = 4x - 8y - 10z \\ y' = -x + 6y + 5z \\ z' = x - 8y - 7z \end{cases}$

21. $\begin{cases} x' = 2x - 2y + z \\ y' = -3x + 3y + z \\ z' = 3x - 2y \end{cases}$

For problems 22–27, (i) find the eigenvalues and eigenvector of the coefficient matrix using Matlab, Maple, or Mathematica; (ii) use Theorem 6.1.2 to classify the stability of the origin (the only equilibrium solution); (iii) use the methods of Sections 6.1 and 6.2 to obtain three linearly independent solutions and write the general solution.

22. a. $\begin{cases} x' = -11x + 4y \\ y' = 8x - 25y \\ z' = -6x + 12y - 9z \end{cases}$ b. $\begin{cases} x' = -8x - 8y - 2z \\ y' = -13x + 14y - z \\ z' = 12x - 15y - 6z \end{cases}$

23. a. $\begin{cases} x' = 2x - 16y + 6z \\ y' = -3x + 4y - 3z \\ z' = -4x + 19y - 8z \end{cases}$ b. $\begin{cases} x' = -13x + 4y - 8z \\ y' = 8x + y + 4z \\ z' = 42x - 6y + 21z \end{cases}$

24. a. $\begin{cases} x' = 11x - 6y + 10z \\ y' = -5x + 4y - 5z \\ z' = -13x + 9y - 12z \end{cases}$ b. $\begin{cases} x' = -x + z \\ y' = -3y + z \\ z' = -y - z \end{cases}$

25. a. $\begin{cases} x' = 4x - 8y - 10z \\ y' = -x + 6y + 5z \\ z' = x - 8y - 7z \end{cases}$ b. $\begin{cases} x' = -2y \\ y' = x + 3y \\ z' = -x - 2y + z \end{cases}$

26. a. $\begin{cases} x' = 2x - 2y + z \\ y' = -3x + 3y + z \\ z' = 3x - 2y \end{cases}$ b. $\begin{cases} x' = x - 2y + 2z \\ y' = -4x + 3y + 2z \\ z' = 4x - 2y - z \end{cases}$

27. a. $\begin{cases} x' = -17x + 4y + 2z \\ y' = 5x - 25y + z \\ z' = 3x + 12y - 12z \end{cases}$ b. $\begin{cases} x' = 6x - 16y + 10z \\ y' = -5x + 4y - 5z \\ z' = -8x + 19y - 12z \end{cases}$

For problems 28–32, (i) find the equilibria of the given system; (ii) use linearization and Theorem 6.4.2 to classify the stability of the equilibria; (iii) use Matlab, Maple, or Mathematica to draw the vector field of the system; (iv) sketch trajectories on the vector field for various initial conditions (either by hand or with the computer). You should verify that your answers from parts (iii) and (iv) agree with your predictions in parts (i) and (ii).

28. $\begin{cases} x' = x + 1 \\ y' = x + y^2 \end{cases}$

29. $\begin{cases} x' = y^2 - 1 \\ y' = x \end{cases}$

30. $\begin{cases} x' = y^3 - y \\ y' = x - y \end{cases}$

31. $\begin{cases} x' = x(y - 1) \\ y' = x - y^2 \end{cases}$

32. Use Matlab, Maple, or Mathematica to examine the Rössler system [34].

$$x' = -y - z$$
$$y' = x + ay$$
$$z' = b + z(x - c)$$

For $a = b = .2, c = 2.5$, determine the equilibria and their stability. Use the computer to draw the phase portrait. Describe what you see. Then repeat these steps for $a = b = .2, c = 3.5$. Again, describe what you see.

PROJECTS FOR CHAPTER 6

Project 1: An MSEIR Model [15]

Consider a model for a disease for which infection confers permanent immunity (e.g., measles, rubella, mumps, and chicken pox) and for which a mother's immunity gives her newborn infant *passive* immunity from the disease; that is, some of the mother's antibodies will protect the newborn from the disease for, say, the first six months of her/his life. Verify that Figure 6.11 gives rise to the following system of differential equations:

$$\frac{dM}{dt} = b(N - S) - \epsilon M - \mu M$$

$$\frac{dS}{dt} = bS + \epsilon M - \beta S \frac{I}{N} - \mu S$$

$$\frac{dE}{dt} = \beta S \frac{I}{N} - \delta E - \mu E$$

$$\frac{dI}{dt} = \delta E - \gamma I - \mu I$$

$$\frac{dR}{dt} = \gamma I - \mu R, \tag{6.63}$$

where $N = M + S + E + I + R$ is the total population.

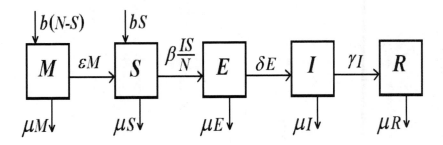

FIGURE 6.11:   Flowchart for $MSEIR$ model with exponentially changing size.

Show that

$$\frac{dN}{dt} = (b - \mu)N,$$

where $N = M + S + E + I + R$ is the total population. Is the total population constant or changing? Rewrite the differential equation in its fractional form (with $m = M/N, s = S/N, e = E/N, \iota = I/N, r = R/N$) and eliminate s from the system to obtain

$$\frac{dm}{dt} = (\mu + q)(e + \iota + r) - \epsilon m$$

$$\frac{de}{dt} = \beta(1 - m - e - \iota - r)\iota - (\delta + \mu + q)e$$

$$\frac{d\iota}{dt} = \delta e - (\gamma + \mu + q)\iota$$

$$\frac{dr}{dt} = \gamma \iota - (\mu + q)r, \tag{6.64}$$

where $q = b - \mu$ is the difference between the birth and death rates. Show that

$$\{(m, e, \iota, r) | 0 \le m, 0 \le e, 0 \le \iota, 0 \le r, m + e + \iota + r \le 1\}$$

is a positively invariant domain. Determine conditions on the stability of the *DFE* and show that the basic reproductive number is given by

$$R_0 = \frac{\beta\delta}{(\gamma + \mu + q)(\delta + \mu + q)}.$$

Interpret the meaning of R_0 in terms of the parameters of the system. Try to find the unique endemic equilibrium of the system that lies in the invariant region. Determine its stability and when it is biologically relevant. For chicken pox, $1/\epsilon = 6$ months, $1/\epsilon = 14$ days, and $1/\gamma = 7$ days; choose reasonable values for the birth and death rates in (6.64). Choose two different β-values that will give qualitatively different behavior in the system (only the *DFE* and then another that gives both an *EE* and a *DFE*). Give plots for these two different situations similar to that of Figure 6.9.

Project 2: Jordan Normal Form
In Example 2 of Section 6.2.1, we examined

$$\mathbf{x}' = \begin{pmatrix} -1 & 2 & -4 \\ 0 & -1 & 0 \\ 0 & 0 & -1 \end{pmatrix} \mathbf{x}.$$

The eigenvalue $\lambda = -1$ was of multiplicity 3 and we found two eigenvectors

$$\mathbf{v}_1 = \begin{pmatrix} 0 \\ 2 \\ 1 \end{pmatrix}, \quad \mathbf{v}_2 = \begin{pmatrix} 1 \\ 0 \\ 0 \end{pmatrix},$$

and one generalized eigenvector from $\mathbf{v}_2$:

$$\mathbf{u}_1 = \begin{pmatrix} 0 \\ 1/2 \\ 0 \end{pmatrix}.$$

In Appendix C.3.1, we saw that when a matrix has a full set of eigenvectors it can be diagonalized as $\boldsymbol{\Lambda} = \mathbf{V}^{-1}\mathbf{A}\mathbf{V}$. For Example 2, create a matrix that has $\mathbf{v}_1, \mathbf{v}_2, \mathbf{u}_1$ as its first, second, and third columns, respectively. Show that

$$\mathbf{J} = \mathbf{V}^{-1}\mathbf{A}\mathbf{V} = \begin{pmatrix} -1 & 0 & 0 \\ 0 & -1 & 1 \\ 0 & 0 & -1 \end{pmatrix}.$$

The lower 2×2 *block matrix* is known as a Jordan block and indicates that we found one generalized eigenvector from $\mathbf{v}_2$. The matrix is said to be in

Jordan normal form, which we can think of as the next best thing to have if we can't diagonalize a matrix. Find the Jordan normal form of

$$
\mathbf{A} = \begin{pmatrix}
7 & -8 & 2 & 5 & 3 & -7 \\
8 & -11 & 3 & 7 & 5 & -10 \\
-9 & 15 & -1 & -8 & -6 & 12 \\
0 & 0 & 0 & 2 & 0 & 2 \\
5 & -8 & 2 & 4 & 5 & -7 \\
-6 & 10 & -2 & -5 & -4 & 10
\end{pmatrix}.
$$

You may use Matlab, Maple, or Mathematica to find the eigenvectors and help you find the generalized eigenvectors. How many Jordan blocks do your have? What is the size of each one? Now experiment with the built-in commands in each of the three that will find the Jordan normal form for you. Is the computer answer different from yours? Explain.

Chapter 7

Laplace Transforms

In this chapter we will develop properties of the Laplace transform. The Laplace transform is very useful in certain applications. Consider the equation

$$m\frac{d^2x}{dt^2} + b\frac{dx}{dt} + kx = F(t)$$

for the position $x(t)$ of a mass on a spring. We saw in Chapter 4 this mass-spring system and considered some of its applications. In all of the problems we considered previously, the forcing function $F(t)$ was continuous. In many practical applications, however, $F(t)$ may have discontinuities. For instance, the mass may be periodically struck, imparting a force $F(t)$ at time t. In this case, the methods we used in Chapter 4 are awkward and the solutions may be cumbersome. It is just these kinds of applications for which the Laplace transform is useful.

The Laplace transform will be formally defined in the next section, but to set the stage and motivate what is about to come, we will let $\mathcal{L}$ represent the Laplace transform. We will see shortly that the Laplace transform of a function $f(t)$ produces a function $F(s)$, that is,

$$F(s) = \mathcal{L}\{f(t)\}$$

where F is a function of a new independent variable s.

After learning some of the basics of computing Laplace transforms of some basic functions, we will see that the Laplace transform converts a *differential* equation in $f(t)$ into an *algebraic* equation in $F(s)$. It is often the case that such an algebraic equation is easier to solve than the differential equation. We will develop a method of converting from the solution of the algebraic equation to the solution of the differential equation; this is the concept of the inverse Laplace transform. The ideas discussed here are schematically shown in the following diagram.

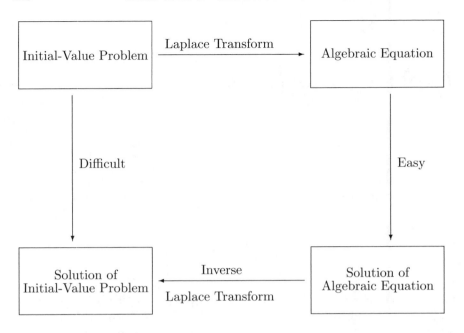

7.1   Fundamentals of the Laplace Transform

We now formally introduce the Laplace transform of a function f.

DEFINITION 7.1   *Let f be a real-valued function of the real variable t, defined for $t > 0$. Consider the function F defined by*

$$F(s) = \int_0^\infty e^{-st} f(t)\, dt \qquad (7.1)$$

for all values of s for which this integral exists. The function F is called the Laplace transform of the function f. We denote the Laplace transform F of f by $\mathcal{L}\{f\}$ and denote $F(s)$ by $\mathcal{L}\{f(t)\}$.

Note that in this definition, the integral is with respect to the variable t, with s a "dummy" variable for the integral. Thus, the Laplace transform of $f(t)$ will produce a function of s. Further, since the integral in (7.1) is improper, it may not exist for every function f for which we might wish to compute the Laplace transform. Conditions will often have to be imposed upon f in order for this integral to exist.

7.1.1 Calculation of Laplace Transforms

Using the definition, it is not difficult to calculate the Laplace transform of familiar functions. We now present several examples to illustrate this concept.

Example 1: For the function $f(t) = 1$ for all $t > 0$, the Laplace transform is

$$\mathcal{L}\{1\} = \int_0^\infty e^{-st} \cdot 1 \, dt = \int_0^\infty e^{-st} \, dt = \lim_{k \to \infty} \left. \frac{-1}{s} e^{-st} \right|_0^k$$

$$= \lim_{k \to \infty} -\frac{1}{s} e^{-sk} - \frac{-1}{s} = \frac{1}{s} \quad \text{for all } s > 0$$

so

$$\mathcal{L}\{1\} = \frac{1}{s}, \quad \text{for} \quad s > 0,$$

which is afunction of s as expected.

Example 2: Let $f(t) = t, \ t > 0$; then integrating by parts gives the Laplace transform as

$$\mathcal{L}\{t\} = \int_0^\infty e^{-st} \cdot t \, dt = \int_0^\infty t \, e^{-st} \, dt$$

$$= \lim_{k \to \infty} \left. -\frac{t}{s} e^{-st} \right|_0^k + \frac{1}{s} \int_0^\infty e^{-st} \, dt$$

$$= \lim_{k \to \infty} \left. -\frac{1}{s^2} e^{-st} \right|_0^k = \frac{1}{s^2}.$$

Hence

$$\mathcal{L}\{t\} = \frac{1}{s^2}, \quad \text{for} \quad s > 0.$$

Example 3: For the function $f(t) = e^{at}$, for $t > 0$, the Laplace transform is found from

$$\mathcal{L}\{e^{at}\} = \int_0^\infty e^{-st} e^{at} \, dt = \int_0^\infty e^{(a-s)t} \, dt$$

$$= \lim_{k \to \infty} \left. \frac{1}{a-s} e^{(a-s)t} \right|_0^k = \lim_{k \to \infty} \frac{1}{a-s} e^{(a-s)k} - \frac{1}{a-s}$$

so that if $s > a$, then $e^{(a-s)k} \to 0$ as $k \to \infty$, thus

$$\mathcal{L}\{e^{at}\} = \frac{1}{s-a} \quad \text{if } s > a.$$

Example 4: For $f(t) = \sin bt, \ t > 0$, the Laplace transform is found from

$$\mathcal{L}\{\sin bt\} = \int_0^\infty e^{-st} \sin bt \, dt.$$

Now recalling (or by integrating by parts twice),

$$\int e^{-st} \sin bt \, dt = \frac{-b \cos bt e^{-st}}{s^2 + b^2} \frac{-s \sin bt \, e^{-st}}{s^2 + b^2} \qquad \text{for } s > 0.$$

Thus

$$\int_0^\infty e^{-st} \sin bt \, dt = \lim_{k \to \infty} \int_0^k e^{-st} \sin bt \, dt$$

$$= \left[\lim_{k \to \infty} \frac{-b \cos bt \, e^{-st}}{s^2 + b^2} - \frac{s \sin bt \, e^{-st}}{s^2 + b^2} \right]_0^k$$

$$= \lim_{k \to \infty} \left[\frac{-b}{s^2 + b^2} (\cos bk \, e^{-sk}) - \frac{s}{s^2 + b^2} \sin bk \, e^{-sk} \right]$$

$$- \left[\frac{-b}{s^2 + b^2} + \frac{0}{s^2 + b^2} \right]$$

$$= \frac{b}{s^2 + b^2},$$

so

$$\mathcal{L}\{\sin bt\} = \frac{b}{s^2 + b^2}, s > 0.$$

A similar calculation shows that

$$\mathcal{L}\{\cos bt\} = \frac{s}{s^2 + b^2} \quad \text{for } s > 0.$$

In each of these examples, we have seen that the integral (7.1) exists for some (possibly infinite) range of values of s. Does the Laplace transform always exist? The answer is "No," as we will shortly see. We will now investigate conditions for which the transform does exist.

7.1.2   Existence of the Laplace Transform

We will introduce a few ideas to help us determine when the Laplace transform exists. The next two definitions will allow us to determine a class of functions for which the transform always exists.

DEFINITION 7.2 *A function f is said to be piecewise continuous on a finite interval $a \le t \le b$ if this interval can be divided into a finite number of subintervals such that*

1. *f is continuous in the interior of each of these subintervals.*

2. *$f(t)$ approaches a limit (finite) as t approaches either endpoint of the subintervals from its interior.*

More specifically, if f is piecewise continuous on $a \le t \le b$ and t_0 is an endpoint of one of the subintervals, then the right and left hand limit exists, that is,

$$f(t_0^-) = \lim_{t \to t_0^-} f(t) \quad \text{and} \quad f(t_0^+) = \lim_{t \to t_0^+} f(t) \text{ are finite and exist.}$$

Example 5: Consider the function f defined by

$$f(t) = \begin{bmatrix} -1 & 0 < t < 2 \\ 1 & t > 2. \end{bmatrix}$$

The graph of $f(t)$ is shown in Figure 7.1. The function f is piecewise continuous, for at $t = 2$ we have

$$f(2^-) = \lim_{t \to 2^-} f(t) = -1$$
$$f(2^+) = \lim_{t \to 2^+} f(t) = 1.$$

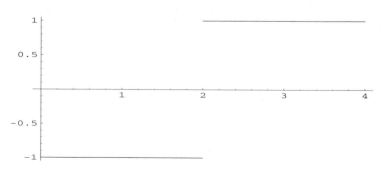

FIGURE 7.1: A piecewise defined function.

The next definition will allow us to determine how fast a function grows.

DEFINITION 7.3   *A function f is said to be of exponential order if there exists a constant α and positive constants t_0 and M such that*

$$e^{-\alpha t}|f(t)| < M$$

for all $t > t_0$ at which $f(t)$ is defined. More explicitly, if f is of exponential order corresponding to some definite constant α, then we say that f is of exponential order $e^{\alpha t}$.

In other words, we say f is of exponential order if a constant α exists such that the product $e^{-\alpha t}|f(t)|$ is bounded for all sufficiently large values of t. Thus, we have

$$|f(t)| < Me^{\alpha t},$$

so the values of $|f(t)|$ cannot become infinite more rapidly than a multiple of some exponential function $e^{\alpha t}$. Note that if f is of exponential order $e^{\alpha t}$, then f is also of exponential order $e^{\beta t}$ for any $\beta > \alpha$. Now clearly every bounded function is of exponential order, with constant $\alpha = 0$. So, for instance, $\sin bt$ and $\cos bt$ are of exponential order. The next few examples consider some combinations of familiar functions.

Example 6: For the function $f(t) = e^{at}\sin bt$, $f(t)$ is of exponential order with constant $\alpha = a$, since

$$e^{-\alpha t}|f(t)| = e^{-at}e^{at}|\sin bt| = |\sin bt|$$

which is bounded for all t.

Example 7: Consider the function $f(t) = t^n$, where $n > 0$. Then

$$e^{-\alpha t}|f(t)| = e^{-\alpha t}t^n.$$

Now for any $\alpha > 0$,

$$\lim_{t\to\infty} e^{-\alpha t}t^n = 0.$$

This means there exists $M > 0$ and $t_0 > 0$ so that

$$e^{-\alpha t}|f(t)| = e^{-\alpha t}t^n < M \quad \text{for } t > t_0.$$

Thus $f(t) = t^n$ is of exponential order, with the constant α equal to any positive number.

Example 8: The function $f(t) = e^{t^2}$ is *not* of exponential order, as

$$e^{-\alpha t}|f(t)| = e^{-\alpha t + t^2}$$

which is unbounded as $t \to \infty$ no matter the value of α.

To apply these ideas, we need a familiar result from calculus. This result gives a condition for the existence of an improper integral.

THEOREM 7.1.1 *(Comparison Test for Improper Integrals) Let g and G be real functions such that*

$$0 \le g(t) \le G(t) \quad on \ a \le t < \infty.$$

Suppose

$$\int_a^\infty G(t)\,dt \quad exists,$$

and g is integrable on every finite closed subinterval of $a \le t < \infty$. Then

$$\int_a^\infty g(t)\,dt \quad exists.$$

The function G is sometimes called a *dominating* or *majorizing* function. The comparison test gives the next result since $g(t) \le |g(t)|$ for any function $g(t)$.

THEOREM 7.1.2 *Suppose the real function g is integrable on every finite closed subinterval of $a \le t < \infty$ and suppose*

$$\int_a^\infty |g(t)|\,dt \quad exists.$$

Then

$$\int_a^\infty g(t)\,dt \quad exists.$$

With these two results in place, we can now state and prove an existence theorem for Laplace transforms.

THEOREM 7.1.3 *Let f be a real function that has the following properties:*

1. *f is piecewise continuous in every finite closed interval $0 \leq t \leq b, b > 0$*

2. *f is of exponential order, that is, there exists $\alpha, M > 0$ and $t_0 > 0$ so that*
$$e^{-\alpha t}|f(t)| < M \quad for \ t > t_0.$$

Then the Laplace transform
$$\mathcal{L}\{f\} = \int_0^\infty e^{-st} f(t)\, dt$$

of f exists for $s > \alpha$.

Proof: We have
$$\int_0^\infty e^{-st} f(t)\, dt = \int_0^{t_0} e^{-st} f(t)\, dt + \int_{t_0}^\infty e^{-st} f(t)\, dt.$$

Now by hypothesis 1, the first integral on the right exists.

Considering the second integral, by hypothesis 2 we have
$$e^{-st}|f(t)| < e^{-st} M e^{-\alpha t} = M e^{-(s-\alpha)t} \text{ for } t > t_0.$$

Also,
$$\int_{t_0}^\infty M e^{-(s-\alpha)t}\, dt = \lim_{k \to \infty} \left. \frac{-M e^{-(s-\alpha)t}}{s - \alpha} \right|_{t_0}^k$$
$$= \lim_{k \to \infty} \left(\frac{M}{s - \alpha} \right) \left(e^{-(s-\alpha)t_0} - e^{-(s-\alpha)k} \right)$$
$$= \left[\frac{M}{s - \alpha} \right] e^{-(s-\alpha)t_0} \text{ if } s > \alpha.$$

Thus
$$\int_{t_0}^\infty M e^{-(s-\alpha)t}\, dt \quad \text{exists for } s > \alpha.$$

Now by hypothesis 1 $e^{-st}|f(t)|$ is integrable on every finite closed subinterval of $t_0 \leq t < \infty$ so applying the comparison test with
$$g(t) = e^{-st}|f(t)|$$

and
$$G(t) = M e^{-(s-\alpha)t}$$

we see that

$$\int_{t_0}^{\infty} e^{-st} |f(t)|\, dt \quad \text{exists for } s > \alpha.$$

That is,

$$\int_{t_0}^{\infty} |e^{-st} f(t)|\, dt \quad \text{exists for } s > \alpha.$$

Thus,

$$\int_{t_0}^{\infty} e^{-st} f(t)\, dt \quad \text{also exists if } s > \alpha.$$

So the Laplace transform of f exists for $s > \alpha$. □

If we look at the proof, we have actually shown that if f is as stated, then

$$\int_{t_0}^{\infty} e^{-st} |f(t)|\, dt \quad \text{exists if } s > \alpha.$$

Further, hypothesis 1 shows that

$$\int_{0}^{t_0} e^{-st} |f(t)|\, dt \text{ exists.}$$

Thus

$$\int_{0}^{\infty} e^{-st} |f(t)|\, dt \quad \text{exists if } s > \alpha,$$

so not only does $\mathcal{L}\{f(t)\}$ exist, but so does $\mathcal{L}\{|f(t)|\}$ for $s > \alpha$. More precisely, we say that $\int_{0}^{\infty} e^{-st} f(t)\, dt$ is *absolutely convergent* for $s > \alpha$.

It should be pointed out that the conditions on f described in the hypothesis of the theorem are not necessary for the existence of $\mathcal{L}\{f\}$; however, there are functions f that *do not* satisfy the hypothesis of the theorem but for which $\mathcal{L}\{f\}$ exists. For instance, suppose we replace hypothesis 1 with the less restrictive condition that f is piecewise continuous in every finite closed interval $a \le t \le b$ where $a > 0$, and is such that $t^n f(t)$ remains bounded as $t \to 0^+$ for some $0 < n < 1$. Then, provided hypothesis 2 remains satisfied, it can be shown that $\mathcal{L}\{f\}$ still exists.

Example 9: Consider the function $f(t) = t^{-\frac{1}{3}}$, $t > 0$; then it can be shown that $\mathcal{L}\{f\}$ exists. For although f does not satisfy the hypothesis of the theorem [$f(t) \to \infty$ as $t \to 0^+$] it *does* satisfy the less restrictive requirement stated above (with $n = \frac{2}{3}$) and is indeed of exponential order.

Problems

In problems 1–10, use the definition of the Laplace transform to find $\mathcal{L}\{f(t)\}$ for each of the given functions $f(t)$. If $\mathcal{L}\{f(t)\}$ exists, give the domain for $F(s) = \mathcal{L}\{f(t)\}$.

1. $f(t) = t^2$
2. $f(t) = t - 3$
3. $f(t) = te^{-t}$
4. $f(t) = \begin{cases} 2 & 0 < t < 4 \\ 0 & \text{otherwise} \end{cases}$
5. $f(t) = |t - 2|$
6. $f(t) = \begin{cases} t & 0 < t < 1 \\ 1 & 1 < t < 2 \\ 0 & \text{otherwise} \end{cases}$
7. $f(t) = \begin{cases} t & 0 < t < 1 \\ t^2 & 1 < t < 2 \\ 0 & \text{otherwise} \end{cases}$
8. $f(t) = \sinh t$
9. $f(t) = te^{t\sqrt{t}}$
10. $f(t) = 28e^{13t}$
11. The gamma function $\Gamma(\alpha)$ is defined as

$$\Gamma(\alpha) = \int_0^\infty x^{\alpha-1}e^{-x}\, dx \quad \text{for } \alpha > 0.$$

We will explore further properties of the gamma function in the next chapter. In this problem, using the definition, show that

$$\mathcal{L}\{t^\alpha\} = \frac{\Gamma(\alpha+1)}{s^{\alpha+1}} \quad \text{for } \alpha > -1.$$

12. Using the results of problem 11, find the following Laplace transforms:

   a. $\mathcal{L}\left\{\frac{1}{\sqrt{t}}\right\}$ b. $\mathcal{L}\{\sqrt{t}\}$ c. $\mathcal{L}\{t^{3/2}\}$.

13. Let $f(t)$ and $g(t)$ be of exponential order. Show that their sum $f(t)+g(t)$ is also of exponential order.

14. Let $f(t)$ and $g(t)$ be piecewise continuous on $a \le t \le b$. Show that their sum $f(t) + g(t)$ is also piecewise continuous on $a \le t \le b$.

15. Using the results of problems 13 and 14 show

$$\mathcal{L}\{f(t) + g(t)\} = \mathcal{L}\{f(t)\} + \mathcal{L}\{g(t)\}.$$

16. Show that the function $f(t) = 3\sin(e^{t^2})$ is of exponential order but that its derivative is not.

7.2   Properties of the Laplace Transforms

Now that we have established the existence of the Laplace transform we will obtain some of its basic properties that we will find very useful.

Using the familiar property of the integral that the integral of a sum is the sum of the integrals, we say that the Laplace transform is *linear*.

THEOREM 7.2.1 *(Linearity) Let f_1 and f_2 be functions whose Laplace transform exists and let C_1 and C_2 be constants. Then,*

$$\mathcal{L}\{C_1 f_1(t) + C_2 f_2(t)\} = C_1 \mathcal{L}\{f_1(t)\} + C_2 \mathcal{L}\{f_2(t)\}.$$

Example 1: The linearity property can be used to compute Laplace transforms of functions such as $\sin^2 at$. That is, since

$$\sin^2 at = \frac{1 - \cos 2at}{2},$$

then

$$\mathcal{L}\{\sin^2 at\} = \mathcal{L}\{\frac{1 - \cos 2at}{2}\} = \frac{1}{2}\left[\mathcal{L}\{1\} - \mathcal{L}\{\cos 2at\}\right].$$

But

$$\mathcal{L}\{1\} = \frac{1}{s} \quad \text{and} \quad \mathcal{L}\{\cos 2at\} = \frac{s}{s^2 + 4a^2},$$

so that

$$\mathcal{L}\{\sin^2 at\} = \frac{1}{2}\left[\frac{1}{s} - \frac{s}{s^2 + 4a^2}\right]$$

$$= \frac{2a^2}{s(s^2 + 4a^2)}.$$

Since we seek to apply Laplace transforms to differential equations, we need to know how they affect the derivative of a function.

THEOREM 7.2.2 *(Differentiation) Let f be a real function that is continuous for $t \geq 0$ and of exponential order $e^{\alpha t}$ and suppose f' is piecewise continuous in every finite closed interval $0 \leq t \leq b$. Then*

$$\mathcal{L}\{f'\} \text{ exists for } s > \alpha \text{ and}$$

$$\mathcal{L}\{f'(t)\} = s\mathcal{L}\{f(t)\} - f(0).$$

Proof: Since

$$\mathcal{L}\{f'(t)\} = \int_0^\infty e^{-st} f'(t)\, dt,$$

integrating by parts gives

$$\mathcal{L}\{f'(t)\} = \lim_{k \to \infty} e^{-st} f(t) \Big|_0^k + s \int_0^\infty e^{-st} f(t)\, dt$$

$$= \lim_{k \to \infty} e^{-sk} f(k) - f(0) + s\mathcal{L}\{f(t)\}$$

$$= s\mathcal{L}\{f(t)\} - f(0) \quad [\text{ since } f \text{ is of exponential order } 1]. \ \Box$$

Example 2: If $f(t) = \sin^2 at$, then $f'(t) = 2a \sin at \cos at$ so that

$$\mathcal{L}\{2a \sin at \cos a\} = s\mathcal{L}\{\sin^2 at\}$$

$$= s \left(\frac{2a^2}{s(s^2 + 4a^2)} \right) = \frac{2a^2}{s^2 + 4a^2}.$$

In a very similar manner, it is also possible to consider higher-order derivatives.

THEOREM 7.2.3 *(Higher-Order Derivatives) Let f be a real function having a continuous $(n-1)^{st}$ derivative $f^{(n-1)}$ for $t \geq 0$ and suppose $f, f', \ldots, f^{(n-1)}$ are all of exponential order $e^{\alpha t}$. Further suppose $f^{(n)}$ is piecewise continuous. Then $\mathcal{L}\{f^{(n)}\}$ exists for $s > \alpha$ and*

$$\mathcal{L}\{f^{(n)}(t)\} = s^n \mathcal{L}\{f(t)\} - s^{(n-1)} f(0)$$
$$- s^{(n-2)} f'(0) - s^{(n-3)} f''(0) - \ldots - f^{(n-1)}(0).$$

The proof of this result follows from induction on the previous theorem.

Example 3: To find $\mathcal{L}\{\sin bt\}$, we proceed as follows:

$$\mathcal{L}\{f''(t)\} = s^2 \mathcal{L}\{f(t)\} - s f(0) - f'(0), \qquad (7.2)$$

so that we have

$$f(t) = \sin bt \text{ so } f(0) = 0.$$

Further,

$$f'(t) = b \cos bt \text{ gives } f'(0) = b$$

and

$$f''(t) = -b^2 \sin bt.$$

This yields

$$\mathcal{L}\{-b^2 \sin bt\} = s^2 \mathcal{L}\{\sin bt\} - b, \quad \text{applying } (7.2).$$

Simplifying

$$(s^2 + b^2)\mathcal{L}\{\sin bt\} = b$$

and solving for $\mathcal{L}\{\sin bt\}$ we have

$$\mathcal{L}\{\sin bt\} = \frac{b}{s^2 + b^2}$$

which agrees with Example 3 of Section 7.1.

An additional useful property of the Laplace transform is given in the next theorem.

THEOREM 7.2.4 *(Translation Property) Suppose f is such that $\mathcal{L}\{f\}$ exists for $s > \alpha$. Then for any constant a,*

$$\mathcal{L}\{e^{at}f(t)\} = F(s - a)$$

for any $s > \alpha + a$, where $F(s) = \mathcal{L}\{f(t)\}$.

Proof: Let

$$F(s) = \mathcal{L}\{f(t)\} = \int_0^\infty e^{-st} f(t)\, dt,$$

so

$$F(s - a) = \int_0^\infty e^{-(s-a)t} f(t)\, dt = \int_0^\infty e^{-st} e^{at} f(t)\, dt = \mathcal{L}\{e^{at}f(t)\}. \quad \square$$

The translation property aids in many calculations of $\mathcal{L}\{f(t)\}$ as can be seen in the following examples.

Example 4: Find $\mathcal{L}\{e^{at}t\}$.

Now

$$\mathcal{L}\{t\} = \frac{1}{s^2} \quad \text{for } s > 0$$

so

$$\mathcal{L}\{e^{at}t\} = \frac{1}{(s - a)^2} \quad \text{for } s > a.$$

Example 5: For $\mathcal{L}\{e^{at}\sin bt\}$, we have

$$\mathcal{L}\{\sin bt\} = \frac{b}{s^2 + b^2}$$

so

$$\mathcal{L}\{e^{at}\sin bt\} = \frac{b}{(s-a)^2 + b^2}.$$

Another useful property concerning the nth derivative of the Laplace transform of a function $f(t)$ is given in the next theorem.

THEOREM 7.2.5 *Suppose f is a function satisfying the hypothesis of Theorem 7.2.4, with the Laplace transform F given by*

$$F(s) = \int_0^\infty e^{-st} f(t)\, dt.$$

Then

$$\mathcal{L}\{t^n f(t)\} = (-1)^n \frac{d^n}{ds^n}[F(s)].$$

Example 6: To find $\mathcal{L}\{t^2 \sin bt\}$, we have

$$\mathcal{L}\{t^2 \sin bt\} = (-1)^2 \frac{d^2}{ds^2} F(s)$$

where

$$F(s) = \mathcal{L}\{\sin bt\} = \frac{b}{s^2 + b^2}$$

and

$$\frac{dF}{ds} = \frac{-2bs}{(s^2 + b^2)^2}$$

$$\frac{d^2 F}{ds^2} = \frac{6bs^2 - 2b^3}{(s^2 + b^2)^3},$$

so that

$$\mathcal{L}\{t^2 \sin bt\} = \frac{6bs^2 - sb^3}{(s^2 + b^2)^3}.$$

Computer Code 7.1:   **Basic Laplace transforms**

Matlab, Maple, Mathematica

```
                              Matlab
>>  syms s t a b
>>  eq1=laplace(1,t,s) %Laplace transform of 1
>>  eq2=laplace(sin(b*t),t,s)
>>  f1=t*cos(b*t)
>>  eq3=laplace(f1,t,s)
>>  f2=exp(-a*t)*sin(b*t)
>>  eq4=laplace(f2,t,s)
```

```
                              Maple
>  with(inttrans):
>  eq1:=laplace(1,t,s); #Laplace transform
>  eq2:=laplace(sin(b*t),t,s);
>  f1:=t*cos(b*t);
>  eq3:=laplace(f1,t,s);
>  f2:=exp(-a*t)*sin(b*t);
>  eq4:=laplace(f2,t,s);
```

```
                           Mathematica
eq1=LaplaceTransform[1,t,s]
eq2=LaplaceTransform[Sin[bt],t,s]
f1=t Cos[b t]
eq3=LaplaceTransform[f1,t,s]
f2 = e^{-at} Sin[b t]  (*note e is entered from the palette*)
eq4=LaplaceTransform[f2,t,s]
```

Problems

In problems 1–10, using the properties of the Laplace transform developed in this section, find the Laplace transform $\mathcal{L}\{f(t)\}$ for each of the given functions $f(t)$.

1. $f(t) = t - 5$
2. $f(t) = |t - 5|$
3. $f(t) = te^{2t}$
4. $f(t) = |t - 5|e^{3t}$
5. $f(t) = \cos^2 t$

6. $f(t) = \cos^2 at$; a, a constant
7. $f(t) = \sin t \sin 2t$
8. $f(t) = \sin at \sin bt$; a, b constants
9. $f(t) = \sin^3 t$
10. $f(t) = \sin^2 t \cos t$

From a table of integrals

$$\int e^{au} \sin bu \, du = e^{au} \frac{a \sin bu - b \cos bu}{a^2 + b^2}$$

$$\int e^{au} \cos bu \, du = e^{au} \frac{a \cos bu + b \sin bu}{a^2 + b^2}.$$

Use the integrals to find the Laplace transform $\mathcal{L}\{f(t)\}$ of the following functions in problems 11–13.

11. $f(t) = \sin \alpha t$
12. $f(t) = \cos \alpha t$
13. $f(t) = \sin(\alpha t - \phi)$

14. Another useful property of Laplace transforms can be given as follows: if $\mathcal{L}\{f(t)\} = F(s)$ exists for some $s > a$ for a function $f(t)$ in which $f(t)/t$ is bounded for $t > 0$, then

$$\mathcal{L}\{\frac{1}{t} f(t)\} = \int_s^\infty F(z) \, dz. \tag{7.3}$$

a. Verify (7.3) by writing

$$F(z) = \int_0^\infty e^{-zt} f(t) \, dt$$

and integrating both sides from $z = s$ to $z = \infty$.

b. Using the definition of the Laplace transform, observe the troubles in evaluating $\mathcal{L}\left\{\frac{\sin t}{t}\right\}$

c. Using (7.3) evaluate $\mathcal{L}\left\{\frac{\sin t}{t}\right\}$.

7.3   Step Functions, Translated Functions, and Periodic Functions

We will now consider the Laplace transform on some basic functions that serve as "building blocks" for other functions. For each real number $a \geq 0$, the *unit step function* U_a is defined by

$$U_a(t) = \begin{cases} 0 & t < a \\ 1 & t \ge a. \end{cases}$$

$U_a(t)$ is also sometimes referred to as a *Heaviside function*. If $a = 0$ then

$$U_0(t) = \begin{cases} 0 & t < 0 \\ 1 & t \ge 0. \end{cases}$$

but since we have defined U_a for t nonnegative, we have

$$U_0(t) = 1 \quad t \ge 0.$$

The unit step function U_a satisfies the conditions of the existence theorem for Laplace transforms so that $\mathcal{L}\{U_a(t)\}$ exists. Thus

$$\mathcal{L}\{U_a(t)\} = \int_0^\infty e^{-st} U_a(t) dt = \int_0^a e^{-st} 0 \, dt + \int_a^\infty e^{-st} \, dt$$

$$= \lim_{k\to\infty} \int_a^k e^{-st} \, dt = \lim_{k\to\infty} \left. -\frac{e^{-st}}{s} \right|_a^k$$

$$= \lim_{k\to\infty} \frac{-e^{-sk} + e^{-sa}}{s} = \frac{e^{-sa}}{s}.$$

That is,

$$\mathcal{L}\{U_a(t)\} = \frac{e^{-as}}{s}, \quad s > 0.$$

We can use unit step functions to express other step functions.

Example 1: Consider

$$f(t) = \begin{cases} 0 & 0 < t < 2 \\ 3 & 2 < t < 5 \\ 0 & t > 5. \end{cases}$$

This is just the linear combination $3U_2(t) - 3U_5(t)$, that is,

$$3U_2(t) - 3U_5(t) = \begin{cases} 0 & t < 2 \\ 3 & t \ge 2 \end{cases} - \begin{cases} 0 & t < 5 \\ 3 & t \ge 5 \end{cases}$$

$$= \begin{cases} 0 & 0 < t < 2 \\ 3 & 2 \le t \le 5 \\ 0 & t > 5 \end{cases} = f(t) \quad .$$

Thus

$$\mathcal{L}\{f(t)\} = \mathcal{L}\{3U_2(t) - 3U_5(t)\} = 3\mathcal{L}\{U_2(t)\} - 3\mathcal{L}\{U_5(t)\}$$

$$= \frac{3}{5} e^{-2t} - \frac{3}{5} e^{-st} = \frac{3}{5} \left(e^{-2t} - e^{-st} \right).$$

Another useful property of the unit step function in connection with the Laplace transform is the translation of a function $f(t)$ in a positive direction. That is, consider a new function defined by

$$\begin{cases} 0 & 0 < t < a \\ f(t-a) & t \geq a. \end{cases}$$

Then since

$$U_a(t) = \begin{cases} 0 & 0 < t < a \\ 1 & t \geq a \end{cases}$$

we have

$$U_a(t)f(t-a) = \begin{cases} 0 & 0 < t < a \\ f(t-a) & t \geq a. \end{cases}$$

Thus for Laplace transforms of this function we have the following result.

THEOREM 7.3.1 *Suppose f is a function which has a Laplace transform F, i.e.,*

$$F(s) = \int_0^\infty e^{-st} f(t)\, dt.$$

Then the translated function

$$U_a(t)f(t-a) = \begin{bmatrix} 0 & 0 < t < a \\ f(t-a) & t >\geq a \end{bmatrix}$$

has Laplace transform

$$\mathcal{L}\{U_a(t)f(t-a)\} = e^{-as}\mathcal{L}\{f(t)\}.$$

Proof: We have

$$\mathcal{L}\{U_a(t)f(t-a)\} = \int_0^\infty e^{-st} U_a(t) f(t-a)\, dt$$

$$= \int_0^a e^{-st} \cdot 0\, dt + \int_a^\infty e^{-st} f(t-a)\, dt$$

$$= \int_a^\infty e^{-st} f(t-a)\, dt.$$

We now change variables, letting $\tau = t - a$. Then

$$\int_a^\infty e^{-st} f(t-a)\, dt = \int_0^\infty e^{-s(\tau+a)} f(\tau)\, d\tau = e^{-sa} \int_0^\infty e^{-s\tau} f(\tau)\, d\tau$$

$$= e^{-sa}\mathcal{L}\{f(t)\}. \;\square$$

Example 2: Find the Laplace transform of

$$g(t) = \begin{bmatrix} 0 & 0 < t < 5 \\ t - 3 & t \geq 5. \end{bmatrix}$$

We must express $t - 3$ for $t \geq 5$ in terms of $t - 5$. Thus $t - 3 = t - 5 + 2$ so

$$g(t) = \begin{bmatrix} 0 & 0 < t < 5 \\ t - 5 + 2 & t \geq 5 \end{bmatrix}$$

which is

$$U_5(t)f(t - 5) = \begin{bmatrix} 0 & 0 < t < 5 \\ (t - 5) + 2 & t \geq 5 \end{bmatrix}$$

where $f(t) = t + 2$. Thus we need

$$\mathcal{L}\{f(t)\} = \mathcal{L}\{t + 2\} = \mathcal{L}\{t\} + \mathcal{L}\{2\}$$
$$= \frac{1}{s^2} + \frac{2}{s}.$$

This gives

$$\mathcal{L}\{g(t)\} = \mathcal{L}\{U_s(t)f(t - 5)\} = e^{-5s}\mathcal{L}\{f(t)\}$$
$$= e^{-5s}\left(\frac{1}{s^2} + \frac{2}{s}\right).$$

Example 3: Find the Laplace transform of

$$g(t) = \begin{bmatrix} 0 & 0 < t < \frac{\pi}{2} \\ \sin t & t \geq \frac{\pi}{2}. \end{bmatrix}$$

We must express $\sin t$ in terms of $t - \frac{\pi}{2}$. Noting that $\sin t = \cos(t - \frac{\pi}{2})$, we have

$$g(t) = \begin{bmatrix} 0 & 0 < t < \frac{\pi}{2} \\ \cos(t - \frac{\pi}{2}) & t \geq \frac{\pi}{2}. \end{bmatrix}$$

Hence

$$U_{\frac{\pi}{2}}f\left(t - \frac{\pi}{2}\right) = \begin{bmatrix} 0 & 0 < t < \frac{\pi}{2} \\ \cos(t - \frac{\pi}{2}) & t \geq \frac{\pi}{2} \end{bmatrix}$$

where $f(t) = \cos t$. Thus, since

$$\mathcal{L}\{\cos t\} = \frac{s}{s^2 + 1}$$

we have

$$\mathcal{L}\{g(t)\} = \mathcal{L}\left\{U_{\frac{\pi}{2}}(t)f\left(t - \frac{\pi}{2}\right)\right\} = e^{-\frac{\pi}{2}s}\left(\frac{s}{s^2 + 1}\right).$$

This example motivates the following theorem for periodic functions. Recall a function $f(x)$ is *periodic* with period p if

$$f(x + p) = f(x) \quad \text{for all} \quad x.$$

THEOREM 7.3.2 *Suppose f is a periodic function of period P for which a Laplace transform $\mathcal{L}\{f(t)\}$ exists. Then*

$$\mathcal{L}\{f(t)\} = \frac{\int_0^P e^{-st} f(t)\, dt}{1 - e^{-Ps}}.$$

Proof: Since

$$\mathcal{L}\{f(t)\} = \int_0^\infty e^{-st} f(t)\, dt = \int_0^P e^{-st} f(t)\, dt + \int_P^\infty e^{-st} f(t)\, dt$$

it follows that we compute

$$\int_P^\infty e^{-st} f(t)\, dt = \int_0^\infty e^{-s(\tau+P)} f(\tau + P)\, d\tau = e^{-sP} \int_0^\infty e^{-st} f(t)\, dt$$

$$= e^{-sP} \mathcal{L}\{f(t)\}.$$

Thus

$$\mathcal{L}\{f(t)\} = \int_0^P e^{-st} f(t)\, dt + e^{-sP} \mathcal{L}\{f(t)\}$$

so that

$$\mathcal{L}\{f(t)\} = \frac{1}{1 - e^{-sP}} \int_0^P e^{-st} f(t)\, dt. \quad \square$$

Example 4: Find the Laplace transform of

$$f(t) = \begin{bmatrix} 1 & 0 \le t < 2 \\ -1 & 2 \le t < 4 \end{bmatrix}$$

and $f(t + 4) = f(t)$ for $t \ge 4$.

This is a periodic function with period $P = 4$ as seen in the graph shown in Figure 7.2. So

$$\mathcal{L}\{f(t)\} = \frac{\int_0^4 e^{-st} f(t)\, dt}{1 - e^{-4s}}$$

$$= \frac{1}{1 - e^{-4s}} \left[\int_0^2 e^{-st}\, dt + \int_2^4 e^{-st}(-1)\, dt \right]$$

$$= \frac{1}{1 - e^{-4s}} \left[\frac{-1}{s} e^{-st} \Big|_0^2 + \frac{e^{-st}}{s} \Big|_2^4 \right]$$

$$= \frac{1}{1 - e^{-4s}} \left(\frac{1}{s}\right) \left[-e^{-2s} + 1 + e^{-4s} - e^{-2s} \right]$$

$$= \frac{1 - e^{-2s}}{s(1 + e^{-2s})}.$$

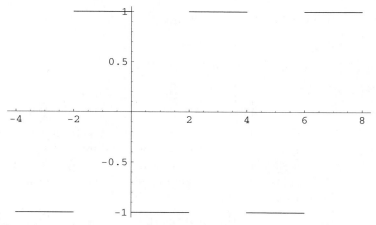

FIGURE 7.2:   A piecewise defined periodic function.

As we mentioned in the introduction to this chapter, another important aspect of the unit step function $U_a(t)$ occurs in connection with differential equations of the form

$$a\,y''(t) + b\,y'(t) + c\,y(t) = f(t)$$

where we think of $f(t)$ as some driving or forcing function. We have earlier considered these ideas in connection with the mass-spring system. However, now we would like to think of $f(t)$ as "large" over a short interval, say $t_0 - \alpha < t < t_0 + \alpha$ for some constants t_0 and α.

That is

$$f(t) = \begin{bmatrix} K & \text{for} \quad t_0 - \alpha \leq t \leq t_0 + \alpha \\ 0 & \text{otherwise} \end{bmatrix} \tag{7.4}$$

with K a constant, so that we can think of f delivering an amount K to the system over the interval $[t_0 - \alpha, t_0 + \alpha]$. In this regard, we write

$$f(t) = K\left(U_{t_0-\alpha}(t) - U_{t_0+\alpha}(t)\right)$$

in terms of the unit step function $U_a(t)$. Let

$$\delta_\alpha(t - t_0) = U_{t_0-\alpha}(t) - U_{t_0+\alpha}(t)$$

and define the impulse delivered by $f(t)$ over the interval $(t_0 - \alpha, t_0 + \alpha)$ as

$$I(t) = \int_{-\infty}^{\infty} f(t)\,dt$$

or equivalently as

$$I(t) = \int_{t_0-\alpha}^{t_0+\alpha} f(t)\,dt.$$

But with $f(t)$ defined as in (7.4), this integral for $I(t)$ becomes

$$I(t) = 2\alpha K.$$

Setting $K = \frac{1}{2\alpha}$, we have the *unit impulse function*

$$I(t) = \begin{bmatrix} 1 & \text{for } t_0 - \alpha \le t \le t_0 + \alpha \\ 0 & \text{otherwise.} \end{bmatrix}$$

We note here that the value of $I(t)$ does not depend upon α as long as $\alpha \ne 0$. Indeed, for every $\alpha \ne 0$, $I(t) = 1$ over the interval $[t_0 - \alpha, t_0 + \alpha]$ regardless of how small α is. This idea allows us to define the *idealized impulse function* by requiring that $\delta_\alpha(t - t_0)$ act on smaller and smaller intervals.

From our calculations, we have

$$\lim_{\alpha \to 0} I(t) = 1$$

while

$$\lim_{\alpha \to 0} \delta_\alpha(t - t_0) = 0, \quad \text{for } t \ne t_0.$$

Thus we have the following definition.

DEFINITION 7.4   *The (idealized) unit impulse function δ is the function that satisfies*

$$\delta(t - t_0) = 0 \quad \text{for } t \ne t_0$$

and

$$\int_{-\infty}^{\infty} \delta(t - t_0)\, dt = 1.$$

The function $\delta(t - t_0)$ is also termed the *Dirac delta function* and is very useful in applications with impulse forcing functions. A useful theorem involving the Dirac delta function is stated here.

THEOREM 7.3.3 *Suppose that $g(t)$ is a bounded continuous function; then*

$$\int_{-\infty}^{\infty} \delta(t - t_0)g(t)\, dt = g(t_0).$$

The Laplace transform of $\delta(t - t_0)$ can easily be found using $\delta_\alpha(t - t_0)$ and L'Hopital's rule as follows:

$$\mathcal{L}\{\delta(t - t_0)\} = \mathcal{L}\{\lim_{\alpha \to 0} \delta_\alpha(t - t_0)\}$$

and we have

$$\mathcal{L}\{\lim_{\alpha \to 0} \delta_\alpha(t - t_0)\} = \lim_{\alpha \to 0} \mathcal{L}\{\delta_\alpha(t - t_0)\}$$

(as can be justified by the bounded convergence theorem). Also,

$$
\begin{aligned}
\mathcal{L}\{\delta_\alpha(t - t_0)\} &= \mathcal{L}\left\{\frac{1}{2\alpha}[U_{t_0-\alpha}(t) - U_{t_0+\alpha}(t)]\right\} \\
&= \frac{1}{2\alpha}[\mathcal{L}\{U_{t_0-\alpha}(t)\} - \mathcal{L}\{U_{t_0+\alpha}(t)\}] \\
&= \frac{1}{2\alpha}\left[\frac{e^{-s(t_0-\alpha)}}{s} - \frac{e^{-s(t_0+\alpha)}}{s}\right] \\
&= e^{-st_0}\left(\frac{e^{\alpha s} - e^{-\alpha s}}{2\alpha s}\right).
\end{aligned}
$$

So

$$
\begin{aligned}
\mathcal{L}\{\delta(t - t_0)\} &= \lim_{\alpha \to 0} \mathcal{L}\{\delta_\alpha(t - t_0)\} \\
&= \lim_{\alpha \to 0} e^{-st_0}\left(\frac{e^{\alpha s} - e^{-\alpha s}}{2\alpha s}\right) \\
&= \lim_{\alpha \to 0} e^{-st_0}\left(\frac{se^{\alpha s} + se^{-\alpha s}}{2s}\right) \quad \text{(by L'Hopital's rule)} \\
&= e^{-st_0} \cdot 1 \\
&= e^{-st_0}.
\end{aligned}
$$

We state this result here for reference.

THEOREM 7.3.4 *For $t_0 \geq 0$,*

$$\mathcal{L}\{\delta(t - t_0)\} = e^{-st_0}.$$

Matlab, Maple, and Mathematica all have built-in Dirac delta and Heaviside functions. This next set of commands shows some of the basic properties of their Laplace transforms and also an interesting relationship between the two functions.

Computer Code 7.2:   Laplace transform of the Dirac delta function; Heaviside function

Matlab, Maple, Mathematica

| Matlab |

```
>>  syms s t
>>  f1=(2*t-3)*heaviside(t-4) %Heaviside function
>>  eq1=laplace(f1,t,s)
>>  f2=dirac(t-3)
>>  eq2=int(f2,t,-inf,inf) %must be 1 by definition
>>  eq3=laplace(f2,t,s)
>>  f3=heaviside(t-3)
>>  eq4=diff(f3,t)
>>  eq5=int(f2,t)
```

| Maple |

```
>  with(inttrans):
>  f1:=(2*t-3)*Heaviside(t-4); #Heaviside function
>  convert(f1,'piecewise');
>  eq1:=laplace(f1,t,s);
>  f2:=Dirac(t-3);
>  convert(f2,'piecewise');
>  eq2:=int(f2,t=-infinity..infinity); #must be 1 by defn
>  eq3:=laplace(f2,t,s);
>  f3:=Heaviside(t-3);
>  eq4:=diff(f3,t);
>  eq5:=int(f2,t);
```

| Mathematica |

```
f1=(2 t-3) UnitStep[t-4]
eq1=LaplaceTransform[f1,t,s]
f2=DiracDelta[t-3]
```
$$eq2=\int_{-\infty}^{\infty} f2 \; dt \; (*entered \; from \; palette*)$$
```
eq3=LaplaceTransform[f2,t,s]
f3=UnitStep[t-3]
eq4=D[f3,t]
```
$$eq5=\int f2 \; dt \; (*entered \; from \; palette*)$$

Problems

Find $\mathcal{L}\{f(t)\}$ for each of the functions $f(t)$ in problems 1–12.

1. $f(t) = \begin{cases} 0 & 0 < t < 4 \\ 3 & t \geq 4 \end{cases}$

2. $f(t) = \begin{cases} 0 & 0 < t < 6 \\ 5 & t \geq 6 \end{cases}$

3. $f(t) = \begin{cases} 4 & 0 \leq t \leq 10 \\ 0 & t > 10 \end{cases}$

4. $f(t) = \begin{cases} 0 & 0 < t < 5 \\ 2 & 5 \leq t \leq 7 \\ 0 & t > 7 \end{cases}$

5. $f(t) = \begin{cases} 0 & 0 < t < 3 \\ -6 & 3 \leq t \leq 9 \\ 0 & t > 9 \end{cases}$

6. $f(t) = \begin{cases} 1 & 0 < t < 2 \\ 2 & 2 \leq t \leq 3 \\ 3 & 3 < t \leq 5 \\ 0 & t > 7 \end{cases}$

7. $f(t) = \begin{cases} 1 & 0 \leq t \leq 2 \\ 0 & 2 < t < 6 \\ 1 & t \geq 6 \end{cases}$

8. $f(t) = \begin{cases} t & 0 \leq t < 3 \\ 3 & t \geq 3 \end{cases}$

9. $f(t) = \begin{cases} 2t & 0 \leq t \leq 5 \\ 10 & t > 5 \end{cases}$

10. $f(t) = \begin{cases} 0 & 0 < t < 1 \\ e^{-t} & t \geq 1 \end{cases}$

11. $f(t) = \begin{cases} 0 & 0 < t < 4 \\ t - 4 & 4 \leq t \leq 7 \\ 3 & t > 7 \end{cases}$

12. $f(t) = \begin{cases} 6 & 0 \leq t \leq 1 \\ 8 - 2t & 1 < t \leq 3 \\ 2 & t > 3 \end{cases}$

Evaluate the following integrals in problems 13–16 involving the Dirac delta function.

13. $\int_0^3 (1 + e^{-t})\delta(t - 2)\, dt$

14. $\int_{-2}^{1}(1+e^{-t})\delta(t-2)\,dt$

15. $\int_{-1}^{2}\cos 2t\delta(t)\,dt$

16. $\int_{-3}^{2}(e^{2t}+t)\delta(t+2)\,dt$

17. Determine a value of t_0 so that

$$\int_{0}^{1}\sin^{2}[\pi(t-t_0)]\delta\left(t-\frac{1}{2}\right)dt=\frac{3}{4}$$

18. If $\int_{1}^{5}t^{n}\delta(t-2)\,dt=8$ what is the value of the exponent n?

19. Sketch the graph of the function $f(t)$ defined by the function

$$f(t)=\int_{0}^{t}\delta(z)\,dz$$

$0<t<\infty$. Can the graph be obtained from a unit step function?

7.4   The Inverse Laplace Transform

Thus far our work has been centered upon the problem: Given a function f, find its Laplace transform $\mathcal{L}\{f(t)\}$. The next question for us to pursue is given a function F, find a function f whose Laplace transform is the given F.

We introduce the notation $\mathcal{L}^{-1}\{F\}$ to denote such a function f, i.e.,

$$f(t)=\mathcal{L}^{-1}\{F(s)\},$$

which means $\mathcal{L}\{f(t)\}=F(s)$. $\mathcal{L}^{-1}$ is called the *inverse Laplace transform*. There are now three questions that we need to consider:

1. Given a function F, does an inverse transform exist?
2. If it exists, is it unique?
3. How do you find it?

These questions have different levels of import in our considerations here, from very theoretical mathematics to very practical. We will first answer question 2.

THEOREM 7.4.1 *(Uniqueness) Let f and g be two functions that are continuous for t ≥ 0 and that have the same Laplace transform, that is,*

$$\mathcal{L}\{f(t)\} = \mathcal{L}\{g(t)\},$$

then

$$f(t) = g(t) \quad \text{for all } t \geq 0.$$

Remark: In probability theory the uniqueness theorem plays an essential role in the theory of moment generating functions (m.g.f.) where the uniqueness is essential for determining the distribution of a random variable.

We must emphasize that in the uniqueness theorem the function is continuous, as the next example clarifies.

Example 1: Consider

$$g(t) = \begin{bmatrix} 1 & 0 < t < 3 \\ 2 & t = 3 \\ 1 & t > 3. \end{bmatrix}$$

Then

$$\mathcal{L}\{g(t)\} = \int_0^\infty e^{-st} g(t)\, dt = \int_0^3 e^{-st}\, dt + \int_3^\infty e^{-st}\, dt$$

$$= \frac{1}{s} \quad \text{if } s > 0.$$

This, however, is also the Laplace transform of 1, i.e.,

$$\mathcal{L}\{1\} = \frac{1}{s} \quad \text{now clearly } g(t) \neq 1 \quad \text{for all } t.$$

Thus, this discontinuous function g is also an inverse transform of F defined by

$$F(s) = \frac{1}{s}.$$

However, the only continuous inverse transform of

$$F(s) = \frac{1}{s}$$

is $f(t) = 1$ for all t so that the *unique* continuous inverse transform is

$$\mathcal{L}^{-1}\left\{\frac{1}{s}\right\} = 1.$$

This is how we address uniqueness. We require continuity of the inverse transform.

Question: How do we compute inverse Laplace transforms?

This is an important question for us. However, we will not directly compute inverse transforms in this text. We will refer to the table in Figure 7.3 to give us some often-used transforms.

As with all tables, certain preliminary manipulations might be required to put a given $F(s)$ into a form present in the table. Often partial fraction decomposition is required. Of course, with Matlab, Maple, or Mathematica, our job is much easier as will be seen at the end of this section.

	$f(t) = \mathcal{L}^{-1}\{F(s)\}$	$F(s) = \mathcal{L}\{f(t)\}$
1.	1	$\dfrac{1}{s}$
2.	e^{at}	$\dfrac{1}{s-a}$
3.	$\sin bt$	$\dfrac{b}{s^2+b^2}$
4.	$\cos bt$	$\dfrac{s}{s^2+b^2}$
5.	$\sinh bt$	$\dfrac{b}{s^2-b^2}$
6.	$\cosh bt$	$\dfrac{s}{s^2-b^2}$
7.	$t^n (n=1,2,\ldots)$	$\dfrac{n!}{s^{n+1}}$
8.	$t^n e^{at} (n=1,2,\ldots)$	$\dfrac{n!}{(s-a)^{n+1}}$
9.	$t \sin bt$	$\dfrac{2bs}{(s^2+b^2)^2}$
10.	$t \cos bt$	$\dfrac{s^2-b^2}{(s^2+b^2)^2}$
11.	$e^{-at} \sin bt$	$\dfrac{b}{(s+a)^2+b^2}$
12.	$e^{-at} \cos bt$	$\dfrac{s+a}{(s+a)^2+b^2}$
13.	$\dfrac{\sin bt - bt\cos bt}{2b^3}$	$\dfrac{1}{(s^2+b^2)^2}$
14.	$\dfrac{t\sin bt}{2b}$	$\dfrac{s}{(s^2+b^2)^2}$

FIGURE 7.3:   Table of Laplace transforms

Example 2: Find $\mathcal{L}^{-1}\{\frac{1}{s^2+6s+13}\}$.

We look for

$$F(s) = \frac{1}{as^2 + bs + c}$$

in the table but there is no such critter; we do, however, find

$$\frac{b}{(s+a)^2 + b^2}.$$

Thus we rewrite

$$s^2 + 6s + 13 = s^2 + 6s + 9 + 4 = (s+3)^2 + 4.$$

Hence

$$\frac{1}{s^2 + 6s + 13} = \frac{1}{(s+3)^2 + 4} = \frac{1}{2}\frac{2}{(s+3)^2 + 2^2}.$$

So

$$\mathcal{L}^{-1}\left\{\frac{1}{s^2 + 6s + 13}\right\} = \frac{1}{2}\mathcal{L}^{-1}\left\{\frac{2}{(s+3)^2 + 2^2}\right\}$$

$$= \frac{1}{2}e^{-3t}\sin 2t.$$

Example 3: Compute $\mathcal{L}^{-1}\{\frac{1}{s(s^2+1)}\}$.

By partial fractions,

$$\frac{1}{s(s^2 + 1)} = \frac{A}{s} + \frac{Bs + c}{s^2 + 1}$$

so that

$$A = 1, \quad B = -1, \quad \text{and } C = 0.$$

Thus

$$\frac{1}{s(s^2 + 1)} = \frac{1}{s} - \frac{s}{s^2 + 1},$$

which means

$$\mathcal{L}^{-1}\left\{\frac{1}{s(s^2 + 1)}\right\} = \mathcal{L}^{-1}\left\{\frac{1}{s}\right\} - \mathcal{L}^{-1}\left\{\frac{s}{s^2 + 1}\right\}$$

$$= 1 - \cos t.$$

Example 4: Find $\mathcal{L}^{-1}\{e^{-4s}(\frac{2}{s^2} + \frac{5}{s})\}$.

This is of the form $\mathcal{L}^{-1}\{e^{-as}F(s)\}$ where $a = 4$ with $F(s) = \frac{2}{s^2} + \frac{5}{s}$. Thus

$$\mathcal{L}^{-1}\{e^{-as}F(s)\} = U_a(t)f(t - a)$$

where

$$f(t) = \mathcal{L}^{-1}\{F(s)\} = \mathcal{L}^{-1}\left\{\frac{2}{s^2} + \frac{5}{s}\right\}$$

$$= 2\mathcal{L}^{-1}\left\{\frac{1}{s^2}\right\} + 5\mathcal{L}^{-1}\left\{\frac{1}{s}\right\}$$

$$= 2t + 5 \cdot 1 = 2t + 5.$$

So $f(t-4) = 2(t-4) + 5 = 2t - 3$. Hence

$$\mathcal{L}^{-1}\{e^{-4s}F(s)\} = U_4(t)f(t-4) = U_4(t)(2t-3)$$

$$= \begin{bmatrix} 0 & 0 < t < 4 \\ 2t - 4 & t > 4. \end{bmatrix}$$

Computer Code 7.3:   Inverse Laplace transforms

Matlab, Maple, Mathematica

Matlab
```
≫  syms s t a b
≫  g1=1/(s^2+6*s+13)
≫  eq1=ilaplace(g1,s,t) %inverse Laplace transform
≫  g2=exp(-4*s)*(2/s^2+5/s)
≫  eq2=ilaplace(g2,s,t)
≫  f1=dirac(t-3)
≫  eq3=laplace(f1,t,s)
≫  eq4=ilaplace(eq3,s,t)
```

Maple
```
>  with(inttrans):
>  g1:=1/(s^2+6*s+13);
>  eq1:=invlaplace(g1,s,t); #inverse Laplace transform
>  g2:=exp(-4*s)*(2/s^2+5/s);
>  eq2:=invlaplace(g2,s,t);
>  convert(eq7,'piecewise');
>  f1:=Dirac(t-3);
>  convert(g1,'piecewise');
>  eq3:=laplace(f1,t,s);
>  eq4:=invlaplace(eq3,s,t);
```

$$\boxed{\textbf{Mathematica}}$$

$g1 = \dfrac{1}{s^2 + 6s + 13}$

eq1=InverseLaplaceTransform[g1,s,t]

FullSimplify[eq1]

$g2 = e^{-4s}\left(\frac{2}{s^2} + \frac{5}{s}\right)$

eq2=InverseLaplaceTransform[g2,s,t]

f1=DiracDelta[t-3]

eq3=LaplaceTransform[f1,t,s]

eq4=InverseLaplaceTransform[eq3,s,t]

Problems

Find $\mathcal{L}^{-1}\{F(s)\}$ for each of the functions $F(s)$ in problems 1–16.

1. $F(s) = \frac{3}{s^2+4}$

2. $F(s) = \frac{5s}{s^2-9}$

3. $F(s) = \frac{5}{(s-3)^2}$

4. $F(s) = \frac{5s}{s^2+4s+4}$

5. $F(s) = \frac{s+2}{s^2+4s+7}$

6. $F(s) = \frac{s+6}{s^2+8s+20}$

7. $F(s) = \frac{s+1}{s^3+2s}$

8. $F(s) = \frac{s+1}{(s^2+9)^2}$

9. $F(s) = \frac{2s+7}{(s+3)^4}$

10. $F(s) = \frac{8(s+1)}{(2s-1)^3}$

11. $F(s) = \frac{s-2}{s^2+5s+6}$

12. $F(s) = \frac{5s+8}{s^2+3s-10}$

13. $F(s) = \frac{5s+6}{s^2+9}e^{-\pi s}$

14. $F(s) = \frac{2s+9}{s^2+4s+13}e^{-3s}$

15. $F(s) = \frac{e^{-4s}-e^{-7s}}{s^2}$

16. $F(s) = \frac{2-e^{-3s}}{s^2+9}$

7.5   Laplace Transform Solution of Linear Differential Equations

We now consider how the Laplace transform may be applied to solve initial-value problems consisting of an nth order linear differential equation with a specified initial condition.

The Method:

Consider the nth order linear differential equation with constant coefficients:

$$a_0 \frac{d^n y}{dt^n} + a_1 \frac{d^{n-1} y}{dt^{n-1}} + \cdots + a_{n-1} \frac{dy}{dt} + a_n y = b(t)$$

which satisfies the initial conditions

$$y(0) = c_0, y'(0) = c_1, \ldots, y^{(n-1)}(0) = c_{n-1}.$$

The existence theorem in Section 4.2 ensures that this initial-value problem has a unique solution. Thus if we now take the Laplace transform of the differential equation,

$$a_0 \mathcal{L}\left\{\frac{d^n y}{dt^n}\right\} + a_1 \mathcal{L}\left\{\frac{d^{n-1} y}{dt^{n-1}}\right\} + \cdots + a_{n-1} \mathcal{L}\left\{\frac{dy}{dt}\right\} + a_n \mathcal{L}\{y(t)\} = \mathcal{L}\{b(t)\}.$$

Each Laplace transform is given as

$$\mathcal{L}\left\{\frac{d^n y}{dt^n}\right\} = s^n \mathcal{L}\{y(t)\} - s^{n-1} y(0) - s^{n-2} y^1(0) - \ldots - y^{(n-1)}(0)$$

$$= s^n \mathcal{L}\{y(t)\} - c_0 s^{n-1} - c_1 s^{n-2} - \ldots - c_{n-1}$$

$$\mathcal{L}\left\{\frac{d^{n-1} y}{dt^{n-1}}\right\} = s^{n-1} \mathcal{L}\{y(t)\} - s^{n-2} y(0) - s^{n-3} y^1(0) - \ldots - y(n-2)(0)$$

$$= s^{n-1} \mathcal{L}\{y(t)\} - c_0 s^{n-2} - c_1 s^{n-3} - \ldots - c_{n-2}$$

$$\vdots$$

$$\mathcal{L}\left\{\frac{dy}{dt}\right\} = s\mathcal{L}\{y(t)\} - y(0) = s\mathcal{L}\{y(t)\} - c_0.$$

Now letting $Y(s) = \mathcal{L}\{y(t)\}$ and $B(s) = \mathcal{L}\{b(t)\}$ we obtain

$$[a_0 s^n + a_1 s^{n-1} + \cdots + a_{n-1} s + a_n] Y(s) - c_0 [a_0 s^{n-1} + a_1 s^{n-2} + \cdots + a_{n-1}]$$

$$- c_1 [a_0 s^{n-2} + a_1 s^{n-3} + \cdots + a_{n-2}]$$

$$-$$

$$\vdots$$

$$- c_{n-2}[a_0 s + a_1] - c_{n-1} a_0 = B(s).$$

This is an algebraic equation in the "unknown" $Y(s)$. We solve this equation for $Y(s)$ so that

$$y(t) = \mathcal{L}^{-1}\{Y(s)\}.$$

The idea presented here is one of the main features of the Laplace transform. It allows a differential equation to be expressed as an algebraic equation that can be solved easily. Once the algebraic equation is solved, the inverse Laplace transform is applied to recover the solution of the differential equation. This is the idea we mentioned in the introduction to this chapter and we repeat the diagram given there.

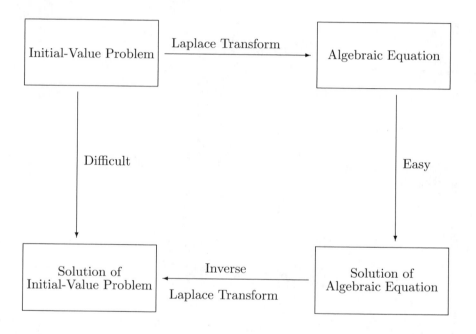

Example 1: Solve the initial-value problem

$$\frac{dy}{dt} - 2y = e^{5t}, \quad y(0) = 3.$$

Taking Laplace transforms of both sides of the differential equation gives

$$\mathcal{L}\left\{\frac{dy}{dt}\right\} - 2\mathcal{L}\{y(t)\} = \mathcal{L}\{e^{5t}\}.$$

Now

$$\mathcal{L}\left\{\frac{dy}{dt}\right\} = s\mathcal{L}\{y(t)\} - y(0),$$

letting
$$Y(s) = \mathcal{L}\{y(t)\}$$

so
$$\mathcal{L}\left\{\frac{dy}{dt}\right\} = sY(s) - 3.$$

Thus
$$sY(s) - 3 - 2Y(s) = \mathcal{L}\{e^{5t}\}$$
$$= \frac{1}{(s-5)},$$

so
$$(s-2)Y(s) - 3 = \frac{1}{s-5}.$$

Solving for $Y(s)$ gives
$$(s-2)Y(s) = \frac{1}{s-5} + 3$$

as
$$Y(s) = \frac{3s - 14}{(s-5)(s-2)}.$$

Now we find $Y(t)$,
$$Y(t) = \mathcal{L}^{-1}\left\{\frac{3s-14}{(s-5)(s-2)}\right\}.$$

Applying partial fractions
$$\frac{3s-14}{(s-5)(s-2)} = \frac{A}{s-5} + \frac{B}{s-2}$$

gives
$$3s - 14 = A(s-2) + B(s-5)$$

so that
$$A = \frac{1}{3}, \quad B = \frac{8}{3}.$$

This implies
$$\mathcal{L}^{-1}\left\{\frac{3s-14}{(s-2)(s-5)}\right\} = \frac{8}{3}\mathcal{L}^{-1}\left\{\frac{1}{s-2}\right\} + \frac{1}{3}\mathcal{L}^{-1}\left\{\frac{1}{s-5}\right\}.$$

Now
$$\mathcal{L}^{-1}\left\{\frac{1}{s-2}\right\} = e^{2t} \quad \text{and} \quad \mathcal{L}^{-1}\left\{\frac{1}{s-5}\right\} = e^{5t}.$$

Thus
$$Y(t) = \frac{8}{3}e^{2t} + \frac{1}{3}e^{5t}$$

so

$$y = \frac{8}{3}e^{2t} + \frac{1}{3}e^{5t}$$

is the solution of the initial-value problem.

Example 2: Solve the initial-value problem.

$$\frac{d^2y}{dt^2} - 2\frac{dy}{dt} - 8y = 0$$

$$y(0) = 3, \quad y'(0) = 6.$$

Taking the Laplace transform of both sides of the differential equation gives

$$\mathcal{L}\left\{\frac{d^2y}{dt^2}\right\} - 2\mathcal{L}\left\{\frac{dy}{dt}\right\} - 8\mathcal{L}\{y\} = \mathcal{L}\{0\}.$$

Now

$$\mathcal{L}\left\{\frac{d^2Y}{dt^2}\right\} = s^2Y(s) - sy(0) - y'(0)$$

where

$$Y(s) = \mathcal{L}\{y(t)\};$$

also

$$\mathcal{L}\left\{\frac{dy}{dt}\right\} = sY(s) - y(0).$$

Thus

$$s^2Y(s) - 3s - 6 - 2(sY(s) - 3) - 8Y(s) = 0,$$

which is

$$[s^2 - 2s - 8]Y(s) - 3s = 0.$$

Solving for $Y(s)$ gives

$$Y(s) = \frac{3s}{s^2 - 2s - 8},$$

which is

$$Y(s) = \frac{3s}{(s-4)(s+2)}.$$

We find $Y(t)$ by applying the inverse transform

$$Y(t) = \mathcal{L}^{-1}\left\{\frac{3s}{(s-4)(s+2)}\right\}.$$

Using partial fractions

$$\frac{3s}{(s-4)(s+2)} = \frac{A}{s-4} + \frac{B}{s+2}$$

gives the solution

$$A = 2, \quad B = 1.$$

Thus

$$\mathcal{L}^{-1}\left\{\frac{3s}{(s-4)(s+2)}\right\} = 2\mathcal{L}^{-1}\left\{\frac{1}{s-4}\right\} + \mathcal{L}^{-1}\left\{\frac{1}{s+2}\right\}$$

$$= 2e^{4t} + e^{-2t}.$$

So the solution of the initial problem is

$$Y(t) = 2e^{4t} + e^{-2t}.$$

Example 3: Solve $y'' + y = \delta(t - \pi)$ with the initial conditions

$$y(0) = 0 \quad \text{and} \quad y'(0) = 0.$$

Here $\delta(t - \pi)$ is the Dirac delta function. Applying the Laplace transform to both sides of the differential equation gives

$$\mathcal{L}\{y''\} + \mathcal{L}\{y\} = \mathcal{L}\{\delta(t - \pi)\}$$

so that if $Y(s) = \mathcal{L}\{y\}$ then

$$s^2 Y(s) - sy(0) - y'(0) + Y(s) = e^{-\pi s},$$

that is,

$$(s^2 + 1)Y(s) = e^{-\pi s}.$$

Here

$$Y(s) = \frac{e^{-\pi s}}{s^2 + 1}$$

so that

$$y(t) = \mathcal{L}^{-1}\left\{\frac{e^{-\pi s}}{s^2 + 1}\right\},$$

but

$$\mathcal{L}^{-1}\left\{\frac{1}{s^2 + 1}\right\} = \sin t$$

thus

$$y(t) = \mathcal{L}^{-1}\left\{\frac{e^{-\pi s}}{s^2 + 1}\right\} = \sin(t - \pi)U_\pi(t).$$

Laplace transforms can sometimes be used to solve initial-value problems involving linear differential equations with nonconstant coefficients. However,

Laplace transforms do not provide a general method for solving equations with nonconstant coefficients.

Example 4: Solve $y'' + 4ty' - 8y = 4$ with the initial conditions $y(0) = 0, y'(0) = 0$.

Applying the Laplace transform to both sides of the equation gives

$$\mathcal{L}\{y''\} + 4\mathcal{L}\{ty'\} - 8\mathcal{L}\{y\} = \mathcal{L}\{4\}. \tag{7.5}$$

Now recall

$$\mathcal{L}\{t^n f(t)\} = (-1)^n \frac{d^n}{ds^n} \mathcal{L}\{f(t)\}$$

so that

$$\mathcal{L}\{ty'\} = -\frac{d}{ds}\mathcal{L}\{y'\}$$
$$= -\frac{d}{ds}[sY(s) - y(0)]$$
$$= -sY'(s) - Y(s).$$

Applying this to equation (7.5), we have

$$s^2 Y(s) - sy(0) - y'(0) + 4[-sY'(s) - Y(s)] - 8Y(s) = \frac{4}{s}.$$

Simplifying, we obtain

$$(s^2 - 12)Y(s) - 4sY'(s) = \frac{4}{s},$$

which is a first-order differential that can be solved for $Y(s)$. Rewriting this equation as

$$Y'(s) + \left(\frac{3}{s} - \frac{s}{4}\right)Y(s) = \frac{-1}{s^2}$$

we see that this is a first-order linear differential equation. An integrating factor for this equation is

$$u(s) = s^3 e^{-s^2/8},$$

and upon applying this integrating factor, we obtain the solution

$$Y(s) = \frac{4}{s^3} + \frac{ce^{s^2/8}}{s^3}.$$

Now if $y(t) = \mathcal{L}^{-1}\{Y(s)\}$ exists, then

$$\lim_{s \to \infty} Y(s) = 0.$$

Thus, here $c = 0$, so

$$Y(s) = \frac{4}{s^3}.$$

Using this

$$y(t) = \mathcal{L}^{-1}\{Y(s)\} = 2\mathcal{L}^{-1}\left\{\frac{2}{s}\right\}$$

$$= 2t^2.$$

The next example gives the general solution for a mass-spring system in terms of the Laplace transform. The latter part of the example considers a discontinuous forcing function.

Example 5: Mass on a Spring (revisited)

As we saw in Chapter 4, the motion of a mass suspended on a spring satisfies the differential equation

$$m\frac{d^2x}{dt^2} + b\frac{dx}{dt} + kx = F(t)$$

where m is the mass, b is a damping coefficient, k the spring constant, and $F(t)$ is a forcing function. The displacement of the mass also satisfies the initial conditions $x(0) = x_0$ and $x'(0) = v_0$. So applying the Laplace transform to the differential equation gives

$$\mathcal{L}\{m\frac{d^2x}{dt^2} + b\frac{dx}{dt} + kx\} = \mathcal{L}\{F(t)\}.$$

So

$$m\mathcal{L}\left\{\frac{d^2x}{dt^2}\right\} + b\mathcal{L}\left\{\frac{dx}{dt}\right\} + k\mathcal{L}\{x\} = \mathcal{L}\{F(t)\}$$

which is

$$m[s^2\mathcal{L}\{x\} - x_0 s - v_0] + b[s\mathcal{L}\{x\} - x_0] + \mathcal{L}\{x\} = \mathcal{L}\{F(t)\}.$$

Letting $X = \mathcal{L}\{x\}$ and rewriting this expression gives

$$(ms^2 + bs + 1)X - mx_0 s - mv_0 - bx_0 = \mathcal{L}\{F(t)\}.$$

Thus

$$X = \frac{\mathcal{L}\{F(t)\} + m(x_0 s + v_0)}{ms^2 + bs + 1}.$$

Applying the inverse transform gives

$$x(t) = \mathcal{L}^{-1}\left\{\frac{\mathcal{L}\{F(t)\} + m(x_0 s + v_0)}{ms^2 + bs + 1}\right\}.$$

Of course, this is a very general expression for the displacement function $x(t)$ of a mass on a spring. Specific examples can be readily calculated.

a) Suppose we had the case of forced undamped motion described by

$$x'' + 4x = \sin 3t$$

with $x(0) = x'(0) = 0$.

Then since $\mathcal{L}\{F(t)\} = \mathcal{L}\{\sin 3t\} = \frac{3}{s^2+9}$ and $b = 0, m = 4, x_0 = 0, v_0 = 0$ we have

$$x(t) = \mathcal{L}^{-1}\left\{ \frac{\frac{3}{s^2+9}}{4s^2+1} \right\} = \mathcal{L}^{-1}\left\{ \frac{3}{(s^2+9)(4s^2+1)} \right\}.$$

Applying partial fractions gives

$$\frac{3}{(s^2+9)(4s^2+1)} = \frac{3}{5}\left(\frac{1}{s^2+4} \right) - \frac{3}{5}\left(\frac{1}{s^2+9} \right)$$

$$= \frac{3}{10}\left(\frac{2}{s^2+4} \right) - \frac{1}{5}\left(\frac{3}{s^2+9} \right).$$

Now

$$\mathcal{L}^{-1}\left\{ \frac{2}{s^2+4} \right\} = \sin 2t \quad \text{and} \quad \mathcal{L}^{-1}\left\{ \frac{3}{s^2+9} \right\} = \sin 3t.$$

Thus

$$x(t) = \frac{3}{10}\sin 2t - \frac{1}{5}\sin 3t$$

is the displacement.

b) Perhaps, more interestingly, we can also consider forced undamped motion described by

$$x'' + 9x = 3\delta(t - \pi)$$

with $x(0) = 1, x'(0) = 0$. Then since $\mathcal{L}\{F(t)\} = 3\mathcal{L}\{\delta(t - \pi)\} = 3e^{-\pi s}$ and $b = 0, m = 9, x_0 = 1, v_0 = 0$ we have

$$x(t) = \mathcal{L}^{-1}\left\{ \frac{3e^{-\pi s} + 9s}{9s^2+1} \right\}.$$

Now using the translation property, we have

$$x(t) = \cos 3t + \sin 3(t - \pi)u(t - \pi)$$

which can be written as

$$x(t) = \begin{bmatrix} \cos 3t & t < \pi \\ \cos 3t - \sin 3t & \pi < t. \end{bmatrix}$$

Problems

For each of the problems 1–15, use Laplace transforms to solve each of the initial-value problems.

1. $\frac{dy}{dt} - y = e^{3t}$, $y(0) = 2$

2. $\frac{dy}{dt} + y = 3\cos t$, $y(0) = -1$

3. $\frac{d^2y}{dt^2} + 4y = 0$, $y(0) = 2$, $y'(0) = 3$

4. $\frac{d^2y}{dt^2} + 16y = 0$, $y(0) = 7$, $y'(0) = 0$

5. $\frac{d^2y}{dt^2} - \frac{dy}{dt} - 6y = 0$, $y(0) = 2$, $y'(0) = 6$

6. $\frac{d^2y}{dt^2} - 5\frac{dy}{dt} + 6y = 0$, $y(0) = 1$, $y'(0) = 3$

7. $\frac{d^2y}{dt^2} + \frac{dy}{dt} - 12y = 0$, $y(0) = 4$, $y'(0) = -1$

8. $\frac{d^2y}{dt^2} + 8y = 4$, $y(0) = 0$, $y'(0) = 6$

9. $\frac{d^2y}{dt^2} + 2\frac{dy}{dt} + 5y = 0$, $y(0) = 2$, $y'(0) = 4$

10. $\frac{d^2y}{dt^2} + 2\frac{dy}{dt} + y = te^{-3t}$, $y(0) = 2$, $y'(0) = 0$

11. $\frac{d^2y}{dt^2} - 8\frac{dy}{dt} + 15y = 6te^{4t}$, $y(0) = 5$, $y'(0) = 4$

12. $\frac{d^2y}{dt^2} - \frac{dy}{dt} - 2y = 13e^{-t}\sin 3t$, $y(0) = 0$, $y'(0) = 3$

13. $\frac{d^2y}{dt^2} + 3\frac{dy}{dt} - 10y = 50e^{2t}\cos t$, $y(0) = -1$, $y'(0) = 1$

14. $\frac{d^3y}{dt^3} - 5\frac{d^2y}{dt^2} + y\frac{dy}{dt} - 3y = 20\sin t$, $y(0) = 0$, $y'(0) = 0$, $y''(0) = -2$

15. $\frac{d^3y}{dt^3} - 6\frac{d^2y}{dt^2} + 11\frac{dy}{dt} - 6y = 19te^{4t}$, $y(0) = -1$, $y'(0) = 0$, $y''(0) = -6$

16. The differential equation

$$v\frac{dy}{dt} = r_i f(t) - r_0 y$$

is used to describe the absorption of a drug by a body organ of volume v. The function $y(t)$ is the concentration of the drug in the organ's fluid at time t, and r_i and r_0 are the rates of fluid flow into and out of the organ, respectively. The function $f(t)$ is the concentration of the drug entering the organ.

a. If $y(0) = 0$ and $r_i = r_0$ find $y(t)$ if $f(t) = t[1 - U_a(t)]$.

b. If $y(0) = 1$ and $r_i \neq r_o$ find $y(t)$ if $f(t) = t[1 - U_a(t)]$.

17. The differential equation for the current $I(t)$ in a series circuit with inductance L and resistance R is

$$LI' + RI = E(t)$$

where $E(t)$ is the applied voltage. If $I(0) = 0$ and E_0 is a constant, solve this equation if

a. $E(t)$ is the square pulse $E(t) = E_0[U_1(t) - U_2(t)]$

b. $E(t)$ is a single pulse of a sine wave $E(t) = E_0[1 - U_\pi(t)]$

18. A particular forced vibration of a mass m at the end of a vertical spring with spring constant k is described by the differential equation

$$m\frac{d^2x}{dt^2} + kx(t) = 1 + U_1(t) - 2U_2(t)$$

a. Describe what $f(t)$ represents concerning the motion of the top of the spring.

b. Solve the differential equation for $m = 1$ and $k = 4$ with the initial conditions $x(0) = x'(0) = 0$.

c. Plot the solution on $0 \leq t \leq 3$. Considering the driving function $1 + U_1(t) - 2U_2(t)$, what do you observe?

7.6 Solving Linear Systems using Laplace Transforms

We can apply the Laplace transform method to find the solution of a first-order system

$$a_1\frac{dx}{dt} + a_2\frac{dy}{dt} + a_3x + a_4y = \beta_1(t)$$

$$b_1\frac{dx}{dt} + b_2\frac{dy}{dt} + b_3x + b_4y = \beta_2(t) \tag{7.6}$$

where $a_1, a_2, a_3, a_4, b_1, b_2, b_3$, and b_4 are constants and $\beta_1(t), \beta_2(t)$ are known functions. In addition, x and y satisfy the initial conditions

$$x(0) = c_1 \quad \text{and} \quad y(0) = c_2$$

where c_1 and c_2 are constants.

The procedure we employ is a straightforward extension of the method we used to solve an ordinary differential equation. Letting

$$X(s) = \mathcal{L}\{x(t)\} \quad \text{and} \quad Y(s) = \mathcal{L}\{y(t)\}$$

we apply the Laplace transform to equation (7.6) to obtain a system of algebraic equations in $X(s)$ and $Y(s)$. This system of algebraic equations is then solved for $X(s), Y(s)$. Once these are obtained, we find (if possible)

$$x(t) = \mathcal{L}^{-1}\{X(s)\} \quad \text{and} \quad y(t) = \mathcal{L}^{-1}\{Y(s)\}.$$

Example 1: Solve the system

$$\frac{dx}{dt} - 6x + 3y = 8e^t$$

$$\frac{dy}{dt} - 2x - y = 4e^t$$

that satisfies the initial conditions

$$x(0) = -1$$
$$y(0) = 0$$

using Laplace transforms.

Applying the Laplace transform to both sides of the system gives

$$\mathcal{L}\left\{\frac{dx}{dt}\right\} - 6\mathcal{L}\left\{x(t)\right\} + 3\mathcal{L}\left\{y(t)\right\} = \mathcal{L}\{8e^t\}$$

$$\mathcal{L}\left\{\frac{dy}{dt}\right\} - 2\mathcal{L}\{x(t)\} - \mathcal{L}\{y(t)\} = \mathcal{L}\{4e^t\}$$

and letting $X(s) = \mathcal{L}\{x(t)\}$ and $Y(s) = \mathcal{L}\{y(t)\}$ gives

$$\mathcal{L}\left\{\frac{dx}{dt}\right\} = sX(s) - x(0) = sX(s) + 1$$

$$\mathcal{L}\left\{\frac{dy}{dt}\right\} = sY(s) - y(0) = sY(s).$$

So, using the Laplace transform table, we have

$$sX(s) + 1 - 6X(s) + Y(s) = \frac{8}{s-1}$$

$$sY(s) - 2X(s) - Y(s) = \frac{4}{s-1}$$

which simplifies to

$$(s-6)X(s) + 3Y(s) = \frac{-s+9}{s-1}$$

$$-2X(s) + (s-1)Y(s) = \frac{4}{s-1}.$$

Now solving this system of algebraic equations for $X(s)$ and $Y(s)$ gives

$$X(s) = \frac{-s+7}{(s-1)(s-4)} \quad \text{and} \quad Y(s) = \frac{2}{(s-1)(s-4)}.$$

Thus

$$x(t) = \mathcal{L}^{-1}\{X(s)\} = \mathcal{L}^{-1}\left\{\frac{-s+7}{(s-1)(s-4)}\right\}$$

and

$$y(t) = \mathcal{L}^{-1}\{Y(s)\} = \mathcal{L}^{-1}\left\{\frac{2}{(s-1)(s-4)}\right\}.$$

Applying partial fractions we have

$$\frac{-s+7}{(s-1)(s-4)} = \frac{A}{s-1} + \frac{B}{s-4}$$

which gives $A = -2$ and $B = 1$, so

$$x(t) = -2\mathcal{L}^{-1}\left\{\frac{1}{s-1}\right\} + \mathcal{L}^{-1}\left\{\frac{1}{s-4}\right\}.$$

Using the Laplace transform table, we find $x(t)$ as

$$x(t) = -2e^t + e^{4t}$$

and in a very similar manner, we find $y(t)$ as

$$y(t) = \frac{-2}{3}e^t + \frac{2}{3}e^{4t}.$$

Example 2: Consider the system

$$x' = -x + y + 25\sin t$$
$$y' = -x - 3y$$

with the initial conditions $x(0) = -1, y(0) = -1$.

If we take the Laplace transform of both sides of this system we obtain

$$\mathcal{L}\{x'\} = -\mathcal{L}\{x\} + \mathcal{L}\{y\} + 25\mathcal{L}\{\sin t\}$$
$$\mathcal{L}\{y'\} = -\mathcal{L}\{x\} - 3\mathcal{L}\{y\}.$$

Letting $X(s) = \mathcal{L}\{x(t)$ and $Y(s) = \mathcal{L}\{y(t)\}$, then this becomes

$$sX(s) + 1 = -X(s) + Y(s) + \frac{25}{s^2+1}$$
$$sY(s) + 1 = -X(s) - 3Y(s).$$

Solving these linear equations for $X(s)$ and $Y(s)$ gives

$$X(s) = -(3+s)Y(s) - 1$$

and

$$Y(s) = -\frac{s}{(s+2)^2} - \frac{25}{(s+2)^2(s^2+1)}.$$

Now applying partial fractions to the second term here,

$$\frac{25}{(s+2)^2(s^2+1)} = \frac{A}{s+2} + \frac{B}{(s+2)^2} + \frac{Cs+D}{(s^2+1)},$$

gives

$$A + C = 0$$
$$2A + B + 4C + D = 0$$
$$A + 4C + 4D = 0$$
$$2A + B + 4D = 25.$$

Solving this system gives $A = 4$, $B = 5$, $C = -4$, and $D = 3$.
  Thus

$$\mathcal{L}^{-1}\left\{\frac{25}{(s+2)^2(s^2+1)}\right\} = 4e^{-2t} + 5te^{-2t} - 4\cos t + 3\sin t.$$

Combining the above yields the explicit form of the solution as

$$y(t) = -e^{-2t}(3t+5) + 4\cos t - 3\sin t.$$

We find $x(t)$ by solving $y' = -x - 3y$ for x as

$$x(t) = -3y - y'$$

so that

$$x(t) = e^{-2t}(3t+8) - 9\cos t + 13\sin t.$$

Problems

  In each of the following problems 1–8, use the Laplace transform to find the
solution of the linear system that satisfies the given initial conditions.

1. $\frac{dx}{dt} - 2y = 0$
   $\frac{dy}{dt} + x - 3y = 2$
   $x(0) = 3, \quad y(0) = 0$

2. $\frac{dx}{dt} + y = 3e^{2t}$
   $\frac{dy}{dt} + x = 0$
   $x(0) = 2, \quad y(0) = 0$

3. $\frac{dx}{dt} - 2x - 3y = 0$
   $\frac{dy}{dt} + x + 2y = t$
   $x(0) = -1, \quad y(0) = 0$

4. $\frac{dx}{dt} - 4x + 2y = 2t$
   $\frac{dy}{dt} - 8x + 4y = 1$
   $x(0) = 3, \quad y(0) = 5$

5. $\frac{dx}{dt} - 5x + 2y = 3e^{4t}$

$\quad \frac{dy}{dt} - 4x + y = 0$

$\quad x(0) = 3, \quad y(0) = 0$

6. $\frac{dx}{dt} + x + y = 5e^{2t}$

$\quad \frac{dy}{dt} - 5x - y = -3e^{2t}$

$\quad x(0) = 3, \quad y(0) = 2$

7. $2\frac{dx}{dt} + \frac{dy}{dt} - x - y = e^{-t}$

$\quad \frac{dx}{dt} + \frac{dy}{dt} + 2x + y = e^{t}$

$\quad x(0) = 2, \quad y(0) = 1$

8. $2\frac{dx}{dt} + \frac{dy}{dt} + x + 5y = 4t$

$\quad \frac{dx}{dt} + \frac{dy}{dt} + 2x + 2y = 2$

$\quad x(0) = 3, \quad y(0) = -4$

In each of the problems 9–12, use Laplace transforms to solve the following initial-value problems.

9. $x' + 2x + y' = 16e^{-2t}$

$\quad 2x' + 3y' + 5y = 15$

$\quad x(0) = 4, \quad y(0) = -3$

10. $2x' - 2x + y' = 1$

$\quad x' - 3x + y' - 3y = 2$

$\quad x(0) = 0, \quad y(0) = 0$

11. $x' - y' = -e^{t}$

$\quad 2x' - 2y' - y = 8$

$\quad x(0) = -1, \quad y(0) = -10$

12. $x'' + 2x - 4y' = 0$

$\quad x' + y'' - 4y = 0$

$\quad x(0) = -4, \quad y(0) = 1$

$\quad x'(0) = 8, \quad y'(0) = 2$

13. **a.** Letting, as usual, $X(s) = \mathcal{L}\{x(t)\}$ and $Y(s) = \mathcal{L}\{(t)\}$, the Laplace transform converts the initial-value problem

$$2x' + y' - y = t$$

$$x' + y' = t^2$$

with $x(0) = 1$ and $y(0) = 0$ into the system

$$2sX + (s-1)Y = 2 + \frac{1}{s^2}$$

$$sX + sY = 1 + \frac{2}{s^3}.$$

b. Solve the system in part (a) for $Y(s)$ as

$$Y(s) = \frac{4-s}{s^3(s+1)}$$

$$= \frac{4}{s^3(s+1)} - \frac{1}{s^2(s+1)}$$

and use partial fractions to show that

$$y(t) = 5 - 5t + 2t^2 - 5e^{-t}.$$

c. Show that $X(s) + Y(s) = \frac{1}{s} + \frac{2}{s^4}$ and solve for $x(t)$.

7.7   The Convolution (Optional)

Sometimes inverse Laplace transforms and the use of the table of transforms can be simplified by using the convolution theorem that we give below. We first define the convolution of two functions f and g.

DEFINITION 7.5   *Let f and g be two functions that are piecewise continuous on every finite closed interval $0 \leq t \leq b$ and of exponential order. The function denoted by $f * g$ and defined by*

$$f(t) * g(t) = \int_0^t f(\tau)g(t-\tau)\, d\tau \qquad (7.7)$$

is called the convolution of the functions f and g.

The convolution operation $*$ is commutative. To see this, we need only change variables in (7.7). Letting $u = t - \tau$, we have

$$f(t) * g(t) = \int_0^t f(\tau)g(t-\tau)\,d\tau$$

$$= -\int_t^0 f(t-\mu)g(u)\,du$$

$$= \int_0^t g(u)f(t-\mu)\,du$$

$$= g(t) * f(t),$$

so $f * g = g * f$.

Example 1: Let $f(t) = e^{-at}$ and $g(t) = \sin bt$. Then we have

$$f(t) * g(t) = \int_0^t e^{-a\tau} \sin b(t-\tau)\,d\tau$$

$$= \int_0^t e^{-a(t-\tau)} \sin b\tau\,d\tau \quad (\text{using } f * g = g * f)$$

$$= e^{-at} \int_0^t e^{at} \sin b\tau\,d\tau$$

$$= \frac{1}{a^2 + b^2}[a \sin bt - b \cos bt + be^{-at}].$$

We now state a useful fact about the convolution of two functions f and g.

PROPOSITION 7.7.1 *If the functions f and g are piecewise continuous on a closed interval $0 \le t \le b$ and of exponential order e^{at}, then the convolution $f * g$ is also piecewise continuous on every finite closed interval $0 \le t \le b$ and of exponential order $e^{(a+\varepsilon)t}$ where $\varepsilon > 0$ is a constant.*

We will not prove this proposition, but we note here that it guarantees the existence of $\mathcal{L}\{f * g\}$ when $\mathcal{L}\{f\}$ and $\mathcal{L}\{g\}$ exist.

THEOREM 7.7.1 *(Convolution) If f and g are piecewise continuous functions on every closed interval $0 \le t \le b$ and are of exponential order e^{at} then*

$$\mathcal{L}\{f * g\} = \mathcal{L}\{f\}\mathcal{L}\{g\}$$

for $s > a$.

Proof: Applying the definition of the Laplace transform, $\mathcal{L}\{f * g\}$ is the function defined by

$$\int_0^\infty e^{-st}\left[\int_0^t f(\tau)g(t-\tau)\,d\tau\right]\,dt$$

which can be written as

$$\int_0^\infty \int_0^t e^{-st}f(\tau)g(t-\tau)\,d\tau\,dt.$$

Now consider the region R_1 shown in Figure 7.4.

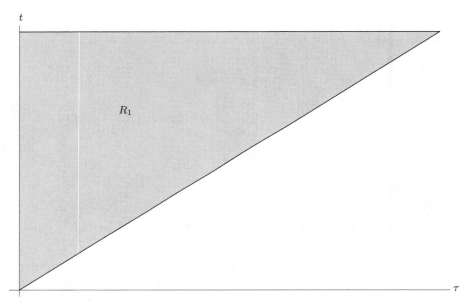

R_1

FIGURE 7.4:   Region of integration R_1.

Thus

$$\int_0^\infty \int_0^t e^{-st}f(\tau)g(t-\tau)\,d\tau\,dt = \int\int_{R_1} e^{-st}f(\tau)g(t-\tau)\,d\tau\,dt.$$

So if we consider the change of variables $u = t - \tau, v = \tau$, which has Jacobian 1, the double integral over R_1 becomes the double integral

$$\int\int_{R_2} e^{-s(u+v)}f(v)g(u)\,du\,dv$$

where R_2 is the quarter plane defined by $u > 0, v > 0$ shown in Figure 7.5. We then have

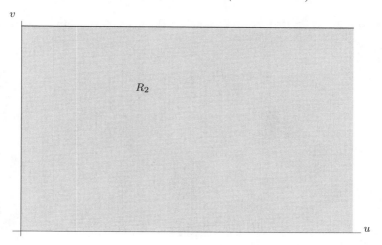

FIGURE 7.5: Region of integration R_2

$$\int\int_{R_2} e^{-s(u+v)} f(u)g(u)\, du\, dv = \int_0^\infty \int_0^\infty e^{-s(u+v)} f(v)g(u)\, du\, dv$$
$$= \int_0^\infty e^{-sv} f(v)\, dv \int_0^\infty e^{-su} g(u)\, du.$$

Indeed, since the integrals involved in this calculation are absolutely convergent for $s > a$, the operations we performed above are valid so we have shown

$$\mathcal{L}\{f * g\} = \mathcal{L}\{f\}\mathcal{L}\{g\} \quad \text{for } s > a. \,\square$$

If we denote $\mathcal{L}\{f\}$ by F and $\mathcal{L}\{g\}$ by G then the conclusion of the last theorem can be written as

$$\mathcal{L}\{f(t) * g(t)\} = F(s)G(s).$$

Thus,

$$\mathcal{L}^{-1}\{F(s)G(s)\} = f(t) * g(t) = \int_0^t f(\tau)g(t-\tau)\, d\tau \qquad (7.8)$$

and since $*$ is a commutative operation, we have

$$\mathcal{L}^{-1}\{F(s)G(s)\} = g(t) * f(t) = \int_0^t g(\tau)f(t-\tau)\, d\tau. \qquad (7.9)$$

These remarks allow another method for computing inverse Laplace transforms. Suppose we are given a function H and need to find $\mathcal{L}^{-1}\{H(s)\}$. If we can express $H(s)$ as a product $F(s)G(s)$ where $\mathcal{L}^{-1}\{F(s)\} = f(t)$ and $\mathcal{L}^{-1}\{G(s)\} = g(t)$ are known, then we can apply (7.8) or (7.9) to determine $\mathcal{L}^{-1}\{H(s)\}$.

Example 2: Find $\mathcal{L}^{-1}\left\{\frac{1}{s(s^2+1)}\right\}$ using the convolution and the transform table.

Writing $\frac{1}{s(s^2+1)}$ as the product $F(s)G(s)$ where

$$F(s) = \frac{1}{s} \quad \text{and} \quad G(s) = \frac{1}{s^2+1}$$

we have by the table

$$f(t) = \mathcal{L}^{-1}\left\{\frac{1}{s}\right\} = 1$$

and

$$g(t) = \mathcal{L}^{-1}\left\{\frac{1}{s^2+1}\right\} = \sin t$$

so

$$\mathcal{L}^{-1}\left\{\frac{1}{s(s^2+1)}\right\} = f(t) * g(t)$$

$$= \int_0^t 1 \cdot \sin(t-\tau)\,d\tau$$

or

$$\mathcal{L}^{-1}\left\{\frac{1}{s(s^2+1)}\right\} = g(t) * f(t)$$

$$= \int_0^t \sin\tau \cdot 1\,d\tau.$$

Here the second integral is (slightly) more simple to evaluate, so we have

$$\mathcal{L}^{-1}\left\{\frac{1}{s(s^2+1)}\right\} = 1 - \cos t.$$

Note that we obtained this result earlier by means of a partial fractions decomposition.

Example 3: Newton's Law of Cooling (revisited)

Recall from Chapter 1 we considered Newton's Law of Cooling:

$$\frac{dT}{dt} = k[T(t) - T_s(t)],$$

which says that the rate of change of the temperature $T(t)$ of a body is proportional to the difference between the temperature of the body and its surroundings $T_s(t)$. Unlike our work in Chapter 1, here we assume T_s is not necessarily a constant. So, applying the Laplace transform to both sides of this differential equation gives

$$\mathcal{L}\left\{\frac{dT}{dt}\right\} = \mathcal{L}\{k[T(t) - T_s(t)]\}$$

which is

$$s\mathcal{L}\{T\} - T(0) = k\mathcal{L}\{T\} - k\mathcal{L}\{T_s\}.$$

Solving this equation for $\mathcal{L}\{T\}$ gives

$$\mathcal{L}\{T\} = \frac{T(0)}{s-k} - \frac{k}{s-k}\mathcal{L}\{T_s\}.$$

Using the convolution theorem we find the inverse Laplace transform as

$$T(t) = T(0)e^{kt} - \int_0^t kT_s(z)e^{k(t-z)}\, dz,$$

that is,

$$T(t) = T(0)e^{kt} - ke^{kt}\int_0^t T_s(z)e^{-kz}\, dz. \tag{7.10}$$

Equation (7.10) is the solution of Newton's Law of Cooling with a nonconstant surrounding temperature. If we take $T_s(t) = T_s$ a constant, as we did in Chapter 1, then equation (7.10) becomes

$$\begin{aligned}
T(t) &= T(0)e^{kt} - ke^{kt}\int_0^t T_s e^{-kz}\, dz \\
&= T(0)e^{kt} - T_s e^{kt} + T_s \\
&= (T(0) - T_s)e^{kt} + T_s
\end{aligned}$$

which is the solution we obtained in Chapter 1; see equation (1.4).

Computer Code 7.4:   **Inverse Laplace transform and convolution**

Matlab, Maple, Mathematica

```
                          Matlab
≫  syms s t a b tau
≫  g1=1/(s^2*(s^2+1))
≫  eq1=ilaplace(g1,s,t)
≫  eq2=int(tau*sin(t-tau),tau,0,t) %using convolution
≫  f1=exp(-a*tau)*sin(b*(t-tau))
≫  eq3=int(f1,tau,0,t)
≫  simplify(laplace(eq3,t,s))
```

```
                           Maple
>  with(inttrans):
>  g1:=1/(s^2*(s^2+1));
>  eq1:=invlaplace(g3,s,t);
>  eq2:=int(tau*sin(t-tau),tau=0..t); #using convolution
>  f1:=exp(-a*tau)*sin(b*(t-tau))
>  eq3:=int(f1,tau=0..t)
>  simplify(laplace(eq3,t,s))
```

```
                      Mathematica
```
$$g1 = \frac{1}{s^2 \left(s^2 + 1\right)} \quad \text{(*entered from palette*)}$$
```
eq1=InverseLaplaceTransform[g1,s,t]
```
$$eq2 = \int_0^t \tau \, \text{Sin}[t - \tau] d\tau \quad \text{(*convolution; entered from palette*)}$$
$$f1[\tau\_] = e^{-a\tau} \, \text{Sin}[b\,(t - \tau)] \quad \text{(*entered from palette*)}$$
$$eq3[t\_] = \int_0^t f1[\tau] d\tau$$
```
FullSimplify[LaplaceTransform[eq3[t],t,s]]
```

Problems

In each of the following problems 1–6, compute the convolution $(f * g)(t)$ for the given pair of functions.

1. $f(t) = 1, g(t) = t^2$
2. $f(t) = \sin 2t, g(t) = e^{-t}$
3. $f(t) = t^3, g(t) = \sin 4t$
4. $f(t) = \cos 2t, g(t) = e^{-st}$
5. $f(t) = t, g(t) = \delta(t)$
6. $f(t) = \sin t, g(t) = \delta\left(t - \frac{\pi}{2}\right)$

Find $f(t) = \mathcal{L}^{-1}\{F(s)\}$ using the convolution theorem for each of the functions $F(s)$ in problems 7–12.

7. $F(s) = \frac{1}{s^2+5s+6}$
8. $F(s) = \frac{1}{s^2+3s-4}$
9. $F(s) = \frac{1}{s(s^2+9)}$
10. $F(s) = \frac{1}{s(s^2+4s+13)}$
11. $F(s) = \frac{1}{s^3(s+3)}$
12. $F(s) = \frac{1}{(s^2+4)(s+7)}$

13. Show that the convolution operation $*$ is *associative*; that is

$$(f * (g * h))(t) = ((f * g) * h)(t)$$

for functions f, g, and h.

14. Show that the convolution operation $*$ satisfies the *distributive property over addition*; that is,

$$(f * (g + h))(t) = (f * g)(t) + (f * h)(t)$$

for functions f, g, and h.

15. Show that for any constant k

$$((kf) * g)(t) = k(f * g)(t).$$

16. Show that $t^{-1/2} * t^{-1/2} = \pi$. Hint: Find $\mathcal{L}\{t^{-1/2}\}$.

17. Compute for a constant k

   a) $\sin kt * \cos kt$

   b) $\sin kt * \sin kt$

   c) $\cos kt * \cos kt$

   d) Using the results, compute $\mathcal{L}^{-1}\left\{\frac{ks}{(s^2+k^2)^2}\right\}$.

7.8   Chapter 7: Additional Problems and Projects

ADDITIONAL PROBLEMS

In problems 1–5, determine whether the statement is true or false. If it is true, give reasons for your answer. If it is false, give a counterexample or other explanation of why it is false.

1. For any function $f(t)$, the Laplace transform $\mathcal{L}\{f(t)\}$ exists.

2. The translated function

$$U_a(t)f(t-a) = \begin{bmatrix} 0 & 0 < t < a \\ f(t-a) & t > a \end{bmatrix}$$

  has Laplace transform

$$\mathcal{L}\{U_a(t)f(t-a)\} = e^{-as}\mathcal{L}\{f(t)\}.$$

3. If $\mathcal{L}\{f(t)\} = F(s)$ exists for some $s > a$ for a function $f(t)$ in which $f(t)/t$ is bounded for $t > 0$, then

$$\mathcal{L}\{\frac{1}{t}f(t)\} = \int_s^\infty F(z)\, dz.$$

4. $\mathcal{L}\{C_1 f_1(t) + C_2 f_2(t)\} = C_1\mathcal{L}\{f_1(t)\}C_2\mathcal{L}\{f_2(t)\}.$

5. If $\mathcal{L}\{f(t)\} = \mathcal{L}\{g(t)\}$, then $f(t) = g(t)$ for all $t \geq 0$.

In problems 6–10, use the definition of the Laplace transform to find $\mathcal{L}\{f(t)\}$ for each of the given functions $f(t)$.

6. $f(t) = \begin{cases} 5 & 0 < t < 7 \\ 0 & \text{otherwise} \end{cases}$

7. $f(t) = 2t^4$

8. $f(t) = t^2 e^{-t}$

9. $f(t) = |t - 2|$

10. $f(t) = e^t \cos t$

11. Using the definition of the Laplace transform, show that

$$\mathcal{L}\{e^{(a+bi)t}\} = \frac{1}{s - (a + bi)} \quad \text{for } s > a.$$

12. Prove that if f and g are piecewise continuous on $[a, b]$, then the product fg is piecewise continuous on $[a, b]$.

13. The *floor* or *unit staircase* function is defined as

$$f(t) = n \text{ if } n - 1 \le t < n,$$

for n a positive integer. Show that

$$f(t) = \sum_{k=0}^{\infty} U(t - k).$$

Using this result, and assuming that the Laplace transform exists for this infinite series, show using a geometric series that

$$\mathcal{L}\{f(t)\} = \frac{1}{s(1 - e^{-s})}.$$

14. Using the method in problem 13, show that the function

$$f(t) = \sum_{k=0}^{\infty} (-1)^k U(t - k)$$

has Laplace transform

$$\mathcal{L}\{f(t)\} = \frac{1}{s(1 + e^{-s})}.$$

15. Using problem 14, show that the *square wave* function

$$g(t) = \begin{cases} \alpha & n - 1 \le t < n, n \text{ an odd integer} \\ \beta & n - 1 \le t < n, n \text{ an even integer} \end{cases}$$

has Laplace transform

$$\mathcal{L}\{f(t)\} = \frac{\alpha + \beta e^{-s}}{s(1 + e^{-s})}.$$

In problems 16–19, evaluate the following integrals involving the Dirac delta function.

16. $\int_0^4 (1 + e^{-t})\delta(t - 3)\, dt$

17. $\int_{-5}^1 (1 + e^{-t})\delta(t - 2)\, dt$

18. $\int_{-1}^2 \sin 2t\delta(t)\, dt$

19. $\int_0^{\frac{\pi}{3}} \tan t\delta(t - \frac{\pi}{4})\, dt$

In problems 20–24, find $\mathcal{L}^{-1}\{F(s)\}$.

20. $F(s) = \frac{5}{s^2+9}$

21. $F(s) = \frac{7s}{s^2-4}$

22. $F(s) = \frac{6}{(s-2)^2}$

23. $F(s) = \frac{5s}{s^2+6s+9}$

24. $F(s) = \frac{s-3}{s^2+5s+10}$

In problems 25–34, use Laplace transforms to solve the initial-value problem.

25. $\frac{dy}{dt} - y = e^{4t}, \quad y(0) = 6$

26. $\frac{d^2y}{dt^2} + 9y = 0, \quad y(0) = 2, \quad y'(0) = 3$

27. $\frac{d^2y}{dt^2} + 81y = 0, \quad y(0) = 1, \quad y'(0) = 2$

28. $\frac{d^2y}{dt^2} - \frac{dy}{dt} - 12y = 0, \quad y(0) = 1, \quad y'(0) = 0$

29. $\frac{d^2y}{dt^2} + 2\frac{dy}{dt} + 5y = 0, \quad y(0) = 6, \quad y'(0) = 0$

30. $\frac{d^2y}{dt^2} + 2\frac{dy}{dt} + y = te^{-2t}, \quad y(0) = 1, \quad y'(0) = 0$

31. $\frac{d^2y}{dt^2} - 8\frac{dy}{dt} + 15y = 5te^{2t}, \quad y(0) = 5, \quad y'(0) = 4$

32. $\frac{d^2y}{dt^2} - \frac{dy}{dt} - 2y = 13e^{-t}\cos 4t, \quad y(0) = 0, \quad y'(0) = 3$

33. $\frac{d^2y}{dt^2} + 3\frac{dy}{dt} - 10y = e^{t}, \quad y(0) = -1, \quad y'(0) = 1$

34. $\frac{d^3y}{dt^3} - 7\frac{dy}{dt} - 6y = 13t, \quad y(0) = 1, \quad y'(0) = 2, \quad y''(0) = -2$

35. Suppose an insect pest population grows at a rate proportional to its size with rate $r = 1/10$ (per day). In an attempt to eliminate the pest, a farmer applies a pesticide. The pesticide kills the insect at a rate that, in one day, decreases linearly with time from the maximum rate $P > 0$ to 0 (in units of 10,000 per day). As this is a rather potent pesticide, regulations allow the farmer to apply it only once. If the initial pest population is approximately 200,000, what dosage P must the farmer use to eliminate the pest population?

36. A forced vibration of a mass m at the end of a vertical spring, with spring constant k, is described by

$$mx''(t) + kx'(t) = f(t)$$

where

$$f(t) = 1 + U(t-1) - 2U(t-2).$$

Solve this differential equation for $m = 1, k = 8, x(0) = x'(0) = 0$. Plot the solution on the interval $0 \le t \le 3$ and discuss what you observe.

In problems 37–40, use Laplace transforms to solve the following initial-value problems.

37. $\begin{cases} x' + 2x + y' = 8e^{-3t} \\ 2x' + 3y' + 5y = 12 \end{cases}$ with $x(0) = 2, y(0) = -3$

38. $\begin{cases} 2x' - 2x + y' = 12\sin t \\ x' - 3x + y' - 3y = 2 \end{cases}$ with $x(0) = 1, y(0) = 3$

39. $\begin{cases} x' - y' = -e^2 t \\ 2x' - 2y' - y = 8 \end{cases}$ with $x(0) = -1, y(0) = -10$

40. $\begin{cases} x' - 2y' = -5e^3 t \\ x' - 2y' - y = 5 \end{cases}$ with $x(0) = -10, y(0) = -1$

In problems 41–45, compute the convolution $(f * g)(t)$ for the given pair of functions.

41. $f(t) = 3, g(t) = t^3$

42. $f(t) = \sin 3t, g(t) = e^{-2t}$

43. $f(t) = 2t^4, g(t) = \cos 4t$

44. $f(t) = \cos at, g(t) = e^{-bt}$

45. $f(t) = \cos t, g(t) = \delta\left(t - \frac{\pi}{2}\right)$

In problems 46–50, find $\mathcal{L}^{-1}\{F(t)\}$.

46. $F(s) = \frac{1}{s^2 + 7s + 12}$

47. $F(s) = \frac{1}{s^2 + 3s - 40}$

48. $F(s) = \frac{1}{s(s^2 + 4)}$

49. $F(s) = \frac{6}{s^2(s^2 + 36)}$

50. $F(s) = \frac{1}{(s^2 + 64)(s + 8)}$

PROJECTS FOR CHAPTER 7

Project 1: Carrier-Borne Epidemics

This project is a follow-up on some of the ideas presented in Section 6.5. It would be a good idea to review that section before working this project. Our discussion follows the work in Daley and Gani [11].

There are certain diseases such as typhoid and tuberculosis in which the infectives who are the source of the infection in a population appear healthy. Since the infectives appear healthy, they often go undetected in the population until the disease spread warrants a mass screening program. We call such infectives *carriers* to distinguish them from susceptibles, who, on being infected, may be quickly recognized by their symptoms and removed from the population. Once carriers are detected, they are removed from contact with the susceptibles.

The situation described here requires a different type of model than presented in Section 6.5. The infection of a susceptible through contact with a carrier results in the removal of the infected susceptible, while the number of carriers remains unaltered. The carrier population declines through a process that is independent of the susceptibles.

If we let $C(t)$ represent the size of the carrier population at time t and $S(t)$ be the size of the susceptible population at time t, then one simple model for this carrier-borne infection is

$$\frac{dS}{dt} = -\beta S(t) C(t)$$

$$\frac{dC}{dt} = -\gamma C(t) \tag{7.11}$$

with initial conditions $S(0) = s_0$ and $C(0) = c_0$, where β and γ are positive constants.

Solve this system of differential equations to find $S(t)$ and $C(t)$. For different values of β and γ and initial conditions s_0, c_0, plot your solutions and discuss the behavior of the susceptible and carrier populations.

In practice, when a carrier-borne disease is recognized in a population where it is not normally present, measures are taken to locate the source or sources of the infection. Suppose then that there are c_0 carriers initially in the population, and at some time later, say at t_0, the disease is recognized in the population. At this time, efforts are increased to remove the carriers from the population so that the carrier removal rate γ is increased to γ'. This modification changes the system (7.11) to the system

$$\frac{dS}{dt} = -\beta S(t) C(t)$$

$$\frac{dC}{dt} = \begin{cases} -\gamma C(t) & 0 < t \le t_0 \\ -\gamma' C(t) & t > t_0 \end{cases} \tag{7.12}$$

with the same initial conditions $S(0) = s_0$ and $C(0) = c_0$ as before.

Using Laplace transforms, solve this system of differential equations to find $S(t)$ and $C(t)$. For the values of β, γ and initial conditions s_0, c_0 you used for the system (7.11), plot your solutions for different values of γ'. Discuss the behavior of the susceptible and carrier populations comparing your results to the results obtained for the system (7.11).

One possible modification of the system (7.12) that can be studied using Laplace transforms arises by allowing there to be a constant influx of new carriers to the carrier population. If we let α be the rate at which new carries arrive in the population (α can be thought of as an immigration rate), and assume that this rate is independent of the populations of carriers and susceptibles, then we have the modified system

$$\frac{dS}{dt} = -\beta S(t)C(t)$$

$$\frac{dC}{dt} = \begin{cases} -\gamma C(t) + \alpha & 0 < t \le t_0 \\ -\gamma' C(t) + \alpha & t > t_0 \end{cases} \qquad (7.13)$$

with the same initial conditions $S(0) = s_0$ and $C(0) = c_0$ as before.

Solve this system using Laplace transforms and perform the same kind of analysis you did on the other systems with this system. What is the effect of the immigration rate α?

As a final extension and challenge, we can extend the carrier-borne model described by system (7.12) in a different manner. Consider the affect of allowing a small proportion k of the susceptibles infected to become carriers. This modification changes the differential equation for the carriers and gives the equation

$$\frac{dC}{dt} = -\gamma C(t) + k\beta S(t)C(t),$$

since the proportion of new carriers introduced to the carrier population arises from the infected susceptible population. Using this, the system (7.12) with the increased detection rate γ' becomes

$$\frac{dS}{dt} = -\beta S(t)C(t)$$

$$\frac{dC}{dt} = -\gamma C(t) + k\beta S(t)C(t). \qquad (7.14)$$

Observe that this system is similar to the SI model, equation (6.44) considered in Section 6.5. Determine conditions on the stability of the *DFE* and find the basic reproductive number R_0. Interpret the meaning of R_0 in terms of the parameters of the system. Try to find the unique endemic equilibrium of the system that lies in the invariant region. Determine its stability and when it is biologically relevant.

If the system has initial conditions $S(0) = s_0$ and $C(0) = c_0$, show that the number of susceptibles $s_\infty = \lim_{t\to\infty} S(t)$ surviving the epidemic is the unique solution of the equation

$$k(s_0 - s_\infty) + c_0 = \frac{\gamma}{\beta} \ln\left(\frac{s_0}{s_\infty}\right).$$

Project 2: Integral Equations

The convolution theorem has many useful theoretical applications in other branches of mathematics. We will now briefly consider an application closely related to differential equations.

An equation in which an unknown function occurs in a definite integral is called an *integral equation*. In some applications, the unknown function occurs not only in the integral, but in differentiated form in other terms of the same equation. In such cases, the equation is called an *integrodifferential equation.* In either case, if the definite integral is of the convolution type, then the convolution theorem may enable us to transform such an equation into an algebraic equation.

To illustrate this discussion, we will consider an integral that is of the convolution type. Consider an equation of the form

$$x(t) = f(t) + \int_0^t g(t - \tau)x(\tau)\,d\tau \qquad (7.15)$$

in which f and g are *known* functions. Our goal is to find a function x for which (7.15) is true.

If we assume that the functions in (7.15) have transforms, we have

$$X(s) = F(s) + G(s)X(s),$$

which we can algebraically solve for $X(s)$ as

$$X(s) = \frac{F(s)}{1 - G(s)}. \qquad (7.16)$$

Equation (7.16) is the transform of the solution function $x(t)$ of (7.15). To find $x(t)$ explicitly, we calculate $\mathcal{L}^{-1}\{X(s)\}$.

For example, we can find $x(t)$ so that

$$x(t) = t^3 + \int_0^t x(\tau)\sin(t - \tau)\,d\tau.$$

Here, in this equation, $f(t) = t^3$ and $g(t) = \sin t$ so that applying Laplace transforms yields the equation

$$X(s) = \frac{6}{s^4} + \frac{X(s)}{s^2 + 1}.$$

Solving this equation for $X(s)$ gives

$$X(s) = \frac{6}{s^4} + \frac{6}{s^6},$$

so that

$$x(t) = \mathcal{L}^{-1}\{X(s)\} = \mathcal{L}^{-1}\left\{\frac{6}{s^4} + \frac{6}{s^6}\right\}$$

$$= t^3 + \frac{t^5}{20}.$$

Solve the given integral equation using Laplace transforms.

1. $x(t) = t - \displaystyle\int_0^t (t-u)x(u)\,du$

2. $x(t) = 2 + \displaystyle\int_0^t x(u)\,du$

3. $x(t) = 5 - \displaystyle\int_0^t x(t-u)u\,du$

As a simple application, consider the current $i(t)$ in a single loop R-L-C circuit which satisfies the equation

$$L\frac{di(t)}{dt} + Ri(t) + \frac{1}{C}\int_0^t i(\tau)\,d\tau = e(t)$$

where R, L, and C are constants and $e(t)$ is a known driving function.

Convert this equation into an algebraic equation in the transform $I(s) = \mathcal{L}\{i(t)\}$ by taking the Laplace transform of both sides. If the constants R, L, and C are given and the function $e(t)$ has a transform, solve for $I(s)$ and find the current $i(t)$ from $\mathcal{L}^{-1}\{I(s)\}$.

For instance, in a single loop R-L-C circuit, we have

$$L = .01, \qquad R = 10, \qquad C = \frac{1}{2500}$$

and a constant driving function $e(t) = 30$. Using these values, find the current $i(t)$ if we assume $i(0) = 0$.

Chapter 8

Series Methods

In this chapter we will explore a different approach to solving an ordinary differential equation. We have developed many methods in the previous chapters for obtaining the solution to an ordinary differential equation, in this chapter we will present a method based upon the power series representation of a function.

In previous chapters, we have not discussed (in general) a method of solution for an ordinary differential equation with variable coefficients; series methods provide one possible method of solution.

8.1 Power Series Representations of Functions

We begin with a discussion of the basic properties of a power series.

A *power series* in $(x - a)$ is an infinite series of the form

$$\sum_{n=0}^{\infty} c_n(x - a)^n = c_0 + c_1(x - a) + c_2(x - a)^2 + \ldots, \tag{8.1}$$

where $c_0, c_1, \ldots$, are constants, called the *coefficients* (what else would they be called?) of the series; a is a constant, called the *center* of the series; and x is an independent variable. In particular, the power series centered at zero $(a = 0)$ has the form

$$\sum_{n=0}^{\infty} c_n x^n = c_0 + c_1 x + c_2 x^2 + \ldots. \tag{8.2}$$

It is a *very* important first observation to note that all polynomials are (finite termed) power series.

A series of the form (8.1) can always be reduced to the form (8.2) by the substitution $X = x - a$. This substitution is merely a translation of the coordinate system. It is easy to see that the behavior of (8.2) near zero is exactly the same as the behavior of (8.1) near a. For this reason we need only study the properties of series of the form (8.2).

We say that (8.2) *converges* at a point $x = x_0$ if the infinite series (of real numbers) $\sum_{n=0}^{\infty} c_n x_0^n$ converges. More specifically, this means that the

$$\lim_{N \to \infty} \sum_{n=0}^{N} c_n x_0^n < \infty.$$

If this limit does not exist as a finite (real) number, then the power series *diverges* at $x = x_0$. It should be clear that the series (8.2) converges at $x = 0$, but are there other values for which the series converges?

Every power series (8.2) has an *interval of convergence* consisting of all values x for which the series converges. The interval of convergence is given in the form $|x| < R$, where R is called the *radius of convergence* of the power series (8.2). On its interval of convergence, a power series (8.2) *converges absolutely*, that is, the series

$$\sum_{n=0}^{\infty} |c_n|\, |x^n|$$

converges; outside the interval of convergence, $|x| > R$, the series (8.2) diverges. When $R = 0$, the interval of convergence consists only of $x = 0$; when $R = \infty$, the power series (8.2) converges absolutely for all x. We can often find the radius of convergence from the familiar ratio test we studied in calculus. We state it here for reference.

THEOREM 8.1.1 *(Ratio Test) Suppose for a numerical series*

$$\sum_{n=1}^{\infty} c_n \qquad\qquad (8.3)$$

that the limit of the ratio of successive terms is

$$\lim_{n \to \infty} \left| \frac{c_{n+1}}{c_n} \right| = L.$$

i) *If $L < 1$ then the series (8.3) converges.*
ii) *If $L > 1$ then the series (8.3) diverges.*
iii) *If $L = 1$, either convergence or divergence is possible. The test is inconclusive.*

Example 1: Determine the radius and interval of convergence for the series

$$\sum_{n=0}^{\infty} \frac{x^n}{(n+1)2^n}.$$

Substituting for each $x = x_0$ in this power series gives a numerical series with coefficients

$$c_n = \frac{x_0^n}{(n+1)2^n}$$

so that

$$L = \lim_{n\to\infty} \left| \frac{c_{n+1}}{c_n} \right|$$

$$= \lim_{n\to\infty} \left| \frac{\frac{x_0^{n+1}}{(n+2)2^{n+1}}}{\frac{x_0^n}{(n+1)2^n}} \right|$$

$$= \frac{|x_0|}{2} \lim_{n\to\infty} \frac{n+1}{n+2}$$

$$= \frac{|x_0|}{2}.$$

Now the ratio test requires $L < 1$ so

$$\frac{|x_0|}{2} < 1,$$

which is $-2 < x_0 < 2$, so the radius of convergence is 2 and the series converges (absolutely) in $(-2, 2)$.

Convergence of the power series is also possible at the endpoints of this interval. If $x = 2$, the series becomes

$$\sum_{n=0}^{\infty} \frac{1}{(n+1)}$$

which is the harmonic series and hence diverges. If $x = -2$, the series becomes

$$\sum_{n=0}^{\infty} \frac{(-1)^n}{(n+1)}$$

which converges by the alternating series test. Thus, the series converges absolutely for $-2 < x < 2$ and converges on $[-2, 2)$.

As we just saw in this last example, a power series may converge or diverge at the endpoints of its interval of convergence. In general, extra care must be taken when determining whether a power series with interval of convergence $|x| < R$ converges at one or both endpoints $x = \pm R$.

Example 2: Determine the radius of convergence of the power series

$$\sum_{n=0}^{\infty} n! x^n.$$

The radius of convergence is found by substituting $x = x_0$ into this power series and calculating the limit of successive terms of the corresponding numerical series. Specifically,

$$L = \lim_{n \to \infty} \left| \frac{c_{n+1}}{c_n} \right|$$

$$= \lim_{n \to \infty} \left| \frac{(n+1)! x_0^{n+1}}{n! x_0^n} \right|$$

$$= |x_0| \lim_{n \to \infty} (n+1)$$

$$= \infty \text{ for } x_0 \neq 0.$$

The only value that will make $L < 1$ is $x_0 = 0$. Thus the power series converges only for $x_0 = 0$.

Example 3: Determine the radius of convergence of the power series

$$\sum_{n=0}^{\infty} \frac{x^n}{n!}.$$

The radius of convergence is found by computing

$$L = \lim_{n \to \infty} \left| \frac{c_{n+1}}{c_n} \right|$$

$$= \lim_{n \to \infty} \left| \frac{\frac{x_0^{n+1}}{(n+1)!}}{\frac{x_0^n}{n!}} \right|$$

$$= |x_0| \lim_{n \to \infty} \frac{n!}{(n+1)!}$$

$$= |x_0| \lim_{n \to \infty} \frac{1}{n+1}$$

$$= 0.$$

Thus, $L < 1$ for all x_0, hence the power series converges for all real x_0.

The interval of convergence for a power series is crucial. For each x in this interval, the power series

$$\sum_{n=0}^{\infty} c_n x^n = c_0 + c_1 x + c_2 x^2 + \dots$$

converges. Thus, the series defines, in a very natural way, a *function* of x. Thus, on the interval of convergence, we have

$$f(x) = \sum_{n=0}^{\infty} c_n x^n$$

where $f(x)$ is the limit of the series at the value x, that is,

$$f(x) = \lim_{N \to \infty} \sum_{n=0}^{N} c_n x^n.$$

Now what is truly great about power series is that on their interval of convergence they behave nicely. In fact, the function

$$f(x) = \sum_{n=0}^{\infty} c_n x^n \tag{8.4}$$

is differentiable on $|x| < R$. Moreover, the derivative is obtained by *term-by-term* differentiation:

$$f'(x) = \left(\sum_{n=0}^{\infty} c_n x^n \right)' = \sum_{n=1}^{\infty} n c_n x^{n-1}, \tag{8.5}$$

and the termwise derivative $f'(x)$ has the *same* radius of convergence as the original series (8.4). This wonderful fact is a consequence of the absolute convergence of the series (8.4) on its interval of convergence $|x| < R$ and forms the basis upon which a method of solution of differential equations will be developed.

The series in (8.5) may again be differentiated term by term to obtain the second derivative $f''(x)$. In fact, we can repeat this process indefinitely, so that the function $f(x)$ defined by (8.4) is *infinitely differentiable*.

Observe that

$$f(0) = c_0,$$
$$f'(0) = c_1,$$
$$f''(0) = 2! c_2,$$

and that in general

$$f^{(n)}(0) = n! c_n.$$

Thus,

$$c_n = \frac{f^{(n)}(0)}{n!},$$

so that (8.4) becomes the *Maclaurin* series of $f(x)$,

$$f(x) = \sum_{n=0}^{\infty} \frac{f^{(n)}(0)}{n!} x^n. \tag{8.6}$$

Any function that has a Maclaurin series (8.6) or, more generally, a *Taylor series*,

$$f(x) = \sum_{n=0}^{\infty} \frac{f^{(n)}(a)}{n!} (x-a)^n, \qquad (8.7)$$

which converges to $f(x)$ in the interval $|x| < R$ for series (8.6) or the interval $|x-a| < R$ for the series (8.7), is said to be *analytic* in the interval $|x| < R$ or $|x-a| < R$, respectively.

Now, applying these ideas it is possible to obtain power series representations of familiar functions.

Example 4: Find the Maclaurin series for

$$f(x) = e^x.$$

Since

$$f^{(n)}(x) = e^x \text{ for all } n \geq 0,$$

then

$$f^{(n)}(0) = 1 \text{ for all } n \geq 0.$$

Thus,

$$e^x = \sum_{n=0}^{\infty} \frac{f^{(n)}(0)}{n!} x^n = \sum_{n=0}^{\infty} \frac{x^n}{n!},$$

which is the series we considered above. We showed that this series converges absolutely for $|x| < \infty$. Thus,

$$e^x = 1 + x + \frac{x^2}{2} + \frac{x^3}{3!} + \cdots . \qquad (8.8)$$

The Maclaurin series (8.8) gives an infinite termed "polynomial-like" representation for e^x. Using this series, the function e^x can be approximated by the first few terms. For instance, using just the first two terms

$$e^x \approx 1 + x.$$

Graphing the functions $e^x 1 + x, 1 + x + x^2/2$ and $1 + x + \frac{x^2}{2} + \frac{x^3}{6}$, as seen in Figure 8.1, higher-order polynomials yield a better approximation.

Now using the same procedure as in Example 4, we can give the following power series which are analytic for all x:

$$\sin x = \sum_{n=0}^{\infty} \frac{(-1)^n x^{2n+1}}{(2n+1)!} = x - \frac{x^3}{3!} + \frac{x^5}{5!} - \frac{x^7}{7!} + \cdots \qquad (8.9)$$

$$\cos x = \sum_{n=0}^{\infty} \frac{(-1)^n x^{2n}}{(2n)!} = 1 - \frac{x^2}{2!} + \frac{x^4}{4!} - \frac{x^6}{6!} + \cdots \qquad (8.10)$$

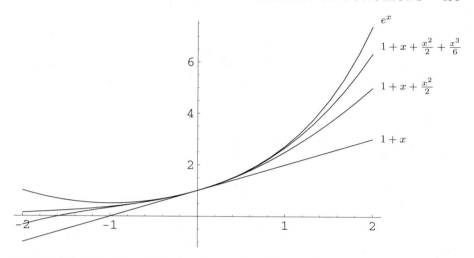

FIGURE 8.1:   Three Maclaurin polynomial approximations for e^x.

$$\sinh x = \sum_{n=0}^{\infty} \frac{x^{2n+1}}{(2n+1)!} = x + \frac{x^3}{3!} + \frac{x^5}{5!} + \frac{x^7}{7!} + \cdots \qquad (8.11)$$

$$\cosh x = \sum_{n=0}^{\infty} \frac{x^{2n}}{(2n)!} = 1 + \frac{x^2}{2!} + \frac{x^4}{4!} + \frac{x^6}{6!} + \cdots . \qquad (8.12)$$

Before we use these series to study differential equations, we consider a few useful properties.

THEOREM 8.1.2 *(Addition of power series) If the power series*

$$c_0 + c_1 x + c_2 x^2 + \ldots$$

and

$$b_0 + b_1 x + b_2 x^2 + \ldots$$

converge for a given value of x, then for that same value of x we have

$$(b_0 + b_1 x + b_2 x^2 + \ldots) + (c_0 + c_1 x + c_2 x^2 + \ldots)$$
$$= (b_0 + c_0) + (b_1 + c_1)x + (b_2 + c_2)x^2 + \ldots .$$

Power series can be scaled by a constant multiple, as given next.

THEOREM 8.1.3 *(Multiplication by a constant) If the power series*

$$c_0 + c_1 x + c_2 x^2 + \cdots$$

converges for a given value of x and a is a real number, then for that same value of x we have

$$a(c_0 + c_1 x + c_2 x^2 + \cdots) = ac_0 + ac_1 x + ac_2 x^2 + \cdots.$$

Using the distributive property, multiplication of power series is very familiar. It is just "infinite-termed" polynomial multiplication.

THEOREM 8.1.4 *(Multiplication of two power series) If the power series*

$$c_0 + c_1 x + c_2 x^2 + \cdots$$

and

$$b_0 + b_1 x + b_2 x^2 + \cdots$$

converge for a given value of $|x| < R$, then for $|x| < R$ we have

$$(b_0 + b_1 x + b_2 x^2 + \ldots)(c_0 + c_1 x + c_2 x^2 + \cdots)$$
$$= b_0 c_0 + (b_0 c_1 + b_1 c_0) x + (b_0 c_2 + b_1 c_1 + b_2 c_0) x^2 + \cdots.$$

We have already noted how to differentiate a power series. Integration is similar.

THEOREM 8.1.5 *(Integration of power series) If the power series*

$$c_0 + c_1 x + c_2 x^2 + \cdots$$

converges for a given value of $|x| < R$, then for $|x| < R$ we have

$$\int (c_0 + c_1 x + c_2 x^2 + \cdots)\, dx = c_0 x + \frac{c_1}{2} x^2 + \frac{c_2}{3} x^3 + \cdots.$$

Let's look at a few examples.

Example 5: In the exercises, you will show that the power series

$$\sum_{n=0}^{\infty} x^n = 1 + x + x^2 + x^3 + \cdots$$

converges to

$$\frac{1}{1-x}$$

for $|x| < 1$. This is the familiar geometric series. Now if we replace x by $-x$, we obtain the power series

$$1 - x + x^2 - x^3 + \cdots = \frac{1}{1+x}. \tag{8.13}$$

Replacing x by x^2 in this representation gives yet another power series

$$1 - x^2 + x^4 - x^6 + \cdots = \frac{1}{1+x^2}. \tag{8.14}$$

Integrating this power series term by term and noting that

$$\int \frac{1}{1+x^2}\,dx = \arctan x$$

gives the power series for the inverse tangent function as

$$\arctan x = x - \frac{x^3}{3} + \frac{x^5}{5} - \frac{x^7}{7} + \cdots. \tag{8.15}$$

Setting $x = 1$ in this power series representation gives a remarkable result

$$\frac{\pi}{4} = 1 - \frac{1}{3} + \frac{1}{5} - \frac{1}{7} + \cdots.$$

This result, known as *Gregory's series*, has been discovered (and rediscovered) by some of history's greatest mathematicians. We should note here that our argument is not yet completely valid. You should show that the series representation for $\arctan x$ converges for $x = 1$.

Computer Code 8.1: Series expansions and summations

Matlab, Maple, Mathematica

Matlab

```
>>  syms x k
>>  f=exp(x)
>>  T=taylor(f,8,0) %expands to O(x^8) about x=0
>>  pretty(T)

>>  g=1/x^k
>>  s1=symsum(g,k,0,inf)
>>  s2=simplify(symsum(g,k,0,5)) %partial sum only
```

Maple

```
>  eq1:=exp(x);
>  eq2:=taylor(eq1,x=0,8);#expands about x=0 to O(x^8)
>  eq3:=mtaylor(eq1,x=0,8); #same as above but w/o order term
    #mtaylor will often be useful if computations must follow
    #mtaylor is the form required for multivariable expansions

>  eqg:=1/x^k;
>  sum(eqg,'k'=0..infinity);
>  simplify(sum(eqg,'k'=0..5)); #partial sum only
```

Mathematica

```
eq1 = e^x  (*e entered from palette*)
eq2=Series[eq1,{x,0,8}]
eq3=Normal[eq2]
eqg[x_,k_]=1/x^k  (*entered from palette*)
Sum[eqg[x,k],{k,0,∞}] (*∞ entered from palette*)
Sum[eqg[x,k],{k,0,5}]
```

Example 6: Show, using a power series approach, that

$$e^{x+y} = e^x e^y.$$

The power series representation for e^x is given by (8.8), as

$$e^x = 1 + x + \frac{x^2}{2} + \frac{x^3}{3!} + \cdots$$

which converges for all real x. Thus, using the product of power series, we have

$$e^x e^y = \left(1 + x + \frac{x^2}{2} + \frac{x^3}{3!} + \cdots\right)\left(1 + y + \frac{y^2}{2} + \frac{y^3}{3!} + \cdots\right)$$

$$= \left(1 + y + \frac{y^2}{2} + \frac{y^3}{3!} + \cdots\right) + x\left(1 + y + \frac{y^2}{2} + \frac{y^3}{3!} + \cdots\right)$$

$$+ \frac{x^2}{2}\left(1 + y + \frac{y^2}{2} + \frac{y^3}{3!} + \cdots\right) + \frac{x^3}{3}\left(1 + y + \frac{y^2}{2} + \frac{y^3}{3!} + \cdots\right) + \cdots$$

$$= 1 + (x + y) + \frac{x^2}{2} + xy + \frac{y^2}{2} + \cdots$$

$$= 1 + (x + y) + \frac{(x + y)^2}{2} + \cdots$$

$$= e^{x+y}.$$

The last equality follows from recognizing the power series expansion for e^{x+y}.

Example 7: Show, using power series, that

$$\sin^2 x + \cos^2 x = 1.$$

Using the power series (8.9) for $\sin x$ we have

$$\sin^2 x = \left(x - \frac{x^3}{3!} + \frac{x^5}{5!} - \frac{x^7}{7!} + \cdots\right)\left(x - \frac{x^3}{3!} + \frac{x^5}{5!} - \frac{x^7}{7!} + \cdots\right)$$

$$= x^2 + \left(-\frac{1}{3!} - \frac{1}{3!}\right)x^4 + \cdots$$

$$= x^2 - \frac{x^4}{3} + \cdots.$$

Similarly,

$$\cos^2 x = \left(1 - \frac{x^2}{2!} + \frac{x^4}{4!} - \frac{x^6}{6!} + \cdots\right)\left(1 - \frac{x^2}{2!} + \frac{x^4}{4!} - \frac{x^6}{6!} + \cdots\right)$$

$$= 1 - x^2 + \frac{x^4}{3} + \cdots.$$

Thus

$$\sin^2 x + \cos^2 x = \left(x^2 - \frac{x^4}{3} + \cdots\right) + \left(1 - x^2 + \frac{x^4}{3} + \cdots\right)$$

$$= 1.$$

In addition to these power series manipulations, power series are also useful for computing limits.

Example 8: We know from calculus that

$$\lim_{x \to 0}\left(\frac{\sin x}{x}\right) = 1.$$

We can verify this result by using a power series. Since

$$\sin x = x - \frac{x^3}{3!} + \frac{x^5}{5!} - \frac{x^7}{7!} + \cdots$$

it follows that

$$\frac{\sin x}{x} = 1 - \frac{x^2}{3!} + \frac{x^4}{5!} - \frac{x^6}{7!} + \cdots \qquad \text{for } x \neq 0.$$

Thus, taking the limit gives

$$\lim_{x \to 0}\left(\frac{\sin x}{x}\right) = \lim_{x \to 0}\left(1 - \frac{x^2}{3!} + \frac{x^4}{5!} - \frac{x^6}{7!} + \cdots\right)$$

$$= 1.$$

As we move through our work with power series, we see their nature as infinite, termed "polynomial-like" functions. Since polynomials are very friendly functions, power series must surely be friendly functions as well!

Before we consider power series methods of solving differential equations, we give one additional useful application.

Many integrals, even simple-looking ones, cannot be calculated in "closed form." Power series, being easy to integrate, offer a useful alternative approach.

Example 9: Approximate

$$\int_0^1 \sin x^2 \, dx.$$

Using the power series (8.9) for $\sin x$ with x^2 replacing x, we have

$$\sin x^2 = x^2 - \frac{x^6}{3!} + \frac{x^{10}}{5!} - \frac{x^{14}}{7!} + \cdots$$

which converges for all x. Thus

$$\int_0^x \sin t^2 \, dt = \frac{x^3}{3} - \frac{x^7}{7 \times 3!} + \frac{x^{11}}{11 \times 5!} - \frac{x^{15}}{15 \times 7!} + \cdots .$$

Now, this power series is not a representation for any elementary function. But, it is a perfectly good antiderivative of $\sin x^2$. Thus,

$$\int_0^1 \sin t^2 \, dt = \frac{x^3}{3} - \frac{x^7}{7 \times 3!} + \frac{x^{11}}{11 \times 5!} - \frac{x^{15}}{15 \times 7!} + \cdots \Big|_0^1$$

$$= \frac{1}{3} - \frac{1}{7 \times 3!} + \frac{1}{11 \times 5!} - \frac{1}{15 \times 7!} + \cdots$$

$$\approx \frac{1}{3} - \frac{1}{7 \times 3!} + \frac{1}{11 \times 5!} - \frac{1}{15 \times 7!}$$

$$\approx 0.3102681578.$$

Using a numerical integration routine, we have

$$\int_0^1 \sin t^2 \, dt \approx 0.31026830172$$

so that the series approximation is very good.

Problems

1. Among the simplest and most useful power series, known as the geometric series, is

$$S(x) = 1 + x + x^2 + x^3 + \ldots = \sum_{n=0}^{\infty} x^n.$$

   a. For which real numbers x does $S(x)$ converge? Why?

   b. What function does the series converge to? Why?

2. Derive the power series representations (8.9)–(8.12). Show that each of these series is analytic for all x.

3. Consider the function

$$f(x) = \begin{cases} e^{-1/x^2}, & x \neq 0 \\ 0, & x = 0. \end{cases}$$

a. Show that f has derivatives of all orders at $x = 0$ and that

$$f'(0) = f''(0) = \ldots = 0.$$

You might want to use the limit

$$f'(0) = \lim_{x \to 0} \frac{f(x) - f(0)}{x}.$$

b. Conclude that $f(x)$ does not have a Taylor series expansion at $x = 0$, even though it is infinitely differentiable there. Thus, explain why f is not analytic at $x = 0$.

4. Find the interval and radius of convergence of the power series

$$\sum_{n=0}^{\infty} (-1)^n \frac{x^{2n+1}}{2n+1} = x - \frac{x^3}{3} + \frac{x^5}{5} - \frac{x^7}{7} + \cdots.$$

Show, using a power series argument, the following identities in problems 5–8.

5. $e^{-x} = \dfrac{1}{e^x}$.

   (Hint: Show $e^{-x}e^x = 1$.)

6. $\sin 2x = 2 \sin x \cos x$.

7. Assume that the power series representations for e^x, $\sin x$, and $\cos x$ are valid for complex numbers $z = x + iy$, where x and y are real and i is the imaginary number. Here $i^2 = -1$. Using power series, derive Euler's formula

$$e^{ix} = \cos x + i \sin x.$$

This is yet another proof of this remarkable result.

8. $\lim\limits_{x \to 0} \dfrac{1 - \cos x}{x} = 0$.

9. Suppose that two functions $f(x)$ and $g(x)$ have Taylor series expansions

$$f(x) = a_0 + a_1(x - c) + a_2(x - c)^2 + \ldots$$

and

$$g(x) = b_0 + b_1(x - c) + b_2(x - c)^2 + \ldots$$

for $|x - c| < r$.

Show that

a. The Wronskian of f and g at c is $a_0 b_1 - a_1 b_0$.

b. f and g are linearly independent on $|x - c| < r$ if $W(f, g)(c) \neq 0$.

8.2 The Power Series Method

The fundamental assumption made in solving a linear differential equation of the form

$$f(x, y, y', y'', \ldots) = 0$$

by the power series method is that *the solution of the differential equation can be expressed in the form of a power series,* say

$$y = \sum_{n=0}^{\infty} c_n x^n = c_0 + c_1 x + c_2 x^2 + \cdots . \tag{8.16}$$

Since a power series can be differentiated, term by term, it follows that $y', y'', \ldots$ can be obtained as

$$y' = \sum_{n=1}^{\infty} n c_n x^{n-1} = c_1 + 2c_2 x + 3c_3 x^2 + \cdots \tag{8.17}$$

$$y'' = \sum_{n=2}^{\infty} n(n-1) c_n x^{n-2} = 2c_2 + 3 \cdot 2c_3 x + \cdots . \tag{8.18}$$

$$\vdots$$

These series can be substituted into the given differential equation and the terms involving like powers of x are then collected. Such an expression would be of the form

$$k_0 + k_1 x + k_2 x^2 + \ldots = \sum_{n=0}^{\infty} k_n x^n = 0, \tag{8.19}$$

where $k_0, k_1, k_2, \cdots$ are expressions involving the unknown coefficients $c_0, c_1, c_2, \cdots$. Now Equation (8.19) must hold for all values of x in the interval of convergence, so all the coefficients $k_0, k_1, k_2, \ldots$ must be zero, that is,

$$k_0 = 0, \quad k_1 = 0, \quad k_2 = 0, \quad \cdots .$$

Using these equations, we may be able to determine the coefficients $c_0, c_1, c_2, \ldots$ of the power series solution y. We illustrate this method in the next several examples.

Example 1: Solve the initial value problem

$$y' = y + x^2$$

with $y(0) = 1$.

Using the equations (8.16) and (8.17) in the differential equation $y' = y + x^2$ gives

$$c_1 + 2c_2 x + 3c_3 x^2 + \ldots = (c_0 + c_1 x + c_2 x^2 + \ldots) + x^2.$$

Collecting like powers of x we obtain

$$(c_1 - c_0) + (2c_2 - c_1)x + (3c_3 - c_2 - 1)x^2 + (4c_4 - c_3)x^3 + \ldots = 0.$$

Equating each of these coefficients to zero gives the equations

$$c_1 - c_0 = 0,$$

$$2c_2 - c_1 = 0,$$

$$3c_3 - c_2 - 1 = 0,$$

$$4c_4 - c_3 = 0,$$

$$\vdots$$

From these equations we have $c_0 = c_1$ and

$$c_2 = \frac{c_1}{2} = \frac{c_0}{2!},$$

$$c_3 = \frac{c_2 + 1}{3} = \frac{c_0 + 2}{3!},$$

$$c_4 = \frac{c_3}{4} = \frac{c_0 + 2}{4!}$$

$$\vdots$$

We now substitute these back into the power series solution (8.16) y to get

$$y = c_0 + c_0 x + \frac{c_0}{2!}x^2 + \frac{c_0 + 2}{3!}x^3 + \frac{c_0 + 2}{4!}x^4 + \cdots .$$

The initial condition $y(0) = 1$ gives

$$1 = c_0 + 0 + 0 + \ldots$$

so the series becomes

$$y = 1 + x + \frac{x^2}{2!} + \frac{3}{3!}x^3 + \frac{3}{4!}x^4 + \cdots .$$

This almost looks like an exponential power series. If you look at it hard enough, you will see (maybe) that adding and subtracting

$$2\left(1 + x + \frac{x^2}{2!}\right)$$

gives

$$y = 3\left(1 + x + \frac{x^2}{2!} + \frac{x^3}{3!} + \ldots\right) - 2\left(1 + x + \frac{x^2}{2!}\right).$$

Now we see that the series is the power series of e^x. Thus, the solution is

$$y = 3e^x - x^2 - 2x - 2.$$

This is our first example; we will not always be so lucky to recognize a power series expansion. Note that this equation is a first-order nonhomogeneous linear differential equation and we could have solved it as such.

Example 2: Solve the differential equation

$$y'' + y = 0.$$

Using the equations (8.16) and (8.18), we have

$$(2c_2 + 3 \cdot 2c_3 x + \ldots) + (c_0 + c_1 x + c_2 x^2 + \ldots) = 0.$$

Collecting like powers of x gives

$$(2c_2 + c_0) + (3 \cdot 2c_3 + c_1)x + (4 \cdot 3c_4 + c_2)x^2 + \ldots = 0.$$

Each of these coefficients must be zero, so

$$2c_2 + c_0 = 0,$$

$$3 \cdot 2c_3 + c_1 = 0,$$

$$4 \cdot 3c_4 + c_2 = 0,$$

$$5 \cdot 4c_5 + c_3 = 0,$$

$$\vdots$$

Thus,

$$c_2 = -\frac{c_0}{2!}, \quad c_3 = -\frac{c_1}{3!}, \quad c_4 = -\frac{c_2}{4 \cdot 3} = \frac{c_0}{4!}, \quad c_5 = -\frac{c_3}{5 \cdot 4} = \frac{c_1}{5!}, \quad \ldots$$

Using these coefficients in the power series (8.16) for y gives

$$y = c_0 + c_1 x - \frac{c_0}{2!}x^2 - \frac{c_1}{3!}x^3 + \frac{c_0}{4!}x^4 + \frac{c_1}{5!}x^5 + \cdots.$$

This can be written as the sum of two power series

$$y = c_0\left(1 - \frac{x^2}{2!} + \frac{x^4}{4!} - \frac{x^6}{6!} + \ldots\right) + c_1\left(x - \frac{x^3}{3!} + \frac{x^5}{5!} - \frac{x^7}{7!} + \ldots\right).$$

Here, again, we recognize each of the power series, the first series is that of $\cos x$ and the second is the one for $\sin x$. Hence, the solution is

$$y = c_0 \cos x + c_1 \sin x.$$

Note that this differential equation is a second-order homogeneous equation with constant coefficients.

   You might now wonder why we did the power series approach on the last two examples, as each was solvable by our earlier methods. The next two examples will reveal one of the reasons we have introduced this method.

Example 3: Solve the differential equation

$$y'' + xy' + y = 0.$$

Using power series written in summation notation (for the sake of compactness) we have

$$\sum_{n=2}^{\infty} n(n-1)c_n x^{n-2} + x \sum_{n=1}^{\infty} nc_n x^{n-1} + \sum_{n=0}^{\infty} c_n x^n = 0. \qquad (8.20)$$

Now, to group coefficients of like powers of x, we need to collect the sums into one single sum. The procedure is akin to change of variables techniques in integration and is as follows. For the first sum

$$S_1 = \sum_{n=2}^{\infty} n(n-1)c_n x^{n-2},$$

we will obtain an exponent k by substituting $k = n - 2$. Thus, $n = k + 2$ and we see that $n = 2$ gives $k = 0$ while $n = \infty$ gives $k = \infty$. Thus we have

$$S_1 = \sum_{k=0}^{\infty} (k+2)(k+1)c_{k+2} x^k.$$

Next, we need to re-index the second series

$$S_2 = x \sum_{n=1}^{\infty} nc_n x^{n-1} = \sum_{n=1}^{\infty} nc_n x^n = \sum_{k=1}^{\infty} kc_k x^k = \sum_{k=0}^{\infty} kc_k x^k.$$

Note here that the last expression follows since $kc_k x^k = 0$ when $k = 0$. Finally, the third series becomes

$$S_3 = \sum_{k=0}^{\infty} c_k x^k.$$

So, the expression (8.20) becomes

$$\sum_{k=0}^{\infty} (k+2)(k+1)c_{k+2}x^k + \sum_{k=0}^{\infty} kc_k x^k + \sum_{k=0}^{\infty} c_k x^k = 0.$$

Gathering like powers of x gives the expression

$$\sum_{k=0}^{\infty} \left[(k+2)(k+1)c_{k+2} + (k+1)c_k \right] x^k = 0.$$

Setting the coefficients in the brackets to zero gives the general *recursion* formula

$$(k+2)(k+1)c_{k+2} + (k+1)c_k = 0.$$

This gives $c_{k+2} = -\dfrac{c_k}{k+2}$ so that

$$c_2 = -\frac{c_0}{2}, \quad c_3 = -\frac{c_1}{3}, \quad c_4 = -\frac{c_2}{4} = \frac{c_0}{2 \cdot 4}, \quad c_5 = -\frac{c_3}{5} = \frac{c_1}{3 \cdot 5}, \quad \cdots.$$

The power series for y can be written in the form

$$y = c_0 + c_1 x - \frac{c_0}{2}x^2 - \frac{c_1}{3}x^3 + \frac{c_0}{2 \cdot 4}x^4 + \frac{c_1}{3 \cdot 5}x^5 - \cdots.$$

Separating the terms that involve c_0 and c_1 gives

$$y = c_0 \left(1 - \frac{x^2}{2} + \frac{x^4}{2 \cdot 4} - \frac{x^6}{2 \cdot 4 \cdot 6} + \cdots \right)$$
$$+ c_1 \left(x - \frac{x^3}{3} + \frac{x^5}{3 \cdot 5} - \frac{x^7}{3 \cdot 5 \cdot 7} + \cdots \right).$$

There is no closed form for either of these series. This *does not* mean that this solution is not useful. Indeed, it is a perfectly good (and natural) solution. For instance, suppose that the differential equation satisfied the initial conditions $y(0) = 1$ and $y'(0) = 1$; then $c_0 = 1$ and $c_1 = 1$. The solution is

$$y = 1 + x - \frac{x^2}{2} - \frac{x^3}{3} + \frac{x^4}{8} - \frac{x^5}{15} + \cdots.$$

Graphing $1 + x$, $1 + x - \frac{x^2}{2}$, $1 + x - \frac{x^2}{2} - \frac{x^3}{3}$ and $1 + x - \frac{x^2}{2} - \frac{x^3}{3} + \frac{x^4}{8}$ as in Figure 8.2 gives an idea of the behavior of this series solution.

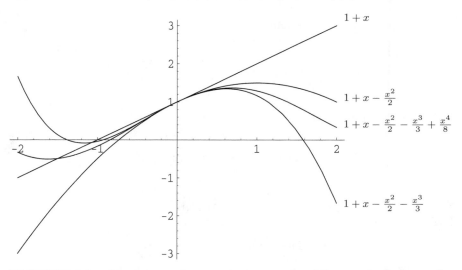

FIGURE 8.2:   Four approximate solutions to the differential equation $y'' + xy' + y = 0$ with initial conditions $y(0) = 1$ and $y'(0) = 1$.

Computer Code 8.2:   **Solving differential equations by series expansion**

Matlab, Maple, Mathematica

```
                            Matlab
>>  syms x c0 c1 c2 c3 c4 c5 c6
>>  y=c0+c1*x+c2*x^2+c3*x^3+c4*x^4+c5*x^5+c6*x^6
>>  eqODE=diff(y,x,2)+x*diff(y,x)+y
>>  eq1=collect(eqODE)
>>  eq2=coeffs(eq1,x) %n must be large enough in y
>>  eq3=solve(eq2(1),eq2(2),eq2(3),eq2(4),c2,c3,c4,c5)
>>  c2=eq3.c2
>>  c3=eq3.c3
>>  c4=eq3.c4
>>  c5=eq3.c5
>>  ysoln=c0+c1*x+c2*x^2+c3*x^3+c4*x^4+c5*x^5
>>  subs(ysoln)
>>  ysoln1=collect(collect(ysoln,c0),c1)
```

Maple

```
> eqODE:=diff(y(x),x$2)+x*diff(y(x),x)+y(x)=0;
> eq1:=y(x)=sum('c[k]*x^k',k=0..n);
> eq1a:=eval(subs(n=6,eq1));
> eq2:=subs(eq1a,eqODE);
> eq3:=collect(eq2,x);
> eq3a:=coeff(lhs(eq3),x,0);
> eq3b:=coeff(lhs(eq3),x,1);
> eq3c:=coeff(lhs(eq3),x,2);
> eq3d:=coeff(lhs(eq3),x,3); #n must be large enough in eq1a
> eq4:=solve({eq3a,eq3b,eq3c,eq3d},{c[2],c[3],c[4],c[5]});
> eqsoln:=subs(eq4,evalf(subs(n=5,eq1)));
> eqsoln1:=collect(collect(eqsoln,c[0]),c[1]);
```

Mathematica

```
y1[x_]=c0 + c1 x + c2 x^2 + c3 x^3 + c4 x^4 + c5 x^5 + c6 x^6
dey[x_]=y''[x]+x y'[x]+y[x]
eq2=Simplify[ReplaceAll[dey[x],{y[x]→ y1[x],y'[x]→ y1'[x],
    y''[x]→ y1''[x]}]]
eq3a=Coefficient[eq2,x,0]
eq3b=Coefficient[eq2,x,1]
eq3c=Coefficient[eq2,x,2]
eq3d=Coefficient[eq2,x,3]
    (*n must be large enough in y1[x_]*)
eq4=Solve[{eq3a==0,eq3b==0,eq3c==0,eq3d==0},{c2,c3,c4,c5}]
eqsoln=ReplaceAll[ReplaceAll[y1[x],eq4],c6→ 0]
eqsoln1=Collect[Collect[eqsoln[[1]],c0],c1]
```

Example 4: Find a power series solution of the initial value problem

$$(x^2 - 1)\frac{d^2y}{dx^2} + 3x\frac{dy}{dx} + xy = 0,$$

with $y(0) = 4$ and $y'(0) = 6$.

Since the initial conditions are prescribed at $x = 0$, we will seek a power series solution about this point. We assume a solution of the form

$$y = \sum_{n=0}^{\infty} c_n x^n.$$

This gives

$$\frac{dy}{dx} = \sum_{n=1}^{\infty} n c_n x^{n-1}$$

and

$$\frac{d^2 y}{dx^2} = \sum_{n=2}^{\infty} n(n-1) c_n x^{n-2}$$

so that substituting these into the differential equation gives

$$(x^2 - 1) \sum_{n=2}^{\infty} n(n-1) c_n x^{n-2} + 3x \sum_{n=1}^{\infty} n c_n x^{n-1} + x \sum_{n=0}^{\infty} c_n x^n = 0.$$

Thus

$$\sum_{n=2}^{\infty} n(n-1) c_n x^n - \sum_{n=2}^{\infty} n(n-1) c_n x^{n-2} + 3 \sum_{n=1}^{\infty} n c_n x^n + \sum_{n=0}^{\infty} c_n x^{n+1} = 0,$$

so that reindexing the second and forth series gives

$$\sum_{n=2}^{\infty} n(n-1) c_n x^n - \sum_{n=0}^{\infty} (n+2)(n+1) c_{n+2} x^n + 3 \sum_{n=1}^{\infty} n c_n x^n + \sum_{n=1}^{\infty} c_{n-1} x^n = 0.$$

Combining like powers of x and writing just a single series gives

$$-2c_2 + (c_0 + 3c_1 - 6c_3)x + \sum_{n=2}^{\infty} [-(n+2)(n+1) c_{n+2} + n(n+2) c_n + c_{n-1}] x^n = 0.$$

Comparing coefficients we have

$$-2c_2 = 0$$
$$c_0 + 3c_1 - 6c_3 = 0$$

and

$$-(n+2)(n+1) c_{n+2} + n(n+2) c_n + c_{n-1} = 0 \quad \text{for } n \geq 2.$$

This gives $c_2 = 0$ and $c_3 = \frac{1}{6} c_0 + \frac{1}{2} c_1$, and the recurrence relation

$$c_{n+2} = \frac{n(n+2) c_n + c_{n-1}}{(n+1)(n+2)} \quad \text{for } n \geq 2.$$

Applying this recurrence relation successively yields

$$c_4 = \frac{8c_2 + c_1}{12} = \frac{1}{12} c_1,$$
$$c_5 = \frac{15 c_3 + c_2}{20} = \frac{1}{8} c_0 + \frac{3}{8} c_1$$

so that

$$y = c_0 + c_1 x + \left(\frac{c_0}{6} + \frac{c_1}{2}\right) x^3 + \frac{c_1}{12} x^4 + \left(\frac{c_0}{8} + \frac{3c_1}{8}\right) x^5 + \cdots .$$

This series can be written as

$$y = c_0 \left(1 + \frac{1}{6} x^3 + \frac{1}{8} x^5 + \cdots\right) + c_1 \left(x + \frac{1}{2} x^3 + \frac{1}{12} x^4 + \frac{3}{8} x^5 + \cdots\right).$$

Now applying the initial condition $y(0) = 4$ gives

$$c_0 = 4$$

and differentiating the series gives

$$y' = c_0 \left(\frac{1}{2} x^2 + \frac{5}{8} x^4 + \cdots\right) + c_1 \left(1 + \frac{3}{2} x^2 + \frac{1}{3} x^3 + \frac{15}{8} x^4 + \cdots\right)$$

so that $y'(0) = 6$ gives

$$c_1 = 6.$$

Thus, the power series solution that satisfies the initial value problem is

$$y = 4 \left(1 + \frac{1}{6} x^3 + \frac{1}{8} x^5 + \cdots\right) + 6 \left(x + \frac{1}{2} x^3 + \frac{1}{12} x^4 + \frac{3}{8} x^5 + \cdots\right)$$

$$= 4 + 6x + \frac{11}{3} x^3 + \frac{1}{2} x^4 + \frac{11}{4} x^5 + \cdots .$$

Suppose now that the initial value problem of Example 4 was replaced by the initial value problem

$$(x^2 - 1)\frac{d^2 y}{dx^2} + 3x\frac{dy}{dx} + xy = 0$$

with $y(5) = 4$ and $y'(5) = 6$.

This problem is virtually the same as Example 4, however the initial values in the problem are prescribed at $x = 5$, instead of $x = 0$. We seek a power series solution of the form

$$y = \sum_{n=0}^{\infty} c_n (x - 5)^n.$$

The simplest way to proceed here is to make the substitution $t = x - 5$. Here $x = t + 5$ and the initial value problem becomes

$$((t + 5)^2 - 1)\frac{d^2 y}{dt} + 3(t + 5)\frac{dy}{dt} + (t + 5)y = 0$$

with $y(0) = 4$ and $y'(0) = 6$. We then seek a solution in powers of t of the form

$$y = \sum_{n=0}^{\infty} c_n t^n.$$

Thus, this problem is now very similar to Example 4 and it is suggested that the reader pursue the calculations to obtain the c_n's. Once the c_n's are found, the initial conditions $y(0) = 4$ and $y'(0) = 6$ can be applied; then replacing t by $x - 5$, the solution to the original initial value problem is obtained.

Taylor and Maclaurin series can also be used to find power series solutions to differential equations. This is a slick method.

Example 5: Again, consider the differential equation

$$y' = y + x^2,$$

with the condition $y(0) = 1$.

Differentiating both sides of the differential equation repeatedly and successively evaluating each derivative at the initial value of $x = 0$, we have

$$y'(0) = (y + x^2)\big|_{x=0} = y(0) = 1,$$

$$y''(0) = (y' + 2x)\big|_{x=0} = y'(0) = 1,$$

$$y'''(0) = (y'' + 2)\big|_{x=0} = y''(0) + 2 = 3,$$

$$y^{(4)}(0) = y'''\big|_{x=0} = y'''(0) = 3.$$

$$\vdots$$

Substituting these values into the Maclaurin series (8.2) gives

$$y(x) = \sum_{n=0}^{\infty} \frac{y^{(n)}(0)}{n!} x^n$$

$$= 1 + x + \frac{x^2}{2!} + \frac{3x^3}{3!} + \frac{3x^4}{4!} + \dots,$$

which we obtained earlier.

Problems

In problems 1–14, solve the following differential equations using the power series method. Also solve the equation using one of the methods we have presented earlier, if possible.

1. $y' = y^2 - x$ with $y(0) = 1$.
2. $y' - 2y = x^2$ with $y(1) = 1$.
3. $y' = y + xe^y$ with $y(0) = 0$.
4. $\frac{dy}{dx} = x + y$ with $y(0) = 1$.
5. $\frac{dy}{dx} - x^2 = y^2$ with $y(0) = 1$.
6. $\frac{dy}{dx} = x + \sin y$ with $y(0) = 0$.
7. $y' - e^x + \sin y$ with $y(0) = 0$.
8. $\frac{dy}{dx} = x^2 + 2y^2$ with $y(0) = 1$.
9. $\frac{dy}{dx} = x + \frac{1}{y}$ with $y(0) = 1$.
10. $\frac{dy}{dx} = 1 + x \sin y$ with $y(0) = 0$.
11. $y' = x^2 + y^3$ with $y(1) = 1$.
12. $\frac{dy}{dx} = x^3 + y^2$ with $y(1) = 1$.
13. $y'' + y = 1 + x + x^2$ with $y(0) = 1$ and $y'(0) = -1$.
14. $y'' - xy' + y^2 = 0$ with $y(0) = 1$ and $y'(0) = -1$.

In problems 15–17, find the general power series solution to the differential equation. Try to recognize the solution in terms of familiar functions. This may not be possible.

15. $(1 + x^2)y'' + xy' + xy = 0$.
16. $xy'' - x^2y' + (x^2 - 2)y = 0$, $y(0) = 0$ and $y'(0) = 1$.
17. $y'' - 2xy' - 2y = x$, $y(0) = 1$ and $y'(0) = -1/4$.

18. Sir George Biddell Airy (1801–1892) was Lucasian professor of mathematics, director of the observatory, and Plumian professor of astronomy at Cambridge University in England until 1835. He was then appointed director of the Greenwich Observatory (Astronomer Royal). He remained there until his retirement in 1881. He did much work in lunar and solar photography, planetary motion, optics, and other areas. The *Airy equation*

$$y'' - xy = 0$$

has applications in the theory of diffraction. Find the general solution of this equation.

19. The *Hermite equation* in honor of Charles Hermite (1882–1901) is

$$y'' - 2xy' + 2py = 0,$$

where p is a constant. This equation arises in quantum mechanics in connection with the *Schrödinger equation* for a harmonic oscillator. Show

that if p is a positive integer, one of the two linearly independent solutions of the Hermite equation is a polynomial, called the *Hermite
polynomial* $H_p(x)$.

8.3   Ordinary and Singular Points

The power series method we have considered in the previous section sometimes fails to yield a solution for one equation while working very well for an
apparently similar equation.

Example 1: Consider the differential equation

$$x^2 y'' + axy' + by = 0.$$

Apply the power series method for each of the following cases:

Case 1: $a = -2$, $b = 2$.

In the differential equation, set

$$y = \sum_{n=0}^{\infty} c_n x^n, \quad y' = \sum_{n=1}^{\infty} n c_n x^{n-1}, \text{ and } y'' = \sum_{n=2}^{\infty} n(n-1)c_n x^{n-2}.$$

Then, since $n(n-1) = 0$ at $n = 0$ and 1,

$$x^2 y'' + axy' + by = x^2 \sum_{n=0}^{\infty} n(n-1)c_n x^{n-2} + ax \sum_{n=0}^{\infty} n c_n x^{n-1} + b \sum_{n=0}^{\infty} c_n x^n$$

$$= \sum_{n=0}^{\infty} n(n-1)c_n x^n + \sum_{n=0}^{\infty} a n c_n x^n + \sum_{n=0}^{\infty} b c_n x^n$$

$$= \sum_{n=0}^{\infty} \left[n(n-1) + an + b \right] c_n x^n$$

$$= 0. \tag{8.21}$$

Now substituting $a = -2$, $b = 2$ into (8.21) gives

$$\sum_{n=0}^{\infty} (n^2 - 3n + 2)c_n x^n = 0.$$

Equating the coefficients to zero gives

$$(n-2)(n-1)c_n = 0,$$

which means that $c_n = 0$ for all $n \neq 1$ or 2. Hence,

$$y = c_1 x + c_2 x^2$$

is the general solution when $a = -2$, $b = 2$.

Case 2: $a = -1$ and $b = 1$.

Substitution into (8.21) gives

$$\sum_{n=0}^{\infty} (n^2 - 2n + 1)c_n x^n = 0,$$

so

$$(n-1)^2 c_n = 0.$$

Thus, $c_n = 0$ for all $n \neq 1$, so that the solution is

$$y = c_1 x.$$

But here we note that the differential equation

$$x^2 y'' + axy' + by = 0$$

is second-order, so that its general solution involves *two* linearly independent solutions. The power series method has given us only one. We use the method of reduction of order to find the other solution.

Let $y = xv$ so that

$$x^2 (xv)'' - x(xv)' + xv = 0,$$

and this equation becomes

$$x^3 v'' + x^2 v' = 0.$$

Now making the substitution $z = v'$, we obtain the first-order separable differential equation

$$x^3 \frac{dz}{dx} + x^2 z = 0,$$

so

$$z = \frac{c}{x}.$$

Thus,

$$v' = \frac{c}{x},$$

which gives

$$v = c \ln |x|.$$

Thus, the general solution is

$$y = Ax + Bx \ln |x|.$$

Note here that $\ln |x|$ is not defined at $x = 0$, so it is not a surprise that the power series method failed to obtain this term as part of the solution.

Case 3: $a = 1$ and $b = 1$.

Equation (8.21) becomes

$$\sum_{n=0}^{\infty} (n^2 + 1)c_n = 0.$$

Equating the coefficients to zero gives

$$(n^2 + 1)c_n = 0,$$

so

$$c_n = 0 \text{ for every } n, \text{ since } n^2 + 1 \neq 0.$$

Thus, the power series method fails completely in helping us find the general solution. The general solution is

$$y = A \cos \ln |x| + B \sin \ln |x|.$$

(Check it!)

We will now present conditions under which the power series method does work. One also notes here that the methods of Chapter 4, Section 3 apply.

THEOREM 8.3.1 *There is a unique Maclaurin series $y(x)$ satisfying the initial value problem*

$$y'' + a(x)y' + b(x)y = 0,$$

with $y(0) = \alpha$ and $y'(0) = \beta$, provided $a(x)$ and $b(x)$ can each be represented by a Maclaurin series converging in an interval $|x| < R$. The power series $y(x)$ also converges in $|x| < R$.

This result guarantees that the power series method works whenever $a(x)$ and $b(x)$ are analytic (have convergent Maclaurin series) at $x = 0$. We need some terminology.

DEFINITION 8.1   *The point $x = x_0$ is called an ordinary point of the differential equation*

$$y'' + a(x)y' + b(x)y = 0 \qquad (8.22)$$

when both $a(x)$ and $b(x)$ are analytic at $x = x_0$. If $x = x_0$ is not an ordinary point, it is called a singular point of the differential equation.

If we write each of the second-order homogeneous equations in the previous example in the form (8.22), we have

$$y'' + \frac{a}{x}y' + \frac{b}{x^2}y = 0.$$

We see here that neither of the terms

$$a(x) = \frac{a}{x} \text{ or } b(x) = \frac{b}{x^2}$$

are defined at $x = 0$, so that they fail to have a power series representation that converges in an open interval containing $x = 0$.

Example 2: Solve the differential equation

$$(1 - x^2)y'' - xy' + 2y = 0.$$

Rewriting this equation gives

$$y'' - \frac{x}{1 - x^2}y' + \frac{2}{1 - x^2}y = 0,$$

for which we see that

$$a(x) = -\frac{x}{1 - x^2} \text{ and } b(x) = \frac{2}{1 - x^2}.$$

Each of these functions has a Maclaurin series

$$a(x) = -x(1 + x^2 + x^4 + \ldots)$$

and

$$b(x) = 2(1 + x^2 + x^4 + \ldots)$$

so that $x = 0$ is an ordinary point and the power series method will yield the general solution. Note here that $x = \pm 1$ are singular points of these functions, so that a Taylor series centered at $x = \pm 1$ may or may not exist.

The previous example is a specific case of a special type of differential equation. The *Čebyšev differential equation* is given as

$$(1 - x^2)y'' - xy' + \mu^2 y = 0 \qquad (8.23)$$

where μ is a constant.

Pafnuty Lvovich Čebyšev (1821–1894) (pronounced and often written as Chebyshev or Tchebyshev) was a mathematics professor at the University of St. Petersburg. He made numerous important contributions in the areas of probability theory, number theory, special functions, and mechanics. The Čebyšev differential equation (8.23) has some very interesting properties, some of which we now consider.

Rewriting the equation as

$$y'' - \frac{x}{1-x^2}y' + \frac{\mu^2}{1-x^2}y = 0$$

gives $x = \pm 1$ as singular points. Thus, on the interval $(-1,1)$ letting

$$y(x) = \sum_{n=0}^{\infty} c_n x^n$$

and substituting into the differential equation gives

$$(1-x^2)\sum_{n=2}^{\infty} c_n n(n-1)x^{n-2} - x\sum_{n=1}^{\infty} c_n n x^{n-1} + \mu^2 \sum_{n=0}^{\infty} c_n x^n = 0.$$

Simplifying this expression and rewriting gives

$$\sum_{n=0}^{\infty}[c_{n+2}(n+2)(n+1) - c_n(n^2 - \mu^2)]x^n = 0.$$

Comparing coefficients gives the recurrence relation

$$c_{n+2} = \frac{(n^2 - \mu^2)c_n}{(n+1)(n+2)} \quad \text{for } n = 0,1,2,\cdots. \qquad (8.24)$$

This recurrence relation yields the even terms as

$$c_2 = \frac{-\mu^2 c_0}{2}, \quad c_4 = \frac{-(4-\mu)^2\mu^2 c_0}{24}, \quad \cdots$$

and the odd terms

$$c_3 = \frac{(1-\mu^2)c_1}{6}, \quad c_5 = \frac{(1-\mu^2)(9-\mu^2)c_1}{120}, \quad \cdots$$

so that the general solution is

$$y(x) = c_0\left(1 + \frac{-\mu^2}{2}x^2 + \frac{-(4-\mu^2)\mu^2}{24}x^4 + \ldots\right)$$

$$+ c_1\left(x + \frac{1-\mu^2}{6}x^3 + \frac{(1-\mu^2)(9-\mu^2)}{120}x^5 + \cdots\right).$$

If μ is an even integer, we see that all even coefficients having index greater than μ are zero. For instance, when $\mu = 4$ we have

$$c_6 = \frac{(4^2 - 4^2)c_4}{30} = 0$$

so that $c_8 = 0, c_{10} = 0 \cdots$. Thus when $\mu = 4$ the recurrence relation gives

$$c_2 = -8c_0, \quad c_4 = -c_2 = 8c_0, \quad c_6 = c_8 = c_{10} \ldots = 0.$$

Using these values, we have the fourth-degree polynomial

$$P(x) = c_0[1 - 8t^2 + 8t^4]$$

as a solution of Čebyšev's equation (8.23).

Similarly, if μ is an odd positive integer then we also obtain a polynomial of (8.23). These polynomial solutions of (8.23) for each nonnegative integer valued μ are known as *Čebyšev (Tchebyshev) polynomials of the first kind* and $T_n(x)$ is usually used to denote these nth degree Čebyšev polynomials. The first few Čebyšev polynomials are

$$T_0(x) = 1, \quad T_1(x) = x, \quad T_2(x) = 2x^2 - 1$$

$$T_3(x) = 4x^3 - 3x, \quad T_4(x) = 8x^4 - 8x^2 + 1,$$

and a graph of $T_2(x)$, $T_3(x)$, and $T_4(x)$ is shown in Figure 8.3. More properties and aspects of Čebyšev polynomials are considered in the exercises.

We next consider a method of solution about a singular point.

8.3.1   Solutions about Singular Points

Consider the second-order homogenous linear differential equation

$$f_0(x)\frac{d^2y}{dx^2} + f_1(x)\frac{dy}{dx} + f_2(x)y = 0 \tag{8.25}$$

and suppose that x_0 is a singular point. In this situation we are not assured that a power series solution

$$y = \sum_{n=0}^{\infty} c_n x^n$$

exists. We need to find a different type of solution, but how should we approach the problem?

To this end, we need to make a definition.

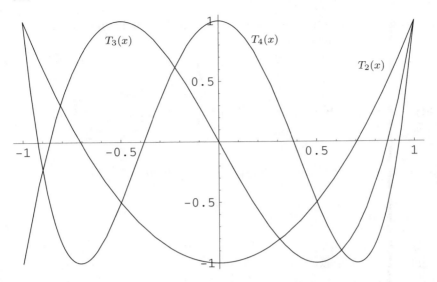

FIGURE 8.3:   The Čebyšev polynomials $T_2(x)$, $T_3(x)$, and $T_4(x)$.

DEFINITION 8.2   *Consider the differential equation*

$$f_0(x)\frac{d^2y}{dx^2} + f_1(x)\frac{dy}{dx} + f_2(x)y = 0$$

and assume that at least one of the functions

$$p(x) = \frac{f_1(x)}{f_0(x)}, \qquad q(x) = \frac{f_2(x)}{f_0(x)}$$

is not analytic at x_0 so that x_0 is a singular point of the differential equation (8.25). If the functions defined by the products

$$(x - x_0)p(x) \quad and \quad (x - x_0)^2 q(x) \tag{8.26}$$

are both analytic at x_0, then x_0 is called a regular singular point of the differential equation (8.25). If either or both of the functions defined by (8.26) are not analytic at x_0, then x_0 is called an irregular singular point of (8.25).

Example 3: The differential equation

$$3x^2\frac{d^2y}{dx^2} - 4x\frac{dy}{dx} + (x + 2)y = 0 \tag{8.27}$$

can be written as

$$\frac{d^2y}{dx^2} - \frac{4}{3x}\frac{dy}{dx} + \frac{(x+2)}{3x^2}y = 0,$$

so that

$$p(x) = \frac{-4}{3x} \quad \text{and} \quad q(x) = \frac{x+2}{3x^2}.$$

Both fail to be analytic at $x = 0$. Thus, $x = 0$ is a singular point of the differential equation (8.27). Now the products are

$$xp(x) = \frac{-4}{3} \quad \text{and} \quad x^2q(x) = \frac{x+2}{3},$$

both of which are analytic at $x = 0$. Thus $x = 0$ is a regular singular point of the differential equation (8.27).

Example 4: The differential equation

$$x^2(x-2)^2\frac{d^2y}{dx^2} + 2(x-2)\frac{dy}{dx} + (x+1)y = 0$$

can be written as

$$\frac{d^2y}{dx^2} + \frac{2}{x^2(x-2)}\frac{dy}{dx} + \frac{(x+1)}{x^2(x-2)^2}y = 0 \qquad (8.28)$$

so that

$$p(x) = \frac{2}{x^2(x-2)} \quad \text{and} \quad q(x) = \frac{x+1}{x^2(x-2)^2}$$

are not analytic at $x = 0$ and $x = 2$. Thus $x = 0$ and $x = 2$ are singular points of the differential equation (8.28).

Now at $x = 0$, we consider the products

$$xp(x) = \frac{2}{x(x-2)} \quad \text{and} \quad x^2q(x) = \frac{x+1}{(x-2)^2};$$

here $x^2q(x)$ is analytic at $x = 0$, while $xp(x)$ is *not* analytic at $x = 0$. Thus $x = 0$ is an irregular singular point of (8.28).

Next considering the point $x = 2$, the products

$$(x-2)p(x) = \frac{2}{x^2} \quad \text{and} \quad (x-2)^2q(x) = \frac{x+1}{x^2}$$

are *both* analytic at $x = 2$ so that $x = 2$ is a regular singular point of (8.28).

These ideas lead us to develop a method of obtaining a series solution when the differential equation has a regular singular point. We will develop this method in the next section.

Problems

For each of the following differential equations in problems 1–6, find all singular points. Determine if these points are regular or irregular.

1. $(x^2 - 3x)y'' + (x + 2)y' + y = 0$.
2. $(x^3 + x^2)\frac{d^2y}{dx^2} + (x^2 - 2x)\frac{dy}{dx} + 4y = 0$.
3. $x^2(x + 2)y'' + xy' - (2x - 1)y = 0$
4. $(x^5 + x^4 - 6x^3)\frac{d^2y}{dx^2} + x^2\frac{dy}{dx} + (x - 2)y = 0$.
5. $(2x + 1)x^2y'' - (x + 2)y' + 2e^x y = 0$.
6. $(x^4 - 2x^3 + x^2)y'' + 2(x - 1)y' + x^2 y = 0$.

7. The Čebyšev differential equation

$$(1 - x^2)y'' - xy' + \mu^2 y = 0$$

   has the polynomial solution $T_n(x)$ when n is a nonnegative integer. Using the recurrence relation (8.24) find $T_5(x)$ and $T_6(x)$.

8. **a.** Show, by substitution, that

$$y(x) = \cos(n \arccos(x))$$

   is a solution of the Čebyšev differential equation

$$(1 - x^2)y'' - xy' + n^2 y = 0$$

   with n a nonnegative integer and $-1 < x < 1$.

   b. Show for $n = 0, 1, 2$ that the function

$$y(x) = \cos(n \arccos(x))$$

   is a polynomial in x and that $T_n(x) = y(x)$. Show that

$$T_n(x) = \begin{cases} \cos(n \arccos(x)) & \text{for } |x| \leq 1 \\ \cosh(n \arccos(x)) & \text{for } |x| > 1. \end{cases}$$

8.4   The Method of Frobenius

We now proceed to give a result concerning solutions about regular singular points.

THEOREM 8.4.1 *If x_0 is a regular singular point of the differential equation*

$$f_0(x)\frac{d^2y}{dx^2} + f_1(x)\frac{dy}{dx} + f_2(x)y = 0,$$

then this differential equation has at least one nontrivial solution of the form

$$|x - x_0|^r \sum_{n=0}^{\infty} c_n(x - x_0)^n,$$

where r is a constant (that can be determined) and this solution is valid in some (deleted) interval $0 < |x - x_0| < R, R > 0$, about x_0.

Example 1: We have seen that $x = 0$ is a regular singular point of the differential equation

$$3x^2\frac{d^2y}{dx^2} - 4x\frac{dy}{dx} + (x + 2)y = 0;$$

the theorem assures us that there is (at least) one nontrivial solution of the form

$$|x|^r \sum_{n=0}^{\infty} c_n x^n$$

which is valid in the interval $0 < |x| < R$ about $x = 0$.

Example 2: We know the differential equation

$$x^2(x - 2)^2\frac{d^2y}{dx^2} + 2(x - 2)\frac{dy}{dx} + (x + 1)y = 0$$

has a regular singular point at $x = 2$ so that there is (at least) one solution of the form

$$|x - 2|^r \sum_{n=0}^{\infty} c_n(x - 2)^n,$$

which is defined for all values in $0 < |x - 2| < R$.

Now, in the case $x = 0$, an irregular singular point of the differential equation, we cannot say that

$$|x|^r \sum_{n=0}^{\infty} c_n x^n$$

is a solution in any interval $0 < |x| < R$.

Question: How do we determine the coefficients c_n and the number r when x_0 is a regular singular point of the differential equation?

The method we consider is very similar to the preceding method of solution for ordinary points. The method we will now develop is often called the *method of Frobenius.*

Georg Frobenius was born in Charlottenburg, Germany in 1849. He received his doctorate at the University of Berlin (awarded with distinction) in 1870 under the supervision of Weierstrass. In 1874, after having taught at the secondary school level for several years, he was appointed to the University of Berlin as an extraordinary professor of mathematics. Frobenius made enormous contributions to group theory, particularly group representations, linear algebra, and analysis. In [14] Haubrich gives the following overview of Frobenius's work:

> *The most striking aspect of his mathematical practice is his extraordinary skill at calculations. In fact, Frobenius tried to solve mathematical problems to a large extent by means of a calculative, algebraic approach. Even his analytical work was guided by algebraic and linear algebraic methods. For Frobenius, conceptual argumentation played a somewhat secondary role. Although he argued in a comparatively abstract setting, abstraction was not an end in itself. Its advantages to him seemed to lie primarily in the fact that it can lead to much greater clearness and precision.*

Frobenius developed a method of determining the coefficients c_n and the number r when x_0 is a regular singular point. We will illustrate this method on the first differential equation we considered.

Example 3: Find a solution to the differential equation

$$3x^2 \frac{d^2 y}{dx^2} - 4x \frac{dy}{dx} + (x+2)y = 0 \qquad (8.29)$$

in some interval $0 < x < R$.

Since $x = 0$ is a regular singular point, we know that

$$y = |x|^r \sum_{n=0}^{\infty} c_n x^n$$

is a solution on $0 < |x| < R$.

Thus for $0 < x < R$, we seek a solution of the form

$$y = x^r \sum_{n=0}^{\infty} c_n x^n$$

$$= \sum_{n=0}^{\infty} c_n x^{n+r}.$$

Now

$$\frac{dy}{dx} = \sum_{n=0}^{\infty} c_n(n+r)x^{n+r-1}$$

and

$$\frac{d^2y}{dx^2} = \sum_{n=0}^{\infty} c_n(n+r)(n+r-1)x^{n+r-2}$$

so that substitution into the differential equation (8.29) gives

$$3x^2\frac{d^2y}{dx^2} - 4x\frac{dy}{dx} + (x+2)y$$

$$= 3x^2 \sum_{n=0}^{\infty} c_n(n+r)(n+r-1)x^{n+r-2}$$

$$-4x \sum_{n=0}^{\infty} c_n(n+r)x^{n+r-1} + (x+2) \sum_{n=0}^{\infty} c_n x^{n+r}$$

$$= 3 \sum_{n=0}^{\infty} c_n(n+r)(n+r-1)x^{n+r}$$

$$-4 \sum_{n=0}^{\infty} c_n(n+r)x^{n+r} + \sum_{n=0}^{\infty} c_n x^{n+r+1} + 2 \sum_{n=0}^{\infty} c_n x^{n+r}$$

$$= 0.$$

Simplification of this expression gives

$$\sum_{n-0}^{\infty}(3(n+r)(n+r-1) - 4(n+r) + 2)c_n x^{n+r} + \sum_{n=1}^{\infty} c_{n-1} x^{n+r} = 0$$

or equivalently

$$(3(r(r-1)-4r+2)c_0 x^r + \sum_{n=1}^{\infty}[(3(n+r)(n+r-1)-4(n+r)+2)c_n+c_{n-1}]x^{n+r} = 0.$$

Assuming $c_0 \neq 0$ and equating the coefficient of the lowest power of x to zero gives

$$(3r(r-1) - 4r + 2)c_0 = 0$$

or

$$3r(r-1) - 4r + 2 = 0.$$

This quadratic equation, known as the *indicial equation*, is

$$3r^2 - 7r + 2 = 0$$

and has roots

$$r_1 = \frac{1}{3} \quad \text{and} \quad r_2 = 2.$$

These roots are known as the *exponents* of the differential equation and are the only possible values for the constant r. Now equating the coefficients of higher powers of x to zero we obtain the recurrence relation

$$(3(n+r)(n+r-1) - 4(n+r) + 2)c_n + c_{n-1} = 0 \quad \text{for} \quad n \geq 1.$$

Now when $r_1 = \frac{1}{3}$ we obtain

$$\left(3\left(n+\frac{1}{3}\right)\left(n-\frac{2}{3}\right) - 4\left(n+\frac{1}{3}\right) + 2\right)c_n + c_{n-1} = 0 \quad \text{for} \quad n \geq 1$$

which simplifies to

$$\left(3n^2 - 5n + \frac{2}{3}\right)c_n + c_{n-1} = 0 \quad \text{for} \quad n \geq 1$$

or

$$c_n = \frac{-c_{n-1}}{3n^2 - 5n + \frac{2}{3}} \quad \text{for} \quad n \geq 1.$$

Thus

$$c_1 = \frac{3c_0}{4},$$

$$c_2 = \frac{-3c_1}{8} = \frac{-9c_0}{32},$$

$$c_3 = \frac{-3c_2}{38} = \frac{27c_0}{1216},$$

$$\vdots$$

so that

$$y_1 = c_0 x^{\frac{1}{3}}\left(1 + \frac{3}{4}x - \frac{9}{32}x^2 + \frac{27}{1216}x^3 - \cdots\right)$$

is the solution corresponding to $r_1 = \frac{1}{3}$.

Similarly, $r_2 = 2$ gives the recurrence relation

$$(3(n+2)(n+1) - 4(n+2) + 2)c_n + c_{n-1} = 0 \quad \text{for} \quad n \geq 1$$

which simplifies as

$$n(3n+5)c_n + c_{n-1} = 0 \quad \text{for} \quad n \geq 1.$$

This gives

$$c_n = \frac{-c_{n-1}}{n(3n+5)} \quad \text{for} \quad n \geq 1$$

so that

$$c_1 = \frac{-c_0}{8}, \quad c_2 = \frac{-c_1}{22} = \frac{c_0}{176}, \quad c_3 = \frac{-c_2}{42} = \frac{-c_0}{7392}, \quad \cdots .$$

This gives

$$y_2 = c_0 x^2 \left(1 - \frac{1}{8}x + \frac{1}{176}x^2 - \frac{1}{7392}x^3 + \dots \right)$$

as the solution corresponding to the root $r_2 = 2$.

These solutions, corresponding to $r_1 = \frac{1}{3}$ and $r_2 = 2$, respectively, are linearly independent, so the general solution of the differential equation (8.29) is

$$y = k_1 x^{\frac{1}{3}} \left(1 + \frac{3}{4}x - \frac{9}{32}x^2 + \frac{27}{1216}x^3 - \dots \right)$$

$$+ k_2 x^2 \left(1 - \frac{1}{8}x + \frac{1}{176}x^2 - \frac{1}{7392}x^3 + \dots \right)$$

where k_1 and k_2 are arbitrary constants.

We now outline the method we presented in this example.

Outline of the Method of Frobenius

For x_0 a regular singular point of the differential equation

$$f_0(x)\frac{d^2 y}{dx^2} + f_1(x)\frac{dy}{dx} + f_2(x)y = 0, \tag{8.30}$$

to find a solution valid in some interval $0 < |x - x_0| < R$,

1. Assume a solution of the form

$$y = (x - x_0)^r \sum_{n=0}^{\infty} c_n (x - x_0)^n$$

$$= \sum_{n=0}^{\infty} c_n (x - x_0)^{n+r}$$

where $c_0 \neq 0$.

2. Differentiate y term by term to obtain

$$\frac{dy}{dx} = \sum_{n=0}^{\infty} c_n (n + r)(x - x_0)^{n+r-1}$$

and

$$\frac{d^2 y}{dx^2} = \sum_{n=0}^{\infty} c_n (n + r)(n + r - 1)(x - x_0)^{n+r-2}$$

and substitute these expressions into the differential equation (8.30) for

$$y, \quad \frac{dy}{dx} \quad \text{and} \quad \frac{d^2 y}{dx^2}.$$

3. Simplify the expression so that it is of the form

$$K_0(x - x_0)^{r+k} + K_1(x - x_0)^{r+k+1} + K_2(x - x_0)^{r+k+2} + \ldots = 0$$

where k is an integer and the coefficients $K_i, i = 0, 1, 2, \ldots$ are functions of r and some of the coefficients c_n.

4. To be valid for all x in $0 < |x - x_0| < R$,

$$K_0 = K_1 = K_2 = \ldots = 0.$$

5. Upon equating to zero the coefficient K_0 of the *lowest* power $r + k$ of $(x - x_0)$, we obtain a quadratic equation in r, called the *indicial equation* of the differential equation (8.30). The two roots are called the *exponents* of the differential equation, denoted by r_1 and r_2. Here r_1 and r_2 can be distinct real numbers, repeated real or complex. For convenience, r_1 is such that

$$\operatorname{Re}(r_1) \geq \operatorname{Re}(r_2).$$

6. Equate the remaining coefficients $K_1, K_2, \ldots$ to obtain a set of equations, involving r, which must be satisfied by the coefficients c_n in the series.

7. Substitute the root r_1 for r obtained in Step **6** and then determine the c_n which satisfy these conditions.

8. If $r_2 \neq r_1$, repeat Step **7** using r_2 instead of r_1.

9. The general solution is obtained as a linear combination of the solution(s) obtained in Steps **7** and **8**.

Example 4: Use the method of Frobenius to find a solution of

$$xy'' + (1 + x)y' - \frac{1}{16x}y = 0.$$

Rewriting this equation as

$$y'' + \frac{1 + x}{x}y' - \frac{1}{16x^2}y = 0$$

so that

$$p(x) = \frac{1 + x}{x} \quad \text{and} \quad q(x) = \frac{-1}{16x^2}.$$

Both fail to be analytic at $x = 0$, so $x = 0$ is a singular point. Now

$$xp(x) = 1 + x \quad \text{and} \quad x^2 q(x) = \frac{-1}{16}$$

are both analytic at $x = 0$, thus $x = 0$ is a regular singular point.
Thus, there is at least one solution of the form

$$y = \sum_{n=0}^{\infty} c_n(x - x_0)^{n+r} \quad \text{with} \quad c_0 \neq 0.$$

Differentiating gives

$$y' = \sum_{n=0}^{\infty} c_n(n + r)(x - x_0)^{n+r-1}$$

and

$$y'' = \sum_{n=0}^{\infty} c_n(n + r)(n + r - 1)(x - x_0)^{n+r-2}$$

which yields upon substitution into the differential equation

$$x \sum_{n=0}^{\infty} c_n(n + r)(n + r - 1)x^{n+r-2}$$

$$+(1 + x) \sum_{n=0}^{\infty} c_n(n + r)x^{n+r-1} - \frac{1}{16x} \sum_{n=0}^{\infty} c_n x^{n+r} = 0.$$

Rearrangement gives

$$\left[r(r - 1) + r - \frac{1}{16} \right] c_0 x^{r-1} + \sum_{n=1}^{\infty} c_n(n + r)(n + r - 1)x^{n+r-1}$$

$$+ \sum_{n=1}^{\infty} c_n(n + r)x^{n+r-1} + \sum_{n=0}^{\infty} c_n(n + r)x^{n+r}$$

$$- \sum_{n=1}^{\infty} \frac{1}{16} c_n x^{n+r-1} = 0,$$

and further simplification gives

$$\left[r(r - 1) + r - \frac{1}{16} \right] c_0 x^{r-1} + \sum_{n=1}^{\infty} \left(\left[(n + r)(n + r - 1) + (n + r) - \frac{1}{16} \right] c_n \right.$$

$$\left. +(n + r - 1)c_{n-1} \right) x^{n+r-1} = 0.$$

Equating the first coefficient to zero with $c_0 \neq 0$ gives the indicial equation

$$r(r - 1) + r - \frac{1}{16} = 0$$

which is

$$r^2 - \frac{1}{16} = 0.$$

This quadratic equation has roots $r_1 = \frac{1}{4}$ and $r_2 = -\frac{1}{4}$. Thus, the differential equation has two solutions

$$y_1 = \sum_{n=0}^{\infty} c_n x^{n+\frac{1}{4}} \quad \text{and} \quad y_2 = \sum_{n=0}^{\infty} c_n x^{n-\frac{1}{4}}.$$

Now, using these series, we can find the coefficients.

Beginning with the larger root $r_1 = \frac{1}{4}$ and equating the series coefficients to zero gives

$$\left[\left(n + \frac{1}{4}\right)\left(n - \frac{3}{4}\right) + \left(n + \frac{1}{4}\right) - \frac{1}{16}\right] c_n + \left(n - \frac{3}{4}\right) c_{n-1} = 0$$

so that the coefficients c_n must satisfy the recurrence relation

$$c_n = \frac{(3 - 4n)c_{n-1}}{2(2n^2 + n)} \quad \text{for} \quad n \geq 1.$$

Hence

$$c_1 = -\frac{1}{6}c_0, \ c_2 = -\frac{1}{4}c_1 = \frac{1}{24}c_0, \ c_3 = -\frac{1}{112}c_0, \ \cdots.$$

Using these gives

$$y_1 = c_0 x^{\frac{1}{4}}\left(1 - \frac{1}{6}x + \frac{1}{24}x^2 - \frac{1}{112}x^3 + \cdots\right).$$

Similarly, we use the root $r_2 = -\frac{1}{4}$ so that

$$\left[\left(n - \frac{1}{4}\right)\left(n - \frac{5}{4}\right) + \left(n - \frac{1}{4}\right) - \frac{1}{16}\right] c_n + \left(n - \frac{5}{4}\right) c_{n-1} = 0.$$

This yields the recurrence relation

$$c_n = \frac{(5 - 4n)c_{n-1}}{2(2n^2 - n)} \quad \text{for} \quad n \geq 1$$

that the coefficients c_n must satisfy. The values are

$$c_1 = \frac{1}{2}c_0, \quad c_2 = -\frac{1}{4}c_1 = -\frac{1}{8}c_0, \quad c_3 = \frac{7}{240}c_0, \cdots.$$

Using these coefficients gives the solution

$$y_2 = c_0 x^{-\frac{1}{4}}\left(1 + \frac{1}{2}x - \frac{1}{8}x^2 + \frac{7}{240}x^3 - \cdots\right).$$

Forming a linear combination of y_1 and y_2 gives a general solution of the differential equation as

$$y = C_1 y_1 + C_2 y_2$$

$$= C_1 x^{\frac{1}{4}} \left(1 - \frac{1}{6}x + \frac{1}{24}x^2 - \frac{1}{112}x^3 + \dots \right)$$

$$+ C_2 x^{-\frac{1}{4}} \left(1 + \frac{1}{2}x - \frac{1}{8}x^2 + \frac{7}{240}x^3 - \dots \right).$$

The previous calculations, while straightforward, can be quite tedious and there is a lot of room for algebraic error. You might imagine that it would be easier to use Matlab, Maple, or Mathematica to find the coefficient of the assumed form of the series even with extra complication of a regular singular point. It will be a bit more complicated as we will have to first deal with the indicial equation. We now give code for the previous example.

Computer Code 8.3: **Series solutions for differential equations with regular singular points**

Matlab, Maple, Mathematica

```
                              Matlab
>>  syms x c0 c1 c2 c3 c4 c5 c6 r
>>  y=c0*x^r+c1*x^(1+r)+c2*x^(2+r)+c3*x^(3+r)+c4*x^(4+r)
>>  eqODE=x*diff(y,x,2)+(x+1)*diff(y,x)-y/(16*x)
>>  eq1=collect(eqnODE)
    %It won't collect with x in denom or r in exponent.
    %We rewrite the expression to get coeffs we need
>>  eq2=coeffs(simplify(eq*x^(1-r)),x)
    %n must be large enough in y
>>  eq3=solve(eq2(1),r)
>>  eq4=solve(eq2(2),eq2(3),eq2(4),c1,c2,c3)
>>  r=eq3(1)
>>  c1=subs(eq3.c1)
>>  c2=subs(eq3.c2)
>>  c3=subs(eq3.c3)
>>  y1=c0*x^r+c1*x^(1+r)+c2*x^(2+r)+c3*x^(3+r)
>>  r=eq3(2)
>>  c1=subs(eq3.c1)
>>  c2=subs(eq3.c2)
>>  c3=subs(eq3.c3)
>>  y2=c0*x^r+c1*x^(1+r)+c2*x^(2+r)+c3*x^(3+r)
```

```
                          ┌─────────┐
                          │  Maple  │
                          └─────────┘
> eqODE:=x*diff(y(x),x$2)+(1+x)*diff(y(x),x)-y(x)/(16*x)=0;
> eq1:=y(x)=sum('c[n]*x^(n+r)', n=0..4);
> eq2:=subs(eq1,eqODE);
> eq3:=combine(expand(eq2),x);
> eq4a:=coeff(lhs(eq3),x^(r-1));
     %n must be large enough in eq1
> eq4b:=solve(eq4a,r);
> eq5a:=coeff(lhs(eq3),x^r);
> eq5b:=coeff(lhs(eq3),x^(r+1));
> eq5c:=coeff(lhs(eq3),x^(r+2));
> eqsoln1:=solve(subs(r=eq4b[1],{eq5a,eq5b,eq5c}),{c[1],c[2],
   c[3]});
> eqsoln2:=solve(subs(r=eq4b[2],{eq5a,eq5b,eq5c}),{c[1],c[2],
   c[3]});
> soln:=sum('c[n]*x^(n+r)', n=0..3);
> soln1:=y[1](x)=subs(eqsoln1,r=eq4b[1],soln);
> soln2:=y[2](x)=subs(eqsoln2,r=eq4b[2],soln);
```

```
                      ┌─────────────┐
                      │ Mathematica │
                      └─────────────┘
yr[x_]=c0 x^r + c1 x^(r+1) + c2 x^(r+2) + c3 x^(r+3) + c4 x^(r+4)

eqODE[x_]=x y''[x]+(x+1) y'[x] - y[x]/(16 x)
eq1=FullSimplify[ReplaceAll[eqODE[x],{y[x]→ yr[x],
   y'[x]→ yr'[x],y''[x]→ yr''[x]}]]
eq4a=Coefficient[eq1,x^(-1+r)]
eq4b=Solve[eq4a==0,r]
eq5a=Coefficient[Coefficient[Expand[eq1],x^r],x,0]
eq5b=Coefficient[Coefficient[Expand[eq1],x^r],x,1]
eq5c=Coefficient[Coefficient[Expand[eq1],x^r],x,2]
eqsoln1= Solve[ReplaceAll[{eq5a==0,eq5b==0,eq5c==0},
   eq4b[[1]]],{c1,c2,c3}]
eqsoln2=Solve[ReplaceAll[{eq5a==0,eq5b==0,eq5c==0},
   eq4b[[2]]],{c1,c2,c3}]
y1[x_]=Collect[ReplaceAll[ReplaceAll[ReplaceAll[yr[x],
   eqsoln1[[1]]]],c4→ 0],eq4b[[1]]],c0]
y2[x_]=Collect[ReplaceAll[ReplaceAll[ReplaceAll[yr[x],
   eqsoln2[[1]]]],c4→ 0],eq4b[[2]]],c0]
```

In the previous examples, we have found the indicial equation via direct substitution of the series solution into the differential equation. The method of Frobenius depends upon the roots of the indicial equation so it is of interest to obtain a general formula for this equation.

Proceeding here, we will assume for simplicity that $x = 0$ is a regular singular point of the differential equation

$$\frac{d^2y}{dx^2} + p(x)\frac{dy}{dx} + q(x)y = 0. \tag{8.31}$$

Otherwise, if $x = x_0 \neq 0$ were a regular singular point, we could then write $X = x - x_0$. Then the functions

$$xp(x) \quad \text{and} \quad x^2q(x)$$

are analytic at $x = 0$. Thus, both $xp(x)$ and $x^2q(x)$ have convergent power series representations.

That is,

$$xp(x) = p_0 + p_1x + p_2x^2 + \cdots$$

and

$$x^2q(x) = q_0 + q_1x + q_2x^2 + \cdots$$

or alternately

$$p(x) = \frac{p_0}{x} + p_1 + p_2x + p_3x^2 + \cdots$$

and

$$q(x) = \frac{q_0}{x^2} + \frac{q_1}{x} + q_2 + q_3x + \cdots.$$

Now substitution of these series into the differential equation (8.31) gives

$$\left(\sum_{n=0}^{\infty} c_n(n+r)(n+r-1)x^{n+r-2}\right) + \left(\frac{p_0}{x} + p_1 + p_2x + p_3x^2 + \ldots\right)$$

$$\times \left(\sum_{n=0}^{\infty} c_n(n+r)x^{n+r-1}\right)$$

$$+ \left(\frac{q_0}{x^2} + \frac{q_1}{x} + q_2 + q_3x + \ldots\right)$$

$$\times \left(\sum_{n=0}^{\infty} c_n x^{n+r}\right) = 0.$$

Expanding a little gives

$$\left(\sum_{n=0}^{\infty} c_n(n+r)(n+r-1)x^{n+r-2}\right)$$

$$+ \left(\sum_{n=0}^{\infty} c_n p_0(n+r)x^{n+r-2}\right)$$

$$+ (p_1 + p_2 x + p_3 x^2 + \ldots)\left(\sum_{n=0}^{\infty} c_n(n+r)x^{n+r-1}\right)$$

$$+ \left(\sum_{n=0}^{\infty} c_n q_0 x^{n+r-2}\right)$$

$$+ \left(\frac{q_1}{x} + q_2 + q_3 x + \ldots\right)\left(\sum_{n=0}^{\infty} c_n x^{n+r}\right) = 0.$$

Thus, with $n = 0$, the coefficient of x^{r-2} is given as

$$-rc_0 + r^2 c_0 + rc_0 p_0 + c_0 q_0 = c_0(r^2 + (p_0 - 1)r + q_0)$$
$$= c_0(r(r-1) + p_0 r + q_0).$$

So for a differential equation of the form

$$\frac{d^2 y}{dx^2} + p(x)\frac{dy}{dx} + q(x)y = 0$$

with $x = 0$ a regular singular point, the indicial equation is given as

$$r(r-1) + p_0 r + q_0 = 0 \tag{8.32}$$

or

$$r^2 + (p_0 - 1)r + q_0 = 0.$$

The values of r that satisfy this equation are called the *exponents of the differential equation* and are found from (8.32) using the quadratic formula. These exponents are

$$r_1 = \frac{1 - p_0 + \sqrt{1 - 2p_0 + p_0^2 - 4q_0}}{2}$$

and

$$r_2 = \frac{1 - p_0 - \sqrt{1 - 2p_0 + p_0^2 - 4q_o}}{2}$$

where

$$p_0 = \lim_{x \to 0} xp(x) \quad \text{and} \quad q_0 = \lim_{x \to 0} x^2 q(x).$$

Example 5: Determine the indicial roots for the following differential equations given in **a**–**d**.

a.

$$\frac{d^2y}{dx^2} - \frac{1}{3x}\frac{dy}{dx} + \frac{1}{3x^2}y = 0.$$

Here we note that $x = 0$ is a singular point of the differential equation which is regular since

$$xp(x) = -\frac{1}{3} \quad \text{and} \quad x^2q(x) = \frac{1}{3}.$$

Now $p_0 = -\frac{1}{3}$ and $q_0 = \frac{1}{3}$ so that the indicial equation is

$$r(r-1) + p_0r + q_0 = r(r-1) - \frac{1}{3}r + \frac{1}{3} = 0.$$

That is,

$$r^2 - \frac{4}{3}r + \frac{1}{3} = 0,$$

which has roots $r_1 = 1$ and $r_2 = \frac{1}{3}$.

b.

$$xy'' + y' - y = 0.$$

Rewriting this differential equation as

$$y'' + \frac{1}{x}y' - \frac{1}{x}y = 0$$

we see that $x = 0$ is a singular point since

$$p(x) = \frac{1}{x} \quad \text{and} \quad q(x) = \frac{-1}{x}$$

fail to be analytic at $x = 0$.

Now $xp(x) = 1$ and $x^2q(x) = -x$ are both analytic at $x = 0$, so $x = 0$ is a regular singular point. Here

$$p_0 = \lim_{x \to 0} xp(x) = 1 \quad \text{and} \quad q_0 = \lim_{x \to 0} x^2q(x) = 0$$

so the indicial equation is

$$r(r-1) + p_0r + q_0 = r(r-1) + r = 0,$$

that is,

$$r^2 = 0.$$

So the roots here are $r_1 = r_2 = 0$.

c.

$$x^2 y'' + (\sin x)y' - (\cos x)y = 0.$$

Rewriting this differential equation as

$$y'' + \frac{\sin x}{x^2}y' - \frac{\cos x}{x^2}y = 0$$

gives that $x = 0$ is a singular point of the differential equation as

$$p(x) = \frac{\sin x}{x^2} \quad \text{and} \quad q(x) = \frac{-\cos x}{x^2}$$

fail to be analytic at $x = 0$.

Now

$$xp(x) = \frac{\sin x}{x} = \frac{1}{x}\left(x - \frac{x^3}{3!} + \frac{x^5}{5!} - \cdots\right)$$

$$= 1 - \frac{x^2}{3!} + \frac{x^4}{5!} - \cdots.$$

So $xp(x)$ is analytic at $x = 0$. Further $x^2 q(x) = -\cos x$ which is (clearly) also analytic at $x = 0$. Thus $x = 0$ is a regular singular point. Now

$$p_0 = \lim_{x \to 0} xp(x) = 1 \quad \text{and} \quad q_0 = \lim_{x \to 0} x^2 q(x) = -1,$$

which gives the indicial equation as

$$r(r - 1) + p_0 r + q_0 = r(r - 1) + r - 1 = 0$$

so

$$r^2 - 1 = 0.$$

The indicial roots for this differential equation are

$$r_1 = 1 \quad \text{and} \quad r_2 = -1.$$

d.

$$x^2 \frac{d^2 y}{dx^2} + \frac{1}{2}(x + \sin x)\frac{dy}{dx} + y = 0.$$

This differential equation can be written as

$$\frac{d^2 y}{dx^2} + \frac{1}{2}\left(\frac{x + \sin x}{x^2}\right)\frac{dy}{dx} + \frac{1}{x^2}y = 0$$

so that $x = 0$ is a singular point as

$$p(x) = \frac{x + \sin x}{2x^2} \quad \text{and} \quad q(x) = \frac{1}{x^2}$$

are not analytic at $x = 0$.

To see that $x = 0$ is a regular singular point expand $xp(x)$ as

$$xp(x) = x\frac{(x + \sin x)}{2x^2} = \frac{x + \sin x}{2x} = \frac{1}{2} + \frac{\sin x}{2x}$$
$$= \frac{1}{2} + \frac{1}{2}\left(1 - \frac{x^2}{3!} + \frac{x^4}{5!} - \cdots\right)$$

so

$$p_0 = \lim_{x \to 0} xp(x) = 1.$$

Further, $x^2 q(x) = 1$ which is analytic for all x. Now the indicial equation is

$$r(r - 1) + p_0 r + q_0 = r(r - 1) + r + 1 = 0$$

which is $r^2 + 1 = 0$ and has (complex!) roots

$$r_1 = i \quad \text{and} \quad r_2 = -i.$$

These examples give some of the possible situations that can arise when finding the indicial roots of the differential equation. The first example we considered has distinct real roots and applying the method of Frobenius is similar to the previous examples we have considered.

In the next example we consider a differential equation with distinct real roots that differ by an integer value. This particular case requires a bit of additional care.

Example 6: Find a general solution of

$$x^2 \frac{d^2 y}{dx^2} + (x^2 - 3x)\frac{dy}{dx} + 3y = 0.$$

In this differential equation

$$p(x) = \frac{x^2 - 3x}{x^2} \quad \text{and} \quad q(x) = \frac{3}{x^2}$$

which fail to be analytic at $x = 0$. However,

$$xp(x) = \frac{x^2 - 3x}{x} = x - 3$$

and

$$x^2 q(x) = 3,$$

which are both analytic at $x = 0$. Hence, $x = 0$ is a regular singular point. Now

$$p_0 = \lim_{x \to 0} xp(x) = -3 \quad \text{and} \quad q_0 = \lim_{x \to 0} x^2 q(x) = 3$$

so the indicial equation is

$$r(r - 1) + p_0 r + q_0 = r(r - 1) - 3r + 3 = 0,$$

that is,

$$r^2 - 4r + 3 = 0.$$

The indicial roots are $r_1 = 3$ and $r_2 = 1$. These roots differ by a positive integer. We assume solutions of the form

$$y_1 = x^3 \sum_{n=0}^{\infty} c_n x^n \quad \text{and} \quad y_2 = x \sum_{n=0}^{\infty} c_n x^n.$$

Now

$$y_1' = \sum_{n=0}^{\infty} (n + 3) c_n x^{n+2}$$

and

$$y_1'' = \sum_{n=0}^{\infty} (n + 3)(n + 2) c_n x^{n+1}$$

so

$$x^2 \sum_{n=0}^{\infty} (n + 3)(n + 2) c_n x^{n+1} + (x^2 - 3x) \sum_{n=0}^{\infty} (n + 3) c_n x^{n+2}$$

$$+ 3 \sum_{n=0}^{\infty} c_n x^{n+3} = 0.$$

Simplifying and equating the higher powers of x to zero give

$$[(n + 3)(n + 2) - 3(n + 3) + 3] c_n + (n + 2) c_{n-1} = 0 \quad \text{for} \quad n \geq 1,$$

which is

$$n(n + 2) c_n + (n + 2) c_{n-1} = 0 \quad \text{for} \quad n \geq 1.$$

Thus, the recurrence relation that the coefficients c_n must satisfy are

$$c_n = -\frac{c_{n-1}}{n} \quad \text{for} \quad n \geq 1$$

so

$$c_1 = -c_0, \quad c_2 = -\frac{c_1}{2} = \frac{c_0}{2!}, \quad c_3 = -\frac{c_2}{3} = -\frac{c_0}{3!}.$$

Indeed, in general,

$$c_n = \frac{(-1)^n c_0}{n!}.$$

so

$$y_1 = c_0 x^3 \left[1 - x + \frac{x^2}{2!} - \frac{x^3}{3!} + \cdots + \frac{(-1)^n x^n}{n!} + \cdots \right]$$
$$= c_0 x^3 e^{-x}.$$

For the root $r_2 = 1$, we obtain the recurrence formula in the same manner as above. Thus, for y_2 the coefficients must satisfy

$$n(n-2)c_n + nc_{n-1} = 0 \quad \text{for} \quad n \geq 1.$$

Now for $n \neq 2$, we have

$$c_n = -\frac{c_{n-1}}{n-2} \quad \text{for} \quad n \geq 1, n \neq 2.$$

So, for $n = 1$

$$c_1 = c_0$$

and for $n = 2$,

$$2 \cdot 0 c_2 + 2c_1 = 0,$$

which is $c_1 = 0$. However, this implies $c_0 = 0$, a contradiction of our assumption. This contradiction implies that there is no solution y_2 of the form $y_2 = \sum_{n=0}^{\infty} c_n x^{n-1}$.

This can be emphasized further by considering the coefficients c_n for $n \geq 3$; from the condition $0 \cdot c_2 + 2c_1 = 0$, c_2 must be arbitrary, so for $n \geq 3$

$$c_3 = -c_2, \quad c_4 = \frac{-c_3}{2} = \frac{c_2}{2!}, \quad c_5 = \frac{-c_4}{3} = \frac{c_2}{3!} \cdots$$

so that using these values of the coefficients gives

$$y_2 = c_2 x \left[x^2 - x^3 + \frac{x^4}{2!} - \frac{x^5}{3!} + \cdots + \frac{(-1)^n x^{n+2}}{n!} + \cdots \right] \qquad (*)$$
$$= c_2 x^3 \left[1 - x + \frac{x^2}{2!} - \frac{x^3}{3!} + \cdots + (-1)^n \frac{x^n}{n!} + \cdots \right]$$
$$= c_2 x^3 e^{-x}$$

which is the solution obtained for y_1.

This really should not be a surprise as the indicial roots r_1 and r_2 in this example differ by 2. [This value of 2 appears as the factor of x^2 in the expansion of y_2 in equation (*).]

So, in the case of the indicial roots differing by an integer, only one solution will arise from the method of Frobenius. To find another (nearly independent) solution of this differential equation, we employ the method of reduction.

Here, let $y = f(x)v$ where $f(x)$ is a known solution of the differential equation. In this case for y_1 we have, with $c_0 = 1$,

$$y = x^3 e^{-x} v.$$

From this, we obtain

$$\frac{dy}{dx} = x^3 e^{-x} \frac{dv}{dx} + (3x^2 e^{-x} - x^3 e^{-x})v$$

and

$$\frac{d^2 y}{dx^2} = x^3 e^{-x} \frac{d^2 v}{dx^2} + 2(3x^2 e^{-x} - x^3 e^{-x})\frac{dv}{dx}$$

$$+(x^3 e^{-x} - 6x^2 e^{-x} + 6x e^{-x})v.$$

Substituting into the differential equation and simplifying give

$$x \frac{d^2 v}{dx^2} + (3 - x)\frac{dv}{dx} = 0.$$

Letting

$$w = \frac{dv}{dx}$$

reduces the order and we obtain

$$x \frac{dw}{dx} + (3 - x)w = 0.$$

This first-order linear differential equation has solution (check it!)

$$w = x^{-3} e^x.$$

Thus

$$v = \int w\, dx = \int x^{-3} e^x\, dx.$$

Hence

$$y = y_2 = x^3 e^{-x} \int x^{-3} e^x\, dx,$$

which is a linearly independent solution from y_1.

Now it is interesting to note the series form of y_2; expanding e^x in a power series gives

$$y_2 = x^3 e^{-x} \int x^{-3} \left(1 + x + \frac{x^2}{2!} + \frac{x^3}{3!} + \frac{x^4}{4!} + \cdots \right) dx$$

$$= x^3 e^{-x} \int \left(x^{-3} + x^{-2} + \frac{1}{2} x^{-1} + \frac{1}{6} + \frac{1}{24} x + \cdots \right) dx$$

$$= x^3 e^{-x} \left[\frac{-1}{2x^2} - \frac{1}{x} + \frac{1}{2}\ln x + \frac{1}{6}x + \frac{1}{48}x^2 + \cdots \right]$$

where we integrated the series, term by term.

Now writing e^{-x} in series form gives

$$y_2 = \left(x^3 - x^4 + \frac{x^5}{2} - \frac{x^6}{6} + \cdots\right)\left(-\frac{1}{2x^2} - \frac{1}{x} + \frac{1}{6}x + \frac{1}{48}x^2 + \cdots\right)$$

$$+ \frac{1}{2}x^3 e^{-x} \ln x$$

$$= \left(-\frac{1}{2}x - \frac{1}{2}x^2 + \frac{3}{4}x^3 - \frac{1}{4}x^4 + \cdots\right) + \frac{1}{2}x^3 e^{-x} \ln x.$$

Note the logarithm here is of the form $cy_1 \ln x$.

The general solution of the differential equation is thus

$$y = C_1 y_1(x) + C_2 y_2(x)$$

where C_1 and C_2 are arbitrary constants.

In this last example, it was fortunate that we were able to express the solution y in closed form. This simplified the computations involved in finding the second solution y_2. The method of reduction still may be applied even if we cannot find a closed form for the first solution y_1. We would carry out the computations term by term in the series for y_1. These computations can be quite spectacular and are often very complicated.

We will now consider (briefly) the case of complex roots of the indicial equation.

Suppose that the indicial equation

$$r(r - 1) + p_0 r + q_0 = 0$$

had complex roots r_1 and r_2. Here r_1 and r_2 occur as complex conjugates so that if $r_1 = a + bi$ then $r_2 = a - bi$ where a, b are real, $b \neq 0$. Now $r_1 - r_2 = 2bi$, so the difference of these roots is not an integer. Therefore, the solution will not contain a logarithmic term as we had previously. Thus, we assume a solution of the form

$$y_1 = x^{r_1} \sum_{n=0}^{\infty} c_n x^n.$$

But how do we treat a complex exponent $x^{r_1} = x^{a+bi}$?

Recalling Euler's formula

$$e^{ix} = \cos x + i \sin x$$

gives

$$x^{a+bi} = x^a x^{bi} = x^a e^{(b \ln x)i}$$

$$= x^a [\cos(b \ln x) + i \sin(b \ln x)] \quad \text{for } x > 0.$$

Thus, the series for y_1 will have coefficients c_n that are complex. Writing

$$c_n = a_n + ib_n \quad \text{for } n \geq 0$$

formally gives the series

$$y_1(x) = x^{r_1} \sum_{n=0}^{\infty} (a_n + ib_n)x^n$$

$$= x^a [\cos(b \ln x) + i \sin(b \ln x)] \left[\sum_{n=0}^{\infty} a_n x^n + i \sum_{n=0}^{\infty} b_n x^n \right]$$

$$= x^a \left[\cos(b \ln x) \sum_{n=0}^{\infty} a_n x^n - \sin(b \ln x) \sum_{n=0}^{\infty} b_n x^n \right]$$

$$+ ix^a \left[\cos(b \ln x) \sum_{n=0}^{\infty} bn x^n + \sin(b \ln x) \sum_{n=0}^{\infty} a_n x^n \right].$$

We note here that both the real and imaginary parts of $y_1(x)$ are real functions. From this analysis, we see that differential equations with complex indicial roots have solutions that are complex valued. The solution $y_2(x)$ corresponding to the complex conjugate root r_2 is the conjugate of $y_1(x)$. It can be shown that $y_1(x)$ and $y_2(x)$ are linearly independent. We will close this discussion by considering an example of this case.

Example 7: Consider the differential equation

$$x^2 y'' + xy' + (1-x)y = 0.$$

Rewriting this equation as

$$y'' + \frac{1}{x}y' + \frac{(1-x)}{x^2}y = 0$$

gives that $x = 0$ is a singular point of the differential equation as

$$p(x) = \frac{1}{x} \quad \text{and} \quad q(x) = \frac{1-x}{x^2},$$

both which fail to be analytic at $x = 0$. However, $xp(x)$ and $xq(x)$ are analytic at $x = 0$, so $x = 0$ is a regular singular point.

Now

$$p_0 = \lim_{x \to 0} xp(x) = 1 \quad \text{and} \quad q_0 = \lim_{x^2 \to 0} xq(x) = 1,$$

so the indicial equation is

$$r(r-1) + p_0 r + q_0 = r(r-1) + r + 1$$
$$= r^2 + 1 = 0.$$

The indicial roots or exponents for this differential equation are

$$r_1 = i \quad \text{and} \quad r_2 = -i.$$

Thus, if we assume a solution of the form

$$y = x^r \sum_{n=0}^{\infty} c_n x^n$$

we have

$$y_1 = x^i \sum_{n=0}^{\infty} c_n x^n \quad \text{and} \quad y_2 = x^{-i} \sum_{n=0}^{\infty} c_n x^n.$$

Formally treating this imaginary exponent of x as one would treat a real exponent, we have upon differentiation of y_1 that

$$y_1' = \sum_{n=0}^{\infty} (n+i) c_n x^{n+i-1}$$

and

$$y_1'' = \sum_{n=0}^{\infty} (n+i)(n+i-1) c_n x^{n+i-2},$$

so substituting into the differential equation gives

$$x^2 \sum_{n=0}^{\infty} (n+i)(n+i-1) c_n x^{n+i-2} + x \sum_{n=0}^{\infty} (n+i) c_n x^{n+i-1} + (1-x) \sum_{n=0}^{\infty} c_n x^{n+i} = 0.$$

Equating coefficients to zero here gives the recurrence relation for the c_n's as

$$c_{n+1} = \frac{1}{(n+1+i)^2 + 1} c_n$$

$$= \frac{(n+1) - 2i}{(n+1)[(n+1)^2 + 4]} c_n \quad \text{for } n \geq 0.$$

Clearly, this recurrence relation will generate complex-valued coefficients in terms of c_0. To simplify our calculations we will take $c_0 = 1 + i$, solely as an illustration. Thus,

$$c_1 = \frac{1 - 2i}{5} c_0 = \left(\frac{1 - 2i}{5}\right)(1 + i) = \frac{3 - i}{2},$$

$$c_2 = \frac{1 - i}{8} c_1 = \left(\frac{1 - i}{8}\right)\left(\frac{3 - i}{2}\right) = \frac{1 - 2i}{20},$$

$$c_3 = \frac{3 - 2i}{39} c_2 = \left(\frac{3 - 2i}{39}\right)\left(\frac{1 - 2i}{20}\right) = \frac{-(1 + 8i)}{780},$$

$$\vdots$$

so that taking the real and imaginary parts we have

$$a_0 = 1, \ a_1 = \frac{3}{5}, \ a_2 = \frac{1}{20}, \ a_3 = \frac{-1}{780}, \ \cdots$$

and

$$b_0 = 1, \ b_1 = \frac{-1}{5}, \ b_2 = \frac{-1}{10}, \ b_3 = \frac{-2}{195}, \ \cdots .$$

Using these we obtain the solution

$$y_1(x) = \cos(\ln x)\left[1 + \frac{3}{5}x + \frac{1}{20}x^2 + \cdots\right] - \sin(\ln x)\left[1 - \frac{1}{5}x - \frac{1}{10}x^2 + \cdots\right]$$

$$+ i\cos(\ln x)\left[1 - \frac{1}{5}x - \frac{1}{10}x^2 + \cdots\right] + \sin(\ln x)\left[1 + \frac{3}{5}x + \frac{1}{20}x^2 + \cdots\right].$$

The solution $y_2(x)$ is found from the conjugate of $y_1(x)$.

Problems

For each of the following differential equations in the problems 1–30, apply the method of Frobenius to obtain the solution.

1. $2x^2\frac{d^2y}{dx^2} + x\frac{dy}{dx} + (x^2 - 1)y = 0.$
2. $2x^2y'' + 2xy' + (x^2 - 5)y = 0.$
3. $5x^2\frac{d^2y}{dx^2} + x\frac{dy}{dx} + (2x^2 - 3)y = 0.$
4. $x^2\frac{d^2y}{dx^2} - x\frac{dy}{dx} + \left(2x^2 + \frac{5}{9}\right)y = 0.$
5. $2x^2y'' + 3xy' - (x + 1)y = 0.$
6. $x^2y'' + xy' + \left(x^2 - \frac{1}{9}\right)y = 0.$
7. $(3x^2 + x^3)y'' - xy' + y = 0.$
8. $3xy'' + 2y' + 2xy = 0.$
9. $3x^2\frac{d^2y}{dx^2} + x\frac{dy}{dx} + \left(x^2 - \frac{1}{9}\right)y = 0.$
10. $x^2y'' + x(x + 1)y' - y = 0.$
11. $xy'' + 2y' + xy = 0.$
12. $x^2y'' + xy' + \left(x^2 - \frac{1}{4}\right)y = 0.$
13. $(x^2 + x)y'' - 2y' - 2y = 0.$
14. $x^2\frac{d^2y}{dx^2} + (x^4 + x)\frac{dy}{dx} - y = 0.$
15. $xy'' - (x^2 + 2)y' + xy = 0.$
16. $xy'' - y = 0.$
17. $x^2\frac{d^2y}{dx^2} + x^2\frac{dy}{dx} - 2y = 0.$
18. $(2x^2 - x)y'' + (2x - 2)y' + (-2x^2 + 3x - 2)y = 0.$

19. $x^2 \frac{d^2 y}{dx^2} - x \frac{dy}{dx} + \frac{3}{4} y = 0.$

20. $x^2 y'' + xy' + (x - 1)y = 0.$

21. $xy'' + y' + y = 0.$

22. $(x^2 + x^3)y'' + (x^2 - x)y' + y = 0.$

23. $(x^2 + x^3)y'' - (x^2 + x)y' + y = 0.$

24. $xy'' - y = 0.$

25. $xy'' - y' + y = 0.$

26. $x^2 y'' + xy' - (2x + 1)y = 0.$

27. $xy'' - xy' - y = 0.$

28. $x^2 y'' + (x^2 - x)y' + 2y = 0.$

29. $(x^2 - x^3)y'' - 3xy' + 5y = 0.$

30. $x^2 y'' + (3x - x^2)y' + (5 - x)y = 0.$

8.5 Bessel Functions

We will now develop the theory of a type of function that occurs in connection with many problems in applied mathematics, physics, and engineering. Its diverse applications range from electric fields to heat conduction to application in abstract probability theory.

The differential equation

$$x^2 \frac{d^2 y}{dx^2} + x \frac{dy}{dx} + (x^2 - p^2)y = 0, \tag{8.33}$$

where p is a parameter, is called *Bessel's equation of order p*. Any solution of Bessel's equation of order p is called a *Bessel function of order p*. The differential equation (8.33) and its solutions are named in honor of Friedrich Wilhelm Bessel (1784–1846) who first studied some of their remarkable properties. [It should be noted here that the solutions of (8.33) are called Bessel functions, not the possessive Bessel's functions. This is common practice with Bessel functions but is not typical of mathematics in general. For example, Green's functions (possessive form) are solutions to another type of differential equation.]

8.5.1 Bessel Functions of Order Zero

If $p = 0$, equation (8.33) becomes

$$x \frac{d^2 y}{dx^2} + \frac{dy}{dx} + xy = 0, \tag{8.34}$$

which is called *Bessel's equation of order zero.* We will first find solutions of this equation in an interval $0 < x < R$.

Noting that $x = 0$ is a regular singular point of (8.34), we assume a solution of the form

$$y = \sum_{n=0}^{\infty} c_n x^{n+r} \tag{8.35}$$

with $c_0 \neq 0$. Differentiating (8.35) twice and substituting into (8.34) gives

$$\sum_{n=0}^{\infty} (n+r)(n+r-1)c_n x^{n+r-1} + \sum_{n=0}^{\infty} (n+r)c_n x^{n+r-1} + \sum_{n=0}^{\infty} c_n x^{n+r+1} = 0.$$

Simplifying this expression, we obtain

$$\sum_{n=0}^{\infty} (n+r)^2 c_n x^{n+r-1} + \sum_{n=2}^{\infty} c_{n-2} x^{n+r-1} = 0,$$

which is

$$r^2 c_0 x^{r-1} + (1+r)^2 c_1 x^r + \sum_{n=2}^{\infty} [(n+r)^2 c_n + c_{n-2}] x^{n+r-1} = 0. \tag{8.36}$$

Thus, upon equating to zero the coefficient of the lowest power of x in (8.36) gives the indicial equation

$$r^2 = 0,$$

which has (clearly) two equal roots $r_1 = r_2 = 0$. This gives, by equating to zero the coefficients of the higher powers of x in (8.36),

$$(1+r)^2 c_1 = 0, \tag{8.37}$$

and

$$(n+r)^2 c_n + c_{n-2} = 0 \quad \text{for } n \geq 2. \tag{8.38}$$

Taking $r = 0$ in (8.37) gives $c_1 = 0$ and, again letting $r = 0$ in (8.38), we have the recurrence

$$n^2 c_n + c_{n-2} = 0 \quad \text{for } n \geq 2,$$

or more usefully

$$c_n = -\frac{c_{n-2}}{n^2} \quad \text{for } n \geq 2.$$

Using this expression, we obtain successively

$$c_2 = \frac{-c_0}{2^2}, \quad c_3 = \frac{-c_1}{3^2} = 0, \quad c_4 = \frac{-c_2}{4^2} = \frac{c_0}{2^2 \cdot 4^2}, \quad c_5 = \frac{-c_3}{5^2} = 0, \cdots$$

so that we see all the odd-termed coefficients are zero and the general form for the even-termed coefficients is

$$c_{2n} = \frac{(-1)^n c_0}{2^2 4^2 6^2 \cdots (2n)^2} = \frac{(-1)^n c_0}{(n!)^2 2^{2n}}, \quad \text{for } n \geq 1.$$

Thus, when $r = 0$ in (8.35) the solution of Bessel's equation (8.33) is

$$y(x) = c_0 \sum_{n=0}^{\infty} \frac{(-1)^n}{(n!)^2} \left(\frac{x}{2}\right)^{2n}.$$

Setting the arbitrary constant $c_0 = 1$, we obtain the particular solution of Bessel's equation (8.33),

$$y(x) = \sum_{n=0}^{\infty} \frac{(-1)^n}{(n!)^2} \left(\frac{x}{2}\right)^{2n}.$$

This particular situation defines a function, denoted by J_0, and is called the *Bessel function of the first kind of order zero*. Thus,

$$J_0(x) = \sum_{n=0}^{\infty} \frac{(-1)^n}{(n!)^2} \left(\frac{x}{2}\right)^{2n} \tag{8.39}$$

is a particular solution of (8.34). A graph of $J_0(x)$ is shown in Figure 8.4; note that $J_0(x)$ has a damped oscillatory behavior.

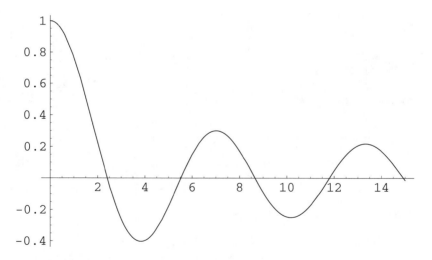

FIGURE 8.4: The Bessel function $J_0(x)$.

Since the indicial equation had two roots that were equal, we know that a solution of (8.34) that is linearly independent of J_0 is of the form

$$y = x \sum_{n=0}^{\infty} c_n^* x^n + J_0(x) \ln x$$

for $0 < x < R$. Using the method of reduction of order, we see that this linearly independent solution y_2 is given by

$$y_2(x) = J_0(x) \int \frac{e^{-\int \frac{1}{x} dx}}{[J_0(x)]^2} dx$$

so that

$$y_2(x) = J_0(x) \int \frac{dx}{x[J_0(x)]^2}.$$

Now

$$[J_0(x)]^2 = \left[1 - \frac{x^2}{4} + \frac{x^4}{64} - \frac{x^6}{2304} + \cdots\right]\left[1 - \frac{x^2}{4} + \frac{x^4}{64} - \frac{x^6}{2304} + \cdots\right]$$

$$= 1 - \frac{x^2}{2} + \frac{3x^4}{32} - \frac{5x^6}{576} + \cdots$$

so

$$\frac{1}{[J_0(x)]^2} = 1 + \frac{x^2}{2} + \frac{5x^4}{32} + \frac{23x^6}{576} + \cdots.$$

This gives

$$y_2(x) = J_0(x) \int \frac{dx}{x[J_0(x)]^2}$$

$$= J_0(x) \int \frac{1}{x}\left[1 + \frac{x^2}{2} + \frac{5x^4}{32} + \frac{23x^6}{576} + \cdots\right] dx$$

$$= J_0(x) \int \left[\frac{1}{x} + \frac{x}{2} + \frac{5x^3}{32} + \frac{23x^5}{576} + \cdots\right] dx$$

$$= J_0(x)\left[\ln x + \frac{x^2}{4} + \frac{5x^4}{128} + \frac{23x^6}{3456} + \cdots\right]$$

$$= J_0(x)\ln x + J_0(x)\left[\frac{x^2}{4} + \frac{5x^4}{128} + \frac{23x^6}{3456} + \cdots\right]$$

$$= J_0(x)\ln x + \left[1 - \frac{x^2}{4} + \frac{x^4}{64} - \frac{x^6}{2304} + \cdots\right]\left[\frac{x^2}{4} + \frac{5x^4}{128} + \frac{23x^6}{3456} + \cdots\right]$$

$$= J_0(x)\ln x + \frac{x^2}{4} - \frac{3x^4}{128} + \frac{11x^6}{13824} + \cdots,$$

so we have found a form of the second solution, but in this form of the expansion we will have trouble finding the general coefficient c_{2n}^*.

Noting that

$$(-1)^2 \frac{1}{2^2(1!)^2}(1) = \frac{1}{2^2} = \frac{1}{4},$$

$$(-1)^3 \frac{1}{2^4(2!)^2}\left(1 + \frac{1}{2}\right) = -\frac{3}{2^4 2^2 \cdot 2} = -\frac{3}{128},$$

$$(-1)^4 \frac{1}{2^6(3!)^2}\left(1 + \frac{1}{2} + \frac{1}{3}\right) = \frac{11}{2^6 6^2 \cdot 6} = \frac{11}{13824}.$$

It seems that, in general,

$$c_{2n}^* = \frac{(-1)^{n+1}}{2^{2n}(n!)^2} \sum_{k=1}^{n} \frac{1}{k} \quad \text{for } n \geq 1.$$

This is indeed the case, as can be shown (c.f. Watson [37]). Using this fact, we have

$$y_2(x) = J_0(x) \ln x + \sum_{n=1}^{\infty} \frac{(-1)^{n+1} x^{2n}}{2^{2n}(n!)^2} \sum_{k=1}^{n} \frac{1}{k}. \tag{8.40}$$

Now $J_0(x)$ and $y_2(x)$ are linearly independent so that the general solution of (8.34) can be written as a linear combination of $J_0(x)$ and $y_2(x)$. However, it is more customary in the theory of Bessel functions to write a "special" linear combination of $J_0(x)$ and $y_2(x)$. This combination is defined as

$$\frac{2}{\pi}[y_2(x) - (\gamma - \ln 2)J_0(x)]$$

where γ is the number

$$\gamma = \lim_{n \to \infty} \left(1 + \frac{1}{2} + \frac{1}{3} + \cdots + \frac{1}{n} - \ln n \right)$$
$$\approx 0.5772,$$

which is known as *Euler's constant*.

This second solution of (8.34) is the function

$$Y_0(x) = \frac{2}{\pi} \left[J_0(x) \ln x + \sum_{n=1}^{\infty} \frac{(-1)^{n+1} x^{2n}}{2^{2n}(n!)^2} \sum_{k=1}^{n} \frac{1}{k} + (\gamma - \ln 2)J_0(x) \right]$$

or

$$Y_0(x) = \frac{2}{\pi} \left[\left(\ln \frac{x}{2} + \gamma \right) J_0(x) + \sum_{n=1}^{\infty} \frac{(-1)^{n+1} x^{2n}}{2^{2n}(n!)^2} \sum_{x=1}^{n} \frac{1}{k} \right]. \tag{8.41}$$

The function $Y_0(x)$ is called the *Bessel function of the second kind of order zero*. The expression (8.41) is known as *Weber's form* of $Y_0(x)$. So, taking $Y_0(x)$ as the second solution of (8.34), the general solution of (8.34) is the linear combination

$$y = C_1 J_0(x) + C_2 Y_0(x) \quad \text{for } 0 < x < R,$$

where C_1 and C_2 are arbitrary constants, with $J_0(x)$ defined by (8.39) and $Y_0(x)$ defined by (8.41).

A graph of $Y_0(x)$ is shown in Figure 8.5, and both $J_0(x)$ and $Y_0(x)$ are shown on the same graph in Figure 8.6. Note that the zeros of $J_0(x)$ separate the zeros of $Y_0(x)$.

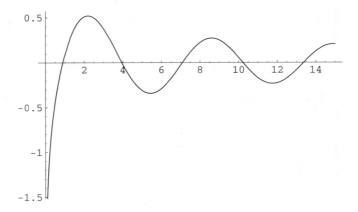

FIGURE 8.5:   The Bessel function $Y_0(x)$.

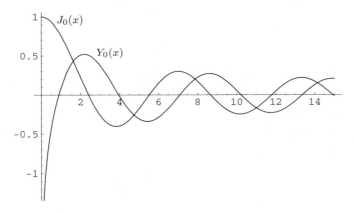

FIGURE 8.6:   The Bessel functions $J_0(x)$ and $Y_0(x)$.

8.5.2   Bessel Functions of Order p

We will now consider solutions of Bessel's equation

$$x^2 \frac{d^2 y}{dx^2} + x \frac{dy}{dx} + (x^2 - p^2)y = 0 \tag{8.42}$$

for real $p > 0$ and valid for $0 < x < R$. Studying equation (8.42), we see that $x = 0$ is a regular singular point, so we can assume a solution

$$y = \sum_{n=0}^{\infty} c_n x^{n+r} \tag{8.43}$$

valid for $0 < x < R$ with $c_0 \neq 0$. Proceeding as we have done previously by differentiating (8.43) and substituting into (8.42) and simplifying, we obtain

$$(r^2 - p^2)c_0 x^r + [(r+1)^2 - p^2]c_1 x^{r+1} + \sum_{n=2}^{\infty} [[(n+r)^2 - p^2]c_n + c_{n-2}]x^{n+r} = 0.$$
(8.44)

Equating the coefficients of (8.44) to zero gives

$$r^2 - p^2 = 0, \quad \text{since } c_0 \neq 0$$
(8.45)

$$[(r+1)^2 - p^2]c_1 = 0$$
(8.46)

and

$$[(n+r)^2 - p^2]c_n + c_{n-2} = 0 \quad \text{for } n \geq 2.$$
(8.47)

Equation (8.45) is, of course, the indicial equation for the differential equation (8.42). The indicial equation has roots $r_1 = p > 0$ and $r_2 = -p$. Now if $r_1 - r_2 = 2p > 0$ is not a positive integer, then we know from the previous section that the differential equation (8.42) has two linearly independent solutions of the form (8.43). On the other hand, if $r_1 - r_2 = 2p$ is a positive integer, we are only certain of a solution in the form (8.43) to exist corresponding to the larger root $r_1 = p$. This is the solution we shall now proceed to obtain.

Thus, letting $r = r_1 = p$ in (8.46) gives

$$(2p+1)c_1 = 0,$$

but since $p > 0$, we have $c_1 = 0$. Letting $r = r_1 = p$ in (8.47) gives the recurrence relation

$$c_n = -\frac{c_{n-2}}{n(n+2p)}, \quad n \geq 2.$$
(8.48)

Now since $c_1 = 0$, (8.48) gives that all odd coefficients are zero, further we find that the even coefficients are given by

$$c_{2n} = \frac{(-1)^n c_0}{[2 \cdot 4 \ldots (2n)][(2+2p)(4+2p) \ldots (2n+2p)]}$$

$$= \frac{(-1)^n c_0}{2^{2n} n![(1+p)(2+p) \ldots (n+p)]} \quad \text{for } n \geq 1.$$

So, the solution of the differential equation (8.42) corresponding to the larger root p is given by

$$y_1(x) = c_0 \sum_{n=0}^{\infty} \frac{(-1)^n x^{2n+p}}{2^{2n} n![(1+p)(2+p) \ldots (n+p)]}.$$
(8.49)

If p is a positive integer then (8.49) can be written as

$$y_1(x) = c_0 2^p p! \sum_{n=0}^{\infty} \frac{(-1)^n}{n!(n+p)!} \left(\frac{x}{2}\right)^{2n+p}.$$
(8.50)

What if p is not a positive integer? In this case, to express $y_1(x)$ in a form similar to (8.50) we need to introduce a function that generalizes the notion of the factorial.

For $\alpha > 0$ the *gamma function* is defined as

$$\Gamma(\alpha) = \int_0^\infty x^{\alpha-1} e^{-x}\, dx \tag{8.51}$$

which is a convergent improper integral for each value of $\alpha > 0$. Integrating (8.51) by parts one time gives the recurrence relation for $\alpha > 0$

$$\Gamma(\alpha) = (\alpha - 1)\Gamma(\alpha - 1), \tag{8.52}$$

so that if α were a positive integer, repeatedly applying (8.52) yields

$$\alpha! = \Gamma(\alpha + 1). \tag{8.53}$$

It is in this sense that if $\alpha > 0$ but not an integer, we use (8.53) to *define* $\alpha!$

Some of these properties may seem surprising. The gamma function, as do the Bessel functions, all belong to a broad class of functions that mathematicians, physicists, and engineers term *special functions*. Special functions are functions, usually defined in terms of a convergent power series or integral, that play a special role in the solution of some problems of practical importance. The gamma and Bessel functions we are studying in this section are perhaps new to you and their properties may seem surprising and unfamiliar. However, the class of special functions contains very familiar functions as well. One only needs to remember that the functions e^x, $\sin x$, and $\cos x$ are all *defined* as convergent power series [although this is *not* how you probably first learned of them], to realize that the "strangeness" of the properties of the gamma and Bessel functions is just an artifact of their newness to your collection of mathematical facts.

Special functions have been extensively studied. Indeed there are some excellent texts that develop the theory and relations of these and other special functions.

Returning now to our discussion of the gamma function, we see that so far, we have defined $\Gamma(\alpha)$ for $\alpha > 0$. It can be shown that

$$\Gamma\left(\frac{1}{2}\right) = \sqrt{\pi}$$

so that using (8.34) we can compute, for instance, $\Gamma(\frac{5}{2})$, that is,

$$\Gamma\left(\frac{5}{2}\right) = \frac{3}{2}\Gamma\left(\frac{3}{2}\right)$$

$$= \left(\frac{3}{2}\right)\left(\frac{1}{2}\right)\Gamma\left(\frac{1}{2}\right)$$

$$= \left(\frac{3}{4}\right)(\sqrt{\pi}) = \frac{3\sqrt{\pi}}{4} \approx 1.3293.$$

In this way, we could say $\left(\frac{3}{2}\right)! = \frac{3\sqrt{\pi}}{4} \approx 1.3293$. For other positive values of α, $\Gamma(\alpha)$ may need to be calculated numerically using (8.51) and a numerical integration routine.

For values of $\alpha < 0$, the integral (8.51) diverges so that $\Gamma(\alpha)$ is not defined for $\alpha < 0$. However, we can extend the definition of $\Gamma(\alpha)$ to $\alpha < 0$ by *demanding* that the recurrence relation (8.52) is valid for all values of α. In this way, $\Gamma(\alpha)$ becomes defined for every noninteger negative value of α; a graph of $\Gamma(x)$ is shown in Figure 8.7.

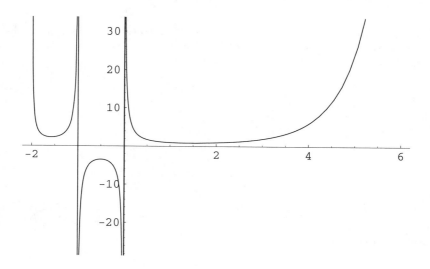

FIGURE 8.7: The gamma function $\Gamma(x)$.

We digressed to the gamma function for a purpose. We are interested in the solution $y_1(x)$ of the differential equation (8.42) when p is not a positive integer. In this case, if we use the recurrence relation (8.52) repeatedly, we obtain

$$\Gamma(n+p+1) = (n+p)(n+p-1)\ldots(p+1)\Gamma(p+1).$$

So, for p not a positive integer, the solution given by (8.49) becomes

$$y_1(x) = c_0\Gamma(p+1)\sum_{n=0}^{\infty} \frac{(-1)^n x^{2n+p}}{2^{2n}n!\Gamma(n+p+1)}$$

$$= c_0 2^p \Gamma(p+1)\sum_{n=0}^{\infty} \frac{(-1)^n}{n!\Gamma(n+p+1)} \left(\frac{x}{2}\right)^{2n+p}. \qquad (8.54)$$

Note that (8.54) reduces to (8.50) when $p = 0$.

If we let c_0 in (8.54) be given as

$$c_0 = \frac{1}{2^p \Gamma(p+1)}$$

we then obtain the particular solution of (8.33) known as the *Bessel function of the first kind of order p*. This function, often denoted J_p, is given as

$$J_p(x) = \sum_{n=0}^{\infty} \frac{(-1)^n}{n!\Gamma(n+p+1)} \left(\frac{x}{2}\right)^{2n+p}. \tag{8.55}$$

Taking $p = 0$ in (8.55), we see that (8.55) reduces to the Bessel function of the first kind of order zero given by (8.39). Now if $p = 1$, then Bessel's equation (8.33) becomes

$$x^2 \frac{d^2y}{dx^2} + x\frac{dy}{dx} + (x^2 - 1) = 0, \tag{8.56}$$

which is Bessel's equation of order 1. Letting $p = 1$ in (8.55) gives the *Bessel function of the first kind of order one*, denoted J_1 and given as

$$J_1(x) = \sum_{n=0}^{\infty} \frac{(-1)^n}{n!(n+1)!} \left(\frac{x}{2}\right)^{2n+1}. \tag{8.57}$$

A graph of $J_1(x)$ is shown in Figure 8.8. Note the damped oscillatory behavior. Graphing both $J_0(x)$ and $J_1(x)$, as shown in Figure 8.9, shows that the positive roots of J_0 and J_1 separate each other.

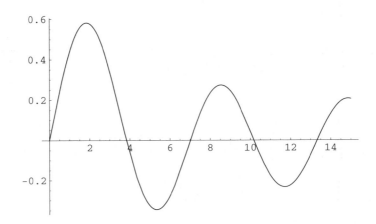

FIGURE 8.8:   The Bessel function $J_1(x)$.

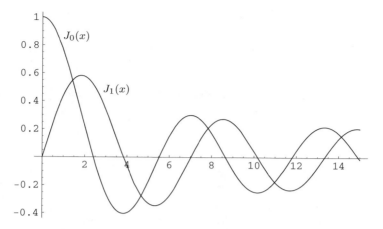

FIGURE 8.9: The Bessel functions $J_0(x)$ and $J_1(x)$.

This is in fact true for the function J_p, $p \geq 0$. It can be shown (c.f. Watson [37]) that J_p has a damped oscillatory behavior as $x \to \infty$ and that the positive roots of J_p and J_{p+1} separate each other.

We have obtained a solution to Bessel's equation (8.33) for every $p \geq 0$. We now find another linearly independent solution of (8.33). We have already given such a solution in the case $p = 0$. This is $Y_0(x)$ defined by (8.41). For $p > 0$, we have noted that if $2p$ is not a positive integer, then the differential equation (8.33) has a linearly independent solution of the form (8.43) corresponding to the smaller root $r_2 = -p$.

Taking $r = r_2 = -p$ in (8.46) gives

$$(-2p + 1)c_1 = 0 \tag{8.58}$$

and in (8.47) we have

$$c_n = -\frac{c_{n-2}}{n(n - 2p)}, \quad n \geq 2, \ n \neq 2p. \tag{8.59}$$

Studying the recurrence relation (8.59), we see that there are three distinct cases for the form of the solution $y = y_2(x)$. These cases are:

1. If $2p$ is not a positive integer then

$$y_2(x) = c_0 x^{-p} \left(1 + \sum_{n=1}^{\infty} \alpha_{2n} x^{2n} \right) \tag{8.60}$$

where c_0 is an arbitrary constant and the $\alpha_{2n} (n = 1, 2, \ldots)$ are (definite) constants.

2. If $2p$ is an odd positive integer then

$$y_2(x) = c_0(x)^{-p}\left(1 + \sum_{n=1}^{\infty}\beta_{2n}x^{2n}\right) + c_{2p}x^p\left(1 + \sum_{n=1}^{\infty}\gamma_{2n}x^{2n}\right) \qquad (8.61)$$

where c_0 and c_{2p} are arbitrary constants and β_{2n}, γ_{2n} $(n = 1, 2, \dots)$ are (definite) constants.

3. If $2p$ is an even positive integer then

$$y_2(x) = c_{2p}x^p\left(1 + \sum_{n=1}^{\infty}\delta_{2n}x^{2n}\right) \qquad (8.62)$$

where c_{2p} is an arbitrary constant and the $\delta_{2n}(n = 1, 2, \dots)$ are (definite) constants.

The solution defined in Case 1 is linearly independent of J_p. In Case 2 the solution defined is linearly independent of J_p if $c_{2p} = 0$. However, in Case 3, the solution is a constant multiple of J_p and hence is *not* linearly independent of J_p. So, if $2p$ is not a positive even integer, there exists a linearly independent solution of the form (8.43) corresponding to the smaller root $-p$. That is, if p is not a positive integer, Bessel's equation (8.33) has a solution of the form

$$y_2(x) = \sum_{n=0}^{\infty}c_{2n}x^{2n-p} \qquad (8.63)$$

which is linearly independent of J_p.

To determine the coefficients c_{2n} in (8.63), we note that (8.59) is obtained if p is replaced by $-p$ in the recurrence relation (8.48). This implies that a solution of the form (8.63) can be obtained from (8.55) by replacing p by $-p$. This gives the solution denoted by J_{-p} as

$$J_{-p}(x) = \sum_{n=0}^{\infty}\frac{(-1)^n}{n!\Gamma(n - p + 1)}\left(\frac{x}{2}\right)^{2n-p}. \qquad (8.64)$$

Thus, if $p > 0$ is not an integer, the general solution of Bessel's equation of order p is given as the linear combination

$$y = c_1 J_p(x) + c_2 J_{-p}(x)$$

where c_1 and c_2 are arbitrary constants.

If p is a positive integer, the solution defined in Case 3 is not linearly independent of J_p as we have already pointed out. So, in this case a solution that is linearly independent of J_p is given by

$$Y_p(x) = x^{-p}\sum_{n=0}^{\infty}c_n^*x^n + CJ_p(x)\ln x$$

where $C \neq 0$. This linearly independent solution can be found using the method of reduction of order, much the same way as we did previously for Y_0. In this manner, we obtain Y_p, so that just as in the case of Bessel's equation of order zero, it is customary to choose a special linear combination as the second solution of (8.33). This special combination, denoted Y_p, is defined as

$$Y_p(x) = \frac{2}{\pi} \left[\left(\ln \frac{x}{2} + \gamma \right) J_p(x) - \frac{1}{2} \sum_{n=0}^{p-1} \frac{(p-n-1)}{n!} \left(\frac{x}{2} \right)^{2n-p} \right.$$
$$\left. + \frac{1}{2} \sum_{n=0}^{\infty} (-1)^{n+1} \left(\sum_{k=1}^{n} \frac{1}{k} + \sum_{k=1}^{n+p} \frac{1}{k} \right) \left[\frac{1}{n!(n+p)!} \left(\frac{x}{2} \right)^{2n+p} \right] \right]$$

$$(8.65)$$

where γ is Euler's constant. The solution Y_p is called the *Bessel function of the second kind of order p*. This formulation of Y_p is sometimes called Weber's form of the Bessel function. A graph of $Y_1(x)$ is shown in Figure 8.10.

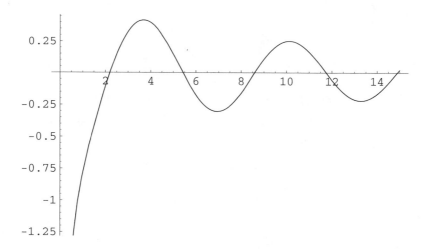

FIGURE 8.10: The Bessel function $Y_1(x)$.

Problems

1. Show directly that the series for $J_0(x)$,

$$J_0(x) = \sum_{n=0}^{\infty} \frac{(-1)^n}{(n!)^2} \left(\frac{x}{2} \right)^{2n},$$

converges absolutely for all x.

2. Show directly that the series for $J_1(x)$,

$$J_1(x) = \sum_{n=0}^{\infty} \frac{(-1)^n}{n!(n+1)!} \left(\frac{x}{2}\right)^{2n+1},$$

converges absolutely for all x and that

$$J_0'(x) = -J_1(x).$$

3. Using the series for $J_p(x)$,

$$J_p(x) = \sum_{n=0}^{\infty} \frac{(-1)^n}{n!\Gamma(n+p+1)} \left(\frac{x}{2}\right)^{2n+p},$$

show directly that

$$\frac{d}{dx}[x^p J_p(kx)] = kx^p J_{p-1}(kx)$$

and

$$\frac{d}{dx}[x^{-p} J_p(kx)] = -kx^{-p} J_{p+1}(kx)$$

where k is a constant.

4. Show that $J_0(kx)$, where k is a constant, satisfies the differential equation

$$x\frac{d^2y}{dx^2} + \frac{dy}{dx} + k^2 xy = 0.$$

5. Show that the transformation

$$y = \frac{u(x)}{\sqrt{x}}$$

reduces the Bessel equation of order p

$$x^2\frac{d^2y}{dx^2} + x\frac{dy}{dx} + (x^2 - p^2)y = 0$$

to the form

$$\frac{d^2u}{dx^2} + \left[1 + \left(\frac{1}{4} - p^2\right)\frac{1}{x^2}\right]u = 0.$$

Using this, show that

$$y_1(x) = \frac{\cos x}{\sqrt{x}} \quad \text{and} \quad y_2(x) = \frac{\sin x}{\sqrt{x}}$$

are solutions of the Bessel equation of order one half.

6. By a suitable change of variables a differential equation sometimes can be transformed into a Bessel equation. Show that a solution of

$$x^2 y'' + (\alpha^2 \beta^2 x^{2\beta} + \frac{1}{4} - p\beta^2)y = 0 \quad \text{for } x > 0$$

is given by $y = \sqrt{x}\, f(\alpha x^\beta)$ where $f(\cdot)$ is a solution of the Bessel equation of order p.

7. Multiply the power series for $e^{xt/2}$ and $e^{-xt/2}$ to show that

$$c^{\frac{x}{2}(t-\frac{1}{t})} = J_0(x) + \sum_{n=1}^{\infty} J_n(x)t^n[1 + (-1)^n].$$

This exponential is called the generating function for the Bessel functions J_n.

8. **a.** Show that the differential equation

$$\frac{d^2 y}{dt^2} + e^{-t}y = 0,$$

which models an aging spring, can be converted to $J_0(t)$ by using the change of variable $x = e^{-t}$.

b. Predict the behavior of the solution by considering the limiting differential equation as $t \to \infty$.

8.6 Chapter 8: Additional Problems and Projects

ADDITIONAL PROBLEMS

In problems 1–5, determine whether the statement is true or false. If it is true, give reasons for your answer. If it is false, give a counterexample or other explanation of why it is false.

1. When $L = 1$, the ratio test gives useful information about the convergence of a series.

2. Power series "behave" a lot like polynomials.

3. Only closed-form solutions are useful as solutions of differential equations.

4. Every ordinary point is a singular point.

5. Every function has a convergent power series representation.

In problems 6–8, show, using a power series argument, the following identities.

6. $e^x e^{-x} = 1$

7. $\lim\limits_{x \to 0} \dfrac{1 - \cos x}{x} = 0$

8. $1 - e^{-x} \approx x$ for small x

In problems 9–20, solve the following differential equations using the power series method.

9. $y' = 2y + 6$ with $y(0) = 3$.

10. $y' - 2y = 2x^3$ with $y(1) = 1$.

11. $y'' + 2y = 4 + 2x + x^2$ with $y(0) = 6$ and $y'(0) = -2$.

12. $y' = x + 2y$ with $y(0) = 3$.

13. $\frac{dy}{dx} = 3 + xy^2$ with $y(0) = 5$.

14. $y' = y + \cos x$ with $y(0) = 1$.

15. $\frac{dy}{dx} = e^{2x} + x \sin y$ with $y(0) = 0$.

16. $y' = x^2 + 5y$ with $y(0) = 6$.

17. $\frac{dy}{dx} = 2x^2 + y^4$ with $y(0) = 2$.

18. $y' = 1 + x \cos y$ with $y(0) = 2$.

19. $y' = x^3 + y^2$ with $y(0) = 4$.

20. $\frac{dy}{dx} = 4x^2 + 2y^3$ with $y(0) = 6$.

In problems 21–30, apply the method of Frobenius to obtain the solution to each of the following differential equations.

21. $2x^2 y'' + xy' + (2x^2 - 5)y = 0$.

22. $x^2 y'' - xy' + \left(x^2 + \frac{8}{9}\right)y = 0$.

23. $x^2 \frac{d^2 y}{dx^2} - 2x \frac{dy}{dx} + \left(x^2 + \frac{3}{4}\right)y = 0$.

24. $2x^2 \frac{d^2 y}{dx^2} - x \frac{dy}{dx} + \left(2x^2 + \frac{7}{9}\right)y = 0$.

25. $x^2 y'' + xy' + \left(x^2 - \frac{1}{9}\right)y = 0$.

26. $3x \frac{d^2 y}{dx^2} + \frac{dy}{dx} + 5y = 0$.

27. $2x \frac{d^2 y}{dx^2} - (x - 3)\frac{dy}{dx} - 3y = 0$.

28. $2xy'' + y' + 2y = 0$.

29. $3xy'' - (x - 2)y' - 2y = 0$.

30. $3x^2 \frac{d^2 y}{dx^2} + (x^4 + x)\frac{dy}{dx} - y = 0$.

PROJECTS FOR CHAPTER 8

Project 1: Asymptotic Series [5]

There will be times when the method of Frobenius fails and we need to consider other approaches in our attempts to find series solutions to a given differential equation. Consider

$$x^2 y'' + (1 + 3x)y' + y = 0. \tag{8.66}$$

Show that there is an irregular singular point at 0. Even though the method of Frobenius should only work at regular singular points, let's try it here. By assuming a solution of the form $y = \sum a_n x^{n+r}$, show that the indicial equation is $r = 0$. By substituting into (8.66), show that the coefficients satisfy $a_{n+1} = -(n+1)a_n$, $n = 0, 1, 2, \cdots$ and thus $a_n = (-1)^n n! a_0$. Conclude that one solution is thus

$$y_1 = a_0 \sum_{n=0}^{\infty} (-1)^n n! x^n. \tag{8.67}$$

Now comes the strange part. Show that this series has a radius of convergence of zero and thus diverges for all $x \neq 0$. Since we assumed that a Frobenius series should have a non-zero radius of convergence, this means that (8.66) actually has no Frobenius series solution!

We often focus on series that converge; that is, a series that approaches the exact function as we add more and more terms. In the remainder of this project we will consider an *asymptotic series*, which has the property that it approaches the function at a specific point even though it is a divergent series. In summary, if $f_n(x)$ is the partial sum approximation to $f(x)$, then

$$\text{Convergent Series: } \lim_{n \to \infty} f_n(x) = f(x),$$
$$\text{Asymptotic Series: } \lim_{x \to x_0} f_n(x) = f(x_0).$$

For two functions $f(x)$ and $g(x)$, we use the notation

$$f(x) \sim g(x), \quad x \to x_0$$

to mean that the relative error between f and g goes to zero as $x \to x_0$ and

$$f(x) \ll g(x), \quad x \to x_0$$

to mean that $f(x)/g(x)$ goes to zero as $x \to x_0$. Our solution (8.67) is an asymptotic series solution to (8.66). We now outline how we can obtain a second solution. Substitute $y = e^S$ into (8.66) to obtain

$$x^2 S'' + x^2 (S')^2 + (1 + 3x)S' + 1 = 0. \tag{8.68}$$

Near an irregular singular point, it is usually the case that $x^2 S''$ can be neglected relative to $x^2 (S')^2$ and that $3xS'$ can be neglected relative to S'. We thus can write

$$x^2 (S')^2 + S' + 1 \sim 0, \quad x \to 0^+.$$

Solve this quadratic for S' and conclude that (because x is small)

$$S' \sim -1, \quad x \to 0^+,$$
$$S' \sim -x^{-2}, \quad x \to 0^+. \tag{8.69}$$

Integrate both of the above asymptotic relations. The first gives the solution already obtained. Use the result from the second integration to conclude that the *controlling factor in the leading behavior* is governed by $e^{1/x}$. To find the full leading behavior, substitute $S(x) = x^{-1} + C(x)$ (with $C(x) \ll x^{-1}$) into (8.68) to obtain

$$x^2 C'' + x^2 (C')^2 - (1 - 3x)C' - x^{-1} + 1 = 0.$$

We can again neglect certain terms as $x \to 0^+$ and obtain

$$c'(x) + x^{-1} \sim 0, \ x \to 0^+.$$

Solve this and conclude that the full leading behavior is given by

$$y_2 = c_2 x^{-1} e^{1/x}, \ x \to 0^+.$$

Choose an initial condition and verify, by numerically solving (8.66), that $y_1 + y_2$ agrees with this solution for small x.

Project 2: Hypergeometric Functions

We will examine solutions of the differential equation

$$x(1 - x)y'' + [c - (a + b + 1)x]y' - aby = 0,$$

for $|x| < 1$ satisfying the initial condition, $y(0) = 1$.

To do so, we first state two helpful definitions. The first uses the gamma function, $\Gamma(\alpha)$, seen earlier in this chapter. We define

$$(\alpha)_n \stackrel{\text{def}}{=} \frac{\Gamma(\alpha + n)}{\Gamma(\alpha)} = \begin{cases} \alpha(\alpha + 1) \cdots (\alpha + n - 1), & n \geq 1 \\ 1, & n = 0. \end{cases}$$

With this definition of $(\alpha)_n$, we can define the following:

DEFINITION 8.3   *Let a, b, and c be real numbers, and suppose c is a nonnegative number. Consider*

$$F(a, b; c; x) \stackrel{\text{def}}{=} \sum_{j=0}^{\infty} \frac{(a)_j (b)_j}{n!(c)_j} x^n$$

on the open interval $(-1, 1)$. *The series is a real analytic function called a* **hypergeometric function.**

THEOREM 8.6.1 *The hypergeometric function, $F(a, b; c; x)$, is the unique real analytic solution to the hypergeometric differential equation*

$$x(1 - x)\frac{\partial^2 y}{\partial x^2} + [c - (a + b + 1)x]\frac{\partial y}{\partial x} - aby = 0$$

for $|x| < 1$, satisfying the initial condition, $y(0) = 1$.

Prove this by letting $y(x) = \sum_{j=1}^{\infty} \alpha_j x^j$ with $\alpha_0 = 1$. It may help you to make the observation

$$(n + 1)(x + n)\alpha_{n+1} = (a + n)(b + n)\alpha_n.$$

Show that the indicial roots of equation (8.6.1) are $r_1 = 0$ and $r_2 = 1 - c$. Using the Frobenius method with the root $r_1 = 0$, obtain a series solution for (8.6.1).

Appendix A

An Introduction to Matlab, Maple, and Mathematica

This appendix is meant as a crash course in Matlab, Maple, and Mathematica or brief refresher, whichever is more applicable to a given reader. All have fairly friendly "help" menus which the reader should consult when questions arise that are not answered here. Both also have *Student Versions* available for about the price of two hardback textbooks. Typically, Matlab is used for numerical problems, whereas Maple and Mathematica are used for symbolic manipulation. We will see throughout the text that this distinction is often blurred.

Initially, we want to plot functions of one variable. Sometimes this will be an explicit representation as in $y = f(x)$. But enough times the separation of variables process (or other method) leaves us with an implicit representation that can be written as $f(x, y) = c$. In order to plot solutions for either case we need to specify initial conditions. As we will see, Matlab, Maple, and Mathematica have methods for handling this.

A.1 Matlab

In Matlab, we execute the commands by either (i) typing them at the command line prompt $\gg$ and pressing <Enter> or (ii) typing the commands in an m-file, saving the m-file, and then typing executing the m-file by typing the name in the command window and pressing <Enter>. (Note that you DO NOT type in $\gg$ for any of these commands.) You cannot go back and correct a line that you typed. However, you can type in the correct one and re-execute it. For example,

```
>> x=0:1:5
>> y=x^2
```

gives the error

```
???  Error using ==> mpower
Matrix must be square.
```

and you need to retype (or use the up arrow to recall the line) and then correct it as

$$\gg \text{ y=x.}^\wedge 2$$

and the last line is the one Matlab will use until you redefine or clear the variable. On occasion, it may confuse variables if you rename them multiple times. The commands

$$\gg \text{ clear y}$$

and

$$\gg \text{ clear all}$$

will wipe the memory clean of the variable y and then of all variables.

As we saw above, we will proceed with the command window input. An important thing to realize is that everything in Matlab is a matrix. If you have not yet encountered this word in your study of mathematics, just think of a matrix as a structure with both rows and columns. A vector is just considered a matrix of either one column or one row. And a scalar is considered to be a matrix with exactly one row and one column.

Another main thing to remember about Matlab is that you have the choice of putting a semicolon ';' at the end of each line or of not putting one. Putting the semicolon suppresses the output, whereas not putting the semicolon shows you the results of the statement you just executed. It does not matter whether you put a semicolon after a statement that plots or does something to an already existing plot. The Matlab commands given in this text will be in `typewriter` font. Anything that follows a percentage sign '%' is ignored by Matlab. Matlab is also case sensitive. That is, the variables t and T are two different variables. Variables also continue to be defined/hold a value until we overwrite them or clear them.

We again note at this point that if you type something incorrectly and get an error message, you can use the up and down arrows on the keyboard to scroll through previous commands that were entered. Sometimes an error is because of a simple typo and using the up arrow to recall your previous command and then correcting the mistake is often quite time saving. Also, if you entered something wrong, you cannot cut the error from the command window—you can only execute the correct command and remember that the most recent assignment of a variable is the one in Matlab's memory.

If we want to plot the function $y = f(x)$ we need to know the x-values that we are going to plug into our function. Then we need to evaluate the function at these values. Sounds simple enough, right? Let's try to plot $f(x) = x^3 - x$ between the x-values of -2 and 2.

```
≫  x=-2:.01:2;
≫  y=x.^3 -x;
≫  plot(x,y)
```

The first line can be read as "go from $x = -2$ to $x = 2$ in steps of .01." The step size of .01 was more or less arbitrary but will give us a smooth plot. The second line can be read as "take the entire vector x and cube *each* entry and then subtract the corresponding entry of x from this." Remember x is a vector (or a matrix) and it doesn't make mathematical sense to cube a vector—we really just want to cube each entry. The '.' is used to tell Matlab to do an operation *elementwise* and this applies to multiplication, division, and raising to a power. (Addition and subtraction are always done elementwise.) If we want to label the axes, title the graph, and change the x-y values that we see in the window to, say, $x \in [-\pi/2, \pi/2]$ and $y \in [-2, 2]$, we could do so with the commands

```
≫  xlabel('x');
≫  ylabel('y');
≫  title('Plot of f(x)=x^3-x');
≫  axis([-pi/2 pi/2 -2 2]);
```

We could also do things like plot the graph in a different color or superimpose another graph on it. Suppose we wanted to also graph the function $f(x) = x^2 - 1$ in the color green and on the same graph. We could use the commands

```
≫  y1=x.^2 -1;
≫  hold on
≫  plot(x,y1,'g')
≫  title('Plot of x^3-x and x^2 -1');
≫  hold off
```

A.1.1 Some Helpful Matlab Commands

Matlab has a help section that can be accessed with the mouse or the keyboard. If you know the command you want to use but forget the proper syntax, the keyboard version is probably the way to go. E.g., if you want to know how to use the built-in Matlab function ode45, you would type help ode45 and information would appear in the command window. We summarize the commands in the list below.

help <command> - specify a command and the info about it will be given
lookfor <topic> - specify a topic and all commands that have related topics will be listed
who - lists current variables

`cd amssi` - changes the working directory to amssi (assuming it exists)
`pwd` - tells you which directory you are currently working in
`dir` - lists files in that directory
`clear all` - clears variables from memory
`format long` - lets you see more non-zero digits
`size(x)` - gives the size of the variable (matrix) x
`length(x)` - gives the length of the vector x
`zeros(n,m)` - an n x m matrix of zeros
`ones(n,m)` - an n x m matrix of ones
`eye(n,m)` - the n x m identity matrix
`rand(n,m)` - an n x m matrix whose entries are random numbers uniformly distributed in (0,1)
`randn(n,m)` - an n x m matrix whose entries are random numbers normally distributed with mean 0 and variance 1
`linspace(x1,x2,n)` - a vector with uniformly spaced points whose beginning value is x1 and the last value is x2 with a total of n points

Plotting in Matlab

Matlab gives numerous options for plotting. Some of the more often used ones are given here.

`figure` - creates a new graph window
`orient tall`, `orient portrait`, `orient landscape` - orients the picture in the desired manner for printing (portrait is the default)

Various line types, plot symbols, and colors can also be implemented in Matlab with the command `plot(x,y,option)` where `option` is a character string made from one element from any or all the following 3 columns:

y	yellow	.	point	-	solid
m	magenta	o	circle	:	dotted
c	cyan	x	x-mark	-.	dashdot
r	red	+	plus	--	dashed
g	green	s	square	p	pentagram
w	white	d	diamond	*	star
k	black	v	triangle (down)	<	triangle (left)
b	blue	^	triangle (up)	>	triangle (right)

As an example, `plot(x,y,'m+:')` plots a magenta dotted line with a plus at each data point, whereas `plot(x,y,'bd')` plots a blue diamond at each data point but does not connect the points with a line.

The reader is encouraged to go through the following example in detail, making sure that the syntax of each line is understood.

Before getting into a detailed example, we pause to discuss Matlab's `subplot` command. In the previous code, we graphed a function that used the entire figure window. There will be many times when this is an inefficient use of space and we would like to restrict ourselves to a smaller portion of the screen. For example, suppose we want to plot x and x^3 separately so that we can see how they appear separately and then plot them together. We can do this conveniently with `subplot`.

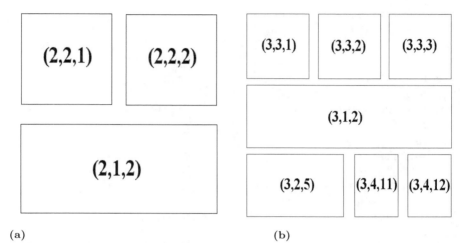

(a)   (b)

FIGURE A.1: Two examples of the subplot command. In (a), we could obtain two plots on top by dividing the figure into 2 rows and 2 columns and then obtain one plot on the bottom by dividing the figure into 2 rows and 1 column. In (b), we again divide up the picture so that it is consistent. E.g., we obtain the middle strip by dividing the picture into 3 rows and 1 column (with the middle strip being picture 2) and we could obtain the lower right one by dividing the picture into 3 rows and 4 columns (with the bottom right one being picture 12).

The command `subplot` simply divides the figure window into smaller sub-windows and then plots what follows in the given smaller window. The syntax for it is

$$\text{subplot}(\text{m,n,p})$$

where `m`, `n`, `p` are positive integers. If `m`, `n`, `p` are single digits, i.e., $0 <$`m`, `n`, `p`< 10, then we could also use the syntax `subplot(mnp)`. The figure is divided into `m` (horizontal) rows and `n` (vertical) columns. The `p` refers to the figure and the numbering starts with 1 as the top left picture. One of the beauties of `subplot` is that we can mix and match how we divide up the window as long as there is no contradiction. See Figure A.1 for two examples of this and the example below for an implementation of `subplot`.

If we wanted to superimpose figures, whether using `subplot` or not, we use the command `hold on` to tell Matlab to superimpose everything that follows until we tell it to `hold off` for the given figure (or subfigure). Example 1 now illustrates some of these additional tools.

Example 1: The intersection of x^2 and x^3

```
>>  x=-10:.05:10;   %don't forget the semicolon!
>>  y=x.^2;   %the . before ^2 is necessary since x is a vector)
>>  plot(x,y);
>>  subplot(2,2,1),plot(x,y);
>>  xlabel('x');
>>  ylabel('x^2');
>>  y1=x.^3;
>>  subplot(2,2,2),plot(x,y1);
>>  subplot(2,1,2),plot(x,y,'r:');
>>  hold on %plots that follow are superimposed
>>  plot(x,y1,'g-')
>>  plot(0,0,'*b')
>>  xlabel('x')
>>  ylabel('y')
>>  legend('x^2','x^3')
>>  title('x^3 vs.  x^2')
>>  axis([-8 8 -100 100]) %restricts the viewing window
>>  grid
>>  subplot(2,2,2),xlabel('x')
>>  subplot(2,2,2),ylabel('x^3')
>>  hold off %anything that follows will not be superimposed
>>  orient tall %makes picture take up full page when you print
```

Unlike in Maple and Mathematica, worksheets are not saved for future execution by default. To save your input in a file, try examining the commands `diary on` and `diary off` to see their usefulness. Alternatively, all the commands may be stored in an m-file to be executed.

There are many *toolboxes* that can accompany Matlab. Although it is not necessary for the reader to have access to any of these in order to use many of the numerical capabilities mentioned here and throughout the text, the *Symbolic Math Toolbox* is often helpful as it will allow symbolic manipulation. For example, we will be able to solve for the roots of a polynomial or calculate a "nice" form of a basis of the nullspace of a matrix. The reader may or may not find it worthwhile to purchase this particular toolbox.

Example 2: Some commands that use Matlab's *Symbolic Math Toolbox*

```
>>  clear all
>>  pi*exp(1) %Note that multiplication is done with *
>>  log(exp(3^2/2))
```

```
>>  syms x r t %defines x, r, t as symbolic variables
>>  eq1=sin(x)+3*x^4
>>  eq2=diff(eq1)
>>  eq3=diff(eq1,x)
>>  eq4=diff(eq1,t) %function doesn't depend on t so answer is 0
>>  eq4a=diff(eq1,x,2) %2nd derivative
>>  eq4b=diff(eq1,x,3) %3rd derivative
>>  eq5=r*x*(1-x)
>>  eq6=solve(eq5,x)
>>  eq6a=eq6(2)
>>  eq7=int(eq5,x)
>>  r=2
>>  eq8=subs(eq5) %evaluates eq5 at the value r=2
>>  eq9=x^2+2*x+3
>>  eq10=solve(eq9,x)
>>  eq10(1)*eq10(2)
>>  expand(eq10(1)*eq10(2)) %multiplies out completely
```

A.1.2   Using m-Files in Matlab

The use of m-files allows one to create functions or simply store a set of commands to be executed. One needs to be careful to save m-files to the appropriate directory and/or be in the right directory in the command window. An m-file is simply a text file that Matlab knows to execute because it has the extension 'm'. Clicking on *File*, then *New*, and then *M-file* will pop up a new window in which you can type.

The easiest use of an m-file is to simply enter a list of commands that would normally be entered in the command window. E.g., we could create an m-file called `ExampleA1.m` and enter the following commands:

```
x=-2:.01:2;
y=x.^3 -x;
plot(x,y)
xlabel('x');
ylabel('y');
title('Plot of f(x)=x^3-x');
axis([-pi/2 pi/2 -2 2]);
y1=x.^2 -1;
hold on
plot(x,y1,'g')
```

In the command window, we would then type

```
>>  ExampleA1;
```

and the result would be the same we obtained at the beginning of this appendix. That is, it would plot the function $x^3 - x$ and $x^2 - 1$ on the same graph and with the various colors, axes, etc. specified.

A more important use of an m-file is to create a function. Matlab has several built-in functions but you can create them as well, and they are stored as an m-file. Suppose we want to create a file that (i) takes two numbers as input, (ii) squares each of them, (iii) takes the fourth root of the sum of their squares, (iv) squares the sum of original numbers, and then (v) returns these values. Let's call this file FourthSq.m. In a text editor, we would enter:

```
function [ans1,ans2]=FourthSq(x,y)

ans1=(x^2+ y^2)^(1/4);
ans2=(x+y)^2;
```

and save the file as FourthSq.m. Then, in the command window we could enter

```
>>  [ans1,ans2]=FourthSq(2,3)
```

We close this part by noting that it is not necessary that the variable use in the m-file agree with the variable use in the command window. Thus we could have also typed

```
>>  [x1,x2]=FourthSq(2,3)
```

and the program would have worked fine. The use of m-files will be very important in the implementation of the Runge-Kutta method given in the book.

Problems

This exercise will make use of many Matlab commands, including zoom, fzero, sqrt, subplot, axis, format long, legend, xlabel, ylabel, title. Turn in both the commands that you entered as well as the output/figures. Please highlight or clearly indicate all requested answers.

1. Determine which of e^π or π^{e^1} is the larger value. Use format long to write your answer correct to 12 decimal places.

2. Plot $y = 3x$, $-2 \le x \le 2$ dashed and green. Label the axes.

3. Plot $y = \sin 2x$, $-3\pi \le x \le 3\pi$ dotted and red. Label the axes.

4. Plot $y = (1-x)^2 - 2$, $-1 \le x \le 3$ dash-dot and cyan. Label the axes.

5. Plot $y = \tan x$, $-\frac{\pi}{2} < x < \frac{\pi}{2}$ with blue stars at the points. Label the axes.

6. Plot $y = e^{-x^2}$, $-5 \leq x \leq 5$ solid and magenta. Label the axes.

7. Plot $x = \ln t$, $.1 \leq t \leq 5$ and only show x-values from -3 to 2. Label the axes.

8. Plot $x = \sqrt{t}, 0 \leq t \leq 4$ and use title and legend. Label the axes.

9. Use `subplot` to plot $y = \sqrt{x^2 - a}$, $\sqrt{a} \leq x \leq \sqrt{a} + 3$ for the values $a = 1, a = 4, a = 9$ in three different subfigures (in the same window).

10. Use `subplot` to plot $y = \sin x, \sin 2x, \cos x, \cos 2x$, $-2\pi \leq x \leq 2\pi$ in four different subfigures (in the same window).

11. Superimpose the plots of $y = \sin x, y = \cos x$ on $0 \leq x \leq \pi$ in the same figure. Use `zoom` to approximate the intersection point to 4 decimal places.

12. Superimpose the plots of $y = \tan x, y = x^2 + 1$ on $0 \leq x \leq \pi/2$ in the same figure. Use `zoom` to approximate the intersection point to 4 decimal places.

13. Use `fzero` to find the intersection point of $y = \tan x, y = x^2 + 1$ on $0 \leq x \leq \pi/2$ to 8 decimal places. (Hint: To find the intersection of $\sin x$ and $\cos x$ at $5\pi/4$ we could type `fzero(inline('sin(x)-cos(x)'),4)` (where 4 is an initial guess "close" to the intersection point) and the intersection point would be displayed numerically).

14. Consider the functions

$$y = x \qquad\qquad (A.1)$$
$$y = r\, x\, (1 - x). \qquad\qquad (A.2)$$

a. Plot the curves (A.1) and (A.2) on the same figure for $r = 1.2$ and find the intersections in the first quadrant (allow both $x = 0$ and $y = 0$) accurate to 3 decimal places for `zoom` and 8 decimal places for `fzero`. Explain any difference between the answers using `zoom` and `fzero`. Use the `solve` command to symbolically solve the equation.

b. Use `subplot` to show the plots in the first quadrant as Matlab gives them and also as you choose to view them (e.g., using the `axis` command). Please do not show more than four pictures on one page. Use different colors and/or symbols to plot the curves. Use the command axis to make sure the display window is okay. Use the command legend to label the graphs.

c. Repeat the above steps with $r = \sqrt{2}$.

Pseudocode answer:

-create a vector that has the x-axis coordinates

-define the curves to be plotted

-use subplot to graph the curves in the arrangement you want

-adjust axes for better viewing, label the curves

-find the intersection points (the syntax for `fzero` is a bit cumbersome; as an example, to find the intersection of $\sin x$ and $\cos x$ at $5\pi/4$ we could type `fzero(inline('sin(x)-cos(x)'),4)` and the intersection point would be displayed numerically)

The remaining problems can only be done if you have the *Symbolic Math Toolbox*.

15. Find the derivative of $t^2 + \cos t$ with respect to t.

16. Find the derivative of $x^2 + \cos x$ with respect to x.

17. Find the second derivative of $t^2 + \cos t$ with respect to t.

18. Find the second derivative of $x^2 + \cos x$ with respect to x.

19. Find the antiderivative of $t^2 + \cos t$ with respect to t.

20. Find the antiderivative of $x^2 + \cos x$ with respect to x.

21. Find the derivative of $\ln(tx) + \cos^2 x + \sin t^2$ with respect to t.

22. Find the derivative of $e^{tx} + \cos^2 x + \sin t^2$ with respect to x.

23. Find the antiderivative of $\ln(tx) + \cos^2 x + \sin t^2$ with respect to t.

24. Find the antiderivative of $e^{tx} + \cos^2 x + \sin t^2$ with respect to x.

25. Find the roots of $x^2 + 3x + 2$.

26. Find the roots of $x^2 + 3x + 4$.

27. Find the roots of $2x^2 - 3x + 4$.

28. Find the roots of $x^2 + 4x + 4$.

29. Find the roots of $x^3 + 3x + 4$.

30. Find the roots of $x^4 + 2x^2 + 1$.

31. Find the roots of $x^4 + 2x^2 + 5$.

32. Find the roots of $x^4 + 2x^2 + x - 5$.

A.2   Maple

In Maple, we execute commands by entering them in the command window and pressing <Enter>. The command prompt for Maple 9 Classic Worksheet and for earlier versions of Maple is >, while the default in Maple 10 is to not show a command prompt.[1] The code in this book was done in Maple 9 Classic Worksheet and you SHOULD NOT type > at the beginning of any line.

---

[1]If you have earlier versions, also be careful of compatibility when saving files, as the later versions will understand the extensions .mw and .mws, while the earlier versions only understand .mws. Most of the code in this book will work with Maple versions 6 through 10.

Any errors that were made can be corrected on the same line that they were typed. For example, the input

```
> x^+1;
```

gives the error

```
Error, '+' unexpected
```

and we can click with the mouse on this same input line and correct it as

```
> x^2+1;
```

and the command will be executed after we press <Enter>.

Maple is a powerful tool for symbolically solving equations and manipulating them. We can integrate, differentiate, solve differential equations exactly, apply various techniques from linear algebra, and much more. We note that Maple 10 allows for entering objects such as fractions and integrals from a palette but since earlier versions of Maple do not have this feature, all code here is entered only from the keyboard.

In Maple, it is a good idea to always start your sessions by typing restart on the first line. If funny things start happening, you can go back to the beginning and re-execute everything, the first line clearing the Maple memory. Also, each command needs to be followed with either a semicolon ';' or a colon ':'. The former allows the output to be seen, whereas the latter suppresses the output. To assign a variable a value, we use a ':=' instead of just an '='. It is often useful to assign each line a name so that you can refer to it later. One common way is to label them eq1, eq2, etc. (even if you technically have an expression instead of an equation). The exception for labeling statements is for any command that plots. If you label one of these, it will return the calculated points and not the desired plot. Sometimes this is desirable but most of the times it is not.

Anything following a # is ignored by Maple. Maple is also case sensitive. That is, the variables t and T are two different variables. Variables also continue to be defined/hold a value until we overwrite them or clear them. In contrast to Matlab, if you execute a command and want to go back and change it, you can go and correct the line where the error occurred or where the change is desired. Once this is re-executed, the new value is the one in Maple's memory.

If we want to plot the function $y = f(x)$ we need to know the x-values that we are going to plug into our function. It is often useful to enter the equation first to make sure there are no typos and then ask Maple to plot it. Let's again try to plot $f(x) = x^3 - x$ between the x-values of -2 and 2.

```
> eq1:=x^3 -x;
> plot(eq1,x=-2..2);
```

The first line defines the equation for us and can be read as "eq1 is assigned the value $x^3 - x$" and the second line says to plot **eq1** from $x = -2$ to $x = 2$. If we want to label the axes, title the graph, and change the x-y values that we see in the window to, say, $x \in [-\pi/2, \pi/2]$ and $y \in [-2, 2]$, we could do so with the commands

```
> plot(eq1,x=-Pi/2..Pi/2,y=-2..2,title="Plot of f(x)=x^3-x");
```

We can also do things like plot the graph in a different color or superimpose another graph on it. Suppose we wanted to also graph the function $f(x) = x^2 - 1$ in the color green and on the same graph. We could do this two different ways. The first way is:

```
> eq2:=x^2 -1;
> plot([eq1,eq2],x=-Pi/2..Pi/2,y=-2..2,color=[blue,green],
    linestyle=[1,2],legend=["x^3-x", "x^2-1"], title="Plot of
    x^3-x and x^2-1");
> #for the linestyle, 1=solid, 2=dotted, 3=dashed, 4=dash-dot
```

A second way to plot the two equations at the same time is with the following commands:

```
> with(plots):
> eq3:=plot(eq1,x=-Pi/2..Pi/2,y=-2..2):
> eq4:=plot(eq2,x=-Pi/2..Pi/2,y=-2..2):
> display([eq3,eq4]);
```

Note that we ended the first two lines with a colon and not a semicolon because we did not want to see all the calculated points.

A.2.1   Some Helpful Maple Commands

Maple also has a help section that can be accessed with the mouse or the keyboard. If you know the command you want to use but forget the proper syntax, the keyboard version is probably the way to go. Suppose the command you have questions about is the **solve** command. In the Maple window, you could enter **?solve** to open the help page for this command, you could enter **??solve** to see the syntax for this command, or you could enter **???solve** to see some examples of this command in use.

Some commands that you might use are:
simplify, expand, evalf, evalc, allvalues, eliminate, coeff, diff, Diff, exp, fsolve, lhs, rhs, plot, roots, solve, subs, sqrt, assume, combine, mtaylor. There are also commands for linear algebra including **evalm, eigenvects, eigenvals, jacobian, transpose, vector, minor, multiply, matrix, trace, inverse, row, col, minor.**

This list is not exhaustive but the commands frequently come up in computations. There are also numerous options for plotting and the reader is again referred to the help menu for instructions. *Some of the above commands, as well as others that are not on the list, require a specific* **Maple package** *to be loaded first.* Three of the common packages that we will take advantage of are `plots`, `DEtools`, `linalg`. E.g., we load the linear algebra package by typing `with(linalg):` and note that it's usually best to end the line with a colon. If we ended with a semicolon instead, we would see all the commands that are included in this package. If we are just beginning, this may be helpful but after a while it's not necessary to see these all the time. See the help menu for instructions on using the various commands and additional packages.

The reader is encouraged to go through the following example in detail, making sure that the syntax of each line is understood.

Example 3: Some sample commands from Maple

```
> restart;
> eqy:=x^2; #defines the function
> plot(eqy, x=-10..10, labels=[x,y]);
> eqy1:=x^3;
> plot(eqy1, x=-10..10, labels=[x,y]);
> plot([eqy,eqy1], x=-8..8, -10..10, labels=[x,y],
   legend=["x^2","x^3"],linestyle=[2,3], color=[red,green]);
> #for the linestyle, 1=solid, 2=dotted, 3=dashed, 4=dash-dot
```

Example 4: Some sample commands from Maple

```
> restart;
> Pi*exp(1);
> evalf(Pi*exp(1)); %gives a decimal answer
> simplify(log(exp(3^2/2)));
> eq1:=sin(x)+3*x^4; #Note that multiplication is done with *
> eq2:=Diff(eq1,x);
> eq3:=diff(eq1,x);
> eq4:=diff(eq1,t); #function doesn't depend on t so answer is 0
> eq4a:=diff(eq1,x$2); #2nd derivative
> eq4b:=diff(eq1,x$3); #3rd derviative
> eq5:=Diff(x,t)=r*x*(1-x);
> eq6:=solve(rhs(eq5),x); #default is to set it equal to 0
> eq6a:=eq6[1];
> eq6b:=eq6[2];
> eq7:=Int(rhs(eq5),x);
> eq8:=int(rhs(eq5),x);
> eq9:=subs(r=2,eq8);
```

```
> plot(eq9,x=-2..3);
> eq10:=x^2+2*x+3;
> eq11:=solve(eq10,x);
> eq12:=eq11[1]*eq11[2];
> eq13:=evalc(eq11[1]*eq11[2]);
   #evalc tells Maple to execute a command on complex numbers
```

Problems

This exercise will make use of many Maple commands, including `plot` (using options such as `title`, `color`), `solve`, `subs`, `evalf`, `simplify`, `sqrt`, `rhs`. Submit only the answers and output requested to the problems given here. (Remember that you can write over an incorrect Maple command and re-execute it, thus saving lots of space due to typos/errors.) Highlight or clearly mark the answers that you are asked to provide.

1. Determine which of e^π or π^{e^1} is the larger value. Use `Digits` and `evalf` to write your answer correct to 12 decimal places.

2. Plot $y = 3x$, $-2 \le x \le 2$ dashed and green. Label the axes.

3. Plot $y = \sin 2x$, $-3\pi \le x \le 3\pi$ dotted and red. Label the axes.

4. Plot $y = (1-x)^2 - 2$, $-1 \le x \le 3$ dash-dot and cyan. Label the axes.

5. Plot $y = \tan x$, $-\frac{\pi}{2} < x < \frac{\pi}{2}$ with blue points. Label the axes.

6. Plot $y = e^{-x^2}$, $-5 \le x \le 5$ solid and magenta. Label the axes.

7. Plot $x = \ln t$, $.1 \le t \le 5$ and only show x-values from -3 to 2. Label the axes.

8. Plot $x = \sqrt{t}, 0 \le t \le 4$ and use title and legend. Label the axes.

9. Superimpose the plots $y = \sqrt{x^2 - a}$, $\sqrt{a} \le x \le \sqrt{a} + 3$ for the values $a = 1, a = 4, a = 9$ in the same figure.

10. Superimpose the plots $y = \sin x, \sin 2x, \cos x, \cos 2x$ on $-2\pi \le x \le 2\pi$ in the same figure.

11. Superimpose the plots of $y = \sin x, y = \cos x$, $0 \le x \le \pi$ in the same figure.

12. Superimpose the plots of $y = \tan x, y = x^2 + 1$ on $0 \le x \le \pi/2$ in the same figure.

13. Use `fsolve` to find the intersection point of $y = \tan x, y = x^2 + 1$ on $0 \le x \le \pi/2$ to 8 decimal places.

14. Find the derivative of $t^2 + \cos t$ with respect to t.

15. Find the derivative of $x^2 + \cos x$ with respect to x.

16. Find the second derivative of $t^2 + \cos t$ with respect to t.

17. Find the second derivative of $x^2 + \cos x$ with respect to x.

18. Find the antiderivative of $t^2 + \cos t$ with respect to t.
19. Find the antiderivative of $x^2 + \cos x$ with respect to x.
20. Find the derivative of $\ln(tx) + \cos^2 x + \sin t^2$ with respect to t.
21. Find the derivative of $e^{tx} + \cos^2 x + \sin t^2$ with respect to x.
22. Find the antiderivative of $\ln(tx) + \cos^2 x + \sin t^2$ with respect to t.
23. Find the antiderivative of $e^{tx} + \cos^2 x + \sin t^2$ with respect to x.
24. Find the roots of $x^2 + 3x + 2$.
25. Find the roots of $x^2 + 3x + 4$.
26. Find the roots of $2x^2 - 3x + 4$.
27. Find the roots of $x^2 + 4x + 4$.
28. Find the roots of $x^3 + 3x + 4$.
29. Find the roots of $x^4 + 2x^2 + 1$.
30. Find the roots of $x^4 + 2x^2 + 5$.
31. Find the roots of $x^4 + 2x^2 + x - 5$.
32. Consider the curve given by

$$y = r\, x\, (1 - x) - \frac{1}{2}x \tag{A.3}$$

where $r \in \mathbb{R}$.
a. Plot the curves for $r = 1.2$ and $\sqrt{2}$ in the same figure/window, labeling them appropriately. Find the roots of this curve as a function of r. For what values of r will there be two nonnegative roots?
b. Find the numerical value of the roots for $r = .5, 1.2$, and $\sqrt{2}$.
c. Find the *exact* value of the roots for these same r-values (remembering, e.g., that $1.2 = 6/5$).
Pseudocode answer:
-enter the equation
-graph the right-hand side of the original equation for the given r values (resize the picture, if necessary, so that it is not too large)
-solve that equation (set equal to 0) for x
-substitute in the given r-values
-use evalf to obtain a numeric answer

A.3 Mathematica

In Mathematica, we execute commands by entering them and holding down <Shift> and then pressing <Enter>. Unlike Matlab and earlier versions of Maple, there is no command prompt (just start typing). Any errors that were made can be corrected on the same line that they were typed. For example, the input

```
y[x-]=x+1;
```

gives the error

Syntax::sntxf: "y[x-" cannot be followed by "]". More...

and we can click with the mouse on this same input line and correct it as

```
y[x_]=x+1;
```

and the command will be executed after pressing <Shift> <Enter>.

Mathematica is comparable to Maple in its ability to symbolically solve and manipulate equations (both are quite good). We can integrate, differentiate, solve differential equations exactly, apply various techniques from linear algebra, and much more.

In correcting input lines, Mathematica's kernel (i.e., computation engine) sometimes gets confused. When this happens, you should go up to `Kernel` $\longrightarrow$ `Quit Kernel` $\longrightarrow$ `Local` and then re-execute the relevant commands. If your line ends with a semicolon ';' then the output is suppressed. If your line has nothing then the output is shown. To assign a variable a value, we use '=' unless it is a function. For functions we must type the variable name followed by an underscore as in the first example above. It is often useful to assign each line a name so that you can refer to it later. One common way is to label them `eq1`, `eq2`, etc. (even if you technically have an expression instead of an equation). Plots can also be labeled, too, which helps when you want to superimpose various plots.

Anything inside (* *) is ignored by Mathematica. Mathematica is also case sensitive. That is, the variables `t` and `T` are two different variables. Variables also continue to be defined/hold a value until we overwrite them or clear them.

If we want to plot the function $y = f(x)$ we need to know the x-values that we are going to plug into our function. It is often useful to enter the equation first to make sure there are no typos and then ask Mathematica to plot it. Let's try to plot $f(x) = x^3 - x$ between the x-values of -2 and 2. We could enter this equation with the keyboard only as

```
eq0[x_]=x^3-x
```

but it's often easier to enter it using the palette as

```
eq1[x_]=x³ - x
Plot[eq1[x],{x,-2,2}]
```

The first line defines the equation as a function of the variable x and can be read as "eq1 is the function defined as $x^3 - x$," while the second line says to plot the function `eq1[x]` from $x = -2$ to $x = 2$. Note that the underscore was needed to define the function but not when it was referred to at a later step. If we want to label the axes, title the graph, and change the x-y values that we see in the window to, say, $x \in [-\pi/2, \pi/2]$ and $y \in [-2, 2]$, we could do so with the commands

$$\texttt{Plot} \left[\texttt{eq1}[\texttt{x}], \left\{\texttt{x}, -\tfrac{\pi}{2}, \tfrac{\pi}{2}\right\}, \texttt{PlotRange} \rightarrow \{-2, 2\}\right]$$

We can also do things like plot the graph in a different color or superimpose another graph on it. Suppose we wanted to also graph the function $f(x) = x^2 - 1$ in the color green and on the same graph. We need to first load the color graphics package:

```
<<Graphics'Colors'
eq2[x_]=x²-1
Plot [{eq1[x],eq2[x]}, {x, -π/2, π/2}, PlotRange→{-2,2},
   PlotStyle→ {{Blue,Dashing[{0}]},{Green,Dashing[{.05}]}}]
p1 = Plot [eq1[x], {x, -π/2, π/2}, PlotRange→{-2,2},
   PlotStyle→ {Blue,Dashing[{0}]}]
p2 = Plot [eq2[x], {x, -π/2, π/2}, PlotRange→{-2,2},
   PlotStyle→ {Green,Dashing[{.05}]}]
Show[p1,p2]
```

A.3.1   Some Helpful Mathematica Commands

Mathematica has a help section that can be accessed with the mouse to find many commands and the proper syntax to use them. Some commands that you might use are:
`Simplify, FullSimplify, Expand, N, ReplaceAll, Coefficient, D, E, Solve, NSolve, NDSolve, FindRoot, Plot, FullExpand ComplexExpand, PowerExpand, Series, Eliminate, Integrate`. There are also commands for linear algebra including `Eigenvalues, Eigenvectors, Eigensystem, CharacteristicPolynomial, TakeRows, MatrixForm, LinearSolve, Tr, Det, SubMatrix`.

This list is not exhaustive but the commands frequently come up in computations. We can sometimes enter a command in two different ways—with the keyboard or with the palette. E.g., `Integrate` can be used or we could click on the integral sign. There are also numerous options for plotting and the reader is again referred to the help menu for instructions. *Some of the above commands, as well as others that are not on the list, require a specific* **Mathematica package** *to be loaded first.* Some of the common packages that we will take advantage of are `LinearAlgebra'MatrixManipulation'`, `Graphics'Colors'`, `Graphics'ImplicitPlot'`; there are many others. See the help menu for instructions on using the various commands and additional packages.

The reader is encouraged to go through the following example in detail, making sure that the syntax of each line is understood.

Example 5: Some sample commands from Mathematica

```
<<Graphics'Colors'
eqy[x_]=x² (*defines the function*)
   (*Note:  power form entered from palette*)
p1=Plot[eqy[x],{x,-10,10},AxesLabel→ {"x","y"},
  PlotStyle→ {Red,Dashing[{0.01}]}]
eqy1[x_]=x³
p2=Plot[eqy1[x],{x,-10,10},AxesLabel→ {"x","y"},
  PlotStyle→ {Green,Dashing[{.05, .01 , .05}]}]
Show[p1,p2]
```

Example 6: Some sample commands from Mathematica

```
π e¹ (*entered from palette*)
N[Pi E^1] (*entered from keyboard*)
N[πe¹] (*gives a decimal answer*)
Log[e^{3²/2}]
eq1[x_]=Sin[x]+3 x⁴
(*Note:  multiplication is done with '*' or with a <Space>*)
eq2=D[eq1[x],x]
eq3=eq1'[x] (*alternate notation*)
D[eq1[x],t] (*function doesn't depend on t so answer is 0*)
D[eq1[x],{x,2}] (*2nd derivative; '' notation okay too*)
eq1'''[x] (*3rd derivative; alternate derivative notation okay*)
eq5[x_,r_]=r x (1-x) (*note a <Space> for multiplication*)
eq6=Solve[eq5[x,r]==0,x] (*'double equal' is needed for solving*)
eq6a=eq6[[1]]
eq6b=eq6[[2]]
eq7[x_,r_]=∫ eq5[x,r]dx (*entered from palette*)
eq8[x_]=ReplaceAll[eq7[x,r],r→ 2]
Plot[eq8[x],{x,-2,3}]
eq9[x_]=x² + 2x + 3
eq11=Solve[eq9[x]==0,x]
Simplify[eq11[[1]][[1]][[2]] eq11[[2]][[1]][[2]]]
   (*note the space for multiplication in above line*)
```

Problems

This exercise will make use of many Mathematica commands, including Plot (using options such as AxesLabel, PlotStyle), Solve, ReplaceAll, Simplify. Submit only the answers and output requested to the problems given here. (Remember that you can write over an incorrect Mathematica command and re-execute it, thus saving lots of space due to typos/errors.) If problems arise, just clear the kernel and re-execute your already-typed commands. Highlight or clearly mark the answers that you are asked to provide.

1. Determine which of e^π or π^{e^1} is the larger value. Use N to write your answer correct to 12 decimal places.

2. Plot $y = 3x$, $-2 \le x \le 2$ dashed and green. Label the axes.

3. Plot $y = \sin 2x$, $-3\pi \le x \le 3\pi$ dotted and red. Label the axes.

4. Plot $y = (1-x)^2 - 2$, $-1 \le x \le 3$ dash-dot and blue. Label the axes.

5. Plot $y = \tan x$, $-\frac{\pi}{2} < x < \frac{\pi}{2}$ with blue points. Label the axes.

6. Plot $y = e^{-x^2}$, $-5 \le x \le 5$ solid and red. Label the axes.

7. Plot $x = \ln t$, $.1 \le t \le 5$ and only show x-values from -3 to 2. Label the axes.

8. Plot $x = \sqrt{t}, 0 \le t \le 4$ and label the axes.

9. Superimpose the plots $y = \sqrt{x^2 - a}$, $\sqrt{a} \le x \le \sqrt{a} + 3$ for the values $a = 1, a = 4, a = 9$ in the same figure.

10. Superimpose the plots $y = \sin x, \sin 2x, \cos x, \cos 2x$ on $-2\pi \le x \le 2\pi$ in the same figure.

11. Superimpose the plots of $y = \sin x, y = \cos x$, $0 \le x \le \pi$ in the same figure.

12. Superimpose the plots of $y = \tan x, y = x^2 + 1$ on $0 \le x \le \pi/2$ in the same figure.

13. Use NSolve to find the intersection point of $y = \tan x, y = x^2 + 1$ on $0 \le x \le \pi/2$ to 8 decimal places.

14. Find the derivative of $t^2 + \cos t$ with respect to t.

15. Find the derivative of $x^2 + \cos x$ with respect to x.

16. Find the second derivative of $t^2 + \cos t$ with respect to t.

17. Find the second derivative of $x^2 + \cos x$ with respect to x.

18. Find the antiderivative of $t^2 + \cos t$ with respect to t.

19. Find the antiderivative of $x^2 + \cos x$ with respect to x.

20. Find the derivative of $\ln(tx) + \cos^2 x + \sin t^2$ with respect to t.

21. Find the derivative of $e^{tx} + \cos^2 x + \sin t^2$ with respect to x.

22. Find the antiderivative of $\ln(tx) + \cos^2 x + \sin t^2$ with respect to t.

23. Find the antiderivative of $e^{tx} + \cos^2 x + \sin t^2$ with respect to x.

24. Find the roots of $x^2 + 3x + 2$.
25. Find the roots of $x^2 + 3x + 4$.
26. Find the roots of $2x^2 - 3x + 4$.
27. Find the roots of $x^2 + 4x + 4$.
28. Find the roots of $x^3 + 3x + 4$.
29. Find the roots of $x^4 + 2x^2 + 1$.
30. Find the roots of $x^4 + 2x^2 + 5$.
31. Find the roots of $x^4 + 2x^2 + x - 5$.
32. Consider the curve given by

$$y = r\,x\,(1 - x) - \frac{1}{2}x \qquad\qquad\text{(A.4)}$$

where $r \in \mathbb{R}$.
a. Plot the curves for $r = 1.2$ and $\sqrt{2}$ in the same figure/window, label-ing them appropriately. Find the roots of this curve as a function of r. For what values of r will there be two nonnegative roots?
b. Find the numerical value of the roots for $r = .5, 1.2$, and $\sqrt{2}$.
c. Find the *exact* value of the roots for these same r-values (remember-ing, e.g., that $1.2 = 6/5$).
Pseudocode answer:
-enter the function
-graph the right-hand side of it for the given r values (resize the picture, if necessary, so that it is not too large)
-solve that equation (set equal to 0) for x
-substitute in the given r-values
-use N to obtain a numeric answer

We hope this was an adequate refresher course or a not-too-brutal crash course into Matlab, Maple, and Mathematica. They are utilized throughout the text and at least one of them should be your close friend. They will definitely aid us in our study of differential equations.

Appendix B

Graphing Factored Polynomials

We give a brief reminder on how to graph factored polynomials quickly. We will not be concerned with maxima, minima, inflection points, etc. Instead we care only about whether a graph is positive or negative between roots, how it passes through a root, and its behavior for large x-values. Knowing how to graph these quickly will be especially helpful in Sections 2.3 and 2.4, which deal with first-order autonomous equations and their applications.

We will assume that we begin with a polynomial in the form

$$P(x) = a_0 x^{k_0} (a_1 x - b_1)^{k_1} \cdots (a_n x - b_n)^{k_n} (c_m x^m + c_{m-1} x^{m-1} + \cdots c_1 x + c_0)^{k_q},$$

(B.1)

where the a_i, b_i, c_i are real numbers and where the last factor $c_m x^m + \cdots + c_0$ is assumed to be a polynomial with no real roots. That is, it does not factor any further over the field of real numbers. Note that $c_m x^m + \cdots + c_0$ can be factored as a product of irreducible quadratic factors, but since our concern will be with the term with the highest power and its leading coefficient, we will typically leave it in this expanded form. For example, it could be written as

$$(x^2 + 1)(x^2 + 6)^2$$

or it could be written out as

$$x^6 + 13x^4 + 48x^2 + 36.$$

The latter form is given in (B.1). To be a bit more explicit, we are requiring

$$a_i \neq 0, \quad i = 0, 1, 2, \ldots n, \qquad c_m \neq 0, \quad c_0 \neq 0.$$

The roots of (B.1) are given by

$$r_i = \frac{b_i}{a_i} \quad \text{with } r_i \neq r_j \text{ if } i \neq j,$$

(B.2)

and each root r_i occurs k_i times where $k_i \geq 1$ is a positive integer. For the $a_0 x^{k_0}$ factor, we should observe that $b_0 = 0$.

In order to sketch the curve, we need to know how many times each root occurs (its multiplicity), the highest power in the polynomial $P(x)$ given in (B.1), and the sign of the coefficient of this highest power. The roots are given by (B.2). For the last two needed items, we multiply the terms with highest power of x from each factor. That is, we consider

$$a_0 x^{k_0} (a_1 x)^{k_1} (a_2 x)^{k_2} \cdots (a_n x)^{k_n} (c_m x^m)^{k_q} \tag{B.3}$$

which can be simplified slightly as

$$a_0 (a_1)^{k_1} (a_2)^{k_2} \cdots (a_n)^{k_n} (c_m)^{k_m} x^{k_0 + k_1 + k_2 + \cdots + k_n + k_q} \equiv A x^K, \tag{B.4}$$

where

$$A = a_0 (a_1)^{k_1} (a_2)^{k_2} \cdots (a_n)^{k_n} (c_m)^{k_q} \tag{B.5}$$

and

$$K = k_0 + k_1 + k_2 + \cdots + k_n + k_q. \tag{B.6}$$

With the knowledge of A, K, and the roots (and their multiplicity), we can sketch the curve with the following two charts; see Figures B.1 and B.2.

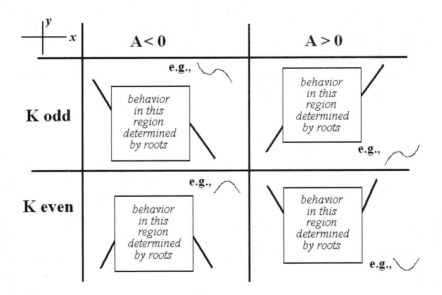

FIGURE B.1:   Behavior of factored polynomial (B.1) for large $|x|$. The curves obey $\lim_{x \to -\infty} P(x)$ and $\lim_{x \to \infty} P(x)$.

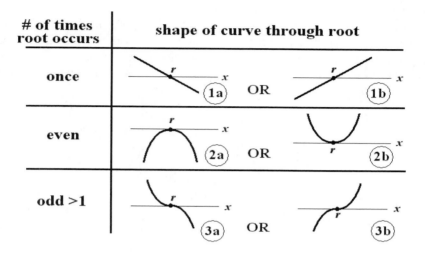

# of times root occurs	shape of curve through root

FIGURE B.2: Behavior near the roots of factored polynomial (B.1).

We finish this appendix with two examples.

Example 1: Sketch the polynomial

$$P(x) = (x-1)^2(2+3x)(4-x)^3(x+2).$$

We see that the polynomial has the following roots and multiplicities:

root	$x = 1$	$x = -2/3$	$x = 4$	$x = -2$
multiplicity	2	1	3	1

Calculating the highest power and its coefficient gives

$$(x)^2(3x)(-x)^3(x) = -3x^7.$$

We are in the case K odd and $A < 0$ and we can identify the long-term behavior from Figure B.1. To find the behavior near the roots, we go from left to right. The first root we encounter is $x = -2$ and it occurs once. Since we are coming from above, we must pass through the root as in 1a of Figure B.2. The second root we encounter is $x = -2/3$ and it occurs once. Since we are coming from below, we must pass through it as in 1b of Figure B.2. The next root we encounter is $x = 1$ and it occurs twice. Since we are coming from above, we must pass through it as in 2b. The final root is $x = 4$ and it occurs three times. Since we are now coming from above, we must pass through it as in 3a of Figure B.2. Note that we can match up with the long-term behavior for large positive x. A sketch is given in Figure B.3. Note that for these rough sketches, we do *not* care about the relative maxima etc. Our goal is to know when the graph is positive vs. negative and how it passes through the roots. Thus, we will not draw a scale on the y-axis.

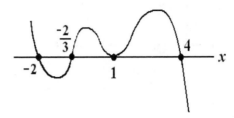

FIGURE B.3:   Sketch of $P(x) = (x - 1)^2(2 + 3x)(4 - x)^3(x + 2)$.

Example 2: Sketch the polynomial

$$P(x) = -x^3(x^2 - 4)(x - 1)(5 - x)(3 + 3x)(4 - 2x)^2(x^2 + 1).$$

Noting that $(x^2 - 4)$ factors into $(x + 2)(x - 2)$, we see that this polynomial has the following roots and multiplicities:

root	$x = 0$	$x = 2$	$x = -2$	$x = 1$	$x = 5$	$x = -1$
multiplicity	3	3	1	1	1	1

Calculating the highest power and its coefficient gives

$$-x^3(x^2)(x)(-x)(3x)(-2x)^2(x^2) = +12x^{12}.$$

We are in the case K even and $A > 0$ and we can identify the long-term behavior from Figure B.1. To find the behavior near the roots, we go from left to right. The first root we encounter is $x = -2$. It occurs only once and since we are coming from above, we must pass through it as in 1a of Figure B.1. The next root we encounter is $x = -1$. We are now coming from below and because it occurs once, we pass through it as in 1b. The next root is $x = 0$ and it occurs three times. Thus we pass through it as in 3a of Figure B.1. The next root is $x = 1$ and it occurs once. Since we are now coming from below we pass through it as in 1b. The next root is $x = 2$ which occurs three times. We are now coming from above and thus pass through it as in 3a. The final root is $x = 5$ and occurs only once. Thus we pass through as in 1b. Note that we can match up with the long-term behavior for large positive x. A sketch is given in Figure B.4.

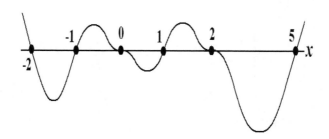

FIGURE B.4: Sketch of
$$P(x) = -x^3(x^2 - 4)(x - 1)(5 - x)(3 + 3x)(4 - 2x)^2(x^2 + 1).$$

Problems

Sketch the graphs of the following equations using the methods of this section.

1. $f(x) = x^2(x - 2)(4 - 3x)$
2. $f(x) = (x^2 - 9)^2(x - 1)$
3. $g(x) = (1 + x)(3 + 2x)(5 - x)^7$
4. $h(x) = x^4(2x - 3)^3(x^2 + 4)(x^2 + 2x + 1)$
5. $P(x) = (x^4 - 16)^2$
6. $x(t) = (16 - 3t^2)(t^2 - 4)^3$
7. $f(x) = (4 - x)^8(2 - x)(4 - 3x^2)^4(2x + 2)(x + 5)$
8. $f(x) = x(4 - 3x)^3(2x + 1)^4(x^3 - 1)$

Appendix C

Selected Topics from Linear Algebra

C.1 A Primer on Matrix Algebra

This section is a concise introduction to matrix algebra. We assume no prior knowledge of matrices.

As a motivation, we consider a system of two first-order differential equations

$$\frac{dx}{dt} = -5x + 8y$$

$$\frac{dy}{dt} = -4x + 7y.$$

We want to find a compact way to write and ultimately solve this system. We can write the coefficients of the variables in a rectangular array. For reasons we will see shortly, the way we do this is

$$\begin{pmatrix} \dfrac{dx}{dt} \\ \dfrac{dy}{dt} \end{pmatrix} = \begin{pmatrix} -5 & 8 \\ -4 & 7 \end{pmatrix} \begin{pmatrix} x \\ y \end{pmatrix}.$$

We say that

$$\begin{pmatrix} \dfrac{dx}{dt} \\ \dfrac{dy}{dt} \end{pmatrix} \quad \text{and} \quad \begin{pmatrix} x \\ y \end{pmatrix}$$

are *vectors* and

$$\begin{pmatrix} -5 & 8 \\ -4 & 7 \end{pmatrix}$$

is the *coefficient matrix* of our original system of equations. If we are not dealing with differential equations, the left-hand side may be constant but we can still write our system in the form of vectors and matrices.

Simply put, a *matrix* is a rectangular array of numbers. The horizontal lists are called rows, and the vertical lists are called columns. For example, let

$$\mathbf{A} = \begin{pmatrix} 2 & 3 & 4 \\ 0 & 1 & -2 \end{pmatrix}.$$

This matrix has two rows: $\begin{pmatrix} 2 & 3 & 4 \end{pmatrix}$ and $\begin{pmatrix} 0 & 1 & -2 \end{pmatrix}$. It has three columns: $\begin{pmatrix} 2 \\ 0 \end{pmatrix}$, $\begin{pmatrix} 3 \\ 1 \end{pmatrix}$, and $\begin{pmatrix} 4 \\ -2 \end{pmatrix}$.

We refer to the size of a matrix as its number of rows by its number of columns. A 2×3 matrix (read "two by three matrix") is a matrix with two rows and three columns. Although we sometimes will talk abstractly about vectors, for now when we say *vector*, we mean the traditional understanding as a matrix with either one row or one column; a vector with only one row is called a *row vector*, and a vector with only one column is called a *column vector*. Our concepts of vector spaces (see Section 3.2.3) apply to these "vectors" as well.

We frequently need to refer to specific elements of a matrix, and we use the row and column information to provide an address for each element. For example, the ijth element is the element in the ith row and the jth column (rows are always listed before columns). Thus we denote a generic matrix by

$$\mathbf{A} = \begin{pmatrix} a_{11} & a_{12} & \cdots & a_{1n} \\ a_{21} & a_{22} & \cdots & a_{2n} \\ \vdots & \vdots & \ddots & \vdots \\ a_{m1} & a_{m2} & \cdots & a_{mn} \end{pmatrix}$$

or more succinctly by $\mathbf{A} = (a_{ij})_{m \times n}$. We call a matrix *square* if the number of rows is equal to the number of columns.

Matrices can be considered to be a number system much as the integers, the real numbers, or the complex numbers are number systems. As with other number systems we define ways to add, subtract, multiply, and divide matrices. Thus we construct an algebra for matrices which we call, appropriately enough, matrix algebra. We again note that a vector (in the traditional multivariable calculus context) is simply a matrix with only one column or one row and thus the statements about matrices apply to vectors as well. Two matrices can be added provided that they have the same size; thus two 2×3 matrices can be added together, but a 2×3 cannot be added to a 3×5 matrix. Addition and subtraction are defined elementwise. Thus $\mathbf{A} + \mathbf{B} = \mathbf{C}$ means that for all i and j, $c_{ij} = a_{ij} + b_{ij}$. Similarly $\mathbf{A} - \mathbf{B} = \mathbf{C}$ means $c_{ij} = a_{ij} - b_{ij}$. For example

$$\begin{pmatrix} a & b \\ c & d \end{pmatrix} + \begin{pmatrix} e & f \\ g & h \end{pmatrix} = \begin{pmatrix} a+e & b+f \\ c+g & d+h \end{pmatrix},$$

$$\begin{pmatrix} a & b \\ c & d \end{pmatrix} - \begin{pmatrix} e & f \\ g & h \end{pmatrix} = \begin{pmatrix} a-e & b-f \\ c-g & d-h \end{pmatrix}.$$

In algebra on the real numbers, 0 is a distinguished number called the additive identity. This means that for any real a, $a + 0 = a$ and $0 + a = a$. There is a matrix (for each size of matrix) playing this same role for matrix arithmetic. It is called the *zero matrix* and is the matrix consisting of all zeros, and denoted by $\mathbf{0}$. It is easy to verify that

$$\mathbf{A} + \mathbf{0} = \mathbf{0} + \mathbf{A} = \mathbf{A}$$

There are actually two number systems involved in matrix algebra: the matrices themselves and the elements that make up the matrices. We distinguish these by calling the former matrices and the latter *scalars*. Technically the 1×1 matrix (a) is a matrix, while a is a scalar. Whenever feasible, matrices are denoted by capital letters and scalars are denoted by lower case letters. We occasionally need to multiply scalars and matrices together. If s is a scalar and $\mathbf{A}$ is a matrix, we define a new matrix $s\mathbf{A}$ to be the matrix that results from multiplying all the entries of $\mathbf{A}$ by s. Thus if $\mathbf{B} = s\mathbf{A}$, then $b_{ij} = sa_{ij}$. For example,

$$s \begin{pmatrix} a & b \\ c & d \end{pmatrix} = \begin{pmatrix} sa & sb \\ sc & sd \end{pmatrix}.$$

Multiplication in matrix algebra is, in general, much stranger than multiplication in the real numbers. While the multiplication by a scalar that we just defined is similar to addition and subtraction of matrices, there is a major conceptual difference. Multiplication of a matrix by a scalar is the multiplication of two different types of objects, namely, scalars and matrices. It is useful to define a multiplication between two matrices. But first we consider the special case of multiplying a matrix and a vector. If we consider an $m \times n$ matrix, the vector we consider must be of size $n \times 1$. It is perhaps most consistent to consider the multiplication of a matrix $\mathbf{A}$ and a vector $\mathbf{x}$ as a *linear combination of the columns of* $\mathbf{A}$. For example,

$$\begin{pmatrix} 1 & 2 & 3 \\ 4 & 5 & 6 \\ 7 & 8 & 9 \end{pmatrix} \begin{pmatrix} 2 \\ 1 \\ -1 \end{pmatrix} = 2 \cdot \begin{pmatrix} 1 \\ 4 \\ 7 \end{pmatrix} + 1 \cdot \begin{pmatrix} 2 \\ 5 \\ 8 \end{pmatrix} - 1 \cdot \begin{pmatrix} 3 \\ 6 \\ 9 \end{pmatrix}$$
$$= \begin{pmatrix} 1 \\ 7 \\ 13 \end{pmatrix}.$$

In order to be able to use the rows of the vector $\mathbf{x}$ as coefficients used to take linear combinations of the columns of $\mathbf{A}$, it is essential that the number of columns of $\mathbf{A}$ be equal to the number of rows of $\mathbf{x}$. Thus, we have the rule

$$\mathbf{A}_{m \times n} \mathbf{x}_{n \times 1} = \mathbf{b}_{m \times 1}, \tag{C.1}$$

which governs the size of the matrices and vectors that we can multiply together. In general, we have

$$
\begin{pmatrix}
a_{11} & a_{12} & \cdots & a_{1n} \\
a_{21} & a_{22} & \cdots & a_{2n} \\
\vdots & \vdots & \ddots & \vdots \\
a_{m1} & a_{m2} & \cdots & a_{mn}
\end{pmatrix}
\begin{pmatrix}
x_1 \\ x_2 \\ \vdots \\ x_n
\end{pmatrix}
$$

$$
= x_1 \begin{pmatrix} a_{11} \\ a_{21} \\ \vdots \\ a_{m1} \end{pmatrix}
+ x_2 \begin{pmatrix} a_{12} \\ a_{22} \\ \vdots \\ a_{m2} \end{pmatrix}
+ \cdots + x_n \begin{pmatrix} a_{1n} \\ a_{2n} \\ \vdots \\ a_{mn} \end{pmatrix}
$$

$$
= \begin{pmatrix}
x_1 a_{11} + x_2 a_{12} + \cdots + x_n a_{1n} \\
x_1 a_{21} + x_2 a_{22} + \cdots + x_n a_{2n} \\
\vdots \\
x_1 a_{m1} + x_2 a_{m2} + \cdots + x_n a_{mn}
\end{pmatrix}
$$

$$
= \begin{pmatrix} b_1 \\ b_2 \\ \vdots \\ b_m \end{pmatrix}. \tag{C.2}
$$

This is simply an alternate way of writing this system of m equations in n unknowns and we refer to the matrix $\mathbf{A}$ in such situations as the *coefficient matrix* of the linear system.

When we consider multiplication of two matrices, $\mathbf{AB}$, we can easily think of doing matrix vector multiplication, with the vectors $\mathbf{x}$ of the previous matrix-vector multiplication now being the columns of $\mathbf{B}$. Let

$$
\mathbf{A} = (a_{ij})_{m \times n} \text{ and } \mathbf{B} = (b_{ij})_{n \times p}.
$$

We can think of the resulting columns of $\mathbf{C}_{m \times p} = \mathbf{AB}$ as

$$
c_j = b_{1j} \begin{pmatrix} a_{11} \\ a_{21} \\ \vdots \\ a_{m1} \end{pmatrix}
+ b_{2j} \begin{pmatrix} a_{12} \\ a_{22} \\ \vdots \\ a_{m2} \end{pmatrix}
+ \cdots + b_{nj} \begin{pmatrix} a_{1n} \\ a_{2n} \\ \vdots \\ a_{mn} \end{pmatrix}. \tag{C.3}
$$

In terms of multiplication by the elements of the matrices, each entry, c_{ij}, of the product $\mathbf{C} = \mathbf{AB}$ is defined by

$$
c_{ij} = a_{i1} b_{1j} + a_{i2} b_{2j} + \cdots + a_{ir} b_{rj}.
$$

In a sense what we have done is multiply the ith row of $\mathbf{A}$ by the jth column of $\mathbf{B}$. Here are a few examples.

$$
\begin{pmatrix} a & b \\ c & d \end{pmatrix}
\begin{pmatrix} e & f \\ g & h \end{pmatrix}
= \begin{pmatrix} ae + bg & af + bh \\ ce + dg & cf + dh \end{pmatrix}
$$

$$\begin{pmatrix} 1 & 2 \\ 3 & 4 \end{pmatrix} \begin{pmatrix} 0 & 1 \\ 1 & 0 \end{pmatrix} = \begin{pmatrix} 2 & 1 \\ 4 & 3 \end{pmatrix}$$

$$\begin{pmatrix} 0 & 1 \\ 1 & 0 \end{pmatrix} \begin{pmatrix} 1 & 2 \\ 3 & 4 \end{pmatrix} = \begin{pmatrix} 3 & 4 \\ 1 & 2 \end{pmatrix}$$

$$\begin{pmatrix} 1 & 3 \\ 2 & 1 \end{pmatrix} \begin{pmatrix} -1 & 3 & 2 \\ 2 & -3 & 4 \end{pmatrix} = \begin{pmatrix} 5 & -6 & 14 \\ 0 & 3 & 8 \end{pmatrix}.$$

From these examples several things are evident: 1) If you multiply an $m \times r$ matrix by an $r \times n$ matrix, the result is an $m \times n$ matrix. 2) You can multiply matrices of different sizes, but the number of columns of the matrix on the left must equal the number of rows of the matrix on the right. 3) In general $\mathbf{AB}$ and $\mathbf{BA}$ are different matrices if, indeed, they both exist. This is often phrased as "matrix multiplication is not commutative," which is different from multiplication of the real numbers so beginners should always use great care when manipulating matrices.

After you have practiced multiplying some matrices, you should become convinced that, while more complicated than real number multiplication, matrix multiplication is not very hard to do. Despite not being very hard, it is cumbersome and not very pleasant if the matrices are large and many need to be multiplied together. These two statements suggest that matrix multiplication would be ideal for a calculator or computer. Indeed many calculators and software packages are able to perform matrix multiplication, and we assume that you have access to one and have mastered its use from this point on.

In what follows, we frequently need to multiply a matrix by itself repeatedly. The analogous notation from real number multiplication carries through. The product $\mathbf{AA}$ is written $\mathbf{A}^2$, $\mathbf{AAA}$ is $\mathbf{A}^3$, and in general the product of n copies of A is A^n.

While dealing with notation for powers of matrices, we should mention the transpose operation. The *transpose* of a matrix is a new matrix which is the same as the old except the rows and columns have been switched. The transpose of an $m \times n$ matrix is then an $n \times m$ matrix. The notation for the transpose of $\mathbf{A}$ is $\mathbf{A}^T$. We always use a capital T for transpose.

Example 1: The transpose of the following two matrices is calculated here.

$$\begin{pmatrix} -1 & 3 & 2 \\ 2 & -3 & 4 \end{pmatrix}^T = \begin{pmatrix} -1 & 2 \\ 3 & -3 \\ 2 & 4 \end{pmatrix}$$

and

$$\begin{pmatrix} -2 & 3 & 1 \\ 5 & 1 & 4 \\ 2 & -3 & -4 \end{pmatrix}^T = \begin{pmatrix} -2 & 5 & 2 \\ 3 & 1 & -3 \\ 1 & 4 & -4 \end{pmatrix}.$$

We note that the transpose can also be useful to simplify the typesetting. When referring to a column vector, we usually write it in transposed form $(x_1 \ x_2 \ \dots \ x_n)^T$.

We can also define two numbers associated with a given square matrix. The *trace* of matrix

$$\text{Tr}(\mathbf{A}) = \begin{pmatrix} a & b \\ c & d \end{pmatrix} = a + d.$$

That is, the trace is defined as the sum of the (upper left to lower right) diagonal elements of a matrix. In terms of notation for an $n \times n$ matrix $\mathbf{A}$, we can write

$$\text{Tr}(\mathbf{A}) = \sum_{j=1}^{n} a_{jj}.$$

Further, there is a number associated with a matrix called its *determinant*. The determinant of a matrix is a very useful theoretical tool that comes up in many different places in this book. The determinant of a 2×2 matrix

$$\mathbf{A} = \begin{pmatrix} a & b \\ c & d \end{pmatrix}$$

is $ad - cb$. Often straight vertical lines are used to denote a determinant. Thus

$$\det(\mathbf{A}) = \begin{vmatrix} a & b \\ c & d \end{vmatrix} = ad - cb.$$

We present two methods of calculating the determinant of a 3×3 matrix. The first is easier to use but does not generalize to higher dimensions, while the latter is sometimes harder for the student to grasp but extends easily to larger square matrices.

For the first way, we write the first two columns of $\mathbf{A}$ as a 4th and 5th column:

$$\begin{array}{ccccc} a & b & c & a & b \\ d & e & f & d & e \\ g & h & i & g & h \end{array}.$$

We then proceed in an analogous way to the 2×2 case. That is, we begin at the top left, going down the diagonal and multiplying the entries. We repeat this for the next two entries in the first row. Then we subtract off the reverse diagonals beginning with the first row, 3rd column entry and moving to the next two:

Subtract these products $= -gec - hfa - idb$

Add these products $= +aei + bfg + cdh$

Then the formula for the 3×3 case is

$$\det \mathbf{A} = aei + bfg + cdh - gec - hfa - idb.$$

The second approach for a 3×3 matrix also holds for larger square matrices: *expansion-by-minors*. To expand on the first row, we take the entry in the $(1, 1)$ position and multiply it by the determinant of the matrix that remains when we cross out the first row and first column (called a *minor*) and subtract the entry in the $(1, 2)$ position times the determinant of the matrix obtained by crossing out the first row and second column, plus the $(1, 3)$ entry times the determinant of the matrix obtained by crossing out the first row and third column, and so on, alternating between adding and subtracting the next term. For example, if

$$\mathbf{A} = \begin{pmatrix} a & b & c \\ d & e & f \\ g & h & i \end{pmatrix},$$

then

$$\det(\mathbf{A}) = a \begin{vmatrix} e & f \\ h & i \end{vmatrix} - b \begin{vmatrix} d & f \\ g & i \end{vmatrix} + c \begin{vmatrix} d & e \\ g & h \end{vmatrix}$$
$$= aei - ahf - bdi + bgf + cdh - cge,$$

which is the same formula obtained with the other method of calculation. Observe that this method is a recursive definition since the determinant of a 4×4 matrix involves computing four determinants of 3×3 matrices, each of which requires the computation of three 2×2 matrices. Here we defined expansion on the first row. We can expand on any row or column and direct the reader to any introductory matrix algebra book for details or for other methods of computing a determinant.

Example 2: Find the determinant of each matrix. Use the first method to find $|B|$ and the second method to find $|C|$.

$$A = \begin{pmatrix} 1 & 3 \\ 0 & 2 \end{pmatrix}, \qquad B = \begin{pmatrix} 1 & 4 & 0 \\ -2 & 3 & 1 \\ 0 & 2 & 1 \end{pmatrix}, \qquad C = \begin{pmatrix} 1 & 2 & 3 \\ 2 & 3 & 1 \\ 3 & 5 & 4 \end{pmatrix}.$$

We apply the relevant definitions of the determinant from above:

$$|A| = 1 \cdot 2 - 0 \cdot 3 = -1,$$

$$\implies |B| = 3 + 0 + 0 - 0 - 2 - (-8) = 9$$

$$|C| = 1\begin{vmatrix} 3 & 1 \\ 5 & 4 \end{vmatrix} - 2\begin{vmatrix} 2 & 1 \\ 3 & 4 \end{vmatrix} + 3\begin{vmatrix} 2 & 3 \\ 3 & 5 \end{vmatrix}$$
$$= 1(12 - 5) - 2(8 - 3) + 3(10 - 9) = 0.$$

The last operation we consider at this time is division. Strictly speaking, division is undefined, but we define an equivalent. The motivation comes from high school (or earlier!) but is useful for putting the current discussion in its proper context. When working with the real numbers we have both 2 and $1/2$. We know these numbers are multiplicative inverses of each other because $2 \times 1/2 = 1$ and $1/2 \times 2 = 1$. In an abstract algebra course, we express the fact that $1/2$ is the multiplicative inverse of 2 by writing it as 2^{-1}. Similarly $2 = (1/2)^{-1}$. Here the symbol " -1" can be interpreted either as physically inverting the number, or as the abstract inverse, i.e., 2^{-1} is the number which when multiplied by 2 gives 1. If you have not thought about it before, notice that all division can be expressed as multiplying by a multiplicative inverse. This is the idea that is carried over into matrix multiplication. The symbol $\mathbf{I}$ will denote any square matrix (same number of rows as columns) with ones on the diagonal and zeros elsewhere and is called an *identity matrix*. For example, the 3×3 identity matrix is the matrix

$$\mathbf{I} = \begin{pmatrix} 1 & 0 & 0 \\ 0 & 1 & 0 \\ 0 & 0 & 1 \end{pmatrix}.$$

This matrix plays the role of "1" in matrix multiplication. You should verify that

$$\mathbf{AI} = \mathbf{A}$$

and

$$\mathbf{IA} = \mathbf{A}.$$

Because of this, the identity matrix is frequently called the multiplicative identity. The *inverse of a matrix* $\mathbf{A}$ is denoted by $\mathbf{A}^{-1}$. The matrix $\mathbf{A}^{-1}$ is defined to be the matrix such that $\mathbf{AA}^{-1} = \mathbf{I}$ and $\mathbf{A}^{-1}\mathbf{A} = \mathbf{I}$ if it exists. If you followed the discussion about 2 and $1/2$, then you should see this as exactly analogous. The existence of $\mathbf{A}^{-1}$ is an issue. By the way $\mathbf{A}^{-1}$ is defined, it needs to be multiplied by $\mathbf{A}$ on both sides, so $\mathbf{A}$ has to be a square matrix for $\mathbf{A}^{-1}$ to exist. A definition and theorem help us at this point.

DEFINITION C.1   *Let* $\mathbf{A}$ *be an* $n \times n$ *matrix.*
a. If $det(\mathbf{A}) \neq 0$, *then* $\mathbf{A}$ *is said to be nonsingular (invertible).*
b. If $det(\mathbf{A}) = 0$, *then* $\mathbf{A}$ *is said to be singular (noninvertible).*

Although we mentioned determinants at an earlier point, they tie in nicely here.

THEOREM C.1.1 *Suppose* $\mathbf{A}$ *is an* $n \times n$ *matrix. Then* $\mathbf{A}^{-1}$ *exists if and only if* $\mathbf{A}$ *is nonsingular.*

Determining if the inverse of a matrix exists is a straightforward, tedious calculation. Finding the inverse of a matrix is a non-trivial, though also completely algorithmic, process. We put the discussion of techniques for finding one by hand in Appendix C.2. As with matrix multiplication, computing inverses is not usually done by hand, and the packages of Matlab, Maple, and Mathematica are quite capable of doing these computations. See Appendix C.1.1 for a discussion of how to compute matrix inverses with the computer.

Example 3: Determine whether each matrix is singular or nonsingular:

$$\mathbf{A} = \begin{pmatrix} 1 & 3 \\ 0 & 2 \end{pmatrix}, \qquad \mathbf{B} = \begin{pmatrix} 1 & 4 & 0 \\ -2 & 3 & 1 \\ 0 & 2 & 1 \end{pmatrix}, \qquad \mathbf{C} = \begin{pmatrix} 1 & 2 & 3 \\ 2 & 3 & 1 \\ 3 & 5 & 4 \end{pmatrix}.$$

From the previous example, we calculated

$$|\mathbf{A}| = -1, \quad |\mathbf{B}| = 9, \quad |\mathbf{C}| = 0$$

and thus Definition C.1 lets us conclude that $\mathbf{A}, \mathbf{B}$ are nonsingular whereas $\mathbf{C}$ is singular.

We will learn how to calculate inverses in Section C.2 but we give the inverses here to illustrate their use. In the previous example, the inverses of $\mathbf{A}$ and $\mathbf{B}$ are

$$\mathbf{A}^{-1} = \frac{1}{2} \begin{pmatrix} 2 & -3 \\ 0 & 1 \end{pmatrix}, \qquad \mathbf{B}^{-1} = \frac{1}{9} \begin{pmatrix} 1 & -4 & 4 \\ 2 & 1 & -1 \\ -4 & -2 & 11 \end{pmatrix}.$$

Multiplication shows that $\mathbf{A}^{-1}\mathbf{A} = \mathbf{I} = \mathbf{A}\mathbf{A}^{-1}$:

$$\mathbf{A}^{-1}\mathbf{A} = \frac{1}{2} \begin{pmatrix} 2 & -3 \\ 0 & 1 \end{pmatrix} \begin{pmatrix} 1 & 3 \\ 0 & 2 \end{pmatrix}$$

$$= \frac{1}{2} \begin{pmatrix} 2 & 0 \\ 0 & 2 \end{pmatrix}$$

$$= \begin{pmatrix} 1 & 0 \\ 0 & 1 \end{pmatrix}$$

and similarly

$$\mathbf{A}\mathbf{A}^{-1} = \begin{pmatrix} 1 & 3 \\ 0 & 2 \end{pmatrix} \frac{1}{2} \begin{pmatrix} 2 & -3 \\ 0 & 1 \end{pmatrix}$$
$$= \frac{1}{2} \begin{pmatrix} 2 & 0 \\ 0 & 2 \end{pmatrix}$$
$$= \begin{pmatrix} 1 & 0 \\ 0 & 1 \end{pmatrix}.$$

We can similarly show that $\mathbf{B}^{-1}$ is indeed the inverse of $\mathbf{B}$.

Matrices can also be used to succinctly write a system of linear equations. For example,

$$a_{11}x_1 + a_{12}x_2 + \cdots + a_{1n}x_n = b_1$$
$$a_{21}x_1 + a_{22}x_2 + \cdots + a_{2n}x_n = b_2$$
$$\vdots \qquad \vdots \qquad \ddots \qquad \vdots \qquad \vdots \qquad (C.4)$$
$$a_{n1}x_1 + a_{n2}x_2 + \cdots + a_{nn}x_n = b_n$$

can be written

$$\mathbf{A}\mathbf{x} = \mathbf{b},$$

where

$$\mathbf{A} = \begin{pmatrix} a_{11} & a_{12} & \cdots & a_{1n} \\ a_{21} & a_{22} & \cdots & a_{2n} \\ \vdots & \vdots & \ddots & \vdots \\ a_{n1} & a_{n2} & \cdots & a_{nn} \end{pmatrix}, \quad \mathbf{x} = \begin{pmatrix} x_1 \\ x_2 \\ \vdots \\ x_n \end{pmatrix}, \quad \text{and } \mathbf{b} = \begin{pmatrix} b_1 \\ b_2 \\ \vdots \\ b_n \end{pmatrix}.$$
$$(C.5)$$

We define $\mathbb{R}^n$ as the set of all vectors with n real entries and thus $\mathbf{x}$ is a vector in $\mathbb{R}^n$. It is of great interest to find a vector $\mathbf{x}$ that will solve the system given a matrix $\mathbf{A}$ and a vector $\mathbf{b}$. In terms of theory, the solution will exist and be unique when $\mathbf{A}^{-1}$ exists and will be given by

$$\mathbf{x} = \mathbf{A}^{-1}\mathbf{b}.$$

Example 4: Find the unique solution of the system

$$x + 3y = -1$$
$$2y = 4$$

by writing the system in matrix-vector form and computing $\mathbf{A}^{-1}\mathbf{b}$.

We can rewrite this system in its equivalent matrix-vector form:

$$\begin{pmatrix} 1 & 3 \\ 0 & 2 \end{pmatrix} \begin{pmatrix} x \\ y \end{pmatrix} = \begin{pmatrix} -1 \\ 4 \end{pmatrix}.$$

We saw previously that $\mathbf{A}^{-1}$ exists and is given by

$$\mathbf{A}^{-1} = \frac{1}{2}\begin{pmatrix} 2 & -3 \\ 0 & 1 \end{pmatrix}.$$

Thus the unique solution is given by

$$\mathbf{x} = \mathbf{A}^{-1}\mathbf{b} = \frac{1}{2}\begin{pmatrix} 2 & -3 \\ 0 & 1 \end{pmatrix}\begin{pmatrix} -1 \\ 4 \end{pmatrix} = \begin{pmatrix} -7 \\ 2 \end{pmatrix}.$$

The reader may question why we would solve the system this way when we could just solve the last equation for y and substitute it into the other. The reason lies in the insight that we will be able to gain by considering the system in matrix-vector form. We will be able to deal with significantly more difficult systems with this same method and solving one equation for a variable and substituting it into another quickly becomes a clumsy way (at best) to solve the system.

Solutions may exist, however, even when $\mathbf{A}^{-1}$ does not exist. It is **very** computationally expensive to calculate the inverse of a matrix and this is never how a vector $\mathbf{x}$ is found. But writing the solution in this way is very useful for conceptually understanding what we are trying to accomplish.

We will conclude our brief introduction with an important theorem:

THEOREM C.1.2 *The following are equivalent characterizations of the* $n \times n$ *matrix* $\mathbf{A}$*:*
a) $\mathbf{A}$ *is invertible.*
b) *The system* $\mathbf{Ax} = \mathbf{b}$ *has a unique solution* $\mathbf{x}$ *for all* $\mathbf{b}$ *in* $\mathbb{R}^n$*.*
c) *The system* $\mathbf{Ax} = \mathbf{0}$ *has* $\mathbf{x} = \mathbf{0}$ *as its unique solution.*
d) $\det(\mathbf{A}) \neq 0$*.*

Although we leave the mechanics of calculating an inverse to Appendix C.2, we know whether or not an inverse exists simply by calculating its determinant.

C.1.1 Linear Algebra with Matlab, Maple, and Mathematica

All three programs are very powerful tools of computing in linear algebra. Although Matlab has packages which can symbolically manipulate matrices, we focus our attention on the numerical examples as this is the area where it excels. We give the following code as an example of some commands.

Computer Code C.1:   Basic linear algebra commands for manipulating matrices and vectors

Matlab, Maple, Mathematica

Matlab
```
>>  clear all
>>  A=[1 3 5;1 7 9;1 2 3] %creates the matrix A
>>  A+A
>>  2*A %same result as A+A
>>  A*A
>>  A^2 %same result as A*A
>>  B=inv(A) %calculates the inverse of A
>>  B*A %verifies that B is the inverse of A
>>  A' %transpose of A
>>  trace(A)
>>  det(A)
>>  x=[8 17 30] %a row vector
>>  x=[8; 17; 30] %a column vector
>>  A*x %matrix-vector multiplication, note the
    %dimensions must match
>>  x'*A %vector-matrix multiplication, note the
    %dimensions must match
>>  b=[0; 1; 3] %a column vector
>>  y=A\b %solution of the system Ay=b
>>  A=[A;2 4 6; 8 10 12] %appends two rows as the 4th
    %and 5th rows of A
>>  A(:,4)=[1;2;3;4;5] %appends one column as the 4th
    %column of A
>>  A(:,3) %selects the 3rd column of A
>>  A(4,:)  %selects the 4th row of A
>>  A(2:4,3:4) %selects the submatrix of A consisting of
    %the elements in rows 2 through 4 and columns 3 through 4
```

Maple
```
> restart;
> with(linalg): #loads the linalg package for linear
    #algebra; note that this is different from the
    #LinearAlgebra package
> A1:=matrix(2,2,[a,b,c,d]); #creates a general 2 × 2 matrix
> B1:=inverse(A1); #calculates the inverse of A1
> eq1:=multiply(A1,B1);#attempts to verify that B1
    #is the inverse of A1
#CODE CONTINUED IN NEXT BOX
```

```
                            Maple

#CODE CONTINUED FROM PREVIOUS BOX
#(Note:  linalg package already loaded)

> eq2:=simplify(eq1); #simplifies the above answer, if
  #necessary
> eq3:=subs(a=3,b=1,c=4,d=1,evalm(A1)); #evalm tells Maple
  #to execute a command on a matrix
> A:=matrix(3,3,[1,3,5,1,7,9,1,2,3]); #creates the matrix A
> evalm(A+A);
> evalm(2*A); #same result as evalm(A+A)
> evalm(A*A); # multiply(A,A) would have also done the same
> evalm(A^2); #same result as evalm(A*A)
> B:=inverse(A); #calculates the inverse of A
> evalm(B&*A); #verifies that B is the inverse of A
> multiply(B,A); #verifies that B is the inverse of A
> transpose(A); #transpose of A
> trace(A); #trace of A
> det(A); #determinant of A
> x:=matrix(1,3,[8,17,30]); #a row "vector"
> x:=matrix(3,1,[8,17,30]); #a column "vector"
> evalm(A&*x); #matrix-matrix multiplication, note the
  #dimensions must match
> evalm(transpose(x)&*A); #matrix-matrix multiplication;
  #dimensions match
> b:=matrix(3,1,[0, 1, 3]); #a column "vector"
> y:=linsolve(A,b) #solution of the system Ay=b
> evalm(A); #shows matrix A again
> col(A,3); #selects 3rd column of A; the result is a vector
> row(A,2); #selects 2nd row of A; result is again a vector
> A:=matrix([row(A,1),row(A,2),row(A,3),evalm([2,4,6]),
  evalm([8,10,12])]); #appends two rows as the 4th
  #and 5th rows of A
> A:=transpose(matrix([col(A,1),col(A,2),col(A,3),
  evalm([1,2,3,4,5])])); #appends one col as 4th col of A
> delcols(A,1..2); #selects the last two columns by deleting
  #the other ones; the result is still a matrix
> delrows(A,1..3); #selects the last row by deleting the
  #other rows; the result is still a matrix
> submatrix(A,2..4,3..4);#selects the submatrix of A
  #consisting of the elements in rows 2 through 4 and
  #columns 3 through 4
```

| Mathematica |

```
<<LinearAlgebra'MatrixManipulation'
A={{1,3,5},{1,7,9},{1,2,3}}
A//MatrixForm (*displays matrix in a nicer visual form*)
MatrixForm[A] (*alternate syntax*)
A+A//MatrixForm (*displays matrix in a nicer visual form*)
2 A//MatrixForm (*same result as A+A*)
A.A//MatrixForm (*displays matrix in a nicer visual form*)
MatrixPower[A,2]//MatrixForm (*same result as A.A*)
B=Inverse[A] (*calculates the inverse of A*)
B//MatrixForm (*displays matrix in a nicer visual form*)
B.A//MatrixForm (*verifies that B is the inverse of A*)
Transpose[A]//MatrixForm (*transpose of A*)
Tr[A]
Det[A]
X = {{8, 17, 30}} (*a row vector*)
X//MatrixForm (*displays vector in a nicer visual form*)
X = {{8}, {17}, {30}} (*column vector*)
X//MatrixForm (*displays vector in a nicer visual form*)
A.X//MatrixForm (*matrix-vector multiplication, note the*)
   (*dimensions must match*)
Transpose[X].A//MatrixForm (*vector-matrix multiplication,*)
   (*note the dimensions must match*)
b={{0},{1},{3}} (*a column vector*)
b//MatrixForm (*displays in a nicer visual form*)
y=LinearSolve[A,b] (*solution of Ay=b*)
  (*Note that to append rows at the bottom of a matrix,*)
  (*you must append entries at the end of the columns*)
A=AppendColumns[A,{{2,4,6}},{{8,10,12}}] (*appends two rows*)
   (*as the 4th and 5th rows of A*)
A//MatrixForm (*nicer visual form*)
A=AppendRows[A,{{1},{2},{3},{4},{5}}] (*appends one column*)
   (*by placing one additional entry at the end of each row*)
A//MatrixForm (*nicer visual form*)
TakeRows[A,{4,4}]//MatrixForm (*selects the 3rd column of A*)
TakeColumns[A,{3,3}]//MatrixForm (*selects the 4th row of A*)
SubMatrix[A,{2,3},{2,2}]//MatrixForm (*selects the*)
   (*submatrix of A consisting of the elements in rows 2*)
   (*through 4 and columns 3 through 4*)
```

Problems

1. Write the following systems using vectors and matrices:

   a. $\begin{cases} 3x + 5y = 7 \\ x + y = -1 \end{cases}$
   b. $\begin{cases} 2x + 5y - z = 2 \\ -x + 2y = -1 \\ x + 4z = 0 \end{cases}$
   c. $\begin{cases} x' = x - 5y \\ y' = 2x + y \end{cases}$.

2. Calculate (by hand) $\mathbf{AB}$ and $\mathbf{BA}$ for the following matrices:

$$\mathbf{A} = \begin{pmatrix} 2 & -3 \\ -1 & 4 \end{pmatrix}, \ \mathbf{B} = \begin{pmatrix} -2 & 2 \\ 1 & -5 \end{pmatrix}$$

3. Calculate (by hand) $\mathbf{AB}$ and $\mathbf{BA}$ for the following matrices:

$$\mathbf{A} = \begin{pmatrix} 1 & 2 & -3 \\ 0 & -1 & 4 \\ 3 & 0 & 2 \end{pmatrix}, \ \mathbf{B} = \begin{pmatrix} 3 & -2 & 2 \\ -1 & 1 & -5 \\ 0 & 2 & 2 \end{pmatrix}$$

In problems 4–11, consider the following matrices:

$$\mathbf{A} = \begin{pmatrix} 3 & 0 \\ -1 & 2 \\ 1 & 1 \end{pmatrix}, \ \mathbf{B} = \begin{pmatrix} 4 & -1 \\ 0 & 2 \end{pmatrix}, \ \mathbf{C} = \begin{pmatrix} 1 & 4 & -2 \\ 6 & 3 & 1 \end{pmatrix}$$

$$\mathbf{D} = \begin{pmatrix} 1 & 5 & 2 \\ -1 & 0 & 3 \\ 4 & 2 & 1 \end{pmatrix}, \ \mathbf{E} = \begin{pmatrix} 1 & -3 & 4 \\ 5 & 1 & 1 \\ 3 & 2 & 1 \end{pmatrix}$$

Compute the following expressions, when possible. If it is not possible, state why it is not possible.

4. a. $\mathbf{D} + \mathbf{E}$ b. $\mathbf{D} - \mathbf{E}$ c. $2\mathbf{D} + 3\mathbf{E}$ d. $2\mathbf{A}$

5. a. $\mathbf{B} + \mathbf{C}$ b. $4\mathbf{E} - 2\mathbf{D}$ c. $\text{Tr}(\mathbf{E})$ d. $\text{Tr}(\mathbf{B})$

6. a. $\text{Tr}(3\mathbf{D})$ b. $\text{Tr}(\mathbf{D} + \mathbf{C})$ c. $\mathbf{A}^T$ d. $\mathbf{D}^T$

7. a. $\mathbf{D}^T + \mathbf{E}^T$ b. $\mathbf{A}^2$ c. $\mathbf{AC}$ d. $\mathbf{DE}$

8. a. $\mathbf{ED}$ b. $\mathbf{AB}$ c. $\mathbf{BA}$ d. $\mathbf{BA}^T$

9. a. $\mathbf{EA}$ b. $\mathbf{AE}$ c. $\mathbf{A}^T\mathbf{E}$ d. $\mathbf{ABC}$

10. a. $\det(\mathbf{B})$ b. $\det(\mathbf{D})$ c. $\det(\mathbf{E})$ d. $\det(\mathbf{B}^2)$

11. a. $\det(\mathbf{DE})$ b. $\det(\mathbf{ED})$ c. $\det(\mathbf{E})\det(\mathbf{D})$ d. $\det(\mathbf{D}^T)$

C.2   Gaussian Elimination, Matrix Inverses, and Cramer's Rule

This section of Appendix C is essential if the course is to be used as a one-quarter course that combines linear algebra and differential equations or if the reader seeks a deeper understanding of the theory of differential equations as seen in Chapters 3 through 5. If neither of these applies, the reader may skip this material unless a brief review of Cramer's rule (used for Section 4.7 on Variation of Parameters) is needed and this part of the current section is self-contained.

We focus on a system of linear equations:

$$
\begin{aligned}
a_{11}x_1 + a_{12}x_2 + \cdots + a_{1n}x_n &= b_1 \\
a_{21}x_1 + a_{22}x_2 + \cdots + a_{2n}x_n &= b_2 \\
\vdots \qquad \vdots \qquad \ddots \qquad \vdots \quad &\quad \vdots \\
a_{m1}x_1 + a_{m2}x_2 + \cdots + a_{mn}x_n, &= b_m.
\end{aligned}
\tag{C.6}
$$

We can use our matrix and vector notation to write

$$
\mathbf{A}\mathbf{x} = \mathbf{b},
$$

where

$$
\mathbf{A} = \begin{pmatrix} a_{11} & a_{12} & \cdots & a_{1n} \\ a_{21} & a_{22} & \cdots & a_{2n} \\ \vdots & \vdots & \ddots & \vdots \\ a_{m1} & a_{m2} & \cdots & a_{mn} \end{pmatrix}, \quad \mathbf{x} = \begin{pmatrix} x_1 \\ x_2 \\ \vdots \\ x_n \end{pmatrix}, \quad \text{and} \ \mathbf{b} = \begin{pmatrix} b_1 \\ b_2 \\ \vdots \\ b_m \end{pmatrix}.
\tag{C.7}
$$

We will typically be given a vector $\mathbf{b}$ in $\mathbb{R}^m$ and a matrix $\mathbf{A}$ in $\mathbb{R}^{m \times n}$ (that is, an $m \times n$ matrix with real entries). Our goal will be to find a vector $\mathbf{x}$ in $\mathbb{R}^n$ that will make the equation $\mathbf{A}\mathbf{x} = \mathbf{b}$ a true statement.

We make note of two areas of active research at this point. Many times, a mathematical model gives rise to a system that has more variables than equations, thus leading to an $m \times n$ matrix where $m < n$ and both are large, say $50,000 \times 100,000$. Although the methods we learn in linear algebra apply, these kinds of matrices are usually best dealt with in the subject of *optimization*. Many other mathematical models of physical systems give rise to very large *square* systems, on the order of a matrix $\mathbf{A}$ that is $100,000 \times 100,000$. Although we will learn how to solve $\mathbf{A}\mathbf{x} = \mathbf{b}$ by hand in the section, we clearly would not want to solve that large of a system by hand. What do we do? Interested students should take a course in *numerical linear algebra* and explore the wealth of algorithms that efficiently give us our desired solution. There is a tremendous number of unsolved problems in these areas and researchers (maybe even your professors!) are actively seeking solutions today.

C.2.1   Gaussian Elimination

That is sufficient motivation for us and we will concentrate on understanding how to solve the equation $\mathbf{A}\mathbf{x} = \mathbf{b}$ by hand, or with a little help from the computer. We will stay small, probably not bigger than 5×5, and our goal will be to understand how a solution is obtained. Let us begin with a simple example.

Example 1: Solve the given equations simultaneously by using multiples of each equation to reduce the system:

$$
\begin{aligned}
-2x \quad\;\; + 4z &= 2 \quad\; \text{(eq1)} \\
4x + 2y - 4z &= -2 \quad \text{(eq2)} \\
2x + 4y + 2z &= -2 \quad \text{(eq3)}
\end{aligned}
\tag{C.8}
$$

to a solvable form.

We want to systematically add and subtract equations from each other to achieve this. It is extremely important to remember that multiplying both sides of an equation by a (non-zero) number does not change the solution just as adding two equations together does not change the solution either. Thus performing the operations below gives us a simpler system to deal with that has the *same solution set* as the original system. Let us fix the first equation and subtract multiples of it from the other two to try to eliminate x. If we multiply 2 times the first equation and add it to the second, we will eliminate the x variable in that equation. Likewise, if we add equations 1 and 3, we will eliminate the x from the third equation. Doing so gives

$$
\begin{aligned}
-2x \quad\;\; + 4z &= 2 & \text{(C.9)} \\
2y + 4z &= 2 \quad (2\text{eq1} + \text{eq2} \to \text{eq2a}) & \text{(C.10)} \\
4y + 6z &= 0. \quad (\text{eq1} + \text{eq3} \to \text{eq3a}) & \text{(C.11)}
\end{aligned}
$$

If we now combine the new second and third equations in a similar way, we will obtain a form that we can easily solve. Thus we multiply the new second equation (eq2a) times -2 and add it to the new third equation (eq3a) and put it in the place of this last equation (now call it eq3b):

$$
\begin{aligned}
-2x \quad\;\; + 4z &= 2 & \text{(C.12)} \\
2y + 4z &= 2 & \text{(C.13)} \\
-2z &= -4. \quad (-2\text{eq2a} + \text{eq3a} \to \text{eq3b}) & \text{(C.14)}
\end{aligned}
$$

We now use *back substitution* to solve the system, starting with the last row and working our way up. Solving the last equation gives us $z = 2$. We substitute this into the second equation and then see that $y = -3$. Substituting into the first equation gives $x = 3$.

We want to reexamine this last example, but now from the point of view of matrices and vectors. We can write the *coefficient matrix* for our system

(C.8) as

$$\mathbf{A} = \begin{pmatrix} -2 & 0 & 4 \\ 4 & 2 & -4 \\ 2 & 4 & 2 \end{pmatrix}. \tag{C.15}$$

The right-hand side of (C.8) is our vector $\mathbf{b}$:

$$\mathbf{b} = \begin{pmatrix} 2 \\ -2 \\ -2 \end{pmatrix}. \tag{C.16}$$

We can then write $\mathbf{Ax} = \mathbf{b}$ and our goal is to find $\mathbf{x}$. To do this, we insert $\mathbf{b}$ as the *fourth* column of $\mathbf{A}$. The resulting matrix is called the augmented matrix:

$$\begin{pmatrix} -2 & 0 & 4 & 2 \\ 4 & 2 & -4 & -2 \\ 2 & 4 & 2 & -2 \end{pmatrix}. \tag{C.17}$$

We then proceed to reduce the augmented matrix, just as we did with the system of equations. We let R_1, R_2, R_3 denote rows 1, 2, and 3, respectively, of our augmented matrix. We will use the notation $2R_1 + R_2 \rightarrow R_2$ to mean that we multiply 2 times row 1, then add it to row 2, and then put the result in row 2 and this new row 2 will now be called R_2. Then beginning with (C.17)

$$2R_1 + R_2 \rightarrow R_2 \ \text{ gives } \ \begin{pmatrix} -2 & 0 & 4 & 2 \\ 0 & 2 & 4 & 2 \\ 2 & 4 & 2 & -2 \end{pmatrix} \tag{C.18}$$

$$R_1 + R_3 \rightarrow R_3 \ \text{ gives } \ \begin{pmatrix} -2 & 0 & 4 & 2 \\ 0 & 2 & 4 & 2 \\ 0 & 4 & 6 & 0 \end{pmatrix} \tag{C.19}$$

$$-2R_2 + R_3 \rightarrow R_3 \ \text{ gives } \ \begin{pmatrix} -2 & 0 & 4 & 2 \\ 0 & 2 & 4 & 2 \\ 0 & 0 & -2 & -4 \end{pmatrix} \tag{C.20}$$

$$\begin{matrix} \frac{-1}{2}R_1 \rightarrow R_1, \ \frac{1}{2}R_2 \rightarrow R_2, \\ \frac{-1}{2}R_3 \rightarrow R_3 \ \text{ gives} \end{matrix} \ \begin{pmatrix} 1 & 0 & -2 & -1 \\ 0 & 1 & 2 & 1 \\ 0 & 0 & 1 & 2 \end{pmatrix}. \tag{C.21}$$

With this last matrix, we can again use back substitution and calculate the solutions for x, y, z. We mentioned that the solution set is not changed if we (i) multiply a row by a constant or (ii) add a multiple of one row to another. The solution set is also not changed if we (iii) *interchange* two rows. These three operations are called *elementary row operations* and are used to reduce the original system to one that is easier to solve.

This process of zeroing out entries in our matrix, ultimately ending up with only zeros below the main diagonal of the matrix (i.e., an upper triangular matrix), is called *Gaussian elimination*. The element in the matrix that we use to "zero out" everything below it is called a *pivot*. Both (C.20) and (C.21) are upper triangular matrices, the only difference being that (C.21) has the number one as the first non-zero entry of each row. A matrix of this form is said to be in *row-echelon form*. In the event that some rows would end up with all zeros, row-echelon form requires that we move these rows to the end and keep the leading entry in this staircase layout. This should be believable because it is equivalent to switching the physical location of two equations and why should this change the solution? (It shouldn't.)

In the case when a unique solution exists, it is also possible to obtain the final answer by continuing the process from (C.21) instead of doing back substitution. In this situation, we reduce the row-echelon matrix to one that has only ones on the main diagonal, with zeros occurring in positions above and below the main diagonal in each column. The last row remains unchanged. The next step in this process, known as *Gauss-Jordan elimination*, is to work from left to right, eliminating all entries in each column that is not on the main diagonal. Normally, we would need to eliminate the entry in the first row and second column of the row-echelon matrix. But (C.21) already has this done and we thus proceed to the third column.

$$-2R_3 + R_2 \to R_2 \text{ gives } \begin{pmatrix} 1 & 0 & -2 & -1 \\ 0 & 1 & 2 & 1 \\ 0 & 0 & 1 & 2 \end{pmatrix} \tag{C.22}$$

$$-2R_3 + R_2 \to R_2 \text{ gives } \begin{pmatrix} 1 & 0 & -2 & -1 \\ 0 & 1 & 0 & -3 \\ 0 & 0 & 1 & 2 \end{pmatrix} \tag{C.23}$$

$$2R_3 + R_1 \to R_1 \text{ gives } \begin{pmatrix} 1 & 0 & 0 & 3 \\ 0 & 1 & 0 & -3 \\ 0 & 0 & 1 & 2 \end{pmatrix}. \tag{C.24}$$

Now the final answer, $z = 2, y = -3, x = 3$, can be immediately seen because (C.24) is in *reduced-row echelon form*.

This example was one in which we have a *unique solution* but we will not always have this situation. Besides this scenario, we could also have *infinite solutions* or *no solution*. If we have either a unique solution or infinite solutions, we say the system of equations is *consistent*, whereas if there is no solution we say the system of equations is *inconsistent*. In terms of two variables, consider the following equations and their graphs in Figure C.1.

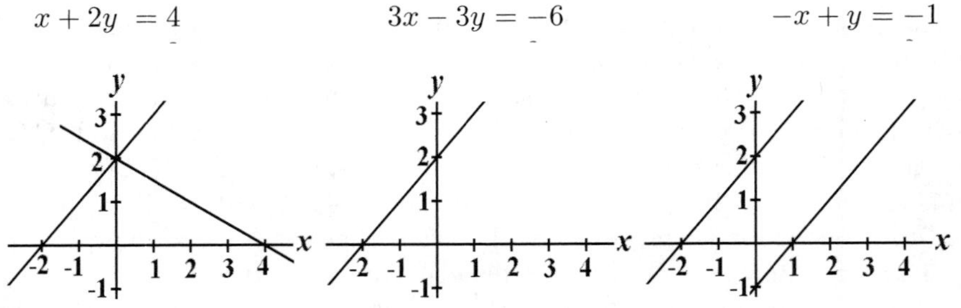

$$x + 2y = 4 \qquad\qquad 3x - 3y = -6 \qquad\qquad -x + y = -1$$

(a)   (b)   (c)

FIGURE C.1:   (a) Consistent, unique solution; (b) consistent, infinite solutions; (c) inconsistent, no solution.

For two variables, the geometric interpretation of unique, infinite, or no
solutions is easily seen. In higher dimensions where the number of equations
is the same as the number of unknown variables, we are also concerned with
finding the intersection of lines, but it is just harder to graph. In performing
Gaussian elimination, there may be times when we get "stuck" at a certain
point in the problem. Sometimes the problem can be overcome by interchang-
ing rows while other times it cannot. If we really are stuck at a certain point
then we will either have infinite solutions or no solution.

Consider the following system written as augmented matrices and observe
the two elementary row operations needed for the first two steps:

$$\begin{pmatrix} 1 & 2 & 1 & 4 \\ 2 & 4 & 1 & 2 \\ -1 & 1 & 2 & 5 \end{pmatrix} \begin{array}{c} -2R_1 + R_2 \to R_2 \\ R_1 + R_3 \to R_3, \end{array} \begin{pmatrix} 1 & 2 & 1 & 4 \\ 0 & 0 & -1 & -6 \\ 0 & 3 & 3 & 9 \end{pmatrix}. \qquad (C.25)$$

We cannot eliminate the '3' in the last row as our next step because the pivot
above it is zero. However, we can simply swap the second and third rows:

$$R_2 \leftrightarrow R_3 \begin{pmatrix} 1 & 2 & 1 & 4 \\ 0 & 3 & 3 & 9 \\ 0 & 0 & -1 & -6 \end{pmatrix}. \qquad (C.26)$$

This takes care of our problem and gives us an upper triangular matrix that
can be solved by back substitution and gives $x = 4, y = -3, z = 6$, and our
solution is unique.

Let us instead consider the following system and first two elementary row
operations needed:

$$\begin{pmatrix} 1 & 2 & 1 & 4 \\ -1 & 1 & 2 & 1 \\ 4 & 2 & -2 & -2 \end{pmatrix} \begin{array}{c} R_1 + R_2 \to R_2 \\ -2R_1 + R_3 \to R_3, \end{array} \begin{pmatrix} 1 & 2 & 1 & 4 \\ 0 & 3 & 3 & 9 \\ 0 & -6 & -6 & -18 \end{pmatrix}. \qquad (C.27)$$

The next step is to eliminate the '−6' in the last row and we do so with

$$2R_2 + R_3 \to R_3 \quad \begin{pmatrix} 1 & 2 & 1 & 4 \\ 0 & 3 & 3 & 9 \\ 0 & 0 & 0 & 0 \end{pmatrix}. \tag{C.28}$$

The last row drops out of the system and we simply need to solve the first two equations. This yields an infinite number of solutions: $x = -2 + z, y = 3 - z$ with z any real number.

Finally, we consider the following system and first two elementary row operations needed:

$$\begin{pmatrix} 1 & 2 & 1 & 4 \\ -1 & 1 & 2 & 5 \\ 2 & -2 & -4 & -2 \end{pmatrix} \begin{matrix} R_1 + R_2 \to R_2 \\ -2R_1 + R_3 \to R_3, \end{matrix} \begin{pmatrix} 1 & 2 & 1 & 4 \\ 0 & 3 & 3 & 9 \\ 0 & -6 & -6 & -10 \end{pmatrix}. \tag{C.29}$$

The next step is to eliminate the '−6' in the last row and we do so with

$$R_2 + R_3 \to R_3 \quad \begin{pmatrix} 1 & 2 & 1 & 4 \\ 0 & 3 & 3 & 9 \\ 0 & 0 & 0 & 8 \end{pmatrix}. \tag{C.30}$$

The last row doesn't make any sense and we thus say that the original system in not consistent and there is no solution.

C.2.2 Matrix Inverse

Equipped with the elementary row operations, we can now find the inverse of a matrix. How? Let us think what Gaussian elimination accomplished for us. We began with

$$\mathbf{Ax} = \mathbf{b}$$

and reduced this to a system

$$\mathbf{Ux} = \mathbf{b_u}$$

where $\mathbf{U}$ is an upper triangular matrix and the $\mathbf{b_u}$ denotes the vector $\mathbf{b}$ after we have performed the same elementary row operations that gave us $\mathbf{U}$. Our answer is the vector $\mathbf{x}$ and we can also write the answer (if it exists, of course) as

$$\mathbf{x} = \mathbf{A}^{-1}\mathbf{b} \quad \text{or equivalently} \quad \mathbf{x} = \mathbf{U}^{-1}\mathbf{b_u}.$$

In a previous example, we considered $\mathbf{Ax} = \mathbf{b}$, where

$$\mathbf{A} = \begin{pmatrix} -2 & 0 & 4 \\ 4 & 2 & -4 \\ 2 & 4 & 2 \end{pmatrix}, \quad \mathbf{b} = \begin{pmatrix} 2 \\ -2 \\ -2 \end{pmatrix}.$$

We reduced the system to row-echelon form, which can equivalently be written as

$$\mathbf{Ux} = \mathbf{b_u} \text{ with } \mathbf{U} = \begin{pmatrix} 1 & 0 & -2 \\ 0 & 1 & 2 \\ 0 & 0 & 1 \end{pmatrix}, \quad \mathbf{b_u} = \begin{pmatrix} -1 \\ 1 \\ 2 \end{pmatrix}.$$

We illustrate below how Gauss-Jordan elimination can also be used to give us the matrix inverses. For now, assume that we are able to calculate the inverses and in doing so we obtain

$$\mathbf{A}^{-1} = \frac{1}{2}\begin{pmatrix} 5 & 4 & -2 \\ -4 & -3 & 2 \\ 3 & 2 & -1 \end{pmatrix}, \text{ or } \mathbf{U}^{-1} = \begin{pmatrix} 1 & 0 & 2 \\ 0 & 1 & -2 \\ 0 & 0 & 1 \end{pmatrix} \text{ with } \mathbf{b_u} = \begin{pmatrix} -1 \\ 1 \\ 2 \end{pmatrix}.$$

The reader should verify that in either case we end up with the desired solution.

In the above example, we were able to obtain $\mathbf{A}^{-1}$, but this will not always be the case as mentioned previously. There are many examples in which our system cannot be solved with the use of our three elementary row operations. In this case, we say the system (or matrix) is *singular* and the inverse of $\mathbf{A}$ does not exist. Assuming that $\mathbf{A}^{-1}$ exists, we were able to proceed with Gauss-Jordan elimination and end up with the system

$$\mathbf{x} = \mathbf{A}^{-1}\mathbf{b}.$$

Thus the inverse of our matrix is related to the elementary row operations of Gauss-Jordan elimination. If we instead wrote our original system as

$$\mathbf{A}\mathbf{x} = \mathbf{I}\mathbf{b} \tag{C.31}$$

and reduced this through Gauss-Jordan elimination to

$$\mathbf{I}\mathbf{x} = \mathbf{A}^{-1}\mathbf{b}, \tag{C.32}$$

we would have the inverse of the matrix on the right-hand side. Note that we did not need to keep track of $\mathbf{b}$ in this case. Let us again consider our previous example given as (C.8) with coefficient matrix

$$\mathbf{A} = \begin{pmatrix} -2 & 0 & 4 \\ 4 & 2 & -4 \\ 2 & 4 & 2 \end{pmatrix}. \tag{C.33}$$

Instead of constructing the augmented matrix with $\mathbf{b}$ as a fourth column, we consider the matrix

$$\mathbf{A} = \begin{pmatrix} -2 & 0 & 4 & | & 1 & 0 & 0 \\ 4 & 2 & -4 & | & 0 & 1 & 0 \\ 2 & 4 & 2 & | & 0 & 0 & 1 \end{pmatrix}. \tag{C.34}$$

According to our previous discussion, if we do Gauss-Jordan elimination on the left half while keeping track of the resulting effect on the right half of the matrix, our final answer will be the inverse of $\mathbf{A}$. We will repeat the elementary row operations from above and will show all intermediate steps:

$$2R_1 + R_2 \rightarrow R_2 \text{ gives} \quad \begin{pmatrix} -2 & 0 & 4 & | & 1 & 0 & 0 \\ 0 & 2 & 4 & | & 2 & 1 & 0 \\ 2 & 4 & 2 & | & 0 & 0 & 1 \end{pmatrix} \quad \text{(C.35)}$$

$$R_1 + R_3 \rightarrow R_3 \text{ gives} \quad \begin{pmatrix} -2 & 0 & 4 & | & 1 & 0 & 0 \\ 0 & 2 & 4 & | & 2 & 1 & 0 \\ 0 & 4 & 6 & | & 1 & 0 & 1 \end{pmatrix} \quad \text{(C.36)}$$

$$-2R_2 + R_3 \rightarrow R_3 \text{ gives} \quad \begin{pmatrix} -2 & 0 & 4 & | & 1 & 0 & 0 \\ 0 & 2 & 4 & | & 2 & 1 & 0 \\ 0 & 0 & -2 & | & -3 & -2 & 1 \end{pmatrix} \quad \text{(C.37)}$$

$$\begin{matrix} \frac{-1}{2}R_1 \rightarrow R_1, \ \frac{1}{2}R_1 \rightarrow R_1, \\ \frac{-1}{2}R_1 \rightarrow R_1 \text{ gives} \end{matrix} \quad \begin{pmatrix} 1 & 0 & -2 & | & \frac{-1}{2} & 0 & 0 \\ 0 & 1 & 2 & | & 1 & \frac{1}{2} & 0 \\ 0 & 0 & 1 & | & \frac{3}{2} & 1 & \frac{-1}{2} \end{pmatrix} \quad \text{(C.38)}$$

$$-2R_3 + R_2 \rightarrow R_2 \text{ gives} \quad \begin{pmatrix} 1 & 0 & -2 & | & \frac{-1}{2} & 0 & 0 \\ 0 & 1 & 0 & | & -2 & \frac{-3}{2} & 1 \\ 0 & 0 & 1 & | & \frac{3}{2} & 1 & \frac{-1}{2} \end{pmatrix} \quad \text{(C.39)}$$

$$2R_3 + R_1 \rightarrow R_1 \text{ gives} \quad \begin{pmatrix} 1 & 0 & 0 & | & \frac{5}{2} & 2 & -1 \\ 0 & 1 & 0 & | & -2 & \frac{-3}{2} & 1 \\ 0 & 0 & 1 & | & \frac{3}{2} & 1 & \frac{-1}{2} \end{pmatrix}. \quad \text{(C.40)}$$

The right half of this matrix is exactly the inverse of **A** as the reader should check. But that was a lot of work! Now you will perhaps understand why it's not even desirable to calculate the inverse with a computer. Nevertheless, it IS instructive to understand the manipulations that are possible. We give sample code that will do these elementary row operations and then the built-in commands that will do the same thing.

C.2.3 Elementary Row Operations with Matlab, Maple, and Mathematica

There exist built-in commands to calculate the inverse of matrices as well as perform Gaussian elimination. As we want to give the reader a hands-on approach, we will describe how to perform the step-by-step calculations of Gauss-Jordan elimination.

We consider the augmented matrix

$$\begin{pmatrix} 2 & 1 & 4 & 1 \\ 2 & 1 & -1 & 0 \\ 4 & 3 & 2 & -1 \end{pmatrix}. \quad \text{(C.41)}$$

Computer Code C.2:   **Elementary row operations for row reduction; built-in solver for Ax = b**

Matlab, Maple, Mathematica

| Matlab |

```
>>  A=[2,1,4,1;2,1,-1,0;4,3,2,-1] %creates augmented matrix
>>  A(2,:)=-A(1,:)+A(2,:)%row 1 of A, mult by -1, then added
    %to row 2 and inserted as the new row 2; A is overwritten
>>  A(3,:)=-2*A(1,:)+A(3,:)  %row 1 of A, mult by -2, then
    %added to row 3 and inserted as the new row 3
>>  A=[A(1,:);A(3,:);A(2,:)]  %swaps 2nd and 3rd rows of A
>>  A(1,:)=A(1,:)/2 %multiplies row 1 by 1/2
>>  A(3,:)=-A(3,:)/5 %multiplies row 3 by -1/5
>>  A(1,:)=-A(2,:)/2+A(1,:)  %row 2 of A, mult by -1/2, then
    %added to row 1 and inserted as the new row 1
>>  A(2,:)=6*A(3,:)+A(2,:)  %row 3 of A, mult by 6, then
    %added to row 2 and inserted as the new row 2
>>  A(1,:)=-5*A(3,:)+A(1,:)  %row 3 of A, mult by -5, then
    %added to row 1 and inserted as the new row 1
>>  %We are now in reduced-row echelon form; alternatively,
>>  B=[2,1,4,1;2,1,-1,0;4,3,2,-1] %re-creates original matrix
>>  rref(B)
```

| Maple |

```
> with(linalg): #loads linalg package for linear algebra;
> A:=matrix(3,4,[2,1,4,1,2,1,-1,0,4,3,2,-1]); #creates
  #the augmented matrix A
> A1:=addrow(A,1,2,-1); #row 1 of A, multiplied by -1,
  #then added to row 2 and inserted as the new row 2
> A2:=addrow(A1,1,3,-2); #row 1 of A1, multiplied by -2,
  #then added to row 3 and inserted as the new row 3
> A3:=swaprow(A2,2,3); #swaps the 2nd and 3rd rows of A2
> A4:=mulrow(A3,1,1/2); #multiplies row 1 by 1/2
> A5:=mulrow(A4,3,-1/5); #multiplies row 3 by -1/5
> A6:=addrow(A5,2,1,-1/2); #row 2 of A5, mult by -1/2,
  #then added to row 1 and inserted as the new row 1
> A7:=addrow(A6,3,2,6); #row 3 of A6, multiplied by 6,
  #then added to row 2 and inserted as the new row 2
> A8:=addrow(A7,3,1,-5); #row 3 of A7, multiplied by -5,
  #then added to row 1 and inserted as the new row 1
> #We are now in reduced-row echelon form; alternatively,
> A9:=gausselim(A);
> backsub(A9);
> gaussjord(A); #or if we just wanted the gauss-jordan form
```

```
                        ┌─────────────┐
                        │ Mathematica │
                        └─────────────┘
<<LinearAlgebra'MatrixManipulation'
A={{2,1,4,1},{2,1,-1,0},{4,3,2,-1}} (*the augmented matrix*)
A//MatrixForm (*visually nicer form*)
A1 ={TakeRows[A,{1,1}][[1]], -1 TakeRows[A,{1,1}][[1]]
   +TakeRows[A,{2,2}][[1]], TakeRows[A,{3,3}][[1]]}
     (*row 1 of A, mult by -1, then added to row 2 and *)
     (*inserted as the new row 2*)
A1//MatrixForm (*visually nicer form*)
A2 = {TakeRows[A1,{1,1}][[1]], TakeRows[A1,{2,2}][[1]],
   -2 TakeRows[A1,{1,1}][[1]]+TakeRows[A1,{3,3}][[1]]}
     (*row 1 of A1, mult by -2, then added to row 3 and*)
     (*inserted as the new row 3*)
A2//MatrixForm (*visually nicer form*)
A3 = {TakeRows[A2,{1,1}][[1]],TakeRows[A2,{3,3}][[1]],
   TakeRows[A2,{2,2}][[1]]}
     (*swaps 2nd and 3rd rows of A2*)
A3//MatrixForm (*visually nicer form*)
A4={1/2 TakeRows[A3,{1,1}][[1]],TakeRows[A3,{2,2}][[1]],
   TakeRows[A3,{3,3}][[1]]}
     (*multiplies row 1 of A3 by 1/2*)
A4//MatrixForm (*visually nicer form*)
A5={TakeRows[A4,{1,1}][[1]],TakeRows[A4,{2,2}][[1]],
   -1/5 TakeRows[A4,{3,3}][[1]]}
     (*multiplies row 3 of A4 by -1/5*)
A5//MatrixForm
A6 = {-1/2 TakeRows[A5,{2,2}][[1]]+TakeRows[A5,{1,1}][[1]],
   TakeRows[A5,{2,2}][[1]],TakeRows[A5,{3,3}][[1]]}
     (*row 2 of A5, mult by -1/2, then added to row 1 and*)
     (*inserted as the new row 1*)
A6//MatrixForm
A7 = {TakeRows[A6,{1,1}][[1]],6 TakeRows[A6,{3,3}][[1]]
   +TakeRows[A6,{2,2}][[1]],TakeRows[A6,{3,3}][[1]]}
     (*row 3 of A6, mult by 6, then added to row 1 and*)
     (*inserted as the new row 1*)
A7//MatrixForm
A8 = {-5 TakeRows[A7,{3,3}][[1]]+ TakeRows[A7,{1,1}][[1]],
   TakeRows[A7,{2,2}][[1]],TakeRows[A7,{3,3}][[1]]}
     (*row 3 of A7, mult by -5, then added to row 1 and*)
     (*inserted as the new row 1*)
A8//MatrixForm
     (*We are now in reduced-row echelon form;*)
     (*alternatively, we could have used:*)
RowReduce[A]//MatrixForm
```

C.2.4   Cramer's Rule

This section of the Appendix can stand alone from the rest, with the possible exception of Appendix C.1, which the reader may need to read parts of in order to take the determinant of a matrix. Cramer's rule is really just another way to solve a system of equations. It is computationally inefficient to use Cramer's rule for even moderately large systems, so we'll skip the theory and just give the formulas for the 2×2, 3×3, and general $n \times n$ cases. We stress that Cramer's rule only works with systems of n equations in n unknowns. In Section 4.7, we need to solve a system of equations and Cramer's rule is a straightforward way to do this. Any of the methods of this section will work, too, but would be a bit more cumbersome because we may be dealing with functions instead of scalars as the entries in our matrices.

Let us consider the 2×2 linear system

$$a_{11}x + a_{12}y = b_1 \tag{C.42}$$

$$a_{21}x + a_{22}y = b_2. \tag{C.43}$$

Cramer's rule states that the solution is

$$x = \frac{\begin{vmatrix} b_1 & a_{12} \\ b_2 & a_{22} \end{vmatrix}}{\begin{vmatrix} a_{11} & a_{12} \\ a_{21} & a_{22} \end{vmatrix}}, \quad y = \frac{\begin{vmatrix} a_{11} & b_1 \\ a_{21} & b_2 \end{vmatrix}}{\begin{vmatrix} a_{11} & a_{12} \\ a_{21} & a_{22} \end{vmatrix}}. \tag{C.44}$$

Mechanically, we obtained the solution to the first variable by replacing the first column of the coefficient matrix by the right-hand side in the numerator (and then took its determinant) and divided by the determinant of the coefficient matrix. We note that the denominator cannot be zero because otherwise our answer will be undefined or indeterminate. But the denominator is the determinant of the coefficient matrix, which is zero only when there is no inverse. In this situation, the system either has no solution or possibly infinitely many solutions. In the case of a non-zero determinant, we will always be able to solve the system.

For the 3×3 linear system

$$a_{11}x + a_{12}y + a_{13}z = b_1 \tag{C.45}$$

$$a_{21}x + a_{22}y + a_{23}z = b_2 \tag{C.46}$$

$$a_{31}x + a_{32}y + a_{33}z = b_3, \tag{C.47}$$

we again replace the first column of the coefficient matrix in the numerator by the right-hand side. For simplicity, let $\mathbf{A}$ denote the coefficient matrix and

$|\mathbf{A}|$ its determinant. Then Cramer's rule states that the solution is

$$x = \frac{\begin{vmatrix} b_1 & a_{12} & a_{13} \\ b_2 & a_{22} & a_{23} \\ b_3 & a_{32} & a_{33} \end{vmatrix}}{|\mathbf{A}|}, \quad y = \frac{\begin{vmatrix} a_{11} & b_1 & a_{13} \\ a_{21} & b_2 & a_{23} \\ a_{31} & b_3 & a_{33} \end{vmatrix}}{|\mathbf{A}|}, \quad z = \frac{\begin{vmatrix} a_{11} & a_{12} & b_1 \\ a_{21} & a_{22} & b_2 \\ a_{31} & a_{32} & b_3 \end{vmatrix}}{|\mathbf{A}|}. \quad (C.48)$$

For the general linear system

$$\begin{aligned} a_{11}x_1 + a_{12}x_2 + \cdots + a_{1n}x_n &= b_1 \\ a_{21}x_1 + a_{22}x_2 + \cdots + a_{2n}x_n &= b_2 \\ \vdots \qquad \vdots \qquad \ddots \qquad \vdots \qquad \vdots \\ a_{n1}x_1 + a_{n2}x_2 + \cdots + a_{nn}x_n &= b_n, \end{aligned} \qquad (C.49)$$

we will again let $\mathbf{A}$ denote the coefficient matrix and Cramer's rule can be stated as

$$x_1 = \frac{\begin{vmatrix} b_1 & a_{12} & \cdots & a_{1n} \\ b_2 & a_{22} & \cdots & a_{2n} \\ \vdots & \vdots & & \vdots \\ b_n & a_{n2} & \cdots & a_{nn} \end{vmatrix}}{|\mathbf{A}|}, \qquad \text{where the first column in the numerator is replaced by } \mathbf{b}, \qquad (C.50)$$

$$x_i = \frac{\begin{vmatrix} a_{11} & \cdots & b_1 & \cdots & a_{1n} \\ a_{21} & \cdots & b_2 & \cdots & a_{2n} \\ \vdots & & \vdots & & \vdots \\ a_{n1} & \cdots & b_n & \cdots & a_{nn} \end{vmatrix}}{|\mathbf{A}|}, \qquad \begin{array}{c} \text{where the } i\text{th column in the} \\ \text{numerator is replaced by } \mathbf{b} \\ \text{for } i = 2, \cdots, n-1, \end{array} \qquad (C.51)$$

$$x_n = \frac{\begin{vmatrix} a_{11} & a_{12} & \cdots & b_1 \\ a_{21} & a_{22} & \cdots & b_2 \\ \vdots & \vdots & & \vdots \\ a_{n1} & a_{n2} & \cdots & b_n \end{vmatrix}}{|\mathbf{A}|}, \qquad \begin{array}{c} \text{where the } n\text{th column in the} \\ \text{numerator is replaced by } \mathbf{b}. \end{array} \qquad (C.52)$$

Example 2: If possible, solve the following system using Cramer's rule:

$$\begin{aligned} 2x + 3y &= 4 \\ -x + 2y = &= 5. \end{aligned} \qquad (C.53)$$

In order to see if the system can be solved by Cramer's rule, we need to calculate the determinant of the coefficient matrix:

$$|\mathbf{A}| = \begin{vmatrix} 2 & 3 \\ -1 & 2 \end{vmatrix} = 4 - (-3) = 7.$$

Because $|\mathbf{A}| \neq 0$, we can solve the system. Applying (C.44) gives

$$x = \frac{\begin{vmatrix} 4 & 3 \\ 5 & 2 \end{vmatrix}}{|\mathbf{A}|} = \frac{-7}{7} = -1, \quad y = \frac{\begin{vmatrix} 2 & 4 \\ -1 & 5 \end{vmatrix}}{|\mathbf{A}|} = \frac{14}{7} = 2.$$

Example 3: If possible, solve the following system using Cramer's rule:

$$2x + 3y = 4$$
$$-x + 2y + z = -5$$
$$7y + 2z = -6. \tag{C.54}$$

In order to see if the system can be solved by Cramer's rule, we need to calculate the determinant of the coefficient matrix:

$$|\mathbf{A}| = \begin{vmatrix} 2 & 3 & 0 \\ -1 & 2 & 1 \\ 0 & 7 & 2 \end{vmatrix} = 2(4 - 7) - 3(-2 - 0) + 0(-7 - 0) = 0.$$

Because $|\mathbf{A}| = 0$, we cannot use Cramer's rule to solve the system. If we instead tried to solve the system by Gaussian elimination, we would obtain, after two steps of elimination,

$$\begin{pmatrix} 2 & 3 & 0 & 4 \\ 0 & 7/2 & 1 & -3 \\ 0 & 0 & 0 & 0 \end{pmatrix}. \tag{C.55}$$

This can be solved to give

$$x = \frac{23 + 3z}{7}, \quad y = \frac{-6 - 2z}{7}, \quad z = \text{anything.}$$

Thus we have infinitely many solutions for this system.

Example 4: If possible, solve the following system using Cramer's rule:

$$2x + 3y = 4$$
$$-x + 2y + z = -5$$
$$3y - 2z = -4. \tag{C.56}$$

In order to see if the system can be solved by Cramer's rule, we need to calculate the determinant of the coefficient matrix:

$$|\mathbf{A}| = \begin{vmatrix} 2 & 3 & 0 \\ -1 & 2 & 1 \\ 0 & 3 & -2 \end{vmatrix} = 2(-4-3) - 3(2-0) + 0(-3-0) = -20.$$

Because $|\mathbf{A}| \neq 0$, we can solve the system. Applying (C.48) gives

$$x = \frac{\begin{vmatrix} 4 & 3 & 0 \\ -5 & 2 & 1 \\ -4 & 3 & -2 \end{vmatrix}}{|\mathbf{A}|}, \quad y = \frac{\begin{vmatrix} 2 & 4 & 0 \\ -1 & -5 & 1 \\ 0 & -4 & -2 \end{vmatrix}}{|\mathbf{A}|}, \quad z = \frac{\begin{vmatrix} 2 & 3 & 4 \\ -1 & 2 & -5 \\ 0 & 3 & -4 \end{vmatrix}}{|\mathbf{A}|}$$

$$= \frac{-70}{-20} = \frac{-7}{2} \qquad\qquad = \frac{20}{-20} = -1 \qquad\qquad = \frac{-10}{-20} = \frac{1}{2}$$

as the solution to our system.

Problems

1. Which of the following 3×3 matrices are in row-echelon form?

   a. $\begin{pmatrix} 1 & 0 & 0 \\ 0 & 1 & 0 \\ 0 & 0 & 1 \end{pmatrix}$
   b. $\begin{pmatrix} 1 & 2 & 0 \\ 0 & 1 & 0 \\ 0 & 0 & 0 \end{pmatrix}$
   c. $\begin{pmatrix} 1 & 0 & 0 \\ 0 & 1 & 0 \\ 0 & 2 & 0 \end{pmatrix}$

2. Which of the following 3×3 matrices are in row-echelon form?

   a. $\begin{pmatrix} 1 & 3 & 4 \\ 0 & 0 & 1 \\ 0 & 0 & 0 \end{pmatrix}$
   b. $\begin{pmatrix} 1 & 5 & -3 \\ 0 & 1 & 1 \\ 0 & 0 & 0 \end{pmatrix}$
   c. $\begin{pmatrix} 1 & 2 & 3 \\ 0 & 0 & 0 \\ 0 & 0 & 1 \end{pmatrix}$

3. Solve each of the following systems by Gauss-Jordan elimination:

   a. $\begin{cases} 2x_1 - 3x_2 = -2 \\ 2x_1 + x_2 = 1 \end{cases}$
   b. $\begin{cases} x_1 + x_2 + 2x_3 = 8 \\ -x_1 - 2x_2 + 3x_3 = 1 \\ 3x_1 - 7x_2 + 4x_3 = 10 \end{cases}$

4. Solve each of the following systems by Gauss-Jordan elimination:

   a. $\begin{cases} 2x_1 + 2x_2 + 2x_3 = 0 \\ -2x_1 + 5x_2 + 2x_3 = 1 \\ 8x_1 + x_2 + 4x_3 = -1 \end{cases}$
   b. $\begin{cases} -2b + 3c = 1 \\ 3a + 6b - 3c = -2 \\ 6a + 6b + 3c = 5 \end{cases}$

5. Solve each of the following systems by Gauss-Jordan elimination:

   a. $\begin{cases} x - 2y + 4z = 2 \\ 2x - 3y + 5z = 3 \\ 3x - 4y + 7z = 7 \end{cases}$
   b. $\begin{cases} 2x + 3y - 2z = 2 \\ x - 2y + 3z = 2 \\ 4x - y + 5z = 1 \end{cases}$

6. Use Cramer's rule to solve the following systems of equations. You must evaluate the determinants by hand. (You may, of course, check you answer with the computer.)

   a. $\begin{cases} 3x + y = 2 \\ 2x - 3y = 5 \end{cases}$
   b. $\begin{cases} x + 4z = 2 \\ 2x - 3y = 3 \\ 3x - 4y + 6z = 0 \end{cases}$
   c. $\begin{cases} 2x + 3y - 2z = 2 \\ x - 2y + 3z = 2 \\ 4x - y + 4z = 1 \end{cases}$

7. Solve using (i) Cramer's rule, (ii) Gaussian elimination, (iii) Gauss-Jordan elimination:

   a. $\begin{cases} 7x_1 - 2x_2 = 3 \\ 3x_1 + x_2 = 5 \end{cases}$

   b. $\begin{cases} 4x + 5y = 2 \\ 11x + y + 2z = 3 \\ x + 5y + 2z = 1 \end{cases}$

8. Solve using (i) Cramer's rule, (ii) Gaussian elimination, (iii) Gauss-Jordan elimination, (iv) built-in commands of Matlab, Maple, or Mathematica:

   a. $\begin{cases} x - 4y + z = 6 \\ 4x - y + 2z = -1 \\ 2x + 2y - 3z = -20 \end{cases}$

   b. $\begin{cases} 4x + y + z + w = 6 \\ 3x + 7y - z + w = 1 \\ 7x + 3y - 5z + 8w = -3 \\ x + y + z + 2w = 3 \end{cases}$

9. Use Cramer's rule to solve $\begin{cases} x = x' \cos\theta - y' \sin\theta \\ y = x' \sin\theta + y' \cos\theta \end{cases}$ for x', y' in terms of x and y.

10. Use the formula for a 2×2 matrix inverse to find the inverse of the following matrices:

$$\mathbf{A}_1 = \begin{pmatrix} 1 & 2 \\ -1 & -1 \end{pmatrix}, \quad \mathbf{A}_2 = \begin{pmatrix} 3 & -4 \\ 1 & 5 \end{pmatrix}$$

11. Use Gauss-Jordan elimination to find the inverses of the following matrices. You may again do this by hand or with the computer but in either case you must show all intermediate steps. Check your answer using the formula for a 2×2 matrix inverse.

$$\mathbf{A}_1 = \begin{pmatrix} 1 & 2 \\ 2 & 7 \end{pmatrix}, \quad \mathbf{A}_2 = \begin{pmatrix} 1 & 5 \\ 3 & -6 \end{pmatrix}$$

12. Use Gauss-Jordan elimination to find the inverses of the following matrices. You may again do this by hand or with the computer but in either case you must show all intermediate steps.

$$\mathbf{A} = \begin{pmatrix} 1 & 2 & -4 \\ -1 & -1 & 5 \\ 2 & 7 & -3 \end{pmatrix}, \quad \mathbf{B} = \begin{pmatrix} 1 & 3 & -4 \\ 2 & 5 & -1 \\ 3 & 13 & -6 \end{pmatrix}$$

C.3 Coordinates and Change of Basis

In this section, we examine how to represent a vector given a basis, and how changing from one basis to another affects this representation. It is best understood after Sections 5.3 and 5.4. In terms of vectors in $\mathbb{R}^n$, we are used to thinking of the *standard basis*, $\left\{ \begin{pmatrix} 1 \\ 0 \\ 0 \end{pmatrix}, \begin{pmatrix} 0 \\ 1 \\ 0 \end{pmatrix}, \begin{pmatrix} 0 \\ 0 \\ 1 \end{pmatrix} \right\}$. For example, if we consider the vector

$$\mathbf{w} = \begin{pmatrix} 3 \\ -2 \\ 5 \end{pmatrix}, \tag{C.57}$$

we probably interpret it as "3 units in the x-direction, -2 units in the y-direction, and 5 units in the z-direction." This is correct if we are using the standard basis (often denoted $\{\mathbf{e}_1, \mathbf{e}_2, \mathbf{e}_3\}$) and we could write

$$\begin{pmatrix} 3 \\ -2 \\ 5 \end{pmatrix} = 3 \begin{pmatrix} 1 \\ 0 \\ 0 \end{pmatrix} - 2 \begin{pmatrix} 0 \\ 1 \\ 0 \end{pmatrix} + 5 \begin{pmatrix} 0 \\ 0 \\ 1 \end{pmatrix} = 3\mathbf{e}_1 - 2\mathbf{e}_2 + 5\mathbf{e}_3. \tag{C.58}$$

Writing $\mathbf{w}$ as we did in (C.57) is called the *coordinate vector of $\mathbf{w}$ relative to the standard basis*. The *coordinates of $\mathbf{x}$ relative to the basis* are exactly the coefficients that we used to write it as a linear combination of the standard basis vectors. We also need to specify that we have an *ordered basis*, which just means that we need to keep the same order of the basis vectors when we refer to the basis. The representation of $\mathbf{w}$ would probably change if we changed the basis.

For example, if our basis B is

$$\begin{pmatrix} 1 \\ 0 \\ 0 \end{pmatrix}, \begin{pmatrix} 1 \\ 1 \\ 0 \end{pmatrix}, \begin{pmatrix} 1 \\ 1 \\ 1 \end{pmatrix}, \tag{C.59}$$

then the vector $\mathbf{w}$ is written

$$[\mathbf{x}]_B = \begin{pmatrix} 5 \\ -7 \\ 5 \end{pmatrix} = 5 \begin{pmatrix} 1 \\ 0 \\ 0 \end{pmatrix} - 7 \begin{pmatrix} 1 \\ 1 \\ 0 \end{pmatrix} + 5 \begin{pmatrix} 1 \\ 1 \\ 1 \end{pmatrix}. \tag{C.60}$$

How did we find the coordinate vector of $\mathbf{w}$ relative to a given basis? We want to find a linear combination of the ordered basis B that gives us the coordinates of $\mathbf{w}$ in the standard basis:

$$c_1 \begin{pmatrix} 1 \\ 0 \\ 0 \end{pmatrix} + c_2 \begin{pmatrix} 1 \\ 1 \\ 0 \end{pmatrix} + c_3 \begin{pmatrix} 1 \\ 1 \\ 1 \end{pmatrix} = \begin{pmatrix} 3 \\ -2 \\ 5 \end{pmatrix}. \tag{C.61}$$

We could have put this as a problem from Appendix C.2, because we simply need to solve a system of three equations in three unknowns. In matrix notation we have

$$\begin{pmatrix} 1 & 1 & 1 \\ 0 & 1 & 1 \\ 0 & 0 & 1 \end{pmatrix} \begin{pmatrix} c_1 \\ c_2 \\ c_3 \end{pmatrix} = \begin{pmatrix} 3 \\ -2 \\ 5 \end{pmatrix}, \tag{C.62}$$

$$\quad P \qquad\quad [\mathbf{w}]_B \qquad [\mathbf{w}]_S \tag{C.63}$$

and we can easily solve this to find the coefficients, which are the coordinates of $\mathbf{w}$ relative to B. Here, we let P denote the *transition matrix from B to S*. The formula

$$P[\mathbf{w}]_S = [\mathbf{w}]_B$$

is the change of basis from B to S. If we wanted the change of basis from S to B, we could use the following theorem:

THEOREM C.3.1 *Let P be the transition matrix from S to B. Then P is invertible and*

$$[\mathbf{w}]_S = P^{-1}[\mathbf{w}]_B$$

is the transition matrix from B to S.

In the above discussion, we could have used any two ordered bases in place of B and S.

It is often of interest to find the matrix that will take an ordered basis B to another ordered basis B', instead of being concerned only with how the coordinate representation of one vector changes.

THEOREM C.3.2 *Let $\{\mathbf{v}_1, \mathbf{v}_2, \cdots, \mathbf{v}_n\}$ be the ordered basis B, and $\{\mathbf{u}_1, \mathbf{u}_2, \cdots, \mathbf{u}_n\}$ be the ordered basis B' for a given vector space V. Then the bases can be related by*

$$\mathbf{v}_1 = c'_{11}\mathbf{u}_1 + c'_{21}\mathbf{u}_2 + \cdots c'_{n1}\mathbf{u}_n \tag{C.64}$$

$$\mathbf{v}_2 = c'_{12}\mathbf{u}_1 + c'_{22}\mathbf{u}_2 + \cdots c'_{n2}\mathbf{u}_n \tag{C.65}$$

$$\vdots = \vdots \tag{C.66}$$

$$\mathbf{v}_n = c'_{1n}\mathbf{u}_1 + c'_{2n}\mathbf{u}_2 + \cdots c'_{n3}\mathbf{u}_n. \tag{C.67}$$

Moreover, applying Gauss-Jordan elimination to the system

$$(B \mid B') = \begin{pmatrix} v_{11} & v_{12} & \cdots & v_{1n} & | & u_{11} & u_{12} & \cdots & u_{1n} \\ v_{21} & v_{22} & \cdots & v_{2n} & | & u_{21} & u_{22} & \cdots & u_{2n} \\ \vdots & \vdots & \ddots & \vdots & | & \vdots & \vdots & \ddots & \vdots \\ v_{n1} & v_{n2} & \cdots & v_{nn} & | & u_{n1} & u_{n2} & \cdots & u_{nn} \end{pmatrix} \tag{C.68}$$

gives us

$$(B \mid B') \longrightarrow (I \mid P^{-1}), \tag{C.69}$$

where P^{-1} is the transition matrix from B to B'.

Example 1: Consider the following two bases in $\mathbb{R}^3$:

$$B = \left\{ \begin{pmatrix} 1 \\ 0 \\ 0 \end{pmatrix}, \begin{pmatrix} 0 \\ 1 \\ 0 \end{pmatrix}, \begin{pmatrix} 0 \\ 0 \\ 1 \end{pmatrix} \right\} \quad \text{and} \quad B' = \left\{ \begin{pmatrix} 2 \\ 1 \\ 0 \end{pmatrix}, \begin{pmatrix} -4 \\ -1 \\ 3 \end{pmatrix}, \begin{pmatrix} 5 \\ 2 \\ -1 \end{pmatrix} \right\}.$$

The transition matrix from B to B' is found by writing the augmented matrix

$$\begin{pmatrix} 2 & -4 & 5 & | & 1 & 0 & 0 \\ 1 & -1 & 2 & | & 0 & 1 & 0 \\ 0 & 3 & -1 & | & 0 & 0 & 1 \end{pmatrix}$$

and row-reducing it via Gauss-Jordan elimination to the form

$$\begin{pmatrix} 1 & 0 & 0 & | & -5 & 11 & -3 \\ 0 & 1 & 0 & | & 1 & -2 & 1 \\ 0 & 0 & 1 & | & 3 & -6 & 2 \end{pmatrix}.$$

The transition matrix from B to B' is thus given by

$$P^{-1} = \begin{pmatrix} -5 & 11 & -3 \\ 1 & -2 & 1 \\ 3 & -6 & 2 \end{pmatrix}.$$

C.3.1 Similarity Transformations

Many times it is convenient to use the standard basis. But when we are trying to understand the qualitative behavior of solutions of differential equations near an equilibrium point (in Sections 5.1 and 6.4), it is often useful to convert to *eigencoordinates* because the resulting matrix will often be diagonal (and at least always in *Jordan form*, where the matrix is *block diagonal*). The eigencoordinates simply give a basis for the space where the eigenvectors are being used as the basis.

In (C.3), we gave the interpretation of matrix-matrix multiplication being equivalent to multiplying the columns of the right matrix with the left matrix to obtain the columns of the result. That is, for $\mathbf{AB} = \mathbf{C}$, we let $\mathbf{b_j}$ and $\mathbf{c_j}$ denote the columns of $\mathbf{B}$ and $\mathbf{C}$, respectively, and we write

$$\mathbf{Ab_j} = \mathbf{c_j}.$$

Using this interpretation we can write our eigenvectors as columns of a matrix $\mathbf{V}$. Then our system becomes

$$\mathbf{AV} = \mathbf{V\Lambda}, \tag{C.70}$$

where

$$\Lambda = \begin{pmatrix} \lambda_1 & 0 & \cdots & 0 \\ 0 & \lambda_2 & \cdots & 0 \\ \vdots & 0 & \ddots & 0 \\ 0 & 0 & \cdots & \lambda_n \end{pmatrix} \quad \text{and} \quad \mathbf{V} = \begin{pmatrix} v_{11} & v_{12} & \cdots & v_{1n} \\ v_{21} & v_{22} & \cdots & v_{2n} \\ \vdots & \vdots & \ddots & \vdots \\ v_{n1} & v_{n2} & \cdots & v_{nn} \end{pmatrix}. \tag{C.71}$$

(Note that the right-hand side is *not* $\mathbf{\Lambda V}$.) If we assume that our matrix $\mathbf{A}$ has n linearly independent eigenvectors, then our matrix $\mathbf{V}$ is invertible and we can write

$$\mathbf{\Lambda} = \mathbf{V}^{-1}\mathbf{AV}. \tag{C.72}$$

Note that $\mathbf{\Lambda}$ is a diagonal matrix. Thus this equation says that left multiplying our original matrix $\mathbf{A}$ with $\mathbf{V}^{-1}$ and right multiplying it with $\mathbf{V}$ yields a diagonal matrix. This process is called a *diagonal similarity transformation*. Our original matrix really did not matter other than we needed it to have n linearly independent eigenvectors, which may or may not happen for a given matrix.

In terms of coordinates and change of bases, we are given a matrix $\mathbf{A}$ relative to the standard basis, S, and we want to find the matrix relative to the basis of the eigenvectors, B'. The transition matrix from B' to S is exactly our matrix $\mathbf{V}$ that has the eigenvectors as the columns. The transition matrix from S to B' is thus $\mathbf{V}^{-1}$. We combine this together to obtain the matrix relative to the basis of eigenvectors as

$$\mathbf{V}^{-1}\mathbf{AV}.$$

Example 1: Diagonalize the matrix

$$\mathbf{A} = \begin{pmatrix} -1 & 4 \\ 3 & 3 \end{pmatrix}$$

by using a diagonal similarity transformation.

We need to find the two eigenvectors of the matrix first. If they are linearly independent, then we will be able to proceed. We could do this by hand or use the computer code from Section 5.4 to find the eigenvalue-eigenvector pairs as

$$\left\{ -3, \begin{pmatrix} -2 \\ 1 \end{pmatrix} \right\} \left\{ 5, \begin{pmatrix} 1 \\ \frac{3}{2} \end{pmatrix} \right\}.$$

We can thus write the matrix of linearly independent eigenvectors and calculate its inverse:

$$\mathbf{V} = \begin{pmatrix} -2 & 1 \\ 1 & \frac{3}{2} \end{pmatrix}, \quad \mathbf{V}^{-1} = \begin{pmatrix} \frac{-3}{8} & \frac{1}{4} \\ \frac{1}{4} & \frac{1}{2} \end{pmatrix}.$$

The original matrix $\mathbf{A}$ can be diagonalized as

$$\mathbf{V}^{-1}\mathbf{AV} = \begin{pmatrix} -3 & 0 \\ 0 & 5 \end{pmatrix}.$$

The above calculations could be done with the following code:

Computer Code C.3: Eigenvalues, eigenvectors, and matrix diagonalization

Matlab, Maple, Mathematica

```
                            Matlab
>> A=[-1, 4; 3, 3]
>> [v,d]=eig(A) %eigenvalues AND eigenvectors of A
>> v %shows the matrix v, which has the eigenvectors as its
   % columns; we should verify that we have lin ind columns
>> inv(v)*A*v %diagonalizes A
```

```
                            Maple
> with(linalg):
> A:=matrix(2,2,[-1,4,3,3]);
> eq1:=eigenvects(A); #eigenvalues AND eigenvectors of A
> #We should verify that above result gives lin ind vectors
> eq2a:=eq1[1][3][1]; #first eigenvector
> eq2b:=eq1[2][3][1]; #second eigenvector
> Vt:= matrix(2,2,[eq2a,eq2b]); #matrix w/eigenvects as rows
> V:=transpose(Vt); #matrix with eigenvectors as columns
> eq3:=multiply(inverse(V),A,V); #diagonalizes A
> simplify(evalm(eq3)); #simplifies above answer, if needed
```

```
                          Mathematica
A = {{-1,4},{3,3}}
A//MatrixForm (*visually nicer form*)
eq1=Eigensystem[A] (*eigenvalues and eigenvectors of A*)
eq2a=eq1[[2]][[1]] (*first eigenvector*)
eq2b=eq1[[2]][[2]] (*second eigenvector*)
Vt = {eq2a,eq2b} (*matrix with eigenvectors as rows*)
Vt//MatrixForm (*visually nicer form*)
V=Transpose[Vt] (*matrix with eigenvectors as columns*)
V//MatrixForm (*visually nicer form*)
eq3 = Inverse[V].A.V (*diagonalizes A*)
eq3//MatrixForm (*visually nicer form*)
```

If we have complex eigenvalues, we usually do *not* diagonalize the matrix simply because this would introduce complex variables into a problem that originally had real numbers. In Section 5.4.1, we considered the system

$$\frac{dx}{dt} = ax - by$$

$$\frac{dy}{dt} = bx + ay, \tag{C.73}$$

where a, b are real and $b \neq 0$. We saw that the eigenvalues are $a \pm ib$ and **we interpreted the action of the matrix as a rotation through an angle θ followed by a stretch by a factor r, where $r = \sqrt{a^2 + b^2}$ and $\tan \theta = b/a$ are the standard polar coordinates**. Thus we can think of (C.73) as being the desired form of a matrix that has complex eigenvalues. Analogous to the case of real eigenvalues, we can use a similarity transformation to convert the matrix to this form.

Example 2: Consider

$$\mathbf{A} = \begin{pmatrix} 3 & 2 \\ -4 & -1 \end{pmatrix}.$$

The eigenvalues can be easily calculated as $1 + 2i, 1 - 2i$ with eigenvectors $(1 \quad -1+i)^T, (1 \quad -1-i)^T$, respectively. We previously used the eigenvectors as the columns of our matrix $\mathbf{V}$. Our eigenvectors are now complex but they still give us the insight that we need. We write one of the eigenvectors in the form $\mathbf{v}_1 + i\mathbf{v}_2$. If we take the first one, this would give

$$\begin{pmatrix} 1 \\ -1+i \end{pmatrix} = \begin{pmatrix} 1 \\ -1 \end{pmatrix} + i \begin{pmatrix} 0 \\ 1 \end{pmatrix}.$$

We then use $\mathbf{v}_1$ and $\mathbf{v}_2$ as the respective columns of our matrix $\mathbf{V}$:

$$\mathbf{V}^{-1}\mathbf{A}\mathbf{V} = \begin{pmatrix} 1 & 0 \\ -1 & 1 \end{pmatrix}^{-1} \begin{pmatrix} 3 & 2 \\ -4 & -1 \end{pmatrix} \begin{pmatrix} 1 & 0 \\ -1 & 1 \end{pmatrix}$$

$$= \begin{pmatrix} 1 & 0 \\ 1 & 1 \end{pmatrix} \begin{pmatrix} 3 & 2 \\ -4 & -1 \end{pmatrix} \begin{pmatrix} 1 & 0 \\ -1 & 1 \end{pmatrix}$$

$$= \begin{pmatrix} 1 & 2 \\ -2 & 1 \end{pmatrix}. \tag{C.74}$$

If we had chosen the second eigenvector, we would have had $\begin{pmatrix} 1 & -2 \\ 2 & 1 \end{pmatrix}$. In either case we see the eigenvalues in the matrix as $1 \pm 2i$.

In general, two $n \times n$ matrices $\mathbf{A}$ and $\mathbf{B}$ are *similar* if there exists an invertible matrix $\mathbf{P}$ such that

$$\mathbf{B} = \mathbf{P}^{-1}\mathbf{A}\mathbf{P}.$$

Based on the previous discussion, we can state the following theorem.

THEOREM C.3.3 *Similar matrices have the same eigenvalues.*

Problems

1. Consider the bases $B = \{\mathbf{u}_1, \mathbf{u}_2\}$ and $B' = \{\mathbf{v}_1, \mathbf{v}_2\}$ for $\mathbb{R}^2$ where

$$\mathbf{u}_1 = \begin{pmatrix} 1 \\ 0 \end{pmatrix}, \qquad \mathbf{u}_2 = \begin{pmatrix} 0 \\ 1 \end{pmatrix}, \qquad \mathbf{v}_1 = \begin{pmatrix} 2 \\ 1 \end{pmatrix}, \qquad \text{and } \mathbf{v}_2 = \begin{pmatrix} -3 \\ 4 \end{pmatrix}.$$

  a. Find the transition matrix from B' to B.
  b. Find the transition matrix from B to B'.
  c. Compute the coordinate matrix $[\mathbf{w}]_B$, where $\mathbf{w} = \begin{pmatrix} 3 \\ -5 \end{pmatrix}$.
  d. Use part c to compute $[\mathbf{w}]_{B'}$.

2. Repeat problem 1 with

$$\mathbf{u}_1 = \begin{pmatrix} 1 \\ 2 \end{pmatrix}, \mathbf{u}_2 = \begin{pmatrix} 4 \\ -1 \end{pmatrix}, \mathbf{v}_1 = \begin{pmatrix} 1 \\ 3 \end{pmatrix}, \mathbf{v}_2 = \begin{pmatrix} -1 \\ -1 \end{pmatrix}, \text{and } \mathbf{w} = \begin{pmatrix} 3 \\ -5 \end{pmatrix}.$$

3. Consider the bases $B = \{\mathbf{u}_1, \mathbf{u}_2, \mathbf{u}_3\}$ and $B' = \{\mathbf{v}_1, \mathbf{v}_2, \mathbf{v}_3\}$ for $\mathbb{R}^3$ where

$$\mathbf{u}_1 = \begin{pmatrix} -3 \\ 0 \\ -3 \end{pmatrix}, \qquad \mathbf{u}_2 = \begin{pmatrix} -3 \\ 2 \\ -1 \end{pmatrix}, \qquad \mathbf{u}_3 = \begin{pmatrix} 1 \\ 6 \\ 1 \end{pmatrix},$$

$$\mathbf{v}_1 = \begin{pmatrix} -6 \\ -6 \\ 0 \end{pmatrix}, \qquad \mathbf{v}_2 = \begin{pmatrix} -2 \\ -6 \\ 4 \end{pmatrix}, \qquad \mathbf{v}_3 = \begin{pmatrix} -2 \\ -3 \\ 7 \end{pmatrix}.$$

  a. Find the transition matrix from B' to B.
  b. Compute the coordinate matrix $[\mathbf{w}]_B$, where $\mathbf{w} = \begin{pmatrix} -5 \\ 8 \\ -5 \end{pmatrix}$.

4. Repeat problem 3 with

$$\mathbf{u}_1 = \begin{pmatrix} 2 \\ 1 \\ 2 \end{pmatrix}, \qquad \mathbf{u}_2 = \begin{pmatrix} 2 \\ -1 \\ 1 \end{pmatrix}, \qquad \mathbf{u}_3 = \begin{pmatrix} 1 \\ 1 \\ 2 \end{pmatrix},$$

$$\mathbf{v}_1 = \begin{pmatrix} 3 \\ 1 \\ -5 \end{pmatrix}, \quad \mathbf{v}_2 = \begin{pmatrix} 1 \\ 1 \\ -3 \end{pmatrix}, \quad \mathbf{v}_3 = \begin{pmatrix} -1 \\ 0 \\ 2 \end{pmatrix}, \quad \mathbf{w} = \begin{pmatrix} -4 \\ 7 \\ -4 \end{pmatrix}.$$

In problems 5–7, if possible, find a matrix $\mathbf{P}$ *that diagonalizes the matrix* $\mathbf{A}$ *and then determine* $\mathbf{P}^{-1}\mathbf{AP}$. *If it is not possible, explain why.*

5. a. $\begin{pmatrix} 2 & 0 \\ 1 & 2 \end{pmatrix}$ b. $\begin{pmatrix} 1 & 0 \\ 7 & -1 \end{pmatrix}$

6. a. $\begin{pmatrix} 2 & -3 \\ 1 & -1 \end{pmatrix}$ b. $\begin{pmatrix} 3 & 0 & 0 \\ 0 & 2 & 0 \\ 0 & 1 & 2 \end{pmatrix}$

7. a. $\begin{pmatrix} 2 & 0 & -2 \\ 0 & 3 & 0 \\ 0 & 0 & 5 \end{pmatrix}$ b. $\begin{pmatrix} 0 & 0 & -2 \\ 1 & 2 & 1 \\ 1 & 0 & 3 \end{pmatrix}$

We have come to the end of the line, our maiden voyage.

Thanks for listening....

Appendix D

Answers to Selected Exercises

Section 1.1

1. $y'(x) = 6x^2 \Rightarrow x(6x^2) = 3(2x^3)$, which is true for all x.

2. $y'(x) = 0 \Rightarrow 0 = x^3(2-2)^2 = 0$, which is true for all x.

4. $y'(x) = e^x - 1 \Rightarrow (e^x - 1) + (e^x - x)^2 = e^{2x} + (1-2x)e^x + x^2 - 1$, which is true for all x.

6. Note the solution does not exist for $x = 3$. $y'(x) = \frac{1}{(x-3)^2} \Rightarrow \frac{1}{(x-3)^2} = \left(\frac{-1}{x-3}\right)^2$ which is true when $x \neq 3$.

8. $y'(x) = \cos x - 2\sin x, y''(x) = -\sin x - 2\cos x \Rightarrow (-\sin x - 2\cos x) + (\sin x + 2\cos x) = 0$, which is true for all x.

9. $y'(x) = 1, y''(x) = 0 \Rightarrow (0) + (x) = x$, which is true for all x.

14. b. e^{-3x}, **c.** xe^{-3x}, and **e.** $2e^{-3x} + xe^{-3x}$ are all solutions; **a.** e^x and **d.** $4e^{3x}$ are not solutions.

17. $y'(x) = re^{rx}, y''(x) = r^2e^{rx} \Rightarrow (r^2e^{rx}) + 3(re^{rx}) + 2(e^{rx} = e^{rx}(r^2 + 3r + 2) = 0 \Rightarrow r = -2, -1$, which is true for all x.

19. a. (i) 2nd order, (ii) linear, (iii) N/A; **b.** (i) 2nd order, (ii) nonlinear, (iii) N/A; **c.** (i) 3rd order, (ii) linear, (iii) IVP; **d.** (i) 1st order, (ii) nonlinear, (iii) IVP; **e.** (i) 2nd order,

(ii) linear, (iii) N/A; **f.** (i) 2nd order, (ii) nonlinear, (iii) BVP.

Section 1.2

2. $y(x) = C\csc^2 x$

4. $y(x) = \tan\left(\frac{\pi}{2} - \arctan(x^2)\right)$

5. $y(x) = C(x+1)e^{-x}$

7. $y(x) = \frac{1}{\ln|x^2-1|+1}$

10. $x(t) = \pm\sqrt{2t - t^2 + C}$

13. Substitute $u = 2y + x$. Solution is $2y - 2\ln|2 + x + 2y| + 4 + 2\ln 2 = 0$

15. $\frac{e^y}{(y+2)^2} = \frac{e^{-1}}{x+4}$

16. $\tan y = 2\sin 2x - 4x + \frac{\pi}{3}$

18. $y(x) = \int_0^x e^{t^2}\,dt$

20. a. $y(x) = \arctan\left(1 - \frac{1}{x}\right)$

24. $y(x) = \frac{Cx^2}{1-Cx}$

26. $y(x) = x\sinh(x+C)$

27. $y(x) = Cxe^x - x$

29. $y = \pm x\sqrt{\ln x + C}$

30. $y = C(y^2 - x^2)$

33. $\frac{1}{\sqrt{y(y-2x)}} = C$

36. $\arctan\left(\frac{y}{x}\right) + \frac{1}{2}\ln\left(1 + \left(\frac{y}{x}\right)^2\right) = -x$

Section 1.3

2. 1.75 sec, 16.3 m, 2 sec, 20 m

3. 1.87 sec, 16.4 m/sec

8. At about 4:48 p.m., (right before dinner!)

10. 40 min

11. 7.8 min

12. $b - \frac{b-a}{60k}\left(1 - e^{-60k}\right)$

13. 10 min

14. 0.5 kg

17. 24 min

19. 200 days

21. 9.75×10^8 years

22. $(c \pm x)y = 2a^2$

23. $b\ln y - y = \pm x + c, 0 < y < b$

25. $y = cx^2$

28. 17.5 min

30. 27 sec

34. 50 sec

36. 98.1%

37. $p \approx e^{-0.12h}$

39. $c\ln(M/m)$

Section 1.4

2. $x^2y + 2y^2 + x = 2$. Only the top curve in this implicit solution passes through the IC.

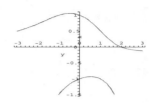

Graph for 1.4#2.

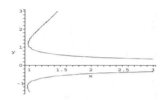

Graph for 1.4#6.

5. $x^2y - \frac{1}{3}y^3 = C$

6. $x^2 - 3x^3y^2 + y^4 = -1$. Only the top curve in this implicit solution passes through the IC.

8. Not exact.

13. $A = -2$; solution is $\frac{y}{x^2} - \frac{y}{x} = C$

16. $N(x, y) = x^2y + \phi(y)$

Section 1.5

2. $y = (1 + x^2)(x - \arctan x + C)$

4. $y = \sin x - \cos x$

6. $y = 4e^x(x + 1)$

10. $y = \frac{\sin x}{x} + x^{-1}$

17. $x = y^2 + Cy$

43. $y^{-3}\sec^3 x = \frac{3}{4}\sin 2x - \frac{3}{2}x + C$

45. $y = (-3x^2 + C|x|^3)^{1/3}$

48. $y = x^2\ln|x| + Cx^2$

49. $y = \dfrac{1}{x^2\sqrt{2e^{-x}(x-1)+C}}$

Chapter 1 Review

1. False. An IC is needed for an IVP.

5. False. Solutions need only be defined for x in the interval (x_0-h, x_0+h).

7. True. This form can always be separated.

9. $y'(x) = Ce^x - 2x - 2 \Rightarrow x^2 + y(x) = x^2 + (Ce^x - x^2 - 2x - 2) = y'(x)$ for all x

12. $y = \pm\sqrt{Ce^{1/x}+2}$

14. $y = \frac{2}{1+Cx^2}$

16. $y = \frac{1}{1+x}$

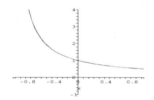

Graph for Chapter 1 Review#16.

17. $y = \ln\left|\dfrac{-\ln|x|+C}{e^{1/x}}\right|$

21. $\arcsin\left(\frac{y}{x}\right) = \ln x + C$

24. $\frac{x^3}{y^2} + x + \frac{5}{y} = C$

Section 2.1

5. d

6. a

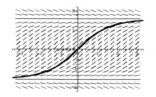

Graph for 2.1#10.

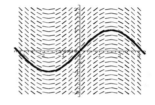

Graph for 2.1#14.

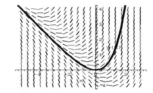

Graph for 2.1#18.

7. b

10. $y' = \cos y$, see figure

14. $y' = \cos x$, see figure

18. $y' = x + y$, see figure

20. $y' = e^{x^2}$, see figure

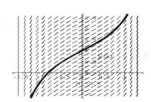

Graph for 2.1#20.

22. $y' = \frac{x^2-1}{y^2+1}$, see figure

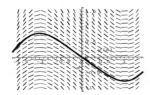

Graph for 2.1#22.

24. $y' = xy(x^2 + 2)$, see figure

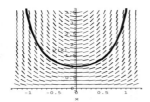

Graph for 2.1#24.

Section 2.2

1. c. (i) Theorem does not guarantee existence or uniqueness of a solution at $(1, 0)$.

2. a. (i) Solutions exist for all (x, y); (ii) solutions are unique for everywhere except possibly when $y = 0$.
b. (i) Solutions exist for all (x, y); (ii) solutions are unique for everywhere except possibly when $y = x$.
c. (i) Solutions exist everywhere except possibly when $y = 0$; (ii) solutions are unique for everywhere except possibly when $y = 0$.

d. (i) Solutions exist everywhere except possibly when $y = -x$; (ii) solutions are unique for everywhere except possibly when $y = -x$.

4. Solutions will exist everywhere and will be unique everywhere except possibly along $y = 0$. Separation of variables $\Rightarrow y = (2x - 2 + \sqrt{3})^2$ passes through $(1, 3)$. There is no problem regarding uniqueness at the given point $(1, 3)$.

5. Solutions will exist everywhere and will be unique everywhere except possibly along $y = 2$. Separation of variables $\Rightarrow y = 2+(2x)^{5/2}$ passes through $(0, 2)$. Since $y = 2$ also passes through $(0, 2)$, the solution is not unique.

7. Solutions will exist everywhere and will be unique everywhere except possibly along $y = 1$. Separation of variables $\Rightarrow y = 1+(3x)^{3/2}$ passes through $(0, 2)$. Since $y = 1$ also passes through $(0, 1)$, the solution is not unique.

Section 2.3

2. (ii) $y^* = -2$ is half-stable; (iii) for $y_0 < -2$, $y \to -2$ as $x \to \infty$; for $y_0 > -2$, $y \to \infty$ as $x \to \infty$;

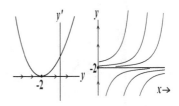

2.3#2 (i), (ii), (iv).

5. (ii) $y^* = -3, 3$ are unstable, $y^* = 2$ is stable; (iii) for $y_0 < -3$, $y \to -\infty$ as $x \to \infty$; for $-3 < y_0 < 2$, $y \to 2$ as $x \to \infty$; for $y_0 > 3$, $y \to \infty$ as $x \to \infty$;

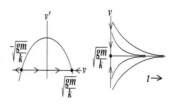

2.3#14 (i), (ii), (iv).

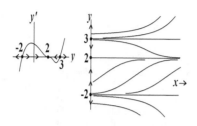

2.3#5 (i), (ii), (iv).

17. (ii) $x^* = 2$ is stable; (iii) for $x_0 \in (-\infty, \infty)$, $x \to 2$ as $t \to \infty$;

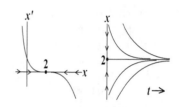

2.3#17 (i), (ii), (iv).

9. (ii) $y^* = 0$ is half-stable, $y^* = 2$ is stable; (iii) for $y_0 < 0$, $y \to 0$ as $x \to \infty$; for $0 < y_0$, $y \to 2$ as $x \to \infty$;

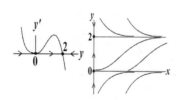

2.3#9 (i), (ii), (iv).

19. (ii) $x^* = 0$ is unstable; $x^* = 1$ is stable; (iii) for $x_0 < 0$, $x \to -\infty$ as $t \to \infty$; for $x_0 > 0$, $x \to 0$ as $t \to \infty$;

14. (ii) $v^* = \sqrt{\dfrac{gm}{k}}$ is stable; $v^* = -\sqrt{\dfrac{gm}{k}}$ is unstable but not physically meaningful (so ignore for rest of problem); (iii) for $v_0 > -\sqrt{\dfrac{gm}{k}}$, $v \to \sqrt{\dfrac{gm}{k}}$ as $t \to \infty$;

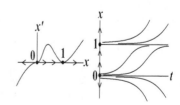

2.3#19 (i), (ii), (iv).

24. $f(y) = y^2 - 1 \Rightarrow f'(y) = 2y$. Equilibria are $y^* = \pm 1$;

$f'(-1) < 0 \Rightarrow y^* = -1$ is stable; $f'(1) > 0 \Rightarrow y^* = 1$ is unstable.

26. $f(y) = y^3 + 1 \Rightarrow f'(y) = 3y^2$. Only real equilibrium is $y^* = -1$; $f'(-1) > 0 \Rightarrow y^* = -1$ is unstable.

Section 2.4

2. $x^* = 0, 1, 6$; $x^* = 0$ is stable; $x^* = 1$ is unstable; $x^* = 6$ is stable.

4. a. $x^* = 0, 1, 4$; $x^* = 0$ is stable; $x^* = 1$ is unstable; $x^* = 4$ is stable. The stability results do not differ from that of the Allee effect.
b. For small x, this model is approximately $x' = -4x^2$ whereas the comparable Allee model would be $x' = -4x$; the population in this model dies off at a much slower rate than in the Allee model.

6. a. $x^* = 0, a, 5$; $x^* = 0$ is unstable; $x^* = a$ is stable; $x^* = 5$ is unstable.
b. For $0 < x_0 < 5$, $x \Rightarrow a$ as $t \Rightarrow \infty$; for $x_0 > 5$, $x \Rightarrow \infty$ as $t \Rightarrow \infty$.
c. The parameter a could be a measure of the health of the individual or the strength of their immune system. If a person had a compromised immune system, then the body will not be able to keep the bacteria in check at low levels but may still be able to keep it in check at higher levels.

8. $x' = x(x-1)(x-6)(x-10)$ is one possibility.

10. a. $x' = x^2(2-x)^2(x-4)$;
b. $x^* = 0, 2, 4$; $x^* = 0$ is half-stable; $x^* = 2$ is half-stable; $x^* = 4$ is unstable.

Section 2.5

2. $y' = x^4 y$, $y(1) = 1$.

x_i	y_i	$y(x_i)$
1.0000	1.0000	1.0000
1.1000	1.1000	1.1299
1.2000	1.2611	1.3467
1.3000	1.5225	1.7205
1.4000	1.9574	2.4004

4. $y' = \frac{\sin x}{y^3}$, $y(\pi) = 2$.

x_i	y_i	$y(x_i)$
3.1416	2.0000	2.0000
3.2416	2.0000	1.9994
3.3416	1.9988	1.9975
3.4416	1.9963	1.9944
3.5416	1.9925	1.9901

7. $y' = y + \cos x$, $y(0) = 0$.

x_i	y_i	$y(x_i)$
0	0	0
0.1000	0.1000	0.1050
0.2000	0.2095	0.2200
0.3000	0.3285	0.3450
0.4000	0.4568	0.4801
0.5000	0.5946	0.6253
0.6000	0.7418	0.7807
0.7000	0.8986	0.9466
0.8000	1.0649	1.1231

10. $y' = x + y$, $y(0) = 0$.

x_i	y_i	$y(x_i)$
0	0	0
0.1000	0	0.0052
0.2000	0.0100	0.0214
0.3000	0.0310	0.0499
0.4000	0.0641	0.0918
0.5000	0.1105	0.1487
0.6000	0.1716	0.2221
0.7000	0.2487	0.3138
0.8000	0.3436	0.4255

Section 2.6

2. $y' = x^4 y$, $y(1) = 1$.

x_i	y_i	$y(x_i)$
1.0000	1.0000	1.0000
1.1000	1.1299	1.1299
1.2000	1.3467	1.3467
1.3000	1.7204	1.7205
1.4000	2.4003	2.4004

4. $y' = \frac{\sin x}{y^3}$, $y(\pi) = 2$.

x_i	y_i	$y(x_i)$
3.1416	2.0000	2.0000
3.2416	1.9994	1.9994
3.3416	1.9975	1.9975
3.4416	1.9944	1.9944
3.5416	1.9901	1.9901

7. $y' = y + \cos x$, $y(0) = 0$.

x_i	y_i	$y(x_i)$
0	0	0
0.1000	0.1050	0.1050
0.2000	0.2200	0.2200
0.3000	0.3450	0.3450
0.4000	0.4801	0.4801
0.5000	0.6253	0.6253
0.6000	0.7807	0.7807
0.7000	0.9466	0.9466
0.8000	1.1231	1.1231

10. $y' = x + y$, $y(0) = 0$.

x_i	y_i	$y(x_i)$
0	0	0
0.1000	0.0052	0.0052
0.2000	0.0214	0.0214
0.3000	0.0499	0.0499
0.4000	0.0918	0.0918
0.5000	0.1487	0.1487
0.6000	0.2221	0.2221
0.7000	0.3138	0.3138
0.8000	0.4255	0.4255

13. $y' = x^3 y - x^2 y^2$, $y(-1) = 1$. Take $h = .001$ viewing window to be $-1 < x < 10, -1 < y < 10$.

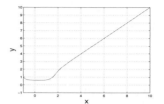

Graph for 2.6#13.

16. $y' = y\sqrt{x^2 + y^2 + 1} + \cos(xy)$, $y(0) = 1$. Take $h = .01$ viewing window to be $-.5 < x < 1, -1 < y < 20$.

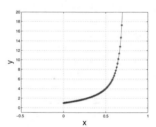

Graph for 2.6#16.

Chapter 2 Review

2. False. RK use four function evaluations to calculate the next step.

4. False. Neither f nor $\frac{\partial f}{\partial y}$ is continuous everywhere.

6. True. Phase line analysis will work and gives information on long-term behavior.

8. a. (i) Solutions exist everywhere; (ii) solutions are unique everywhere except possibly along $y = 0$. **b.** (i) Solutions exist everywhere; (ii) solutions are unique everywhere. **c.** (i) Solutions exist everywhere except possibly when $y = \frac{\pi}{2} \pm n\pi$ for

$n = 0, 1, 2, \cdots$; (ii) solutions are unique everywhere except possibly when $y = \frac{\pi}{2} \pm n\pi$ for $n = 0, 1, 2, \cdots$.
d. (i) Solutions exist everywhere except possibly along $x = -1$; (ii) solutions are unique except possibly along $x = -1$.

15. Left picture matches d; right picture matches b.

17. $y' = \frac{1-y^2}{2x}$, $y(1) = \pi$.

x_i	RK:y_i	Euler:y_x
1.0000	3.1416	3.1416
1.1000	2.7742	2.6981
1.2000	2.5144	2.4127
1.3000	2.3210	2.2118
1.4000	2.1713	2.0621
1.5000	2.0522	1.9459
1.6000	1.9550	1.8531
1.7000	1.8743	1.7770
1.8000	1.8061	1.7135

20. $y' = \frac{y-y^2}{x+1}$, $y(2) = 0$.
All answers are 0 because $y_0 = 0$ is an equilibrium point.

23. $y' = \frac{1-y^2}{2x}$, $y(1) = \pi$.

x_i	RK:y_i	Euler:y_x
1.4142	1.0000	1.0000
1.5142	1.1251	1.1207
1.6142	1.2592	1.2495
1.7142	1.4030	1.3868
1.8142	1.5571	1.5332
1.9142	1.7222	1.6891
2.0142	1.8992	1.8552
2.1142	2.0889	2.0321
2.2142	2.2924	2.2207

26. $y = Cx^2 e^{-3/x}$

27. $y = 0$, $\frac{(x \pm C)^2}{4x^2}$

30. $y = \ln\left(\frac{-1}{\ln(x-2)+C}\right)$

32. $y = \frac{\pm 1}{\sqrt{1+Ce^{x^2}}}$

33. $y = \frac{x - \ln x + C}{(x-1)^2}$

34. $y = \frac{1}{-2 + C\sqrt{x^2-1}}$

36. $y = \pm\sqrt{1 + Ce^{-2x}(1+x)^2}$

38. $y = \pm\frac{\sqrt{6\sin^3 x + 9C}}{3\sin x}$

41. $y^{2/3} - \frac{x}{3} - \frac{1}{6} - Ce^{2x}$

42. $y = \pm\frac{\sqrt{e^{-2x}(x^2+C)}}{e^{-2x}}$

47. $y = \frac{x \pm \sqrt{x^2 - 4C}}{2C}$

49. $\arctan(x+y) = y + C$

59. $y = x\tan(\ln x + C)$

63. $x - \frac{y^3 - 3y + C}{(y-1)^2(y+1)^2} = 0$

Section 3.1

2. A unique solution is not guaranteed because $a_0(x) = x$ is zero at the IC. Substitution gives $c_1 = c_2 = 0$.

5. A unique solution is guaranteed because coefficients are constant. Substitution gives $c_1 = c_2 = 2$.

6. a. Solutions are guaranteed for all (x, y).
b. Solutions are guaranteed everywhere except when $x = \frac{\pi}{2} + n\pi$, $n = 0, 1, 2, \cdots$.
c. Solutions are guaranteed everywhere except when $x = 0$.

Section 3.2

2. By definition: Set $c_1 x + c_2(2x) = 0$ for all x. E.g., choose $c_1 = -2, c_2 = 1$.

5. By Wronskian: $W(x) =$
$\det \begin{pmatrix} e^x & x+1 \\ e^x & 1 \end{pmatrix} = e^x(x+2) \neq 0.$

7. Linearly independent.

9. Linearly dependent.

12. Linearly dependent.

14. Linearly independent.

16. Linearly independent.

18. Linearly independent.

22. Set $c_1 x + c_2 |x| = 0$ for all x.
a. On $[0,1]$, we have $|x| = x$ so choose $c_1 = -c_2$. Thus set is linearly dependent on $[0,1]$.
b. On $[-1,0]$, we have $|x| = -x$ so choose $c_1 = c_2$. Thus set is linearly dependent on $[-1,0]$.
c. On $[-1,1]$, we need results of both **a** and **b** to hold, which can only happen when $c_1 = c_2 = 0$. Thus set is linearly independent on $[-1,1]$.
d. $W(x) = \det \begin{pmatrix} x & -x \\ 1 & -1 \end{pmatrix} = 0$
on $[-1,0]$. $W(x) = \det \begin{pmatrix} x & x \\ 1 & 1 \end{pmatrix} = 0$
on $[0,1]$. Thus $W(x) = 0$ for all x even though $\{x, |x|\}$ is linearly independent.

29. (i) There are the same number of functions as the order of the DE; (ii) each of the two functions are solutions; (iii) the two solutions are linearly independent.

40. Yes, it's a fundamental set of solutions.

43. No, it's not a fundamental set of solutions because $\sin x$ is not a solution.

45. No, it's not a fundamental set of solutions because there are not enough functions to consider.

47. Yes, it's a fundamental set of solutions.

49. Yes, it's a fundamental set of solutions.

52. Yes, it's a fundamental set of solutions.

56. (i) There are the same number of vectors as the dimension of the vector space; (ii) each of the two vectors is in the correct space, $\mathbb{R}^2$ in this case; (iii) the two vectors are linearly independent.

64. No, it's not a basis because vectors are linearly dependent.

66. Yes, it's a basis.

68. No, it's not a basis because vectors aren't in $\mathbb{R}^3$.

70. No, it's not a basis because there aren't enough vectors to span.

Section 3.3

3. $y = \sin 5x$

5. $y = xe^{3x}$

7. $y = e^{-3x}$

9. $y = 1$

11. $y = x$

14. $y = x + 1$

17. $y'' - y = 0 \Rightarrow e^{-x}, e^x$ are solutions.

19. $y'' - 3y' + 2y = 0 \Rightarrow e^{2x}, e^x$ are solutions.

Section 3.4

2. a. 0, **b.** $\cos x + 4 + \sin x$, **c.** $-e^{-2x}$

4. a. $8x + x^3$, **b.** 0, **c.** $-3\sin 2x$

6. a. $4e^x$, **b.** 0, **c.** $2\cos x$

8. $Q(D)P(D)(y) = -5\cos x + 3\sin x - 8x + 3x^2$; $P(D)Q(D)(y) = -5\cos x + 3\sin x - 8x + 3x^2$

10. $Q(D)P(D)(y) = -3\sin x + 3x\cos x$; $P(D)Q(D)(y) = 3x\cos x$

12. $Q(D)P(D)(y) = 5x^3 + 15x^2 + 21x + 21$; $P(D)Q(D)(y) = 5x^3 + 12x^2 + 21x + 16$

17. a. $(D-3)(D+1)$, **b.** already irreducible quadratic, **c.** $(D^2+1)^2$

20. a. $(D+8)(D+2)(y) = 0$, **b.** $D(D^2+2)^2(y) = 0$

22. a. $(D-2)(D^2+2D+4)(y) = 0$, **b.** $D(D+3)^2(y) = 0$

Section 3.5

2. (i) $u_1' = u_2$, $u_2' = -\frac{4}{7}u_2 + \frac{3}{7}u_1$, IC: $u_1(0) = 0, u_2(0) = 1$; **(ii)** with $h = .01$, $y(5) = 5.96$.

4. (i) $u_1' = u_2$, $u_2' = -x^2u_2 - 12u_1$, IC: $u_1(0) = 0, u_2(0) = 7$; **(ii)** with $h = .01$, $y(5) = .0104$.

5. (i) $u_1' = u_2$, $u_2' = -4u_2 - 3\sin u_1 + 1$, IC: $u_1(0) = -1, u_2(0) = \pi$; **(ii)** with $h = .01$, $y(5) = .6846$.

8. (i) $u_1' = u_2$, $u_2' = -2u_2 - 10u_1 + \sin x$, IC: $u_1(\pi) = e, u_2(\pi) = 1$; **(ii)** with $h = .01$, $y(4.99) = .0733$.

11. (i) $u_1' = u_2$, $u_2' = u_3, u_3' = -\frac{1}{8}u_3$, IC: $u_1(0) = 1, u_2(0) = 0, u_3(0) = 2$; **(ii)** with $h = .01$, $y(5) = 21.51$.

14. (i) $u_1' = u_2$, $u_2' = u_3, u_3' = \frac{1}{8}u_3$, IC: $u_1(1) = 1, u_2(1) = 1, u_3(1) = 0$; **(ii)** with $h = .01$, $y(5) = 45.65$.

16. (i) $u_1' = u_2$, $u_2' = u_3, u_3' = u_4, u_4' = \frac{1}{x+2}(u_4 - 6u_3 - 3u_2 + 2u_1$, IC: $u_1(-1) = 3, u_2(-1) = -1, u_3(-1) = 0, u_4(-1) = 1$; **(ii)** with $h = .01$, $y(5) = 7.333$.

Chapter 3 Review

1. False. We only can conclude that this is a possibility.

4. False. Only constant-coefficient differential operators commute.

7. It is guaranteed to have a unique solution through the given IC. The constants are $c_1 = 1 - e^{-1}, c_2 = e^{-1}$.

10. Linearly independent.

12. Linearly independent.

14. Linearly dependent.

17. Yes, it forms a fundamental set of solutions.

19. No, it is not a fundamental set of solutions because e^{-x} is not a solution.

21. No, it is not a fundamental set of solutions because neither is a solution.

24. Yes, it forms a basis.

26. Yes, it forms a basis.

28. No, it is not a basis because vectors are linearly independent.

31. No, it is not a basis because vectors are linearly independent.

35. $y = e^{5x}$

37. $y = e^{-5x}$

40. a. $(D - 4)(D - 3)(y) = 0$, **b.** $D(D - 4)(D + 4)$

Section 4.1

1. $y(x) = c_1 e^{-6x} + c_2 e^{-2x}$

3. $y(x) = c_1 + c_2 x + c_3 e^{\frac{-x}{8}}$

5. $y(x) = e^{-4x}(-1 + e^{7x})$

7. $y(x) = xe^{-2x}$

9. $y(x) = c_1 e^{\frac{-x}{2}} + c_2 x e^{\frac{-x}{2}}$

11. $y(x) = c_1 e^{2x} \sin x + c_2 e^{2x} \cos x$

14. $y(x) = \cos(\sqrt{5}(\pi - x))$ $- \frac{1}{\sqrt{5}} \sin(\sqrt{5}(\pi - x))$

15. $y = \frac{1}{6}e^{2x} - \frac{1}{6}e^{-x} \cos(\sqrt{3}x)$ $+ \frac{1}{6}\sqrt{3}e^{-x} \sin(\sqrt{3}x)$

16. $y(x) = c_1 e^x + c_2 e^{-x} + c_3 \sin x + c_4 \cos x$

19. $y(x) = c_1 e^{-4x} + c_2 x e^{-4x} + c_3 e^{-x} + c_4 e^{2x}$

22. $y(x) = c_1 e^{-3x} + c_2 e^{-x} + c_3 e^x + c_4 e^{3x} + c_5$

25. $y(x) = c_1 e^x + c_2 x e^x + c_3 x^2 e^x$

27. $y(x) = c_1 e^{-2x} + c_2 e^{-x} + c_3 e^x + c_4 e^{2x}$

30. $y(x) =$ $e^{\sqrt{2}x}(c_1 \sin \sqrt{2}x + c_2 \cos \sqrt{2}x)$ $+ e^{-\sqrt{2}x}(c_1 \sin \sqrt{2}x + c_2 \cos \sqrt{2}x)$

32. $D^2(D + 2)(y) = 0$

34. $D^4((D-2-3i)(D-2+3i))^3(y) = 0 \Rightarrow$ $D^4(D^2 - 4D + 13)^3(y) = 0$

Section 4.2

1. $x(t) = \frac{1}{6} \cos 16t$; $\frac{1}{6}$ ft; $\frac{\pi}{8}$ sec; $\frac{8}{\pi}$ oscillations/sec

3. $x(t) = \frac{1}{4} \sin 8t$; $\frac{1}{4}$ ft; $\frac{\pi}{4}$ sec; $\frac{4}{\pi}$ oscillations/sec

6. $\frac{1}{2\pi}\sqrt{K(\frac{1}{I_1} + \frac{1}{I_2})}$

7. $A = \frac{B}{1 - \frac{m}{k}\omega^2}$

8. $I = \frac{V}{R}(1 - e^{\frac{Rt}{L}})$

11. $I = \frac{q}{\omega CL}e^{-\frac{Rt}{2L}} \sin \omega t$, $CR^2 < 4L$

14. Underdamped.

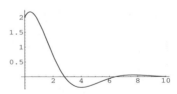

Graph for Chapter 4, Section 4.2#14.

16. Overdamped.

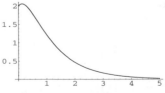

Graph for Chapter 4, Section 4.2#16.

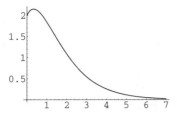

Graph for Chapter 4, Section 4.2#18.

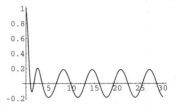

Graph for Chapter 4, Section 4.4#7, $\omega = 1$.

18. Critically damped.

Section 4.3

1. $y(x) = c_1 x^{-2} + c_2 x^{-1}$

3. $y(x) = c_1 \sin(2 \ln x) + c_2 \cos(2 \ln x)$

5. $y(x) = c_1 x^{-3} + c_2 x^{-7}$

7. $y(x) = c_1 x + c_2 x^{3/2}$

9. $y(x) = c_1 x^2 + c_2 x^2 \ln x$

10. $y(x) = c_1 x^{-1+\sqrt{3}} + c_2 x^{-1-\sqrt{3}}$

11. $y(x) = c_1 x^{-2+\sqrt{7}} + c_2 x^{-2-\sqrt{7}}$

13. $y(x) = c_1 x^{1/7} \sin\left(\frac{\sqrt{6}}{7} \ln x\right)$
$+ c_2 x^{1/7} \cos\left(\frac{\sqrt{6}}{7} \ln x\right)$

15. $y(x) = c_1 x^{-5/3} \sin\left(\frac{2\sqrt{2}}{3} \ln x\right)$
$+ c_2 x^{-5/3} \cos\left(\frac{2\sqrt{2}}{3} \ln x\right)$

Section 4.4

1. $y_p = \frac{e^x}{2}$

3. $y_p = 2x \sin 3x + \frac{2}{3} x \cos 3x$

5. $y_p = 2(\frac{1}{6}) - 12(\frac{x}{6} + \frac{5}{36}) + 6(\frac{e^x}{2})$, now simplify.

7. Resonant frequency $\omega = 2$, amplitude 0.2.

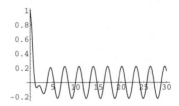

Graph for Chapter 4, Section 4.4#7, $\omega = 2$.

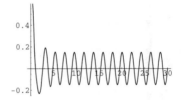

Graph for Chapter 4, Section 4.4#7, $\omega = 3$.

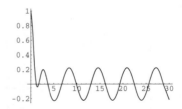

Graph for Chapter 4, Section 4.4#11, $\omega = 1$.

11. Resonant frequency $\omega = \sqrt{3}$, amplitude 0.3.

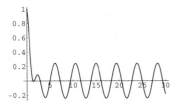

Graph for Chapter 4, Section 4.4#11, $\omega = \sqrt{2}$.

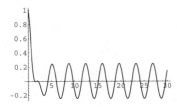

Graph for Chapter 4, Section 4.4#11, $\omega = \sqrt{3}$.

Section 4.5

1. $y_p = A + Bx + Cx^2 + Ee^{-x}$

3. $y_p = Ax \sin x + Bx \cos x + Ce^x \sin x + Ee^x \cos x$

5. $y_p = Ae^{2x} \sin x + Be^{2x} \cos x + C \sin x + E \cos x$

7. $y_p = A \sin x + B \cos x$

9. $A = \frac{1}{9}$

11. $A = \frac{-1}{5}, B = \frac{-3}{5}$

14. $A = \frac{1}{20}, B = \frac{2}{17}$

16. $y(x) = c_1 e^x + c_2 e^{2x} + 2x^2 + 6x + 7$

17. $y(x) = c_1 e^{-2x} + c_2 e^{4x} - 3e^{-3x} - \frac{1}{2} e^{2x}$

19. $y(x) = c_1 e^{-3x} \sin x + c_2 e^{-3x} \cos x + c_3 e^{-4x} + \frac{1}{4} x^2 e^{-4x} + \frac{1}{2} xe^{-4x} + \frac{1}{4} e^{-4x} + \frac{3}{4} e^{-3x} \cos x + \frac{1}{4} e^{-3x} \sin x.$

21. $y(x) = c_1 e^x \sin x + c_2 e^x \cos x + e^x + (\frac{x}{5} + \frac{2}{25}) \cos x - \frac{2}{25}(5x + 7) \sin x$

25. $y(x) = c_1 e^x \sin 2x + c_2 e^x \cos 2x + \frac{1}{2} xe^x + \frac{1}{16} e^x - \frac{1}{4} xe^x \cos 2x$

27. $y(x) = (1 - x)e^x$

21. $y(x) = c_1 e^x \sin x + c_2 e^x \cos x + e^x + (\frac{x}{5} + \frac{2}{25}) \cos x - \frac{2}{25}(5x + 7) \sin x$

25. $y(x) = c_1 e^x \sin 2x + c_2 e^x \cos 2x + \frac{1}{2} xe^x + \frac{1}{16} e^x - \frac{1}{4} xe^x \cos 2x$

27. $y(x) = (1 - x)e^x$

29. $y(x) = c_1 e^{-3x} + c_2 e^{-x} + \frac{e^{-x}}{16}(4x - 2 + e^{2x})$

33. a. $y(x) = c_1 e^{4x} - \frac{1}{4}(x^2 + \frac{x}{2} + \frac{1}{8})$;
c. $y(x) = c_1 e^x + \frac{1}{3} e^{4x}$

Section 4.6

1. $y_p = A + Bx + Cx^2 + Ee^{-x}$

3. $y_p = Ax \sin x + Bx \cos x + Ce^x \sin x + Ee^x \cos x$

5. $y_p = Ae^{2x} \sin x + Be^{2x} \cos x + C \sin x + E \cos x$

7. $y_p = A \sin x + B \cos x$

9. $A = \frac{1}{9}$

11. $A = \frac{-1}{5}, B = \frac{-3}{5}$

14. $A = \frac{1}{20}, B = \frac{2}{17}$

16. $y(x) = c_1 e^x + c_2 e^{2x} + 2x^2 + 6x + 7$

17. $y(x) = c_1 e^{-2x} + c_2 e^{4x} - 3e^{-3x} - \frac{1}{2} e^{2x}$

19. $y(x) = c_1 e^{-3x} \sin x + c_2 e^{-3x} \cos x + c_3 e^{-4x} + \frac{1}{4} x^2 e^{-4x} + \frac{1}{2} xe^{-4x} + \frac{1}{4} e^{-4x} + \frac{3}{4} e^{-3x} \cos x + \frac{1}{4} e^{-3x} \sin x.$

29. $y(x) = c_1 e^{-3x} + c_2 e^{-x} + \frac{e^{-x}}{16}(4x - 2 + e^{2x})$

33. a. $y(x) = c_1 e^{4x} - \frac{1}{4}(x^2 + \frac{x}{2} + \frac{1}{8})$;
c. $y(x) = c_1 e^x + \frac{1}{3} e^{4x}$

Section 4.7

1. $y(x) = c_1 + c_2 x + x \arctan x - \frac{1}{2}\ln(x^2 + 1)$

3. $y(x) = \cos x + \ln(\cos x)\cos x + 2\sin x + x \sin x$

5. $y(x) = -1 + \cos x + \sin x + \ln(\cos(\frac{x}{2}) + \sin(\frac{x}{2}))\sin x - \ln(\cos(\frac{x}{2}) - \sin(\frac{x}{2}))\cos x$

9. $y(x) = \frac{e^{2x}}{6}(8 - 3e^x + e^{3x})$

12. $y(x) = \frac{1}{6}(3 + 2\cos x + \cos 2x)$

15. $y(x) = \frac{e^{3x}}{36} + c_1 \sin 3\sqrt{3}x + c_2 \cos 3\sqrt{3}x$

21. $y(x) = \frac{x}{2} + c_1 \sin(\ln x) + c_2 \cos(\ln x)$

22. $y(x) = \frac{1}{4}x(\ln x - 1) + \frac{c_1}{x} + c_2 x + c_3$.

Chapter 4 Review

1. False, this is true with constant coefficients.

2. True.

4. False.

6. $y(x) = c_1 e^{2x} + c_2 e^{3x}$

8. $y(x) = (3x + 1)e^{-3x}$

9. $y(x) = \frac{1}{2}e^{-4x}(3e^{2x} - 1)$

11. $y(x) = \frac{1}{3}e^{2(x - \pi/2)}(5\cos 3x - 6\sin 3x)$

13. $y(x) = c_1 e^{-x} + c_2 + c_3 x + c_4 e^{4x}$

15. $y(x) = c_1 e^{-x} + c_2 x e^{-x} + c_3 x^2 e^{-x} + c_4$

19. $y(x) = c_1 e^{-7x} + c_2 e^{-2x} + c_3 + c_4 e^{2x}$

21. $y(x) = \frac{c_1}{x} + \frac{c_2}{\sqrt{x}}$

23. $y(x) = c_1 \sin\left(\sqrt{2}\ln x\right) + c_2 \cos\left(\sqrt{2}\ln x\right)$

25. $y(x) = c_1 e^{4x} + c_2 - \frac{1}{17}(4\sin x + \cos x)$

27. $y(x) = c_1 e^{2x} + c_2 x e^{2x} + \frac{1}{16}e^{-2x}$

29. $y(x) = c_1 \sin(\frac{5x}{2}) + c_2 \cos(\frac{5x}{2}) + \frac{2}{29}e^{-x}$

31. $y(x) = c_1 e^{-5x} + c_2 e^{2x} - \frac{1}{100}(10 + 3e^{3x} + 10x e^{3x})$.

33. Overdamped.

35. Overdamped.

37. Critically damped.

38. Underdamped.

45. a. $y(x) = (3/4)e^{-x} + (1/2)xe^{-x} + (1/4)e^x$, **b.** $y(4) = 13.70$

47. a. $y = (7/6)\sin(2x) + \cos(2x) + (1/3)\sin(x)$, **b.** $y(4) = .7565$

50. a. $y(x) = \frac{c_1}{x} + \frac{c_2}{\sqrt{x}}$

52. $y(x) = c_1 \sin\left(\sqrt{2}\ln x\right) + c_2 \cos\left(\sqrt{2}\ln x\right)$

Section 5.1

1. a. $\lambda_{1,2} = 3, -1$, **b.** $\lambda_{1,2} = 3, 2$; see figure

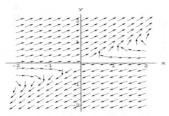

Graph for 5.1#1a.

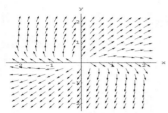

Graph for 5.1#1b.

3. a. $\lambda_{1,2} = -1 \pm \sqrt{6}$, **b.** $\lambda_{1,2} = 3, -1$

Section 5.2

1. $\begin{pmatrix} \cos t & e^t \\ 2t & 3 \end{pmatrix}$, $\begin{pmatrix} -\cos t & e^t \\ \frac{1}{3}t^3 & \frac{3}{2}t^2 \end{pmatrix}$; $\begin{pmatrix} 2\sin t \cos t & e^{-t} - te^{-t} \\ 0 & 1 \end{pmatrix}$

3. $\begin{pmatrix} 0 & -e^{-t} & 3e^{3t} \\ 1 & 0 & 6e^{3t} \\ e^t & -e^{-t} & 9e^{3t} \end{pmatrix}$, $\begin{pmatrix} t & -e^{-t} & \frac{1}{3}e^{3t} \\ \frac{1}{2}t^2 & 0 & \frac{2}{3}e^{3t} \\ e^t & -e^{-t} & e^{3t} \end{pmatrix}$

13. a. Linearly independent, c. linearly dependent.

15. a. Linearly dependent, b. linearly dependent.

Section 5.3

3. b. $\begin{pmatrix} \frac{1}{2} - \frac{\sqrt{3}}{4} \\ \frac{\sqrt{3}}{2} - \frac{3}{2} \end{pmatrix} \approx \begin{pmatrix} 3.0981 \\ -.63397 \end{pmatrix}$

c. $\begin{pmatrix} \frac{-1}{2} - \frac{3\sqrt{3}}{2} \\ \frac{\sqrt{3}}{2} - \frac{3}{2} \end{pmatrix} \approx \begin{pmatrix} -3.09807 \\ -.63397 \end{pmatrix}$

d. $\begin{pmatrix} \frac{1}{4} - \frac{3\sqrt{3}}{4} \\ \frac{\sqrt{3}}{4} - \frac{9}{4} \end{pmatrix} \approx \begin{pmatrix} -1.04904 \\ -1.81699 \end{pmatrix}$

4. b. $\begin{pmatrix} -1 - \frac{3\sqrt{3}}{2} \\ -\sqrt{3} + \frac{3}{2} \end{pmatrix} \approx \begin{pmatrix} -3.59808 \\ -.23205 \end{pmatrix}$

c. $\begin{pmatrix} 1 + \frac{3\sqrt{3}}{2} \\ -\sqrt{3} + \frac{3}{2} \end{pmatrix} \approx \begin{pmatrix} 3.59808 \\ -.23205 \end{pmatrix}$

d. $\begin{pmatrix} \frac{-1}{2} - \frac{3\sqrt{3}}{4} \\ \frac{-\sqrt{3}}{2} + \frac{9}{4} \end{pmatrix} \approx \begin{pmatrix} .79904 \\ 1.38397 \end{pmatrix}$

Section 5.4

1. a. $\lambda_1 = 2, \lambda_2 = -1$, $\mathbf{v}_1 = \begin{pmatrix} 5 \\ 2 \end{pmatrix}, \mathbf{v}_2 = \begin{pmatrix} 1 \\ 1 \end{pmatrix}$

3. b. $\lambda_1 = -2, \lambda_2 = 2$, $\mathbf{v}_1 = \begin{pmatrix} 1 \\ -1 \end{pmatrix}, \mathbf{v}_2 = \begin{pmatrix} 1 \\ 1 \end{pmatrix}$

5. b. $\lambda_1 = 5, \lambda_2 = 1$, $\mathbf{v}_1 = \begin{pmatrix} 1 \\ 1 \end{pmatrix}, \mathbf{v}_2 = \begin{pmatrix} 1 \\ -3 \end{pmatrix}$

6. c. $\lambda_1 = 1, \lambda_2 = 1$, $\mathbf{v}_1 = \begin{pmatrix} 1 \\ 1 \end{pmatrix}, \mathbf{v}_2 = \begin{pmatrix} 0 \\ 0 \end{pmatrix}$

12. b. $\lambda_1 = 2, \lambda_2 = 1, \lambda = -4$, $\mathbf{v}_1 = \begin{pmatrix} 1 \\ 2 \\ 0 \end{pmatrix}, \mathbf{v}_2 = \begin{pmatrix} 0 \\ 0 \\ 1 \end{pmatrix} \mathbf{v}_3 = \begin{pmatrix} 1 \\ -1 \\ 1 \end{pmatrix}$

14. a. $\lambda_1 = -1, \lambda_2 = 3, \lambda = 3$, $\mathbf{v}_1 = \begin{pmatrix} 0 \\ 0 \\ 1 \end{pmatrix}, \mathbf{v}_2 = \begin{pmatrix} 1 \\ 2 \\ 0 \end{pmatrix} \mathbf{v}_3 = \begin{pmatrix} 0 \\ -3 \\ 1 \end{pmatrix}$

Section 5.5

1. a. $\begin{pmatrix} e^{3t} & \frac{2}{9}(e^{-6t} - e^{3t}) \\ 0 & e^{-6t} \end{pmatrix}$

3. b. $\begin{pmatrix} e^{3t}\cos 2t & e^{3t}\sin 2t \\ -e^{3t}\sin 2t & e^{3t}\cos 2t \end{pmatrix}$

Chapter 5 Review

1. True.

3. True.

4. True.

6. False.

17. $\lambda_{1,2,3} = -2, 2, 3,\ v_1 = \begin{pmatrix} 2 \\ -1 \\ 2 \end{pmatrix}$,

$v_2 = \begin{pmatrix} 1 \\ -1 \\ 1 \end{pmatrix},\ v_3 = \begin{pmatrix} 2 \\ -1 \\ 1 \end{pmatrix}$

19. $\lambda_{1,2,3} = -1, 1, 5,\ v_1 = \begin{pmatrix} -1 \\ -1 \\ 1 \end{pmatrix}$,

$v_2 = \begin{pmatrix} 1 \\ 1 \\ 1 \end{pmatrix},\ v_3 = \begin{pmatrix} 1 \\ -1 \\ 1 \end{pmatrix}$

22. $\lambda_{1,2,3} = -27, -18, -9,\ v_1 = $ $\begin{pmatrix} 1 \\ -4 \\ 3 \end{pmatrix},\ v_2 = \begin{pmatrix} -2 \\ -1 \\ 3 \end{pmatrix},\ v_3 = \begin{pmatrix} 2 \\ 1 \\ 6 \end{pmatrix}$

39. a. $\begin{pmatrix} e^{2t} & -e^{t}+e^{2t} \\ 0 & e^{t} \end{pmatrix}$

Section 6.1

1. a. $\lambda_{1,2} = 1, -2$ **b.** $\lambda_{1,2} = 1, -1$

3. a. $\lambda_{1,2} = 4, 2$ **b.** $\lambda_{1,2} = 5, 2$

5. b. $\lambda_{1,2} = -2, -3$

9. a. $\lambda_{1,2} = 3 \pm i$ **b.** $\lambda_{1,2} = \pm 2i$

11. a. $\lambda_{1,2} = 1 \pm 2i$ **b.** $\lambda_{1,2} = -1 \pm 2i$

14. a. $x(t) = \frac{c_1}{3}e^{2t}(2+e^{3t})+\frac{c_2}{3}e^{2t}(-1+e^{3t})$, $y(t) = \frac{2c_1}{3}e^{2t}(-1+e^{3t})+\frac{c_2}{3}e^{2t}(1+2e^{3t})$, $x(t) = \frac{c_1}{3}e^{2t}(2+e^{3t})+\frac{c_2}{3}e^{2t}(-1+e^{3t})$, $z(t) = \frac{c_1}{12}e^{t}(-33+32e^{t}+e^{4t})+\frac{c_2}{12}e^{t}(15-16e^{t}+e^{4t})+c_3e^{t}$

Section 6.2

1. a. $\lambda_{1,2} = 3 \pm i$ **b.** $\lambda_{1,2} = \pm 2i$

3. a. $\lambda_{1,2} = 1 \pm 2i$ **b.** $\lambda_{1,2} = -1 \pm 2i$

5. a. $\lambda_{1,2} = -8 \pm 2i$ **b.** $\lambda_{1,2} = 3 \pm i$

7. a. $\lambda_{1,2} = 3, 2;\ \begin{pmatrix} 1 \\ 2 \end{pmatrix}$

b. $\lambda_{1,2} = -1, 2;\ \begin{pmatrix} 1 \\ 2 \end{pmatrix}$

9. a. $\lambda_{1,2} = 1, 1;$

b. $\lambda_{1,2} = -1, 2;\ \begin{pmatrix} 1 \\ 0 \end{pmatrix}$

11. $\lambda_{1,2,3} = 1, -1 \pm 2i$

13. $\lambda_{1,2,3} = -1, -1 \pm i$

Section 6.3

1. $x(t) = \frac{-c_1}{5}e^{-2t}(-6 + e^{5t})$ $-\frac{2c_2}{5}e^{-2t}(-1+e^{5t})$, $y(t) = \frac{3c_1}{5}e^{-2t}(-1+e^{5t})$ $+\frac{c_2}{5}e^{-2t}(-1+6e^{5t})$

3. $x(t) = c_1 e^{4t}$, $y(t) = \frac{-c_1}{2}e^{2t}(-1+e^{2t})$ $+c_2 e^{2t}$

5. $x(t) = c_1 e^{t} - c_3 e^{-t}(-1+e^{2t})$, $y(t) = c_1 e^{t}(-1+e^{t}) + c_2 e^{2t}$ $+c_3 e^{-t}(-2+e^{2t}+e^{3t})$, $z(t) = c_3 e^{-t}$

7. $x(t) = \frac{c_1}{4}e^{-t}(3+e^{4t})$ $+\frac{c_2}{4}e^{-t}(-3+2e^{2t}+e^{4t})$ $+\frac{c_3}{4}e^{t}(-1+e^{2t})$, $y(t) = \frac{c_1}{4}e^{-t}(-1+e^{4t})$ $+\frac{c_2}{4}e^{-t}(1+e^{2t})^2+\frac{c_3}{4}e^{t}(-1+e^{2t})$, $z(t) = \frac{c_1}{2}e^{-t}(-1+e^{4t})$ $+\frac{c_2}{2}e^{-t}(1+e^{2t})^2+\frac{c_3}{2}e^{t}(-1+e^{2t})$

Section 6.4

2. (i) $(x^*, y^*) = (1,1), (-1,1)$; (ii) $(1,1)$ is a saddle, $(-1,1)$ is a stable spiral.

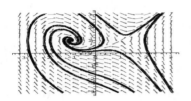

6.4#2 (iii)-(iv).

4. (i) $(x^*, y^*) = (0,0), (9,3)$; (ii) $(0,0)$ is a stable node, $(9,3)$ is a saddle.

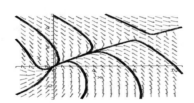

6.4#4 (iii)-(iv).

7. (i) $(x^*, y^*) = (-1,1), (1,1)$; (ii) $(1,1)$ is a stable spiral, $(1,1)$ is a saddle.

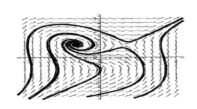

6.4#7 (iii)-(iv).

10. (i) $(x^*, y^*) = (0,0), (3,0), (0,2), (1,1)$; (ii) $(0,0)$ is an unstable node, $(3,0)$ is a stable node, $(0,2)$ is a stable node, $(1,1)$ is a saddle.

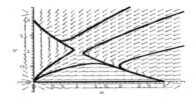

6.4#10 (iii)-(iv).

16. For $r > 0$, $(x^*, y^*) = (\sqrt{r}, 0)$, $(-\sqrt{r}, 0)$; for $r > 0$, $(\sqrt{r}, 0)$ is a stable node, $(-\sqrt{r}, 0)$ is a saddle; for $r = 0$, we have $(x^*, y^*) = (0,0)$ is the only equilibrium and its stability cannot be determined by linearization; for $r < 0$ there are no equilibria. The system undergoes a saddle-node bifurcation.

18. For $r > 0$, $(x^*, y^*) = (\sqrt{r}, 0)$, $(-\sqrt{r}, 0), (0,0)$; for $r > 0$, $(\pm\sqrt{r}, 0)$ are both stable nodes, $(0,0)$ is a saddle; for $r = 0$, we have $(x^*, y^*) = (0,0)$ is the only equilibrium and it is again a stable node; the system undergoes a supercritical pitchfork bifurcation.

21. See figures for $r = -.25, .25$.

Section 6.5

2. We're given $\frac{1}{\alpha} = 3.5$ days. Thus $\beta \approx \frac{1.41}{3.5} \approx .403$.

9. $N'(t) = \Lambda - \mu N$;
$N(t) = \left(N(0) - \frac{\Lambda}{\mu}\right) e^{-\mu t} + \frac{\Lambda}{\mu}$;
$N(t) \to \frac{\Lambda}{\mu}$ as $t \to \infty$

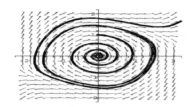

6.4#21 with $r = -.25$; an unstable limit cycle is present because one trajectory spirals into the origin while the other spirals away from the origin.

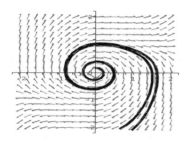

6.4#21 with $r = .25$; there is no unstable limit cycle and both trajectories spiral out of the origin.

10. $R_0 = \frac{\beta}{\mu+\alpha}\frac{\mu}{\mu+\phi}$ and this reduces to the previous expression $R_0 = \frac{\beta}{\mu+\alpha}$ when $\nu = 0$ (as it should).

11. Consider S, E, I for the variables in the reduced system. $R_0 = \frac{\beta}{\mu+\alpha}\frac{\delta}{\delta+\mu}$. Average life-span for the infectious class is $\frac{1}{\mu+\alpha}$, so $\frac{\beta}{\mu+\alpha}$ represents the total adequate contact number of a typical infection. The quantity $\frac{\delta}{\delta+\mu}$ is the probability that an individual in the exposed class becomes infected. Thus R_0 represents the total number of new cases generated by a typical infection during its lifetime.

Chapter 6 Review

7. a. (i) $\lambda_1 = 2, \lambda_2 = -2$,
$$\mathbf{v}_1 = \begin{pmatrix} 1 \\ 1 \end{pmatrix}, \mathbf{v}_2 = \begin{pmatrix} 1 \\ -3 \end{pmatrix};$$
(ii) saddle;
(iv) $x(t) = c_1 e^{2t} + c_2 e^{-2t}$,
$y(t) = c_1 e^{2t} - 3c_2 e^{-2t}$
b. (iv) $x(t) = 2c_1 \cos 4t + 2c_2 \sin 4t$,
$y(t) = -c_1 \sin 4t + c_2 \cos 4t$

9. a. (i) $\lambda_1 = 2, \lambda_2 = 3$,
$$\mathbf{v}_1 = \begin{pmatrix} -1 \\ 1 \end{pmatrix}, \mathbf{v}_2 = \begin{pmatrix} 1 \\ -2 \end{pmatrix};$$ (ii) unstable node;
b. (i) $\lambda_1 = 4i, \lambda_2 = -4i$,
$$\mathbf{v}_1 = \begin{pmatrix} 1 \\ -4i \end{pmatrix}, \mathbf{v}_2 = \begin{pmatrix} 1 \\ 4i \end{pmatrix}$$ (ii) center

11. a. (i) $\lambda_1 = -2, \lambda_2 = -1$,
$$\mathbf{v}_1 = \begin{pmatrix} 1 \\ -2 \end{pmatrix}, \mathbf{v}_2 = \begin{pmatrix} -1 \\ 1 \end{pmatrix};$$ (ii) stable node;
$x(t) = -c_1 e^{-t} + c_2 e^{-2t}, y(t) = c_1 e^{-t} - 2c_2 e^{-2t}$ **b.** (i) $\lambda_1 = 2+i, \lambda_2 = 2-i$,
$$\mathbf{v}_1 = \begin{pmatrix} -3-i \\ 1 \end{pmatrix}, \mathbf{v}_2 = \begin{pmatrix} -3+i \\ 1 \end{pmatrix}$$
(ii) unstable spiral

13. a. (i) $\lambda_1 = -5, \lambda_2 = -1$,
$$\mathbf{v}_1 = \begin{pmatrix} 1 \\ 1 \end{pmatrix}, \mathbf{v}_2 = \begin{pmatrix} -1 \\ 5 \end{pmatrix};$$ (ii) stable node;
$x(t) = c_1 e^{-5t} + c_2 e^{-t}, y(t) = c_1 e^{-5t} + 5c_2 e^{-t}$ **b.** (ii) unstable spiral;
$x(t) = c_1 e^{4t}(\cos(\sqrt{3}t) + \sqrt{3}\sin(\sqrt{3}t)) + c_2 e^{4t}(\sin(\sqrt{3}t) - \sqrt{3}\cos(\sqrt{3}t))$
$y(t) = 2c_1 e^{4t}\cos(\sqrt{3}t) + 2c_2 e^{4t}\sin(\sqrt{3}t)$

15. a. (i) $\lambda_{1,2} = 1$,
$$\mathbf{v}_1 = \begin{pmatrix} 1 \\ 3/2 \end{pmatrix};$$ (ii) unstable node;
b. (i) $\lambda_{1,2} = 2$,
(ii) unstable node

18. $x(t) = c_1 e^{2t} + c_2 e^{-2t}, y(t) = c_1 e^{2t} - 3c_2 e^{-2t}$

19. $x(t) = c_1 e^t + c_2 e^{5t}, y(t) = 5c_1 e^t + c_2 e^{5t}$

23. a. (i) $\lambda_1 = -4, \lambda_{2,3} = 1$,
$$\mathbf{v}_1 = \begin{pmatrix} -1 \\ 0 \\ 1 \end{pmatrix}, \mathbf{v}_2 = \begin{pmatrix} -2 \\ 1 \\ 3 \end{pmatrix}; \text{ (ii) saddle;}$$

b. (i) $\lambda_1 = 3, \lambda_2 = 3+6i, \lambda_3 = 3-6i$,
$$\mathbf{v}_1 = \begin{pmatrix} 1 \\ -2 \\ -3 \end{pmatrix}, \mathbf{v}_2 = \begin{pmatrix} -2 \\ 1 \\ \frac{9}{2} + \frac{3}{2}i \end{pmatrix} \mathbf{v}_3 = \begin{pmatrix} -2 \\ 1 \\ \frac{9}{2} - \frac{3}{2}i \end{pmatrix}; \text{ (ii) unstable;}$$

26. a. (i) $\lambda_1 = 1, \lambda_2 = -1, \lambda_3 = 5$,
$$\mathbf{v}_1 = \begin{pmatrix} 1 \\ 1 \\ 1 \end{pmatrix}, \mathbf{v}_2 = \begin{pmatrix} 1 \\ 1 \\ -1 \end{pmatrix}, \mathbf{v}_2 = \begin{pmatrix} 1 \\ -1 \\ 1 \end{pmatrix};$$
(ii) saddle;

b. (i) $\lambda_1 = -3, \lambda_2 = 5, \lambda_3 = 1$,
$$\mathbf{v}_1 = \begin{pmatrix} -1 \\ -1 \\ 1 \end{pmatrix}, \mathbf{v}_2 = \begin{pmatrix} 1 \\ -1 \\ 1 \end{pmatrix} \mathbf{v}_3 = \begin{pmatrix} 1 \\ 1 \\ 1 \end{pmatrix};$$
(ii) saddle

29. (i) $(x^*, y^*) = (0, 1), (0, -1)$; (ii) $(0, 1)$ is a saddle, $(0, -1)$ is a center

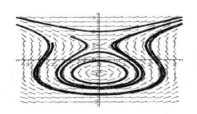

Chapter 6 Review #29 (iii)-(iv).

31. (i) $(x^*, y^*) = (0, 0), (1, 1)$; (ii) $(1, 1)$ is a saddle, $(0, 0)$ is an indeterminate form since linearization gives a zero eigenvalue; graphically, $(0, 0)$ is stable from above but unstable from below.

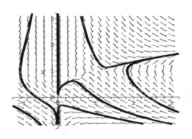

Chapter 6 Review #31 (iii)-(iv).

Section 7.1

1. $F(s) = \frac{2}{s^3}$

2. $F(s) = \frac{1}{s^2} - \frac{3}{s}$

5. $F(s) = \frac{2e^{-2s}}{s^2} - \frac{1-2s}{s^2}$

7. $F(s) = \frac{e^{-2s}(e^{2s}s + e^s(2+s) - 2(1+2s(1+s)))}{s^3}$

8. $F(s) = \frac{1}{s^2-1}$

14. Use the linearity property of the Laplace transform.

Section 7.2

1. $F(s) = \frac{1}{s^2} - \frac{5}{s}$

3. $F(s) = \frac{1}{(s-2)^2}$

5. $F(s) = \frac{s^2+2}{s^3+4s}$

7. $F(s) = \frac{4s}{s^4+10s^2+9}$

9. $F(s) = \frac{6}{s^4+10s^2+9}$

12. $F(s) = \frac{\alpha}{s^2+\alpha^2}$

Section 7.3

1. $F(s) = \frac{e^{-4s}}{s}$

3. $F(s) = \frac{4}{s}(1 - e^{-10s})$

5. $F(s) = \frac{6}{s}(e^{-9s} - e^{-3s})$

7. $F(s) = \frac{1}{s}(1 - e^{-2s} + e^{-6s})$

9. $F(s) = \frac{2}{s^2}(1 - (5s + 1)e^{-5s}) + \frac{10e^{-5s}}{s}$

11. $F(s) = \frac{e^{-7s}(-1+e^{3s}-3s)}{s^2} + \frac{3e^{-7s}}{s}$

13. $1 - e^{-2}$

15. 1

17. $t_0 = \pm\frac{1}{6}$

Section 7.4

1. $f(t) = 3\sin t \cos t$

3. $f(t) = 5te^{3t}$

5. $f(t) = e^{-2t}\cos\sqrt{3}t$

7. $f(t) = \frac{1}{2}(1-\cos\sqrt{2}t+\sqrt{2}\sin\sqrt{2}t)$

9. $f(t) = \frac{1}{2}t(t + 4)e^{-3t}$

11. $f(t) = e^{-3t}(5 - 4e^t)$

13. $f(t) = (2\sin 3(t-\pi)+5\cos 3(t-\pi))U_\pi(t)$

15. $f(t) = (t - 4)U_4(t) - (t - 7)U_7(t)$

Section 7.5

1. $y(t) = \frac{1}{2}e^t(3 + e^{2t})$

3. $y(t) = \frac{1}{2}(4\cos 2t + 3\sin 2t)$

5. $y(t) = 2e^{3t}$

7. $y(t) = \frac{1}{7}e^{-4t}(15e^{7t} + 13)$

9. $y(t) = 2e^{-t}(\cos 2t + 2\sin 2t)$

11. $y(t) = \frac{-1}{2}(12te^t + 5e^{2t} - 15)e^{3t}$

13. $y(t) = \frac{1}{7}e^{-5t}(49e^{7t}\sin t - 7e^{7t}\cos t + 4e^{7t} - 4)$

15. $y(t) = \frac{1}{36}e^t(114te^{3t} - 209e^{3t} + 198e^{2t} + 153e^t - 178)$

Section 7.6

1. $x(t) = 2e^t - e^{2t} + 2, y(t) = e^t(1 - e^t)$

3. $x(t) = -e^{-t}(3te^t+1), y(t) = e^{-t}(1 - e^t + 2te^t)$

5. $x(t) = e^t(5e^{3t}-2), y(t) = 4e^t(e^{3t} - 1)$

7. $x(t) = 2\cos t+8\sin t, y(t) = \cos t - 13\sin t + \sinh t$

10. $x(t) = 2e^{-2t}(e^t + 1), y(t) = 3 + 2e^{-2t}(e^t - 4)$

12. $x(t) = 4(\sin 2t - \cos 2t), y(t) = \sin 2t + \cos 2t$

Section 7.7

1. $\frac{t^3}{3}$

3. $\frac{1}{128}((4t8t^2 - 3) + 3\sin 4t)$

5. $t + tU_0(t)$

7. $f(t) = e^{-3t}(e^t - 1)$

9. $f(t) = \frac{2}{9}\sin^2(\frac{3t}{2})$

11. $f(t) = \frac{1}{54}(9t^2 - 6t - 2e^{-3t} + 2)$

Chapter 7 Review

1. False. See Theorem 7.1.3.

2. True. See Theorem 7.3.1.

4. False. $\mathcal{L}\{C_1 f_1(t) + C_2 f_2(t)\} = C_1 \mathcal{L}\{f_1(t)\} + C_2 \mathcal{L}\{f_2(t)\}$.

5. True. Uniqueness of Laplace transforms.

7. $F(s) = \frac{48}{s^5}$

8. $F(s) = \frac{2}{(1+s)^3}$

16. $1 + e^{-3}$

17. 0

19. 1

21. $\frac{7}{2}e^{-2t}(e^{4t} + 1)$

23. $5(1 - 3t)e^{-3t}$

25. $y(t) = \frac{1}{3}e^t(e^{3t} + 17)$

27. $y(t) = \frac{2}{9}\sin 9t + \cos 9t$

29. $y(t) = 3e^{-t}(\sin 2t + 2\cos 2t)$

31. $y(t) = \frac{-1}{9}e^{2t}(47e^{3t} - 72e^t - 15t - 20)$

34. $y(t) = \frac{1}{180}e^{-2t}(-390te^{2t} + 67e^{5t} + 455e^{2t} - 135e^t - 207)$

37. $x(t) = \frac{-1}{63}e^{-10t}(77e^{9t} - 144e^{7t} - 59), y(t) = \frac{1}{315}e^{-10t}(756e^{10t} - 385e^{9t} - 1080e^{7t} - 236)$

39. $x(t) = \frac{1}{2}(3 - 5e^{2t}), y(t) = -2(4 + e^{2t})$

41. $\frac{3t^4}{4}$

43. $\frac{1}{64}(4t(8t^2 - 3) + 3\sin 4t)$

46. $f(t) = e^{-4t}(e^t - 1)$

48. $f(t) = \frac{1}{2}\sin^2 t$

50. $f(t) = \frac{1}{128}(e^{-8t} + \sin 8t - \cos 8t)$

Section 8.1

1. $|x| < 1$

4. $|x| < 1$

Section 8.2

1. $y = 1 + x + \frac{x^2}{2} + \frac{2x^3}{3} + \frac{7x^4}{12} + \cdots$

3. $y = \frac{x^2}{2} + \frac{x^3}{6} + \frac{x^4}{6} + \cdots$

5. $y = 1 + x + x^2 + \cdots$

7. $y = x + x^2 + \frac{x^3}{2} + \cdots$

9. $y = 1 + x + \frac{x^3}{3} - \frac{x^4}{3} + \cdots$

11. $y = 1 + 2(x - 1) + 4(x - 1)^2 - \frac{25}{3}(x - 1)^3 + \cdots$

14. $y = 1 + 2x - \frac{x^2}{2} - \frac{x^3}{3} - \cdots$

15. $y = c_0\left(1 - \frac{x^3}{6} + \frac{3x^5}{40} + \cdots\right) + c_1\left(x - \frac{x^3}{6} - \frac{x^4}{12} + \frac{3x^5}{40} + \cdots\right)$

17. $y = \frac{1}{4}(4e^{x^2} - x)$

Section 8.3

1. $x = 0$ and $x = 3$ are regular singular points

3. $x = -2$ is a regular singular point; $x = 0$ is an irregular singular point.

5. $x = \frac{-1}{2}$ is a regular singular point; $x = 0$ is an irregular singular point.

6. $x = 1$ is a regular singular point; $x = 0$ is an irregular singular point.

Section 8.4

1. $y = c_1 x\left(1 - \frac{x^2}{14} + \frac{x^4}{616} - \cdots\right) + c_2 x^{\frac{-1}{2}}\left(1 - \frac{x^2}{2} + \frac{x^4}{40} - \cdots\right)$

5. $y = 3c_1 x^{\frac{1}{2}} \sum_{n=0}^{\infty} \frac{2^{n+1}(n+1)}{(2n+3)!} x^n$
$+ c_2 x^{-1} \left(1 - \sum_{n=1}^{\infty} \frac{2^{n-1}(n-1)!}{(2n-2)!} x^n \right)$

7. $y = 3c_1 x^{\frac{1}{3}} \left(1 + \frac{2x}{9} \right.$
$\left. -2\sum_{n=2}^{\infty} (-1)^n \frac{1\cdot4\cdot7\dots(3n-5)}{9^n n!} x^n + c_2 x \right)$

10. $y = 2c_1 x \sum_{n=0}^{\infty} (-1)^n \frac{x^n}{(n+2)!}$
$+ c_2 x^{-1}(1-x)$

13. $y = c_1 \frac{x^3}{1+x} + c_2(1-x+x^2)$

16. $y = c_1 \sum_{n=0}^{\infty} \frac{x^n}{(n+1)!}$

21. $y_1(x) = c_1 \sum_{n=0}^{\infty} (-1)^n \frac{x^n}{(n!)^2}$,
$y = c_1 y_1(x) + c_2 \left(y_1(x) \ln x \right.$
$\left. -2\sum_{n=2}^{\infty} (-1)^n \frac{1+\frac{1}{2}+\dots+\frac{1}{n}}{(n!)^2} x^n \right)$

23. $y_1(x) = x + x^2$,
$y = c_1 y_1(x) + c_2 \left(y_1(x) \ln x \right.$
$\left. -2x^2 - \sum_{n=2}^{\infty} (-1)^n \frac{x^{n+1}}{n(n-1)} \right)$

27. $y_1(x) = xe^x$,
$y = c_1 y_1(x) + c_2 \left(y_1(x) \ln x \right.$
$\left. 1 - \sum_{n=2}^{\infty} \frac{1+\frac{1}{2}+\dots+\frac{1}{n}}{(n-1)!} x^n \right)$

28. $p(x) = 1 - \frac{3}{5}x + \frac{1}{5}x^2 + \dots$,
$q(x) = \frac{1}{5}x - \frac{3}{20}x^2 + \dots$,
$y = c_1 x[p(x)\cos \ln x - q(x)\sin \ln x]$
$+ c_2 x[q(x)\cos \ln x + p(x)\sin \ln x]$

Chapter 8 Review

1. False. See Theorem 8.1.1.

2. True, this is just one reason they are useful.

3. If you think this is true, then you have missed the point of this chapter.

5. False, check out Taylor's theorem.

9. $y(x) = 3(2e^{2x} - 1)$

11. $y(x) = \frac{1}{2} \left(3 + 2x + x^2 + 9\cos\sqrt{2}x - 3\sqrt{2}\sin\sqrt{2}x \right)$

14. $y(x) = \frac{1}{2} \left(3e^x + \sin x - \cos x \right)$

22. $y = c_1 x^{\frac{4}{3}} \left(1 - \frac{3x^2}{16} + \frac{9x^4}{896} - \dots \right)$
$+ c_2 x^{\frac{2}{3}} \left(1 - \frac{3x^2}{8} + \frac{9x^4}{320} - \dots \right)$

25. $y = c_1 x^{\frac{1}{3}} \left(1 - \frac{3x^2}{16} + \frac{9x^4}{896} - \dots \right)$
$+ c_2 x^{\frac{-1}{3}} \left(1 - \frac{3x^2}{8} + \frac{9x^4}{320} - \dots \right)$

29. $y = c_1 \left(1 + x + \frac{3x^2}{10} + \dots \right)$
$+ c_2 x^{\frac{1}{3}} \left(1 + \frac{7x}{12} + \frac{5x^2}{36} - \dots \right)$

Appendix C.1

2. AB $\begin{pmatrix} -7 & 19 \\ 6 & -22 \end{pmatrix}$

3. BA $= \begin{pmatrix} 9 & 8 & -13 \\ -16 & -3 & -3 \\ 6 & -2 & 12 \end{pmatrix}$

4. b. D $-$ E $= \begin{pmatrix} 0 & 8 & -2 \\ -6 & -1 & 2 \\ 1 & 0 & 0 \end{pmatrix}$

5. a. B $+$ C is not possible because the matrices are different sizes.
c. $\mathrm{Tr}\mathbf{E} = 3$

6. c. A^T $= \begin{pmatrix} 3 & -1 & 1 \\ 0 & 2 & 1 \end{pmatrix}$

8. b. AB $= \begin{pmatrix} 12 & -3 \\ -4 & 5 \\ 4 & 1 \end{pmatrix}$
c. BA is not possible because rows of **A** are not equal to the columns of **B**.

10. c. detE $= 33$

Appendix C.2

3. a. $\begin{pmatrix} 12 & -3 & -2 \\ 4 & 1 & 1 \end{pmatrix}$

$\Rightarrow \begin{pmatrix} 1 & 0 & \frac{1}{8} \\ 0 & 1 & \frac{3}{4} \end{pmatrix}$

$\Rightarrow x_1 = \frac{1}{8}, x_2 = \frac{3}{4}$

b. $\begin{pmatrix} 1 & 1 & 2 & 8 \\ -1 & -2 & 3 & 1 \\ 3 & -7 & 4 & 10 \end{pmatrix}$

$\Rightarrow \begin{pmatrix} 1 & 0 & 0 & 3 \\ 0 & 1 & 0 & 1 \\ 0 & 0 & 1 & 2 \end{pmatrix}$

$\Rightarrow x_1 = 3, x_2 = 1, x_3 = 2$

5. b. $\begin{pmatrix} 2 & 3 & -2 & 2 \\ 1 & -2 & 3 & 2 \\ 4 & -1 & 5 & 1 \end{pmatrix}$

$\Rightarrow \begin{pmatrix} 1 & 0 & 0 & 5 \\ 0 & 1 & 0 & -6 \\ 0 & 0 & 1 & -5 \end{pmatrix}$

$\Rightarrow x = 5, y = -6, z = -5$

6. b. $x = \dfrac{\begin{pmatrix} 2 & 0 & 2 \\ 3 & -3 & 0 \\ 0 & -4 & 7 \end{pmatrix}}{-17} = \dfrac{90}{17},$

$y = \dfrac{43}{17}, z = \dfrac{-14}{17}$

7. a.(i) $x_1 = \dfrac{\begin{pmatrix} 3 & -2 \\ 5 & 1 \end{pmatrix}}{\begin{pmatrix} 7 & -2 \\ 3 & 1 \end{pmatrix}} = \dfrac{13}{13} =$

$1, x_2 = \dfrac{26}{13} = 2$

(ii)-(iii). $\begin{pmatrix} 7 & -2 & 3 \\ 3 & 1 & 5 \end{pmatrix}$

$\Rightarrow \begin{pmatrix} 7 & -2 & 3 \\ 0 & \frac{13}{7} & \frac{26}{7} \end{pmatrix}$

$\Rightarrow \begin{pmatrix} 1 & 0 & 1 \\ 0 & 1 & 2 \end{pmatrix}$

$\Rightarrow x_1 = 1, x_2 = 2$

10. $\mathbf{A_1}^{-1} = \begin{pmatrix} -1 & -2 \\ 1 & 1 \end{pmatrix}$

11. $(\mathbf{A_2}|\mathbf{I}) = \begin{pmatrix} 1 & 5 & 1 & 0 \\ 3 & -6 & 0 & 1 \end{pmatrix}$

$\Rightarrow \begin{pmatrix} 1 & 0 & \frac{7}{3} & \frac{-2}{3} \\ 0 & 1 & \frac{-2}{3} & \frac{1}{3} \end{pmatrix}$

$\Rightarrow \mathbf{A_2}^{-1} = \frac{1}{3}\begin{pmatrix} 7 & -2 \\ -2 & 1 \end{pmatrix}$

12. $\mathbf{B}^{-1} = \dfrac{1}{34}\begin{pmatrix} 17 & 34 & -17 \\ -9 & -6 & 7 \\ -11 & 4 & 1 \end{pmatrix}$

References

[1] M.L. Abell and J.P. Braselton. *Modern Differential Equations: Theory, Applications, Technology.* Saunders College Publishing, Fort Worth, Texas, 1996.

[2] L. Almada, R. Rodriguez, M. Thompson, L. Voss, L. Smith, and E.T. Camacho. *Deterministic and Small-World Network Models of College Drinking Patterns*, Technical Report, Department of Mathematics & Statistics, California State Polytechnic University, Pomona, 2006 (<www.amssi.org>).

[3] B.W. Banks. *Differential Equations with Graphical and Numerical Methods.* Prentice Hall, Englewood Cliffs, New Jersey, 2001.

[4] E. Beltrami. *Mathematical Models in the Social and Biological Sciences.* Jones and Bartlett Publishers, Boston, 1993.

[5] C.M. Bender and S.A. Orszag. *Advanced Mathematical Methods for Scientists and Engineers.* McGraw-Hill, New York, 1978.

[6] R.L. Bewernick, J.D. Dewar, E. Gray, N.Y. Rodriguez, and R.J. Swift. *Population Processes*, Technical Report, Department of Mathematics & Statistics, California State Polytechnic University, Pomona, 2005 (<www.amssi.org>).

[7] W.E. Boyce and R.C. DiPrima. *Elementary Differential Equations*, John Wiley & Sons, New York, 4th ed., 1986.

[8] F. Brauer and C. Castillo-Chavez. *Mathematical Models in Population Biology and Epidemiology*, Springer, New York, 2001.

[9] R.L. Burden and J.D. Faires. *Numerical Analysis*, Brooks/Cole, Pacific Grove, 2001.

[10] C. Corduneanu. *Principles of Differential and Integral Equations*, Chelsea Publishing Co., New York, 1977.

[11] D.J. Daley and J. Gani. *Epidemic Modelling*, Cambridge University Press, U.K., 1999.

[12] L. Edelstein-Keshet. *Mathematical Models in Biology.* Birkhäuser Mathematics Series. McGraw-Hill, New York, 1988.

656   *References*

[13] C.A.S. Hall. An assessment of several of the historically most influential theoretical models used in ecology and of the data provided in their support. *Ecol. Model.*, 43:5–31, 1988.

[14] R. Haubrich. Frobenius, Schur, and the Berlin algebraic tradition. *Mathematics in Berlin.* Berlin, 83–96, 1998.

[15] H. Hethcote. The mathematics of infectious diseases. *SIAM Rev*, 42: 599–653, 2000.

[16] M. Hirsch and S. Smale. *Differential Equations, Dynamical Systems, and Linear Algebra.* Academic Press, London, 1974.

[17] J.H. Hubbard and B.H. West. *Differential Equations: A Dynamical Systems Approach, Part I, One Dimensional Equations.* Springer-Verlag, New York, 1990.

[18] W. Kermack and A. McKendrick. Contributions to the mathematical theory of epidemics—I. *Proc. R. Soc.*, 115A:700, 1927.

[19] I. Kyprianidis. Dynamics of a Nonlinear Electrical Oscillator Described by Duffing's Equation, http://www.math.upatras.gr/~crans

[20] A.C. Lazer and P.J. McKenna. Large amplitude periodic oscillations in suspension bridges: some new connections with nonlinear analysis. *SIAM Rev.*, 32 (December):537–578, 1990.

[21] N.N. Lebedev, *Special Functions & Their Applications.* Dover, New York, 1972.

[22] G.N. Lewis, The Collapse of the Tacoma Narrows Suspension Bridge, in *Differential Equations with Boundary-Value Problems*, 5th ed. by D.G. Zill and M.R. Cullen, 2000.

[23] D.O. Lomen and D.Lovelock. *Differential Equations: Graphics, Models, Data*, Wiley, 1998.

[24] D. Ludwig, D.D. Jones, and C.S. Holling. Qualitative analysis of insect outbreak systems: the spruce budworm and forest. *J. Animal Ecol.*, 47:315–332, 1978.

[25] M. Martelli, *Introduction to Discrete Dynamical Systems and Chaos* John Wiley & Sons, Inc., New York, 1999.

[26] R.E. Mickens. *Applications of Nonstandard Finite Difference Schemes.* World Scientific, Singapore, 2000.

[27] D.D. Mooney, and R.J. Swift, *A Course in Mathematical Modeling.* The Mathematical Association of America, Washington D.C., 1999.

[28] C.R. Nave. *HyperPhysics website at Georgia State University* http://hyperphysics.phy-astr.gsu.edu/hbase/hframe.html, 2000.

[29] M. Olinick. *An Introduction to Mathematical Models in the Social and Life Sciences.* Addison-Wesley Publishing Company, Reading, Massachusetts, 1978.

[30] L.F. Richardson. Generalized foreign policy. *Br. J. Psychol. Monogr. Suppl.*, 23:130–148, 1939.

[31] C.C. Ross. *Differential Equations: An Introduction with Mathematica.* Springer-Verlag, New York, 1995.

[32] J.T. Sandefur. *Discrete Dynamical Systems: Theory and Applications.* Clarendon Press, Oxford, 1990.

[33] G. Strang. *Linear Algebra and Its Applications.* Harcourt Brace Jovanovich Publishers, San Diego, 1988.

[34] S.H. Strogatz. *Nonlinear Dynamics and Chaos, with Applications to Physics, Biology, Chemistry, and Engineering.* Addison-Wesley Publishing Company, Reading, Massachusetts, 1994.

[35] K.R. Symon. *Mechanics.* Addison-Wesley Publishing Company, Reading, Massachusetts, 1971.

[36] S. Utida. Cyclic fluctuations of population density intrinsic to the host-parasite system. *Ecology*, 38:442–449, 1957.

[37] G.N. Watson. *A Treatise on the Theory of Bessel Functions.* Cambridge at the University Press, London, second edition, 1966.

Index